CONTENTS

Student Solutions Manual

Laurel Technical Services

ELEMENTARY ALGEBRA

for College Students

Fourth Edition

ALLEN R. ANGEL

PRENTICE HALL, Upper Saddle River, NJ 07458

Production Editor: *Carole Suraci*
Acquisitions Editor: *Melissa Acuna*
Supplement Acquisitions Editor: *Audra Walsh*
Production Coordinator: *Ben Smith*

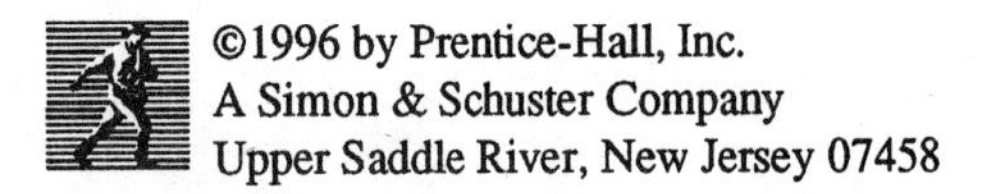

A Simon & Schuster Company
Upper Saddle River, New Jersey 07458

Printed in the United States of America

10 9 8 7 6 5 4 3

ISBN: 0-13-367590-4

Prentice-Hall International (UK) Limited, *London*
Prentice-Hall of Australia Pty. Limited, *Sydney*
Prentice-Hall Canada Inc., *Toronto*
Prentice-Hall Hispanoamericana, S.A., *Mexico*
Prentice-Hall of India Private Limited, *New Delhi*
Prentice-Hall of Japan, Inc., *Tokyo*
Simon & Schuster Asia Pte. Ltd., *Singapore*
Editora Prentice-Hall do Brasil, Ltda., *Rio de Janeiro*

CHAPTER 1

Exercise Set 1.2

1. The greatest common factor of 4 and 16 is 4. $\frac{4}{16}=\frac{4\div4}{16\div4}=\frac{1}{4}$

3. The greatest common factor of 10 and 15 is 5. $\frac{10}{15}=\frac{10\div5}{15\div5}=\frac{2}{3}$

5. The greatest common factor of 15 and 30 is 15. $\frac{15}{30}=\frac{15\div15}{30\div15}=\frac{1}{2}$

7. The greatest common factor of 15 and 35 is 5. $\frac{15}{35}=\frac{15\div5}{35\div5}=\frac{3}{7}$

9. The greatest common factor of 40 and 64 is 8. $\frac{40}{64}=\frac{40\div8}{64\div8}=\frac{5}{8}$

11. 9 and 14 have no common factors other than 1. Therefore, the fraction is in its lowest terms.

13. The greatest common factor of 96 and 72 is 24. $\frac{96}{72}=\frac{96\div24}{72\div24}=\frac{4}{3}$;

$$\begin{array}{r} 1 \\ 3\overline{)\ 4} \\ \underline{-3} \\ 1 \end{array}$$ so $\frac{4}{3}=1\frac{1}{3}$, Answer is $\frac{4}{3}$ or $1\frac{1}{3}$

15. The greatest common factor of 50 and 35 is 5. $\frac{50}{35}=\frac{50\div5}{35\div5}=\frac{10}{7}$;

$$\begin{array}{r} 1 \\ 7\overline{)10} \\ \underline{-7} \\ 3 \end{array}$$ so $\frac{10}{7}=1\frac{3}{7}$, Answer is $\frac{10}{7}$ or $1\frac{3}{7}$

17. (a); $\frac{3}{5}\cdot\frac{10}{11}=\frac{3}{1}\cdot\frac{2}{11}$ Divide both 10 and 5 by 5. This process can be used only when multiplying fractions and so cannot be used for (b), (c) or (d).

19. (b); $6\cdot\frac{5}{12}=\frac{6}{1}\cdot\frac{5}{12}=\frac{5}{2}$ Divide both 6 and 12 by 6. This process can be used only when multiplying fractions and so cannot be used for (a), (c), or (d).

21. $\frac{1}{2}\cdot\frac{3}{4}=\frac{1\cdot3}{2\cdot4}=\frac{3}{8}$

23. $\frac{5}{4}\cdot\frac{2}{7}=\frac{5\cdot1}{2\cdot7}=\frac{5}{14}$

25. $\frac{3}{8}\cdot\frac{2}{9}=\frac{1\cdot1}{4\cdot3}=\frac{1}{12}$

27. $\frac{1}{4}\div\frac{1}{5}=\frac{1}{4}\cdot\frac{5}{1}=\frac{1\cdot5}{4\cdot1}=\frac{5}{4}$ or $1\frac{1}{4}$

29. $\frac{5}{12}\div\frac{4}{3}=\frac{5}{12}\cdot\frac{3}{4}=\frac{5\cdot1}{4\cdot4}=\frac{5}{16}$

31. $\frac{10}{3}\div\frac{5}{9}=\frac{10}{3}\cdot\frac{9}{5}=\frac{2\cdot3}{1\cdot1}=\frac{6}{1}=6$

33. $\frac{5}{12}\cdot\frac{16}{15}=\frac{1\cdot4}{3\cdot3}=\frac{4}{9}$

35. $\frac{4}{15} \div \frac{13}{12} = \frac{4}{15} \cdot \frac{12}{13} = \frac{4 \cdot 4}{5 \cdot 13} = \frac{16}{65}$

37. $\frac{12}{7} \cdot \frac{19}{24} = \frac{1 \cdot 19}{7 \cdot 2} = \frac{19}{14}$ or $1\frac{5}{14}$

39. $1\frac{4}{5} = \frac{5+4}{5} = \frac{9}{5}$

so, $1\frac{4}{5} \cdot \frac{20}{3} = \frac{9}{5} \cdot \frac{20}{3} = \frac{3 \cdot 4}{1 \cdot 1} = 12$

41. $1\frac{2}{3} = \frac{3+2}{3} = \frac{5}{3}$

so, $\left(\frac{3}{5}\right)\left(1\frac{2}{3}\right) = \frac{3}{5} \cdot \frac{5}{3} = 1$

43. $3\frac{2}{3} = \frac{9+2}{3} = \frac{11}{3}$ and $1\frac{5}{6} = \frac{6+5}{6} = \frac{11}{6}$

so, $3\frac{2}{3} \div 1\frac{5}{6} = \frac{11}{3} \div \frac{11}{6} = \frac{11}{3} \cdot \frac{6}{11} = \frac{2}{1} = 2$

45. $\frac{2}{7} + \frac{3}{7} = \frac{2+3}{7} = \frac{5}{7}$

47. $\frac{7}{12} - \frac{5}{12} = \frac{7-5}{12} = \frac{2}{12} = \frac{2 \cdot 1}{2 \cdot 6} = \frac{1}{6}$

49. $\frac{5}{14} + \frac{9}{14} = \frac{5+9}{14} = \frac{14}{14} = 1$

51. $\frac{19}{26} - \frac{5}{26} = \frac{19-5}{26} = \frac{14}{26} = \frac{2 \cdot 7}{2 \cdot 13} = \frac{7}{13}$

53. $\frac{2}{5} = \frac{2}{5} \cdot \frac{6}{6} = \frac{12}{30}$ and $\frac{5}{6} = \frac{5}{6} \cdot \frac{5}{5} = \frac{25}{30}$

$\frac{2}{5} + \frac{5}{6} = \frac{12}{30} + \frac{25}{30} = \frac{37}{30}$ or $1\frac{7}{30}$

55. $\frac{4}{12} = \frac{4 \cdot 1}{4 \cdot 3} = \frac{1}{3} = \frac{1}{3} \cdot \frac{5}{5} = \frac{5}{15}$

$\frac{4}{12} - \frac{2}{15} = \frac{5}{15} - \frac{2}{15} = \frac{3}{15} = \frac{3 \cdot 1}{3 \cdot 5} = \frac{1}{5}$

57. $\frac{2}{10} = \frac{2 \cdot 1}{2 \cdot 5} = \frac{1}{5} = \frac{1}{5} \cdot \frac{3}{3} = \frac{3}{15}$

$\frac{2}{10} + \frac{1}{15} = \frac{3}{15} + \frac{1}{15} = \frac{4}{15}$

59. $\frac{5}{8} = \frac{5}{8} \cdot \frac{7}{7} = \frac{35}{56}$ and $\frac{4}{7} = \frac{4}{7} \cdot \frac{8}{8} = \frac{32}{56}$

$\frac{5}{8} - \frac{4}{7} = \frac{35}{56} - \frac{32}{56} = \frac{35-32}{56} = \frac{3}{56}$

61. $\frac{5}{6} = \frac{5}{6} \cdot \frac{4}{4} = \frac{20}{24}$

$\frac{5}{6} + \frac{9}{24} = \frac{20}{24} + \frac{9}{24} = \frac{29}{24}$ or $1\frac{5}{24}$

63. $\frac{4}{7} = \frac{4}{7} \cdot \frac{4}{4} = \frac{16}{28}$ and $\frac{1}{4} = \frac{1}{4} \cdot \frac{7}{7} = \frac{7}{28}$

$\frac{4}{7} - \frac{1}{4} = \frac{16}{28} - \frac{7}{28} = \frac{9}{28}$

65. $4\frac{1}{4} = \frac{16+1}{4} = \frac{17}{4} = \frac{17}{4} \cdot \frac{5}{5} = \frac{85}{20}$

$\frac{2}{5} = \frac{2}{5} \cdot \frac{4}{4} = \frac{8}{20}$

$4\frac{1}{4} + \frac{2}{5} = \frac{85}{20} + \frac{8}{20} = \frac{93}{20}$ or $4\frac{13}{20}$

67. $2\frac{1}{2} = \frac{4+1}{2} = \frac{5}{2} = \frac{5}{2} \cdot \frac{3}{3} = \frac{15}{6}$

$1\frac{1}{3} = \frac{3+1}{3} = \frac{4}{3} = \frac{4}{3} \cdot \frac{2}{2} = \frac{8}{6}$

$2\frac{1}{2} + 1\frac{1}{3} = \frac{15}{6} + \frac{8}{6} = \frac{23}{6}$ or $3\frac{5}{6}$

69. $4\frac{2}{3}=\frac{12+2}{3}=\frac{14}{3}=\frac{14}{3}\cdot\frac{5}{5}=\frac{70}{15}$

$1\frac{1}{5}=\frac{5+1}{5}=\frac{6}{5}=\frac{6}{5}\cdot\frac{3}{3}=\frac{18}{15}$

$4\frac{2}{3}-1\frac{1}{5}=\frac{70}{15}-\frac{18}{15}=\frac{52}{15}$ or $3\frac{7}{15}$

71. $1\frac{4}{5}=\frac{5+4}{5}=\frac{9}{5}=\frac{9}{5}\cdot\frac{4}{4}=\frac{36}{20}$

$\frac{3}{4}=\frac{3}{4}\cdot\frac{5}{5}=\frac{15}{20}$, $3=\frac{3}{1}\cdot\frac{20}{20}=\frac{60}{20}$

$1\frac{4}{5}-\frac{3}{4}+3=\frac{36}{20}-\frac{15}{20}+\frac{60}{20}$

$=\frac{36-15+60}{20}=\frac{81}{20}$ or $4\frac{1}{20}$

73. $1-\frac{1}{6}=\frac{6}{6}-\frac{1}{6}=\frac{6-1}{6}=\frac{5}{6}$; $\frac{5}{6}$ of the earth's fresh water is elsewhere.

75. $1-\frac{1}{12}=\frac{12}{12}-\frac{1}{12}=\frac{12-1}{12}=\frac{11}{12}$; The chance of an individual in this tax bracket not being audited is $\frac{11}{12}$.

77. $6\frac{3}{4}=\frac{24+3}{4}=\frac{27}{4}$,

$3\left(6\frac{3}{4}\right)=\frac{3}{1}\cdot\frac{27}{4}=\frac{3\cdot 27}{1\cdot 4}=\frac{81}{4}$ or $20\frac{1}{4}$ Paul needs $20\frac{1}{4}$ yds of material.

79. $16\frac{3}{4}=\frac{64+3}{4}=\frac{67}{4}=\frac{67}{4}\cdot\frac{4}{4}=\frac{268}{16}$

$3\frac{1}{16}=\frac{48+1}{16}=\frac{49}{16}$

$16\frac{3}{4}-3\frac{1}{16}=\frac{268}{16}-\frac{49}{16}=\frac{219}{16}$ or $13\frac{11}{16}$

The length of the remaining piece is $13\frac{11}{16}$ in.

81. $3\frac{3}{8}=\frac{24+3}{8}=\frac{27}{8}=\frac{27}{8}\cdot\frac{2}{2}=\frac{54}{16}$

$5\frac{1}{16}=\frac{80+1}{16}=\frac{81}{16}$

$3\frac{3}{8}+5\frac{1}{16}=\frac{54}{16}+\frac{81}{16}=\frac{135}{16}$ or $8\frac{7}{16}$

The total length of the two pieces of pipe is $8\frac{7}{16}$ ft.

83. $20\frac{3}{4}=\frac{80+3}{4}=\frac{83}{4}=\frac{83}{4}\cdot\frac{2}{2}=\frac{166}{8}$

$8\frac{7}{8}=\frac{64+7}{8}=\frac{71}{8}$

$20\frac{3}{4}-8\frac{7}{8}=\frac{166}{8}-\frac{71}{8}=\frac{95}{8}$ or $11\frac{7}{8}$

The water level fell by $11\frac{7}{8}$ ft.

85. $13\frac{1}{2}=\frac{26+1}{2}=\frac{27}{2}$

$22\left(13\frac{1}{2}\right)=\frac{22}{1}\cdot\frac{27}{2}=\frac{594}{2}=297$

The turkey should be baked for 297 min. or 4 hr 57 min.

87. $3\frac{1}{8}=\frac{24+1}{8}=\frac{25}{8}$

$3\frac{1}{8}\div\frac{2}{1}=\frac{25}{8}\cdot\frac{1}{2}=\frac{25}{16}$

Each piece is $\frac{25}{16}$ or $1\frac{9}{16}$ in.

89. $\frac{1}{16}\cdot\frac{80}{1}=\frac{1}{1}\cdot\frac{5}{1}=5$ Mr. Duncan should be given 5 mg of the drug.

91. **(a)** Total height of computer + monitor

$=7\frac{1}{2}$ in. $+14\frac{3}{8}$ in.

$$7\frac{1}{2} = \frac{15}{2} = \frac{15}{2} \cdot \frac{4}{4} = \frac{60}{8},$$

$$14\frac{3}{8} = \frac{112+3}{8} = \frac{115}{8}$$

$$7\frac{1}{2} + 14\frac{3}{8} = \frac{60}{8} + \frac{115}{8} = \frac{175}{8}$$

Total height of computer and monitor is $\frac{175}{8}$ or $21\frac{7}{8}$ in., so there is sufficient room.

(b) $22\frac{1}{2} = \frac{44+1}{2} = \frac{45}{2} = \frac{45}{2} \cdot \frac{4}{4} = \frac{180}{8}$

$$22\frac{1}{2} - 21\frac{7}{8} = \frac{180}{8} - \frac{175}{8} = \frac{5}{8}$$

There will be $\frac{5}{8}$ in. of extra height.

93. **(a)** $\frac{5}{6} \cdot \frac{3}{8} = \frac{5 \cdot 1}{2 \cdot 8} = \frac{5}{16}$

(b) $\frac{5}{6} \div \frac{3}{8} = \frac{5}{6} \cdot \frac{8}{3} = \frac{5 \cdot 4}{3 \cdot 3} = \frac{20}{9}$ or $2\frac{2}{9}$

(c) $\frac{5}{6} = \frac{5}{6} \cdot \frac{4}{4} = \frac{20}{24}$, $\frac{3}{8} = \frac{3}{8} \cdot \frac{3}{3} = \frac{9}{24}$

$$\frac{5}{6} + \frac{3}{8} = \frac{20}{24} + \frac{9}{24} = \frac{29}{24} \text{ or } 1\frac{5}{24}$$

(d) $\frac{5}{6} - \frac{3}{8} = \frac{20}{24} - \frac{9}{24} = \frac{11}{24}$

95. a general term for any collection of numbers, variables, grouping symbols, and operations

97. (b); In part (a) the numerator of one fraction and the denominator of the <u>second</u> fraction are divided by a common factor.

99. Multiply numerators, and multiply denominators.

101. Write fractions with a common denominator, add or subtract numerators, keep the same denominator.

103. Divide the numerator by the denominator. The quotient is the whole number and the remainder is the numerator. The denominator of the mixed number is the same as the denominator of the original fraction.

Group Activity/Challenge Problems

1. **(a)**

1. Divide '2 servings' by 2 to get the amount for 1 serving. Then add this amount to the '2 servings' recipe.
2. Add the '2 servings' and the '4 servings' together and divide by 2.
3. Divide '4 servings' by 4 to get the amount for 1 serving. Subtract this amount from the '4 servings' recipe.

(b)

1. Rice and water:

$$\left(\frac{2}{3}\right) \div 2 + \frac{2}{3} = \frac{2}{3} \cdot \frac{1}{2} + \frac{2}{3} = \frac{1}{3} + \frac{2}{3} = 1$$

Salt: $\frac{1}{4} \div 2 + \frac{1}{4} =$

$$\frac{1}{4} \cdot \frac{1}{2} + \frac{1}{4} = \frac{1}{8} + \frac{2}{8} = \frac{3}{8}$$

Butter: $1 \div 2 + 1 = 1 \cdot \frac{1}{2} + 1 = \frac{1}{2} + \frac{2}{2} = \frac{3}{2}$

or $1\frac{1}{2}$

2. Rice and water:

$$\left(\frac{2}{3} + 1\frac{1}{3}\right) \div 2 = \left(\frac{2}{3} + \frac{4}{3}\right) \cdot \frac{1}{2} = \frac{6}{3} \cdot \frac{1}{2} = 1$$

Salt:
$$\left(\frac{1}{4}+\frac{1}{2}\right)\div 2=\left(\frac{1}{4}+\frac{2}{4}\right)\cdot\frac{1}{2}=\frac{3}{4}\cdot\frac{1}{2}=\frac{3}{8}$$

Butter: $(1+2)\div 2=3\cdot\frac{1}{2}=\frac{3}{2}$ or $1\frac{1}{2}$

3. Rice and water:
$$1\frac{1}{3}-\left(1\frac{1}{3}\div 4\right)=\frac{4}{3}-\frac{1}{3}=1$$

Salt: $\frac{1}{2}-\left(\frac{1}{2}\div 4\right)=\frac{1}{2}-\frac{1}{8}=\frac{4}{8}-\frac{1}{8}=\frac{3}{8}$

Butter: $2-(2\div 4)=2-\frac{1}{2}=1\frac{1}{2}$

(c) Yes, 1 cup of rice and water, $\frac{3}{8}$ tsp salt, and $1\frac{1}{2}$ tsp butter.

3. $\dfrac{\frac{1}{2}+\frac{3}{4}}{\frac{3}{4}-\frac{1}{3}}=\dfrac{\frac{2}{4}+\frac{3}{4}}{\frac{9}{12}-\frac{4}{12}}=\dfrac{\frac{2+3}{4}}{\frac{9-4}{12}}=\dfrac{\frac{5}{4}}{\frac{5}{12}}=\frac{5}{4}\cdot\frac{12}{5}=3$

5. $\left(\frac{5}{12}+\frac{3}{5}\right)\div\left(\frac{5}{7}\cdot\frac{3}{10}\right)=\left(\frac{25}{60}+\frac{36}{60}\right)\div\frac{3}{14}=$

$\frac{61}{60}\cdot\frac{14}{3}=\frac{427}{90}$ or $4\frac{67}{90}$

Exercise Set 1.3

1. $\{\ldots,-3,-2,-1,0,1,2,3,\ldots\}$

3. $\{1,2,3,4,\ldots\}$

5. $\{1,2,3,4,\ldots\}$

7. True; The negative integers are $\{-1,-2,-3,\ldots\}$.

9. True; The integers are $\{\ldots-3,-2,-1,0,1,2,3,\ldots\}$.

11. False; The integers are $\{\ldots-3,-2,-1,0,1,2,3,\ldots\}$.

13. False; $\sqrt{2}$ cannot be expressed as the quotient of two integers.

15. True; $-\frac{3}{5}$ is a quotient of two integers.

17. True; $-4\frac{1}{3}$ can be expressed as a quotient of two integers, $-\frac{13}{3}$.

19. False; $-\frac{5}{3}$ is rational since it is a quotient of two integers.

21. False; Positive numbers are greater than zero.

23. True; All are names for the set $\{1,2,3,\ldots\}$.

25. True; definition of negative

27. True; Every integer can be expressed as a quotient of two integers.

29. False; Irrational numbers are real but not rational.

31. True; Irrational numbers are real numbers which are not rational .

33. True

35. True

37. False; Any positive irrational number is a counterexample.

39. (a) 7, 9

(b) 7, 0, 9; Any positive integer or zero

(c) $-6, 7, 0, 9$

(d) $-6, 7, 12.4, -\frac{9}{5}, -2\frac{1}{4}, 0, 9, 0.35$; Any number that can be expressed as a quotient of two integers.

(e) $\sqrt{3}, \sqrt{7}$; The numbers which cannot be expressed as a quotient of two integers.

(f) $-6, 7, 12.4, -\frac{9}{5}, -2\frac{1}{4}, \sqrt{3}, 0, 9, \sqrt{7}, 0.35$; The real numbers include the rational and the irrational numbers.

41. (a) 5

(b) 5

(c) -300

(d) $5, -300$

(e) $\frac{1}{2}, 4\frac{1}{2}, \frac{5}{12}, -1.67, 5, -300, -9\frac{1}{2}$;

Any number that can be expressed as a quotient of two integers.

(f) $\sqrt{2}, -\sqrt{2}$; The numbers which cannot be expressed as a quotient of two integers.

(g) $\frac{1}{2}, \sqrt{2}, -\sqrt{2}, 4\frac{1}{2}, \frac{5}{12}, -1.67, 5, -300, -9\frac{1}{2}$; The real numbers include the rational and irrational numbers.

43. $\frac{1}{2}, 6.4, -2.6$; Numbers that can be expressed as a quotient of two integers but which are not integers.

45. $\sqrt{2}, -\sqrt{2}, \pi$; Real numbers that cannot be expressed as a quotient of two integers.

47. $5, -8, 4$; Any integer is also a rational number.

49. $-1, -7, -24$; Any negative integer is also a rational number.

51. $1.5, 3, 6\frac{1}{4}$; Any positive number that can be expressed as a quotient of two integers.

53. $-7, 1, 5$; Any real number that can be expressed as a quotient of two integers.

55. Individual Answer

Cumulative Review Exercises 1.3

56. $4\frac{2}{3} = \frac{4 \cdot 3 + 2}{3} = \frac{12+2}{3} = \frac{14}{3}$

57.
$$\begin{array}{r} 5 \\ 3\overline{)\ 16} \\ \underline{-15} \\ 1 \end{array} \quad \frac{16}{3} = 5\frac{1}{3}$$

58. $\frac{3}{5} = \frac{3}{5} \cdot \frac{8}{8} = \frac{24}{40}$, $\frac{5}{8} \cdot \frac{5}{5} = \frac{25}{40}$

$\frac{3}{5} + \frac{5}{8} = \frac{24}{40} + \frac{25}{40} = \frac{49}{40}$ or $1\frac{9}{40}$

59. $4\frac{2}{3} = \frac{12+2}{3} = \frac{14}{3}$

$\left(\frac{5}{9}\right)\left(4\frac{2}{3}\right) = \frac{5}{9} \cdot \frac{14}{3} = \frac{70}{27}$ or $2\frac{16}{27}$

Group Activity/Challenge Problems

1. **(a)** The set of natural numbers does not have a last number.

   (b) It is an infinite set.

3. **(a)** and **(b)** There are an infinite number of fractions between any 2 numbers.

5. $A \cup B = \{a, b, c, d, g, h, i, j, m, p\}$

   $A \cap B = \{b, c, d\}$

Exercise Set 1.4

1. $|4| = 4$

3. $|-15| = 15$

5. $|0| = 0$

7. $-|-8| = -(8) = -8$

9. $-|65| = -(65) = -65$

11. $2 < 3$; 2 is to the left of 3 on the number line.

13. $-3 < 0$; -3 is to the left of 0 on the number line.

15. $\frac{1}{2} > -\frac{2}{3}$; $\frac{1}{2}$ is to the right of $-\frac{2}{3}$ on the number line.

17. $0.2 < 0.4$; 0.2 is to the left of 0.4 on the number line.

19. $-\frac{1}{2} > -1$; $-\frac{1}{2}$ is to the right of -1 on the number line.

21. $4 > -4$; 4 is to the right of -4 on the number line.

23. $-2.1 < -2$; -2.1 is to the left of -2 on the number line.

25. $\frac{5}{9} > -\frac{5}{9}$; $\frac{5}{9}$ is to the right of $-\frac{5}{9}$ on the number line.

27. $-\frac{3}{2} < \frac{3}{2}$; $-\frac{3}{2}$ is to the left of $\frac{3}{2}$ on the number line.

29. $0.49 > 0.43$; 0.49 is to the right of 0.43 on the number line.

31. $5 > -7$; 5 is to the right of -7 on the number line.

33. $-0.006 > -0.007$; -0.006 is to the right of -0.007 on the number line.

35. $-5 < -2$; -5 is to the left of -2 on the number line.

37. $-\frac{2}{3} > -3$; $-\frac{2}{3}$ is to the right of -3 on the number line.

39. $-\frac{1}{2} > -\frac{3}{2}$; $-\frac{1}{2}$ is to the right of $-\frac{3}{2}$ on the number line.

41. $8 > |-7|$ since $|-7| = 7$

43. $|0| < \frac{2}{3}$ since $|0| = 0$

45. $|-3| < |-4|$ since $|-3| = 3$ and $|-4| = 4$

47. $4 < \left|-\frac{9}{2}\right|$ since $\left|-\frac{9}{2}\right| = \frac{9}{2}$ or $4\frac{1}{2}$

49. $\left|-\frac{6}{2}\right| > \left|-\frac{2}{6}\right|$ since $\left|-\frac{6}{2}\right| = \frac{6}{2} = 3$

and $\left|-\frac{2}{6}\right| = \frac{2}{6} = \frac{1}{3}$

51. $\frac{2}{3} + \frac{2}{3} + \frac{2}{3} + \frac{2}{3} = 4 \cdot \frac{2}{3}$ since

$\frac{2}{3} + \frac{2}{3} + \frac{2}{3} + \frac{2}{3} = \frac{2+2+2+2}{3} = \frac{8}{3}$ and

$4 \cdot \frac{2}{3} = \frac{4}{1} \cdot \frac{2}{3} = \frac{8}{3}$

53. $\frac{1}{2} \cdot \frac{1}{2} < \frac{1}{2} \div \frac{1}{2}$ since $\frac{1}{2} \cdot \frac{1}{2} = \frac{1 \cdot 1}{2 \cdot 2} = \frac{1}{4}$

and $\frac{1}{2} \div \frac{1}{2} = \frac{1}{2} \cdot \frac{2}{1} = \frac{1}{1} \cdot \frac{1}{1} = 1$

55. $\frac{5}{8} - \frac{1}{2} < \frac{5}{8} \div \frac{1}{2}$ since $\frac{5}{8} - \frac{1}{2} = \frac{5}{8} - \frac{4}{8} = \frac{1}{8}$

and $\frac{5}{8} \div \frac{1}{2} = \frac{5}{8} \cdot \frac{2}{1} = \frac{5}{4}$

57. $4, -4$ since $|4| = |-4| = 4$

59. $2, -2$ since $|2| = |-2| = 2$

61. The distance between the number and 0 on the number line.

Cumulative Review Exercises: 1-4

62. $1\frac{2}{3} = \frac{3+2}{3} = \frac{5}{3} = \frac{5}{3} \cdot \frac{8}{8} = \frac{40}{24}$,

$\frac{3}{8} = \frac{3}{8} \cdot \frac{3}{3} = \frac{9}{24}$

$1\frac{2}{3} - \frac{3}{8} = \frac{40}{24} - \frac{9}{24} = \frac{31}{24}$ or $1\frac{7}{24}$

63. $\{0, 1, 2, 3, \ldots\}$

64. $\{1, 2, 3, 4, \ldots\}$

65. **(a)** 5

(b) $5, 0$; The whole numbers include 0.

(c) $5, -2, 0$; Integers can be negative or positive and include 0.

(d) $5, -2, 0, \frac{1}{3}, -\frac{5}{9}, 2.3$; Any number that can be expressed as a quotient of two integers is a rational number.

(e) $\sqrt{3}$; Any real number that is not rational is irrational.

(f) $5, -2, 0, \frac{1}{3}, \sqrt{3}, -\frac{5}{9}, 2.3$; The real numbers include the rational and irrational numbers.

Group Activity/Challenge Problems

1. Less than; The factors will always be a fraction with 1 in the numerator and a number larger than 1 in the denominator. When they are multiplied together, the denominator will have a number which is larger in magnitude, therefore making the fraction smaller.

3. 3, –3

$|3| = |-3| = 3$

5. **(a)** If $x \geq 0$, then $|x| = x$.

(b) If $x < 0$, then $|x| = -x$.

(c) $|x| = \begin{cases} x, & x \geq 0 \\ -x, & x < 0 \end{cases}$

Exercise Set 1.5

1. The opposite of 12 is -12 since $12 + (-12) = 0$.

3. The opposite of –40 is 40 since $-40 + 40 = 0$.

5. The opposite of 0 is 0 since $0 + 0 = 0$.

7. The opposite of $\frac{5}{3}$ is $-\frac{5}{3}$ since

$\frac{5}{3} + \left(-\frac{5}{3}\right) = 0$.

9. The opposite of $\frac{3}{5}$ is $-\frac{3}{5}$ since

$\frac{3}{5} + \left(-\frac{3}{5}\right) = 0$.

11. The opposite of 0.63 is –0.63 since $0.63 + (-0.63) = 0$.

13. The opposite of $3\frac{1}{5}$ is $-3\frac{1}{5}$ since

$3\frac{1}{5} + \left(-3\frac{1}{5}\right) = 0$.

15. The opposite of -3.1 is 3.1 since

$-3.1 + 3.1 = 0$.

17. Numbers have same sign, so add absolute values.

$|4| + |3| = 4 + 3 = 7$

Numbers are positive so sum is positive. Thus 4 + 3 = 7.

19. Numbers have different signs so find difference between larger and smaller absolute values.

$|4| - |-3| = 4 - 3 = 1$.

$|4|$ is greater than $|-3|$ so the sum is positive. So $4 + (-3) = 1$.

21. Numbers have same sign, so add absolute values.

$|-4| + |-2| = 4 + 2 = 6$.

Numbers are negative, so sum is negative. Thus $-4 + (-2) = -6$.

23. Numbers have different signs, so find difference between absolute values.

$|6| - |-6| = 6 - 6 = 0$, so $6 + (-6) = 0$.

25. Numbers have different signs, so find difference between absolute values.

$|-4| - |4| = 4 - 4 = 0$, so $-4 + 4 = 0$

27. Numbers have same sign, so add absolute values. $|-8|+|-2|=8+2=10$ Numbers are negative, so sum is negative. Thus $-8+(-2)=-10$.

29. Numbers have different signs, so find difference between absolute values.

$|-3|-|3|=3-3=0$ so $-3+3=0$

31. Numbers have same sign, so add absolute values.

$|-3|+|-7|=3+7=10$

Numbers are negative, so sum is negative. Thus, $-3+(-7)=-10$.

33. $0+0=0$

35. $-6+0=-6$

37. Numbers have different signs, so find difference between larger and smaller absolute values.
$|22|-|-19|=22-19=3$

$|22|$ is greater than $|-19|$ so sum is positive. Thus $22+(-19)=3$.

39. Numbers have different signs, so find difference between larger and smaller absolute values. $|-45|-|36|=45-36=9$

$|-45|$ is greater than $|36|$ so sum is negative. Thus $-45+36=-9$.

41. Numbers have different signs, so find difference between larger and smaller absolute values. $|18|-|-9|=18-9=9$ $|18|$ is greater than $|-9|$ so sum is positive. Thus $18+(-9)=9$

43. Numbers have same sign, so add absolute values.

$|-14|+|-13|=14+13=27$ Numbers are negative, so sum is negative. Thus $-14+(-13)=-27$.

45. Numbers have same sign, so add absolute values.

$|-35|+|-9|=35+9=44$ Numbers are negative, so sum is negative. Thus $-35+(-9)=-44$.

47. Numbers have different signs, so find difference between larger and smaller absolute values. $|-30|-|4|=30-4=26$

$|-30|$ is greater than $|4|$ so sum is negative. Thus $4+(-30)=-26$

49. Numbers have different signs, so find difference between larger and smaller absolute values. $|40|-|-35|=40-35=5$
$|40|$ is greater than $|-35|$ so sum is positive. Thus $-35+40=5$.

51. Numbers have different signs, so find difference between larger and smaller absolute values.

$|-200|-|180|=200-180=20$. $|-200|$ is greater than $|180|$ so sum is negative. Thus $180+(-200)=-20$.

53. Numbers have different signs, so find difference between larger and smaller absolute values.

$|-105|-|74|=105-74=31$

$|-105|$ is greater than $|74|$ so sum is negative. Thus $-105+74=-31$.

55. Numbers have different signs, so find difference between larger and smaller absolute values.

$|184| - |-93| = 184 - 93 = 91$

$|184|$ is greater than $|-93|$ so sum is positive. Thus $184 + (-93) = 91$.

57. Numbers have different signs, so find difference between larger and smaller absolute values.

$|-452| - |312| = 452 - 312 = 140$

$|-452|$ is greater than $|312|$ so sum is negative. Thus $-452 + 312 = -140$.

59. Numbers have same sign, so add absolute values.

$|-60| + |-38| = 60 + 38 = 98$

Numbers are negative so sum is negative. Thus $-60 + (-38) = -98$.

61. **(a)** $|463|$ is greater than $|-197|$ so sum will be positive.

(b) $463 + (-197) = 266$ (by calculator)

(c) Yes; By part (a) we expect a positive sum. The magnitude of the sum is the difference between the larger and smaller absolute values.

63. **(a)** Negative number: The sum of 2 negative numbers is always negative.

(b) $-84 + (-289) = -373$ (by calculator)

(c) Yes; The sum of 2 negative numbers should be (and is) a larger negative number.

65. **(a)** $|-947|$ is greater than $|495|$ so sum will be negative.

(b) $-947 + 495 = -452$ (by calculator)

(c) Yes; By part (a) we expect a negative sum. Magnitude of the sum is the difference between the larger and smaller absolute values.

67. **(a)** Negative number: The sum of 2 negative numbers is always negative.

(b) $-496 + (-804) = -1300$ (by calculator)

(c) Yes; The sum of 2 negative numbers should be (and is) a larger negative number.

69. **(a)** $|-285|$ is greater than $|263|$ so sum will be negative.

(b) $-285 + 263 = -22$ (by calculator)

(c) Yes; By part (a) we expect 0 negative sum. The magnitude of the sum is the difference between the larger and smaller absolute values.

71. **(a)** Negative number; The sum of 2 negative numbers is always negative.

(b) $-1833 + (-2047) = -3880$ (by calculator)

(c) Yes; The sum of 2 negative numbers should be (and is) a larger negative number.

73. **(a)** $|-6060|$ is greater than $|4793|$ so sum will be negative.

(b) $4793+(-6060)=-1267$ (by calculator)

(c) Yes; By part (a) we expect a negative sum. Magnitude of sum is difference between larger and smaller absolute values.

75. **(a)** Negative number: The sum of 2 negative numbers is always negative.

(b) $-1025+(-1025)=-2050$ (by calculator)

(c) Yes; The sum of 2 negative numbers should be (and is) a larger negative number.

77. **(a)** Negative numbers; The sum of 2 negative numbers is always negative.

(b) $-8276+(-6283)=-14,559$ (by calculator)

(c) Yes; The sum of 2 negative numbers should be (and is) a larger negative number.

79. **(a)** $|-9042|$ is greater than $|7827|$ so sum will be negative.

(b) $-9042+7827=-1215$ (by calculator)

(c) Yes; By part (a) we expect a negative sum. The magnitude of the sum is the difference between the larger and smaller absolute values.

81. **(a)** $|-8540|$ is greater than $|1046|$ so sum will be negative.

(b) $1046+(-8540)=-7494$ (by calculator)

(c) Yes; By part (a) we expect a negative sum. The magnitude of the sum is the difference between the larger and smaller absolute values.

83. **(a)** $|8364|$ is greater than $|-906|$ so sum will be positive.

(b) $8364+(-906)=7458$ (by calculator)

(c) Yes; By part (a) we expect a positive sum. The magnitude of the sum is the difference between the larger and smaller absolute values.

85. True

87. False; The sum has the sign of the number with the larger absolute value.

89. False; The sum has the sign of the number with the larger absolute value.

91. Amount owed can be represented as $67+107$. $|67|+|107|=67+107=174$

Both numbers are positive, so the sum is positive. Thus Mr. Yelserp owed the bank $174.

93. Total tax can be represented as $1424+503$. $|1424|+|503|=1424+503$

$=1927$ Both numbers are positive, so the sum will be positive. Thus, her total tax was $1927.

95. Depth of well can be represented as $22+32$. $|22|+|32|=22+32=54$

Both numbers are positive so the sum is positive. Thus the depth of the well is 54 ft.

97. Change in population can be represented as 141 million + (49) million.

$|141|-|-49|=141-49=92$. $|141|$ is greater than $|-49|$ so sum is positive. Thus the change in population is an increase of 92 million.

99. Individual Answer

Cumulative Review Exercises: 1.5

101. $1\frac{2}{3}=\frac{3+2}{3}=\frac{5}{3}$

So $\left(\frac{3}{5}\right)\left(1\frac{2}{3}\right)=\left(\frac{3}{5}\right)\left(\frac{5}{3}\right)=\frac{1}{1}\cdot\frac{1}{1}=1$

102. $3=\frac{3}{1}\cdot\frac{16}{16}=\frac{48}{16}$

So $3-\frac{5}{16}=\frac{48}{16}-\frac{5}{16}=\frac{48-5}{16}$

$=\frac{43}{16}$ or $2\frac{11}{16}$

103. $|-3|>2$ since $|-3|=3$

104. $8>|-7|$ since $|-7|=7$

Group Activity/Challenge Exercises

1. $(-4)+(-6)+(-12)=(-10)+(-12)$ $=-22$

2. $5+(-7)+(-8)=(-2)+(-8)=-10$

3. $29+(-46)+37=(-17)+37=20$

4. $1+2+3+\cdots+10=(1+10)+(2+9)\cdots$ $+(5+6)=(5)(11)=55$

5. $1+2+3+\cdots+20=(1+20)+(2+19)+$ $\cdots+(10+11)=(10)(21)=210$

6. $1+2+3+\cdots+100=(1+100)+$ $(2+99)+\cdots+(50+51)=(50)(101)$ $=5050$

Exercise Set 1.6

1. $6-3=6+(-3)=3$

3. $8-9=8+(-9)=-1$

5. $3-3=3+(-3)=0$

7. $(-7)-(-4)=-7+4=-3$

9. $-3-3=-3+(-3)=-6$

11. $3-(-3)=3+3=6$

13. $0-6=0+(-6)=-6$

15. $0-(-6)=0+6=6$

17. $-3-5=-3+(-5)=-8$

19. $-5+7=2$

21. $5-3=5+(-3)=2$

23. $6-(-3)=6+3=9$

25. $8-8=8+(-8)=0$

27. $-8-10=-8+(-10)=-18$

29. $-4-(-2)=-4+2=-2$

31. $(-4)-(-4)=-4+4=0$

33. $6-6=6+(-6)=0$

35. $9-9=9+(-9)=0$

37. $4-5=4+(-5)=-1$

39. $-2-3=-2+(-3)=-5$

41. $-25-16=-25+(-16)=-41$

43. $37-40=37+(-40)=-3$

45. $-100-80=-100+(-80)=-180$

47. $-20-90=-20+(-90)=-110$

49. $-50-(-40)=-50+40=-10$

51. $130-(-90)=130+90=220$

53. $87-87=87+(-87)=0$

55. $-53-(-7)=-53+7=-46$

57. $-15-3=-15+(-3)=-18$

59. $-8-8=-8+(-8)=-16$

61. $18-8=18+(-8)=10$

63. $-5-(-3)=-5+3=-2$

65. $9-(-4)=9+4=13$

67. $18-18=18+(-18)=0$

69. $8-12=8+(-12)=-4$

71. $-4-(-15)=-4+15=11$

73. $45-(-36)=45+36=81$

75. **(a)** Positive: $296-197=296+(-197)$

$|296|$ is greater than $|-197|$ so the sum will be positive.

(b) $296+(-197)=99$

(c) Yes; By part (a) we expect a positive sum. The size of the sum is the sum of the absolute values of the 2 numbers.

77. (a) Negative: $102 - 697 = 102 + (-697)$

$|-697|$ is greater than $|102|$ so sum will be negative.

(b) $102 + (-697) = -595$

(c) Yes; By part (a) we expect a negative sum. The size of the sum is the difference between the larger and smaller absolute values.

79. (a) Positive: $349 - (-498) = 349 + 498$.

The sum of two positive numbers is always positive.

(b) $349 + 498 = 847$

(c) Yes; By part (a) we expect a positive answer. The size of the answer is the sum of the absolute values of the two numbers.

81. (a) Positive: $950 - (-762) = 950 + 762$.

The sum of two positive numbers is always positive.

(b) $950 + 762 = 1712$

(c) Yes; By part (a) we expect a positive answer. The size of the answer is the sum of the absolute values of the two numbers.

83. (a) Positive:

$-408 - (-604) = -408 + 604$.

$|604|$ is greater than $|-408|$ so the sum will be positive.

(b) $-408 + 604 = 196$

(c) Yes; By part (a) we expect a positive answer. The size of the answer is the difference between the larger and smaller absolute values.

85. (a) Negative:

$-1024 - (-576) = -1024 + 576$

$|-1024|$ is greater than $|576|$ so the sum will be negative

(b) $-1024 + 576 = -448$

(c) Yes; By part (a) we expect a negative answer. The size of the answer is the difference between the larger and the smaller absolute values.

87. (a) Positive:

$165.7 - 49.6 = 165.7 + (-49.6)$

$|165.7|$ is greater than $|-49.6|$ so the sum will be positive.

(b) $165.7 + (-49.6) = 116.1$

(c) Yes; By part (a) we expect a negative answer. The size of the answer is the difference between the larger and the smaller absolute values.

89. (a) Zero:

$-37.2 - (-37.2) = -37.2 + 37.2$

Since the two numbers being added are opposite in sign and equal in magnitude, the answer will be zero.

(b) $-37.2+37.2=0$

(c) Yes; the answer is zero as expected.

91. **(a)** Negative: $295-364=295+(-364)$

Since $|-364|$ is greater than $|295|$ the answer will be negative.

(b) $295+(-364)=-69$

(c) Yes; By part (a) we expect a negative answer. The size of the answer is the difference between the larger and the smaller absolute values.

93. **(a)** Negative:

$-1023-647=-1023+(-647)$.

The sum of two negative numbers is always negative.

(b) $-1023+(-647)=-1670$

(c) Yes; The sum of two negative numbers should be (and is) a larger negative number.

95. **(a)** Positive:

$89.7-(-7.62)=89.7+7.62$.

The sum of two positive numbers is always positive.

(b) $89.7+7.62=97.32$

(c) Yes; By part (a) we expect a positive answer. The size of the answer is the sum of the absolute values of the two numbers.

97. $6+5-(+4)=6+5+(-4)$

$=11+(-4)=7$

99. $-3+(-4)+5=-7+5=-2$

101. $-13-(+5)+3=-13+(-5)+3$

$=-18+3=-15$

103. $-9-(-3)+4=-9+3+4=-6+4=-2$

105. $5-(+3)+(-2)=5+(-3)+(-2)$

$=2+(-2)=0$

107. $25+(+12)-(-6)=25+12+6$

$=37+6=43$

109. $-4-7+5=-4+(-7)+5=-11+5=-6$

111. $-4+7-12=-4+7+(-12)=3+(-12)$

$=-9$

113. $45-3-7=45+(-3)+(-7)$

$=42+(-7)=35$

115. $-9-4-8=-9+(-4)+(-8)$

$=-13+(-8)=-21$

117. $-4-13+5=-4+(-13)+5$

$=-17+5=-12$

119. $-9-3-(-4)+5=-9+(-3)+4+5$

$=-12+4+5=-8+5=-3$

121. $32+5-7-12=32+5+(-7)+(-12)$
$=37+(-7)+(-12)=30+(-12)=18$

123. $6-9-(-3)+12=6+(-9)+3+12$
$=-3+3+12=0+12=12$

125. $19+4-20-25=19+4+(-20)+(-25)$
$=23+(-20)+(-25)=3+(-25)=-22$

127. The number of boxes the troop needs to order can be represented as $1246-920$.
$1246-920=1246+(-920)=326$
Thus the troop must order 326 more boxes.

129. Mike's gain can be represented as $+750$.
Kirk's loss can be represented as -496.
$750-(-496)=750+496=1246$
Thus the difference in their performance is \$1246.

131. $44-(-56)=44+56=100$
Thus the temperature dropped by 100°F.

133. **(a)** To subtract -2 from 6, add the opposite of -2, which is 2, to 6.
(b) $6-(-2)=6+2=8$

135. **(a)** Distance can be represented as
$80-(-68)=80+68=148$
In one hour they will be 148 mi apart.
(b) Distance can be represented as
$80-68=80+(-68)=12$
In one hour, they will be 12 mi apart.

Cumulative Review Exercises: 1.6

137. The integers are $\{..., -2, -1, 0, 1, 2, ...\}$.

138. The set of rational numbers together with the set of irrational numbers forms the set of real numbers.

139. $|-3| > -5$ since $|-3| = 3$

140. $|-6| < |-7|$ since $|-6| = 6$ and $|-7| = 7$

Group Activity/Challenge Problems

1. $1-2+3-4+5-6+7-8+9-10$
$=(1-2)+(3-4)+(5-6)+(7-8)$
$+(9-10)=(-1)+(-1)+(-1)+(-1)$
$+(-1)=-5$

3. $-1+2-3+4-5+6-\cdots-99+100$
$=(-1+2)+(-3+4)+(-5+6)+\cdots$
$+(-99+100)=1+1+1+\cdots+1=50$

4. **(a)** 7 units
(b) $5-(-2)=7$

5. $(128-62)+128+59=66+128+59$
$=253$
The rocket traveled 253 feet.

7. $-5+4-3+2-1+5-4+3-2+1$
$=(5-5)+(4-4)+(3-3)+(2-2)+$
$(1-1)=0$

Exercise Set 1.7

1. Since the numbers have like signs, the product is positive. $(-4)(-3)=12$

3. Since the numbers have unlike signs, the product is negative. $3(-3)=-9$

5. Since the numbers have unlike signs, the product is negative. $(-4)(8)=-32$

7. Since the numbers have unlike signs, the product is negative. $9(-1)=-9$

9. Since the numbers have like signs, the product is positive. $-4(-3)=12$

11. Since the numbers have like signs, the product is positive. $-9(-4)=36$

13. Since the numbers have like signs, the product is positive. $8(12)=96$

15. Since the numbers have like signs, the product is positive. $-9(-9)=81$

17. Since the numbers have unlike signs, the product is negative. $-2(5)=-10$

19. Since there are two negative numbers (an even number), the product will be positive. $(-6)(2)(-3)=(-12)(-3)=36$

21. Zero multiplied by any real number equals zero. $0(3)(8)=0(8)=0$

23. Since there are three negative numbers (an odd number), the product will be negative. $(-1)(-1)(-1)=(1)(-1)=-1$

25. Since there are three negative numbers (an odd number), the product will be negative.
$-5(-3)(8)(-1)=(15)(8)(-1)$
$=(120)(-1)=-120$

27. Since there are two negative numbers (an even number), the product will be positive.
$(-4)(3)(-7)(1)=(-12)(-7)(1)$
$=(84)(1)=84$

29. Since there is one negative number (an odd number), the product will be negative.
$(-3)(2)(5)(3)=(-6)(5)(3)$
$=(-30)(3)=-90$

31. Since there are four negative numbers (an even number), the product will be positive.
$(-5)(-6)(-3)(-4)=(30)(-3)(-4)$
$=(-90)(-4)=360$

33. $\left(\frac{-1}{2}\right)\left(\frac{3}{5}\right)=\frac{(-1)\cdot(3)}{2\cdot 5}=\frac{-3}{10}=-\frac{3}{10}$

35. $\left(\frac{-8}{9}\right)\left(\frac{-7}{12}\right)=\frac{(-8)\cdot(-7)}{9\cdot 12}=\frac{(-2)\cdot(-7)}{9\cdot 3}$
$=\frac{14}{27}$

37. $\left(\frac{6}{-3}\right)\left(\frac{4}{-2}\right)=\left(\frac{2}{-1}\right)\left(\frac{2}{-1}\right)=\frac{(2)\cdot(2)}{(-1)\cdot(-1)}$

$=\frac{4}{1}=4$

39. $\left(\frac{5}{-7}\right)\left(\frac{6}{8}\right)=\left(\frac{5}{-7}\right)\left(\frac{3}{4}\right)=\frac{5\cdot 3}{(-7)\cdot(4)}$

$=\frac{15}{-28}=-\frac{15}{28}$

41. Since the numbers have like signs, the quotient is positive. $\frac{6}{2}=3$

43. Since the numbers have like signs, the quotient is positive. $\frac{-16}{-4}=4$

45. Since the numbers have like signs, the quotient is positive. $\frac{-36}{-9}=4$

47. Since the numbers have unlike signs, the quotient is negative. $\frac{-16}{4}=-4$

49. Since the numbers have unlike signs, the quotient is negative. $\frac{18}{-1}=-18$

51. Since the numbers have like signs, the quotient is positive. $\frac{-15}{-3}=5$

53. Since the numbers have like signs, the quotient is positive. $\frac{-6}{-1}=6$

55. Since the numbers have like signs, the quotient is positive. $\frac{-25}{-5}=5$

57. Since the numbers have unlike signs, the quotient is negative. $\frac{1}{-1}=-1$

59. Since the numbers have unlike signs, the quotient is negative. $\frac{-48}{12}=-4$

61. Zero divided by any nonzero number is zero. $\frac{0}{1}=0$

63. Since the numbers have like signs, the quotient is positive. $\frac{-64}{-4}=16$

65. Zero divided by any nonzero number is zero. $\frac{0}{4}=0$

67. Since the numbers have unlike signs, the quotient is negative. $\frac{30}{-10}=-3$

69. Since the numbers have unlike signs, the quotient is negative. $\frac{-180}{30}=-6$

71. Since the numbers have like signs, the quotient is positive. $\frac{-25}{-5}=5$

73. $\frac{5}{12} \div \left(\frac{-5}{9}\right) = \frac{5}{12} \cdot \left(\frac{9}{-5}\right) = \frac{1}{4} \cdot \left(\frac{3}{-1}\right)$
$= \frac{(1)\cdot(3)}{(4)\cdot(-1)} = \frac{3}{-4} = -\frac{3}{4}$

75. $\frac{3}{-10} \div (-8) = \frac{3}{-10} \cdot \frac{1}{-8} = \frac{(3)\cdot(1)}{(-10)\cdot(-8)}$
$= \frac{3}{80}$

77. $\frac{-15}{21} \div \left(\frac{-15}{21}\right) = \frac{-15}{21} \cdot \frac{21}{-15} = \frac{-1}{1} \cdot \frac{1}{-1}$
$= \frac{(-1)\cdot(1)}{(1)\cdot(-1)} = \frac{-1}{-1} = 1$

79. $-12 \div \frac{5}{12} = \frac{-12}{1} \cdot \frac{12}{5} = \frac{(-12)\cdot(12)}{(1)\cdot(5)}$
$= \frac{-144}{5} = -\frac{144}{5}$

81. Since the numbers have unlike signs, the product is negative. $-6 \cdot 5 = -30$

83. Since the numbers have like signs, the quotient is positive. $\frac{-18}{-2} = 9$

85. Since the numbers have like signs, the quotient is positive. $\frac{-50}{-10} = 5$

87. Since the numbers have like signs, the product is positive. $-5(-12) = 60$

89. Zero divided by any nonzero number is zero. $\frac{0}{5} = 0$

91. Since there are three negative numbers (an odd number), the product is negative.
$(-1)(-5)(-9) = (5)(-9) = -45$

93. Since the numbers have unlike signs, the quotient is negative. $\frac{-100}{5} = -20$

95. Since the numbers have unlike signs, thc quotient is negative. $\frac{60}{-60} = -1$

97. Zero divided by any nonzero number is zero. $\frac{0}{6} = 0$

99. Any nonzero number divided by zero is undefined. $\frac{5}{0}$ is undefined.

101. Zero divided by any nonzero number is zero. $\frac{0}{1} = 0$

103. Zero divided by any nonzero number is zero. $\frac{0}{6} = 0$

105. Zero divided by any nonzero number is zero. $\frac{0}{-6} = 0$

107. Any nonzero number divided by zero is undefined. $\frac{3}{0}$ is undefined.

109. **(a)** Since the numbers have unlike signs, the product will be negative.

(b) $96(-15) = -1440$

(c) Yes; As expected the product is negative.

111. **(a)** Since the numbers have unlike signs, the quotient will be negative.

(b) $\frac{-168}{42} = -4$

(c) Yes; As expected the quotient is negative.

113. **(a)** Since the numbers have unlike signs, the quotient will be negative.

(b) $\frac{-240}{15} = -16$

(c) Yes; As expected the quotient is negative.

115. **(a)** Since the numbers have unlike signs, the quotient will be negative.

(b) $\frac{243}{-27} = -9$

(c) Yes; As expected the quotient is negative.

117. **(a)** Since the numbers have like signs, the product will be positive.

(b) $(-15)(-170) = 2550$

(c) Yes; As expected the product is negative.

119. **(a)** Since the numbers have like signs, the product will be positive.

(b) $(-406)(-42) = 17{,}052$

(c) Yes; As expected the product is positive.

121. **(a)** The product will be zero: zero multiplied by any real number is zero.

(b) $(1530)(0) = 0$

(c) Yes; as expected the product is zero.

123. **(a)** Since the numbers have unlike signs, the product will be negative.

(b) $(-19)(10.5) = -199.5$

(c) Yes; as expected the product is negative.

125. **(a)** Undefined; any nonzero number divided by 0 is undefined.

(b) $\frac{7.2}{0}$ is undefined

(c) Yes

127. **(a)** Zero divided by any nonzero number is zero, so quotient is zero.

(b) $\frac{0}{-5260} = 0$

(c) Yes; as expected the answer is 0.

129. **(a)** Since there are three negative numbers (an odd number), the product will be negative.

(b) $(-90)(-1.2)(-1.6) = -172.8$

(c) Yes; as expected the product is negative.

131. **(a)** Since there are two negative numbers (an even number), the product will be positive.

(b) $(9.6)(-12.2)(-60) = 7027.2$

(c) Yes; As expected the product is positive.

133. True; The product of two numbers with unlike signs is a negative number.

135. True; The quotient of two numbers with like signs is a positive number.

137. True

139. False; Zero divided by one zero.

141. False; One divided by zero is undefined.

143. True; Any nonzero number divided by zero is undefined.

145. True; Any nonzero number divided by zero is undefined.

147. Each pair of negative numbers has a positive product.

Cumulative Review Exercises: 1.7

149. $\frac{5}{7} \div \frac{1}{5} = \frac{5}{7} \cdot \frac{5}{1} = \frac{5 \cdot 5}{7 \cdot 1} = \frac{25}{7}$ or $3\frac{4}{7}$

150. $-20 - (-18) = -20 + 18 = -2$

151. $6 - 3 - 4 - 2 = 3 - 4 - 2 = -1 - 2 = -3$

152. $5 - (-2) + 3 - 7 = 5 + 2 + 3 - 7$
$= 7 + 3 - 7 = 10 - 7 = 3$

Group Activity/Challenge Problems

1. $3^4 = 3 \cdot 3 \cdot 3 \cdot 3 = 9 \cdot 3 \cdot 3 = 27 \cdot 3 = 81$

3. $\left(\frac{2}{3}\right)^3 = \frac{2}{3} \cdot \frac{2}{3} \cdot \frac{2}{3} = \frac{2 \cdot 2 \cdot 2}{3 \cdot 3 \cdot 3} = \frac{8}{27}$

9. $\frac{6 \cdot 5 \cdot 4 \cdot 3 \cdot 2 \cdot 1}{(-6)(-5)(-4)(-3)(-2)(-1)} = \frac{720}{720} = 1$

Exercise Set 1.8

1. $5^2 = 5 \cdot 5 = 25$

3. $2^3 = 2 \cdot 2 \cdot 2 = 8$

5. $3^3 = 3 \cdot 3 \cdot 3 = 27$

7. $6^3 = 6 \cdot 6 \cdot 6 = 216$

9. $(-2)^3 = (-2)(-2)(-2) = -8$

11. $(-1)^3 = (-1)(-1)(-1) = -1$

13. $3^3 = 3 \cdot 3 \cdot 3 = 27$

15. $-6^2 = -(6)(6) = -36$

17. $(-6)^2 = (-6)(-6) = 36$

19. $2^4 = 2 \cdot 2 \cdot 2 \cdot 2 = 16$

21. $4^1 = 4$

23. $(-2)^4 = (-2)(-2)(-2)(-2) = 16$

25. $-2^4 = -(2)(2)(2)(2) = -16$

27. $(-4)^3 = (-4)(-4)(-4) = -64$

29. $5^2 \cdot 3^2 = 5 \cdot 5 \cdot 3 \cdot 3 = 225$

31. $5\left(4^2\right) = 5 \cdot 4 \cdot 4 = 80$

33. $2^1 \cdot 4^2 = 2 \cdot 4 \cdot 4 = 32$

35. $3\left(-5^2\right) = 3(-(5)(5)) = 3(-25) = -75$

37. $x \cdot x \cdot y \cdot y = x^2y^2$

39. $xyyyz = xy^3z$

41. $yyzzz = y^2z^3$

43. $xyxyz = x^2y^2z$

45. $a \cdot x \cdot a \cdot x \cdot y = a^2x^2y$

47. $x \cdot y \cdot y \cdot z \cdot z \cdot z = xy^2z^3$

49. $3xyy = 3xy^2$

51. $x^2y = xxy$

53. $xy^3 = xyyy$

55. $xy^2z^3 = xyyzzz$

57. $3^2yz = 3 \cdot 3yz$

59. $2^3x^3y = 2 \cdot 2 \cdot 2xxxy$

61. $(-2)^2y^3z = (-2)(-2)yyyz$

63. Substitute 3 for x

(a) $x^2 = 3^2 = 3 \cdot 3 = 9$

(b) $-x^2 = -3^2 = -(3)(3) = -9$

65. Substitute 4 for x

(a) $x^2 = 4^2 = 4 \cdot 4 = 16$

(b) $-x^2 = -4^2 = -(4)(4) = -16$

67. Substitute -2 for x

(a) $x^2 = (-2)^2 = (-2)(-2) = 4$

(b) $-x^2 = -(-2)^2 = -(-2)(-2) = -4$

69. Substitute 7 for x

(a) $x^2 = 7^2 = 7 \cdot 7 = 49$

(b) $-x^2 = -7^2 = -(7)(7) = -49$

71. Substitute -1 for x

(a) $x^2 = (-1)^2 = (-1)(-1) = 1$

(b) $-x^2 = -(-1)^2 = -(-1)(-1) = -1$

73. Substitute $-\frac{1}{2}$ for x

(a) $x^2 = \left(-\frac{1}{2}\right)^2 = \left(-\frac{1}{2}\right)\left(-\frac{1}{2}\right) = \frac{1}{4}$

(b) $-x^2 = -\left(-\frac{1}{2}\right)^2 = -\left(-\frac{1}{2}\right)\left(-\frac{1}{2}\right)$

$= -\frac{1}{4}$

75. **(a)** Positive: A positive number raised to any power is positive.

(b) $3^5 = 243$

(c) Yes; As expected the answer is positive.

77. **(a)** Negative: A negative number raised to an odd power is negative.

(b) $(-2)^3 = -8$

(c) Yes; As expected the answer is negative.

79. **(a)** Negative: 2^5 is positive so $-\left(2^5\right)$ is negative.

(b) $-2^5 = -32$

(c) Yes; As expected the answer is negative.

81. **(a)** Negative: 5^6 is positive so $-\left(5^6\right)$ is negative.

(b) $-5^6 = -15625$

(c) Yes; As expected the answer is negative.

83. **(a)** Positive: A negative number raised to an even power is positive.

(b) $(-6)^4 = 1296$

(c) Yes; As expected the answer is positive.

85. **(a)** Positive: A positive number raised to any power is positive.

(b) $(8.4)^3 = 592.704$

(c) Yes; As expected the answer is positive.

87. **(a)** Negative: A negative number raised to an odd power is negative.

(b) $(-2.3)^3 = -12.167$

(c) Yes; As expected the answer is negative.

89. **(a)** Positive: A negative number raised to an even power is positive.

(b) $\left(-\frac{1}{2}\right)^4 = 0.0625$

Yes; as expected the answer is positive.

91. **(a)** Positive: A positive number raised to any power is positive.

(b) $\left(\frac{2}{5}\right)^4 = 0.0256$

(c) Yes; As expected the answer is positive.

93. **(a)** Positive: A negative number raised to an even power is positive.

(b) $\left(-\frac{2}{3}\right)^4 = 0.197530864$

(c) Yes; As expected the answer is positive.

95. False: A negative number raised to an even power is positive.

97. False: $(-3)^{15}$ is negative since a negative number raised to an odd power is negative.

Therefore $-(-3)^{15}$ is positive.

99. True: $x^2y = x^2y^1 = xxy$

101. True: $2x^5y = 2^1x^5y^1 = 2xxxxxy$

103. True

105. Any nonzero number will be positive when squared.

107. Positive; An even number of negative numbers are being multiplied.

109. Any real number except 0 raised to the zero power is 1.

Cumulative Review Exercises: 1.8

110. $12-(-6)=12+6=18$

111. $-4-3+9-7=-7+9-7=2-7=-5$

112. $-4672-5692=-10,364$

113. $\left(\frac{-5}{7}\right)\div\left(\frac{-3}{14}\right)=\left(\frac{-5}{7}\right)\cdot\left(\frac{14}{-3}\right)=\frac{-5}{1}\cdot\frac{2}{-3}$

$=\frac{(-5)(2)}{(1)(-3)}=\frac{10}{3}$ or $3\frac{1}{3}$

114. $\frac{0}{4}=0$ since zero divided by any nonzero number is 0.

Group Activity/Challenge Problems

1. **(a)** $2^2\cdot 2^3 = 2^{2+3} = 2^5$

(b) $3^2\cdot 3^3 = 3^{2+3} = 3^5$

(c) $2^3\cdot 2^4 = 2^{3+4} = 2^7$

(d) $x^m\cdot x^n = x^{m+n}$

3. **(a)** $\left(2^3\right)^2 = 2^{3\cdot 2} = 2^6$

(b) $\left(3^3\right)^2 = 3^{3\cdot 2} = 3^6$

(c) $(c)\left(4^2\right)^2 = 4^{2\cdot 2} = 4^4$

(d) $\left(x^m\right)^n = x^{m\cdot n} = x^{mn}$

Exercise Set 1.9

1. $3+4\cdot 5=3+20=23$

3. $6-6+8=0+8=8$

5. $1+3\cdot 2^2=1+3\cdot 4=1+12=13$

7. $-4^2+6=-16+6=-10$

9. $(4-3)\cdot(5-1)^2=(1)\cdot(4)^2=1\cdot 16=16$

11. $3\cdot 7+4\cdot 2=21+8=29$

13. $[1-(4\cdot 5)]+6=[1-20]+6$
$=-19+6=-13$

15. $4^2-3\cdot 4-6=16-3\cdot 4-6$
$=16-12-6=4-6=-2$

17. $-2[-5+(3-4)]=-2[-5+(-1)]$
$=-2(-6)=12$

19. $(6\div 3)^3+4^2\div 8=(2)^3+4^2\div 8$
$=8+16\div 8=8+2=10$

21. $-4^2+8\div 2\cdot 5+3=-16+4\cdot 5+3$
$=-16+20+3=4+3=7$

23. $3+\left(4^2-10\right)^2-3=3+(16-10)^2-3$
$=3+(6)^2-3=3+36-3=39-3=36$

25. $[6-(-2-3)]^2=[6-(-5)]^2=[6+5]^2$
$=(11)^2=121$

27. $\left(3^2-1\right)\div(3+1)^2=(9-1)\div(3+1)^2$
$=8\div(4)^2=8\div 16=\frac{8}{16}=\frac{1}{2}$

29. $-[(56\div 7)-6\div 2]=-[8-6\div 2]$
$=-[8-3]=-(5)=-5$

31. $2\left[3\left(8-2^2\right)-6\right]=2[3(8-4)-6]$
$=2[3(4)-6]=2[12-6]=2(6)=12$

33. $10-[8-(3+4)]^2=10-[8-7]^2$
$=10-(1)^2=10-1=9$

35. $\left[4+\left((5-2)^2\div 3\right)^2\right]^2$
$=\left[4+\left((3)^2\div 3\right)^2\right]^2$
$=\left[4+(9\div 3)^2\right]^2=\left[4+(3)^2\right]^2$
$=[4+9]^2=(13)^2=169$

37. $\left[-3(4-2)^2\right]^2-\left[-3(3-5)^2\right]$

$=\left[-3(2)^2\right]^2-\left[-3(-2)^2\right]$

$=[-3(4)]^2-[-3(4)]$

$=(-12)^2-[-12]=144-[-12]$

$=144+12=156$

39. $(14\div 7\cdot 7\div 7-7)^2=(2\cdot 7\div 7-7)^2$

$=(14\div 7-7)^2=(2-7)^2=(-5)^2=25$

41. $(8.4+3.1)^2-(3.64-1.2)$

$=(11.5)^2-(2.44)=132.25-2.44$

$=129.81$

43. $(4.3)^2+2(5.3)-3.05$

$=18.49+10.6-3.05$

$=29.09-3.05=26.04$

45. $\left(\frac{2}{7}+\frac{3}{8}\right)-\frac{3}{112}=\left(\frac{16}{56}+\frac{21}{56}\right)-\frac{3}{112}$

$=\frac{37}{56}-\frac{3}{112}=\frac{74}{112}-\frac{3}{112}=\frac{71}{112}$

47. $\frac{3}{4}-4\cdot\frac{5}{40}=\frac{3}{4}-\frac{4}{1}\cdot\frac{5}{40}=\frac{3}{4}-\frac{4}{8}$

$=\frac{3}{4}-\frac{2}{4}=\frac{1}{4}$

49. $2\left(3+\frac{2}{5}\right)\div\left(\frac{3}{5}\right)^2=2\left(\frac{15}{5}+\frac{2}{5}\right)\div\frac{9}{25}$

$=2\left(\frac{17}{5}\right)\div\frac{9}{25}=\frac{34}{5}\div\frac{9}{25}$

$=\frac{34}{5}\cdot\frac{25}{9}=\frac{170}{9}$

51. $6\cdot 3$ Multiply 6 by 3

$(6\cdot 3)-4$ Subtract 4 from the product

$[(6\cdot 3)-4]-2$

Subtract 2 from the difference

Evaluate: $[(6\cdot 3)-4]-2=[18-4]-2$

$=14-2=12$

53. $20\div 5$ Divide 20 by 5

$(20\div 5)+12$ Add 12 to the quotient

$[(20\div 5)+12]-8$

Subtract 8 from the sum

$9[[(20\div 5)+12]-8]$

Multiply the difference by 9

Evaluate:

$9[[(20\div 5)+12]-8]=9[[4+12]-8]$

$=9[16-8]=9(8)=72$

55. $\frac{4}{5}+\frac{3}{7}$ Add $\frac{4}{5}$ to $\frac{3}{7}$

$\left(\frac{4}{5}+\frac{3}{7}\right)\cdot\frac{2}{3}$ Multiply the sum by $\frac{2}{3}$

Evaluate:

$$\left(\frac{4}{5}+\frac{3}{7}\right)\cdot\frac{2}{3}=\left(\frac{28}{35}+\frac{15}{35}\right)\cdot\frac{2}{3}$$

$$=\left(\frac{43}{35}\right)\cdot\left(\frac{2}{3}\right)=\frac{86}{105}$$

57. Substitute -2 for x in the expression.

$x+4=-2+4=2$

59. Substitute 4 for x in the expression.

$3x-2=3(4)-2=12-2=10$

61. Substitute -3 for x in the expression.

$x^2-6=(-3)^2-6=9-6=3$

63. Substitute 1 for x in the expression.

$-3x^2-4=-3(1)^2-4$

$=-3(1)-4=-3-4=-7$

65. Substitute -3 for each x in the expression.

$-4x^2-2x+5=-4(-3)^2-2(-3)+5$

$=-4(9)-(-6)+5=-36+6+5$

$=-30+5=-25$

67. Substitute 7 for the x in the expression.

$3(x-2)^2=3(7-2)^2$

$=3(5)^2=3(25)=75$

69. Substitute 1 for each x in the expression.

$2(x-3)(x+4)=2(1-3)(1+4)$

$=2(-2)(5)=(-4)(5)=-20$

71. Substitute 2 for x and 4 for y in the expression.

$-6x+3y=-6(2)+3(4)=-12+12=0$

73. Substitute -2 for x and -3 for y in the expression.

$x^2-y^2=(-2)^2-(-3)^2=4-9=-5$

75. Substitute 2 for x and -3 for y in the expression.

$4(x+y)^2+4x-3y$

$=4(2+(-3))^2+4(2)-3(-3)$

$=4(-1)^2+8-(-9)=4(1)+8+9$

$=4+8+9=12+9=21$

77. Substitute 4 for each a and -1 for each b in the expression.

$3(a+b)^2+4(a+b)-6$

$=3(4+(-1))^2+4(4+(-1))-6$

$=3(3)^2+4(3)-6=3(9)+12-6$

$=27+12-6=39-6=33$

79. Substitute 2 for each x and 3 for each y in the expression.

$x^2y-6xy+3x=(2)^2(3)-6(2)(3)+3(2)$

$=(4)(3)-12(3)+6=12-36+6$

$=-24+6=-18$

81. Substitute 2 for each x and –3 for each y in the expression.

$6x^2 + 3xy - y^2$

$= 6(2)^2 + 3(2)(-3) - (-3)^2$

$= 6(4) + (6)(-3) - 9 = 24 + (-18) - 9$

$= 6 - 9 = -3$

83. Substitute –2 for x and –1 for y in the expression.

$5(2x-3)^2 - 4(6-y)^2$

$= 5(2(-2)-3)^2 - 4(6-(-1))^2$

$= 5(-4-3)^2 - 4(6+1)^2$

$= 5(-7)^2 - 4(7)^2 = 5(49) - 4(49)$

$= 245 - 196 = 49$

85. Substitute 2 for each x in the expression.

$x^2 + 3x - 5 = (2)^2 + 3(2) - 5$

$= 4 + 6 - 5 = 5$

87. Substitute –3 for each x in the expression.

$x^2 - 4x + 7 = (-3)^2 - 4(-3) + 7$

$= 9 - (-12) + 7 = 9 + 12 + 7$

$= 21 + 7 = 28$

89. Substitute 3 for each x in the expression.

$-x^2 + 6x - 5 = -(3)^2 + 6(3) - 5$

$= -9 + 18 - 5 = 9 - 5 = 4$

91. Substitute –3 for each x in the expression.

$-x^2 - 2x - 5 = -(-3)^2 - 2(-3) - 5$

$= -(9) - (-6) - 5 = -9 + 6 - 5$

$= -3 - 5 = -8$

93. Substitute 5 for each x in the expression.

$2x^2 - 4x - 10 = 2(5)^2 - 4(5) - 10$

$= 2(25) - 20 - 10 = 50 - 20 - 10$

$= 30 - 10 = 20$

95. Substitute 5 for each x in the expression.

$-x^2 - 6x + 8 = -(5)^2 - 6(5) + 8$

$= -25 - 30 + 8 = -55 + 8 = -47$

97. Substitute 5 for each x in the expression.

$x^2 - 16x + 5 = (5)^2 - 16(5) + 5$

$= 25 - 80 + 5 = -55 + 5 = -50$

99. Substitute 4 for each x in the expression.

$x^2 + 8x - 10 = (4)^2 + 8(4) - 10$

$= 16 + 32 - 10 = 48 - 10 = 38$

101. Individual Answer

103. **(a)** Individual Answer

(b) $-4x^2 + 3x - 6 = -4(5)^2 + 3(5) - 6$

$= -4(25) + 3(5) - 6$

$= -100 + 15 - 6 = -91$

Cumulative Review Exercises: 1.9

104. $(-2)(-4)(6)(-1)(-3) = (8)(6)(-1)(-3)$
$= (48)(-1)(-3) = (-48)(-3) = 144$

105. **(a)** $x^2 = (-5)^2 = (-5)(-5) = 25$

(b) $-x^2 = -(-5)^2 = -(-5)(-5) = -25$

106. $(-2)^4 = (-2)(-2)(-2)(-2) = 16$

107. $-2^4 = -(2)(2)(2)(2) = -16$

Group Activity/Challenge Problems

1. $4\left(\left[3(4-2)^2\right]+4\right) = 4\left([3\cdot 2]^2+4\right)$
$= 4\left(6^2+4\right) = 4(36+4) = 4(40) = 160$

2. $\left[(3-6)^2+4\right]^2 + 3\cdot 4 - 12 \div 3$
$= \left[(-3)^2+4\right]^2 + 12 - 4$
$= (9+4)^2 + 12 - 4 = (13)^2 + 12 - 4$
$= 169 + 12 - 4 = 177$

3. $-2\left[\left(3(-2)^2+4\right)^2 - \left(3(-2)^2-2\right)^2\right]$
$= -2\left[(3\cdot 4+4)^2 - (3\cdot 4-2)^2\right]$
$= -2\left[(12+4)^2 - (12-2)^2\right]$
$= -2[256-100] = -2(156) = -312$
$= -2\left[(-16)^2 - (10)^2\right]$

5. $12 - (4-6) + 10 = 24$

7. $30 + (15 \div 5) + 10 \div 2 = 38$

Exercise Set 1.10

1. Distributive property

3. Commutative property of multiplication

5. Distributive property

7. Associative property of multiplication

9. Distributive property

11. $3 + 4 = 4 + 3$

13. $(-6\cdot 4)\cdot 2$

15. $(y)(6)$

17. $1\cdot x + 1\cdot y$ or $x + y$

19. $3y + 4x$

21. $5(x+y)$

23. $3(x+2)$

25. $3x + (4+6)$

27. $(x+y)3$

29. $4x + 4y + 12$

31. Commutative property of addition

33. Distributive property

35. Commutative property of addition

37. Distributive property

39. The order does not affect the outcome so the process is commutative.

41. The order affects the outcome, so the process is not commutative.

43. Individual Answer

Cumulative Review Exercises: 1.10

44. $2\frac{3}{5} = \frac{13}{5} = \frac{3}{3} \cdot \frac{13}{5} = \frac{39}{15}, \frac{2}{3} = \frac{2}{3} \cdot \frac{5}{5} = \frac{10}{15}$

$2\frac{3}{5} + \frac{2}{3} = \frac{39}{15} + \frac{10}{15} = \frac{49}{15}$ or $3\frac{4}{15}$

45. $3\frac{5}{8} = \frac{29}{8} = \frac{58}{16}, 2\frac{3}{16} = \frac{35}{16}$

$3\frac{5}{8} - 2\frac{3}{16} = \frac{58}{16} - \frac{35}{16} = \frac{23}{16}$ or $1\frac{7}{16}$

46. $12 - 24 \div 8 + 4 \cdot 3^2 = 12 - 24 \div 8 + 4 \cdot 9$

$= 12 - 3 + 36 = 9 + 36 = 45$

47. Substitute 2 for x and –3 for y

$-4x^2 + 6xy + 3y^2$

$= -4(2)^2 + 6(2)(-3) + 3(-3)^2$

$= -4(4) + 12(-3) + 3(9)$

$= -16 + (-36) + 27 = -52 + 27 = -25$

Group Activity/Challenge Problem

1. $2 + (3 + 4) = (3 + 4) + 2$

Commutative Property of Addition

Chapter 1 Review Exercises

1. $\frac{3}{5} \cdot \frac{5}{6} = \frac{1}{1} \cdot \frac{1}{2} = \frac{1 \cdot 1}{1 \cdot 2} = \frac{1}{2}$

2. $\frac{2}{5} \div \frac{10}{9} = \frac{2}{5} \cdot \frac{9}{10} = \frac{1}{5} \cdot \frac{9}{5} = \frac{1 \cdot 9}{5 \cdot 5} = \frac{9}{25}$

3. $\frac{5}{12} \div \frac{3}{5} = \frac{5}{12} \cdot \frac{5}{3} = \frac{5 \cdot 5}{12 \cdot 3} = \frac{25}{36}$

4. $\frac{5}{6} + \frac{1}{3} = \frac{5}{6} + \frac{1}{3} \cdot \frac{2}{2} = \frac{5}{6} + \frac{2}{6} = \frac{7}{6}$ or $1\frac{1}{6}$

5. $\frac{3}{8} - \frac{1}{9} = \frac{3}{8} \cdot \frac{9}{9} - \frac{1}{9} \cdot \frac{8}{8} = \frac{27}{72} - \frac{8}{72} = \frac{19}{72}$

6. $2\frac{1}{3} = \frac{6+1}{3} = \frac{7}{3} = \frac{7}{3} \cdot \frac{5}{5} = \frac{35}{15}$

$1\frac{1}{5} = \frac{5+1}{5} = \frac{6}{5} = \frac{6}{5} \cdot \frac{3}{3} = \frac{18}{15}$

Thus, $2\frac{1}{3} - 1\frac{1}{5} = \frac{35}{15} - \frac{18}{15} = \frac{17}{15}$ or $1\frac{2}{15}$

7. $\{1, 2, 3, ...\}$

8. $\{0, 1, 2, 3, ...\}$

9. $\{..., -3, -2, -1, 0, 1, 2, ...\}$

10. The set of rational numbers is the set of all numbers which can be expressed as the quotient of two integers, denominator not zero.

Rational numbers = {quotient of two integers, denominator not 0}.

11. The set of real numbers consists of all numbers that can be represented on the real number line.

Real numbers = {all numbers that can be represented on the real number line}.

12. **(a)** 3, 426

(b) 3, 0, 426; The whole numbers include 0.

(c) 3, −5, −12, 0, 426; Integers include positive and negative integers and zero.

(d) $3, -5, -12, 0, \frac{1}{2}, -0.62, 426, -3\frac{1}{4}$; Any number that can be represented as a quotient of two integers.

(e) $\sqrt{7}$; Any number that is not rational.

(f) $3, -5, -12, 0, \frac{1}{2}, -0.62, \sqrt{7}, 426, -3\frac{1}{4}$; The real numbers include the rational and irrational numbers.

13. **(a)** 1

(b) 1

(c) −8, −9

(d) −8, −9, 1

(e) $-2.3, -8, -9, 1\frac{1}{2}, 1, -\frac{3}{17}$; Any number that can be represented as a quotient of two integers.

(f) $-2.3, -8, -9, 1\frac{1}{2}, \sqrt{2}, -\sqrt{2}, 1, -\frac{3}{17}$; The real numbers include the rational and irrational numbers.

14. $-3 > -5$; -3 is to the right of -5 on the number line.

15. $-2 < 1$; -2 is to the left of 1 on the number line.

16. $0.6 > -1.3$; 0.6 is to the right of -1.3 on the number line.

17. $-2.6 > -3.6$; -2.6 is to the right of -3.6 on the number line.

18. $0.50 < 0.509$; 0.50 is to the left 0.509 on the number line.

19. $4.6 > 4.06$; 4.6 is to the right of 4.06 on the number line.

20. $-3.2 < -3.02$; -3.2 is to the left of -3.02 on the number line.

21. $5 > |-3|$ since $|-3|$ equals 3

22. $-3 < |-7|$ since $|-7|$ equals 7

23. $|-2.5| = \left|\frac{5}{2}\right|$ since $|-2.5| = \left|\frac{5}{2}\right| = 2.5$

24. $-3+6=3$

25. $-4+(-5)=-9$

26. $-6+6=0$

27. $4+(-9)=-5$

28. $0+(-3)=-3$

29. $-10+4=-6$

30. $-8-(-2)=-8+2=-6$

31. $-9-(-4)=-9+4=-5$

32. $4-(-4)=4+4=8$

33. $0-2=-2$

34. $-8-1=-9$

35. $2-12=-10$

36. $7-2=5$

37. $2-7=-5$

38. $0-(-4)=0+4=4$

39. $-7-5=-12$

40. $6-4+3=2+3=5$

41. $-5+7-6=2-6=-4$

42. $-5-4-3=-9-3=-12$

43. $-2+(-3)-2=-5-2=-7$

44. $-(-4)+5-(+3)=4+5-3=9-3=6$

45. $7-(+4)-(-3)=7-4+3=3+3=6$

46. $5-2-7+3=3-7+3=-4+3=-1$

47. $4-(-2)+3=4+2+3=6+3=9$

48. Since the numbers have unlike signs, the product is negative; $-4(7)=-28$

49. Since the numbers have like signs, the product is positive; $(-9)(-3)=27$

50. Since the numbers have unlike signs, the product is negative; $4(-9)=-36$

51. Since the numbers have unlike signs, the product is negative; $-2(3)=-6$

52. $\left(\frac{3}{5}\right)\left(\frac{-2}{7}\right)=\frac{3(-2)}{5\cdot 7}=\frac{-6}{35}=-\frac{6}{35}$

53. $\left(\frac{10}{11}\right)\left(\frac{3}{-5}\right)=\frac{2}{11}\cdot\frac{3}{-1}=\frac{2\cdot 3}{(11)\cdot(-1)}$

$=\frac{6}{-11}=-\frac{6}{11}$

54. $\left(\frac{-5}{8}\right)\left(\frac{-3}{7}\right)=\frac{(-5)\cdot(-3)}{8\cdot 7}=\frac{15}{56}$

55. Zero multiplied by any real number is zero. $0\cdot\frac{4}{9}=0$

56. Since there are two negative numbers (an even number), the product is positive.
$4(-2)(-6)=(-8)(-6)=48$

57. Since there are two negative numbers (an even number), the product is positive.
$(-1)(-3)(4)=(3)(4)=12$

58. Since there is one negative number (an odd number), the product is negative.
$-5(2)(7)=(-10)(7)=-70$

59. Since there are three negative numbers (an odd number), the product is negative.
$(-3)(-4)(-5)=(12)(-5)=-60$

60. Since there are three negative numbers (an odd number), the product is negative.
$-1(-2)(3)(-4)=(2)(3)(-4)$
$=6(-4)=-24$

61. Since there are four negative numbers (an even number), the product is positive.
$(-4)(-6)(-2)(-3)=(24)(-2)(-3)$
$=(-48)(-3)=144$

62. Since the numbers have unlike signs, the quotient is negative; $\frac{15}{-3}=-5$

63. Since the numbers have unlike signs, the quotient is negative; $\frac{6}{-2}=-3$

64. Since the numbers have unlike signs, the quotient is negative; $\frac{-20}{5}=-4$

65. Since the numbers have like signs, the quotient is positive; $\frac{-36}{-2}=18$

66. Zero divided by any nonzero number is zero; $\frac{0}{4}=0$

67. Zero divided by any nonzero number is zero; $\frac{0}{-4}=0$

68. Since the numbers have unlike signs, the quotient is negative; $\frac{72}{-9}=-8$

69. Since the numbers have like signs, the quotient is positive; $\frac{-40}{-8}=5$

70. $-4\div\left(\frac{-4}{9}\right)=\frac{-4}{1}\cdot\frac{9}{-4}=\frac{-1}{1}\cdot\frac{9}{-1}=\frac{-9}{-1}=9$

71. $\frac{15}{32}\div(-5)=\frac{15}{32}\cdot\frac{1}{-5}=\frac{3}{32}\cdot\frac{1}{-1}$
$=\frac{3}{-32}=-\frac{3}{32}$

72. $\frac{3}{8}\div\left(\frac{-1}{2}\right)=\frac{3}{8}\cdot\frac{2}{-1}=\frac{3}{4}\cdot\frac{1}{-1}$
$=\frac{3}{-4}=-\frac{3}{4}$

73. $\frac{28}{-3} \div \frac{9}{-2} = \left(\frac{28}{-3}\right) \cdot \left(\frac{-2}{9}\right) = \frac{-56}{-27} = \frac{56}{27}$

74. $\frac{14}{3} \div \left(\frac{-6}{5}\right) = \frac{14}{3} \cdot \left(\frac{5}{-6}\right) = \frac{7}{3} \cdot \frac{5}{-3}$

$= \frac{35}{-9} = -\frac{35}{9}$

75. $\left(\frac{-5}{12}\right) \div \left(\frac{-5}{12}\right) = \left(\frac{-5}{12}\right) \cdot \left(\frac{12}{-5}\right) = \frac{-1}{1} \cdot \frac{1}{-1}$

$= \frac{-1}{-1} = 1$

76. Zero divided by any nonzero number is zero; $\frac{0}{4} = 0$

77. Zero divided by any nonzero number is zero; $\frac{0}{-6} = 0$

78. Any real number divided by zero is undefined; $\frac{8}{0}$ is undefined.

79. Any real number divided by zero is undefined; $\frac{-4}{0}$ is undefined.

80. Any real number divided by zero is undefined; $\frac{8}{0}$ is undefined.

81. Zero divided by any nonzero number is zero; $\frac{0}{-5} = 0$

82. $-4(2-8) = -4(-6) = 24$

83. $2(4-8) = 2(-4) = -8$

84. $(3-6)+4 = -3+4 = 1$

85. $(-4+3)-(2-6) = (-1)-(-4)$

$= -1+4 = 3$

86. $[4+3(-2)]-6 = [4+(-6)]-6$

$= -2-6 = -8$

87. $(-4-2)(-3) = (-6)(-3) = 18$

88. $[4+(-4)]+(6-8) = (0)+(-2) = -2$

89. $9[3+(-4)]+5 = 9(-1)+5 = -9+5 = -4$

90. $-4(-3)+[4 \div (-2)] = (12)+(-2) = 10$

91. $(-3 \cdot 4) \div (-2 \cdot 6) = -12 \div (-12) = 1$

92. $(-3)(-4)+6-3 = 12+6-3$

$= 18-3 = 15$

93. $[-2(3)+6]-4 = [-6+6]-4$

$= 0-4 = -4$

94. $4^2 = (4)(4) = 16$

95. $6^2 = (6)(6) = 36$

96. $9^3 = (9)(9)(9) = 729$

97. $1^5 = (1)(1)(1)(1)(1) = 1$

98. $3^4 = (3)(3)(3)(3) = 81$

99. $2^4 = (2)(2)(2)(2) = 16$

100. $(-3)^3 = (-3)(-3)(-3) = -27$

101. $(-1)^9$
$= (-1)(-1)(-1)(-1)(-1)(-1)(-1)(-1)(-1)$
$= -1$

102. $(-2)^5 = (-2)(-2)(-2)(-2)(-2) = -32$

103. $\left(\frac{2}{7}\right)^2 = \left(\frac{2}{7}\right)\left(\frac{2}{7}\right) = \frac{4}{49}$

104. $\left(\frac{-3}{5}\right)^2 = \left(\frac{-3}{5}\right)\left(\frac{-3}{5}\right) = \frac{9}{25}$

105. $\left(\frac{2}{5}\right)^3 = \left(\frac{2}{5}\right)\left(\frac{2}{5}\right)\left(\frac{2}{5}\right) = \frac{8}{125}$

106. $xxy = x^2y$

107. $xyy = xy^2$

108. $xxyyx = x^3y^2$

109. $yyzz = y^2z^2$

110. $2 \cdot 2 \cdot 3 \cdot 3 \cdot 3xyy = 2^2 \cdot 3^3xy^2$

111. $5 \cdot 7 \cdot 7 \cdot xxy = 5 \cdot 7^2x^2y$

112. $xyxyz = x^2y^2z$

113. $x^2y = xxy$

114. $xz^3 = xzzz$

115. $y^3z = yyyz$

116. $2x^3y^2 = 2xxxyy$

117. Substitute 3 for x;
$-x^2 = -3^2 = -(3)(3) = -9$

118. Substitute –4 for x;
$-x^2 = -(-4)^2 = -(-4)(-4) = -16$

119. Substitute 3 for x;
$-x^3 = -3^3 = -(3)(3)(3) = -27$

120. Substitute –2 for x;
$-x^4 = -(-2)^4 = -(-2)(-2)(-2)(-2)$
$= -16$

121. $3 + 5 \cdot 4 = 3 + 20 = 23$

122. $7 - 3^2 = 7 - 9 = -2$

123. $3 \cdot 5 + 4 \cdot 2 = 15 + 8 = 23$

124. $(3-7)^2+6=(-4)^2+6=16+6=22$

125. $6+4\cdot 5=6+20=26$

126. $8-36\div 4\cdot 3=8-9\cdot 3=8-27=-19$

127. $6-3^2\cdot 5=6-9\cdot 5=6-45=-39$

128. $2-(8-3)=2-5=-3$

129. $[6-(3\cdot 5)]+5=[6-15]+5$
$=-9+5=-4$

130. $3[9-(4^2+3)]\cdot 2=3[9-(16+3)]\cdot 2$
$=3[9-19]\cdot 2=3\cdot(-10)\cdot 2$
$=-30\cdot 2=-60$

131. $(-3^2+4^2)+(3^2\div 3)$
$=(-9+16)+(9\div 3)=(7)+(3)=10$

132. $2^3\div 4+6\cdot 3=8\div 4+18$
$=2+18=20$

133. $(4\div 2)^4+4^2\div 2^2=(2)^4+16\div 4$
$=16+16\div 4=16+4=20$

134. $(15-2^2)^2-4\cdot 3+10\div 2$
$=(15-4)^2-12+10\div 2$
$=(11)^2-12+10\div 2=121-12+10\div 2$
$=121-12+5=109+5=114$

135. $4^3\div 4^2-5(2-7)\div 5$
$=64\div 16-5(-5)\div 5$
$=64\div 16-(-25)\div 5$
$=4+25\div 5=4+5=9$

136. Substitute 5 for x;
$4x-6=4(5)-6=20-6=14$

137. Substitute 2 for x;
$8-3x=8-3(2)=8-6=2$

138. Substitute –5 for x;
$6-4x=6-4(-5)=6-(-20)$
$=6+20=26$

139. Substitute 6 for x;
$x^2-5x+3=(6)^2-5(6)+3$
$=36-30+3=6+3=9$

140. Substitute –1 for y;
$5y^2+3y-2=5(-1)^2+3(-1)-2$
$=5(1)-3-2=5-3-2=2-2=0$

141. Substitute 2 for x;
$-x^2+2x-3=-2^2+2(2)-3$
$=-4+4-3=0-3=-3$

142. Substitute –2 for x;
$-x^2+2x-3=-(-2)^2+2(-2)-3$
$=-4+(-4)-3=-8-3=-11$

143. Substitute 1 for x;
$-3x^2 - 5x + 5 = -3(1)^2 - 5(1) + 5$
$= -3(1) - 5 + 5 = -3 - 5 + 5$
$= -8 + 5 = -3$

144. Substitute 3 for x and 4 for y;
$3xy - 5x = 3(3)(4) - 5(3)$
$= 9(4) - 15 = 36 - 15 = 21$

145. Substitute –3 for x;
$-x^2 - 8x - 12 = -(-3)^2 - 8(-3) - 12$
$= -9 - (-24) - 12 = -9 + 24 - 12$
$= 15 - 12 = 3$

146. **(a)** $158 + (-493) = -335$

(b) $|-493|$ is greater than $|158|$ so the sum should be (and is) negative.

147. **(a)** $324 - (-29.6) = 324 + 29.6 = 353.6$

(b) The sum of two positive numbers is always positive. As expected, the answer is positive.

148. **(a)** $\frac{-17.28}{6} = -2.88$

(b) Since the numbers have unlike signs, the quotient is negative, as expected.

149. **(a)** $(-62)(-1.9) = 117.8$

(b) Since the numbers have like signs, the product is positive, as expected.

150. **(a)** $5^7 = 78{,}125$

(b) A positive number raised to any power is positive. As expected, the answer is positive.

151. **(a)** $(-3)^6 = 729$

(b) A negative number raised to an even power is positive. As expected, the answer is positive.

152. **(a)** $-(4.2)^3 = -74.088$

(b) Since $(4.2)^3$ is positive, $-(4.2)^3$ should be (and is) negative.

153. Substitute 5 for x;
$3x^2 - 4x + 3 = 3(5)^2 - 4(5) + 3$
$= 3(25) - 20 + 3 = 75 - 20 + 3$
$= 55 + 3 = 58$

154. Substitute –2 for x;
$-2x^2 - 6x - 3 = -2(-2)^2 - 6(-2) - 3$
$= -2(4) - (-12) - 3 = -8 + 12 - 3$
$= 4 - 3 = 1$

155. Associative property of addition

156. Commutative property of multiplication

157. Distributive property

158. Commutative property of multiplication

159. Commutative property of addition

160. Associative property of addition

161. Commutative property of addition

Chapter 1 Practice Test

1. **(a)** 42

   (b) 42, 0; The whole numbers include 0.

   (c) –6, 42, 0, –7,.–1; Integers include positive integers, negative integers, and zero.

   (d) $-6, 42, -3\frac{1}{2}, 0, 6.52, \frac{5}{9}, -7, -1$; Any number that can be expressed as a quotient of two integers.

   (e) $\sqrt{5}$; Any number that is not rational.

   (f) $-6, 42, -3\frac{1}{2}, 0, 6.52, \sqrt{5}, \frac{5}{9}, -7, -1$; Integers include positive integers, negative integers, and zero.

2. $-6 < -3$; –6 is to the left of –3 on the number line.

3. $|-3| > |-2|$ since $|-3| = 3$ and $|-2| = 2$.

4. $-4 + (-8) = -12$

5. $-6 - 5 = -11$

6. $4 - (-12) = 4 + 12 = 16$

7. $5 - 12 - 7 = -7 - 7 = -14$

8. $(-4 + 6) - 3(-2) = (2) - (-6) = 2 + 6 = 8$

9. $(-4)(-3)(2)(-1) = (12)(2)(-1)$
   $= (24)(-1) = -24$

10. $\left(\frac{-2}{9}\right) \div \left(\frac{-7}{8}\right) = \frac{-2}{9} \cdot \frac{8}{-7} = \frac{-16}{-63} = \frac{16}{63}$

11. $\left(-12 \cdot \frac{1}{2}\right) \div 3 = \left(\frac{-12}{1} \cdot \frac{1}{2}\right) \div 3$
    $= \left(\frac{-6}{1} \cdot \frac{1}{1}\right) \div 3 = -6 \div 3 = -2$

12. $3 \cdot 5^2 - 4 \cdot 6^2 = 3 \cdot 25 - 4 \cdot 36$
    $= 75 - 144 = -69$

13. $(4 - 6^2) \div [4(2+3) - 4]$
    $= (4 - 36) \div [4(5) - 4]$
    $= (-32) \div [20 - 4] = -32 \div 16 = -2$

14. $-6(-2-3) \div 5 \cdot 2 = -6(-5) \div 5 \cdot 2$
    $= 30 \div 5 \cdot 2 = 6 \cdot 2 = 12$

15. $(-3)^4 = (-3)(-3)(-3)(-3) = 81$

16. $\left(\frac{3}{5}\right)^3 = \left(\frac{3}{5}\right)\left(\frac{3}{5}\right)\left(\frac{3}{5}\right) = \frac{27}{125}$

17. $2 \cdot 2 \cdot 5 \cdot 5 \cdot yyzzz = 2^2 5^2 y^2 z^3$

18. $2^2 3^3 x^4 y^2 = 2 \cdot 2 \cdot 3 \cdot 3 \cdot 3xxxxyy$

19. Substitute –4 for x;
    $2x^2 - 6 = 2(-4)^2 - 6 = 2(16) - 6$
    $= 32 - 6 = 26$

20. Substitute 3 for x and –2 for y;
    $6x - 3y^2 + 4 = 6(3) - 3(-2)^2 + 4$
    $= 18 - 3(4) + 4 = 18 - 12 + 4 = 6 + 4 = 10$

21. Substitute –2 for each x in the expression;
    $-x^2 - 6x + 3 = -(-2)^2 - 6(-2) + 3$
    $= -4 - (-12) + 3 = -4 + 12 + 3$
    $= 8 + 3 = 11$

22. Commutative property of addition

23. Distributive property

24. Associative property of addition

25. Commutative property of multiplication

CHAPTER 2

Exercise Set 2.1

1. Since $2 + 3 = 5$, then $2x + 3x = 5x$

3. Since $4 - 5 = -1$, then $4x - 5x = -x$

5. The 12 and –3 are like terms.
Rearranging terms: $x + 12 - 3$
Combining like terms: $x + 9$

7. Since $-2 + 5 = 3$, then $-2x + 5x = 3x$

9. The x and $3x$ are like terms.
Combining like terms:
$x + 3x - 7 = 4x - 7$

11. The 6 and –3 are like terms.
Rearranging terms: $2x + 6 - 3$
Combining like terms: $2x + 3$

13. The –4 and 12 are like terms.
Rearranging terms: $5x - 4 + 12$
Combining like terms: $5x + 8$

15. The $2y$ and y are like terms.
Rearranging terms: $5x + 2y + y + 3$
Combining like terms: $5x + 3y + 3$

17. The $4x$ and $-2x$ are like terms.
The 3 and –7 are like terms.
Combining like terms:
$4x - 2x + 3 - 7 = 2x - 4$

19. The 5 and 3 are like terms.
Rearranging terms: $x + 5 + 3$
Combining like terms: $x + 8$

21. The $-3x$ and $-5x$ are like terms.
Rearranging terms: $-3x - 5x + 2$
Combining like terms: $-8x + 2$

23. The $2x$ and $-4x$ are like terms.
The 5 and 6 are like terms.
Rearranging terms: $2x - 4x + 5 + 6$
Combining like terms: $-2x + 11$

25. The x and $2x$ are like terms.
The –2 and –4 are like terms.
Rearranging terms: $x + 2x - 2 - 4$
Combining like terms: $3x - 6$

27. The $-3x$ and $-2x$ are like terms.
The 2 and 1 are like terms.
Rearranging terms: $-3x - 2x + 2 + 1$
Combining like terms: $-5x + 3$

29. The $2y$ and $4y$ are like terms.
Combining like terms:
$2y + 4y + 6 = 6y + 6$

31. The x and $3x$ are like terms.
The –6 and –4 are like terms.
Rearranging terms: $x + 3x - 6 - 4$
Combining like terms: $4x - 10$

33. The $-x$ and $4x$ are like terms.
The 4 and –8 are like terms.
Rearranging terms: $-x + 4x + 4 - 8$
Combining like terms: $3x - 4$

35. The $\frac{3}{4}$ and $-\frac{1}{3}$ are like terms.
Combining like terms:
$$x + \frac{3}{4} - \frac{1}{3} = x + \frac{9}{12} - \frac{4}{12} = x + \frac{5}{12}$$

37. The $68.2x$ and $19.7x$ are like terms.
Combining like terms:
$$68.2x - 19.7x + 8.3 = 48.5x + 8.3$$

39. The $\frac{1}{2}y$ and $-\frac{3}{8}y$ are like terms.
Combining like terms:
$$x + \frac{1}{2}y - \frac{3}{8}y = x + \frac{4}{8}y - \frac{3}{8}y = x + \frac{1}{8}y$$

41. The –3.1 and –5.2 are like terms.
Combining like terms:
$$-4x - 3.1 - 5.2 = -4x - 8.3$$

43. The x and $-3x$ are like terms.
The 1 and 6 are like terms.
Rearranging terms: $x - 3x + 1 + 6$
Combining like terms: $-2x + 7$

45. The $3x$ and $4x$ are like terms.
The –7 and –9 are like terms.
Rearranging terms: $3x + 4x - 7 - 9$
Combining like terms: $7x - 16$

47. The $4x$ and $3x$ are like terms.
The 6 and –7 are like terms.
Rearranging terms: $4x + 3x + 6 - 7$
Combining like terms: $7x - 1$

49. The –4, –6, and 2 are like terms.
Rearranging terms: $x - 4 - 6 + 2$
Combining like terms: $x - 8$

51. The $40.02x$ and $-18.3x$ are like terms
The –19.36 and 12.25 are like terms.
Rearranging terms:
$40.02x - 18.3x - 19.36 + 12.25$
Combining like terms: $21.72x - 7.11$

53. The $\frac{3}{5}x$ and $-\frac{7}{4}x$ are like terms.
The –3 and –2 are like terms.
Rearranging terms: $\frac{3}{5}x - \frac{7}{4}x - 3 - 2$
Combining like terms:
$$\frac{12}{20}x - \frac{35}{20}x - 5 = -\frac{23}{20}x - 5$$

55. $2(x + 6) = 2x + 2(6) = 2x + 12$

57. $5(x + 4) = 5x + 5(4) = 5x + 20$

59. $-2(x-4)=-2[x+(-4)]=-2x+(-2)(-4)=-2x+8$

61. $-\frac{1}{2}(2x-4)=-\frac{1}{2}[2x+(-4)]=-\frac{1}{2}(2x)+\left(-\frac{1}{2}\right)(-4)=-x+2$

63. $1(-4+x)=(1)(-4)+(1)(x)=-4+x=x-4$

65. $\frac{1}{4}(x-12)=\frac{1}{4}[x+(-12)]=\frac{1}{4}x+\frac{1}{4}(-12)=\frac{1}{4}x+(-3)=\frac{1}{4}x-3$

67. $-0.6(3x-5)=-0.6[3x+(-5)]=(-0.6)(3x)+(-0.6)(-5)=-1.8x+3$

69. $\frac{1}{2}(-2x+6)=\frac{1}{2}(-2x)+\frac{1}{2}(6)=-x+3$

71. $0.4(2x-0.5)=0.4[2x+(-0.5)]=0.4(2x)+0.4(-0.5)=0.8x+(-0.2)=0.8x-0.2$

73. $-(-x+y)=-1(-x+y)=-1(-x)+(-1)(y)=x+(-y)=x-y$

75. $-(2x-6y+8)=-1[2x+(-6y)+8]=-1(2x)+(-1)(-6y)+(-1)(8)=-2x+6y+(-8)$
$=-2x+6y-8$

77. $4.6(3.1x-2.3y+1.8)=4.6[3.1x+(-2.3y)+1.8]=4.6(3.1x)+4.6(-2.3y)+4.6(1.8)$
$=14.26x+(-10.58y)+8.28=14.26x-10.58y+8.28$

79. $2\left(\frac{1}{2}x-4y+\frac{1}{4}\right)=2\left[\frac{1}{2}x+(-4y)+\frac{1}{4}\right]=2\left(\frac{1}{2}x\right)+2(-4y)+2\left(\frac{1}{4}\right)=x+(-8y)+\frac{1}{2}=x-8y+\frac{1}{2}$

81. $(x+3y-9)=1[x+3y+(-9)]=1(x)+(1)(3y)+(1)(-9)=x+3y+(-9)=x+3y-9$

83. $-(-x+4+2y)=-1(-x+4+2y)=-1(-x)+(-1)(4)+(-1)(2y)=x+(-4)+(-2y)=x-4-2y$

85. $4(x-2)-x = 4x-8-x$ Use the distributive property.

$= 3x-8$ Combine like terms.

87. $-2(3-x)+1 = -6+2x+1$ Use the distributive property.

$= 2x-5$ Combine like terms.

89. $6x+2(4x+9) = 6x+8x+18$ Use the distributive property.

$= 14x+18$ Combine like terms.

91. $2(x-y)+2x+3 = 2x-2y+2x+3$ Use the distributive property.

$= 2x+2x-2y+3$ Rearrange terms.

$= 4x-2y+3$ Combine like terms.

93. $(2x+y)-2x+3 = 2x+y-2x+3$ Use the distributive property.

$= 2x-2x+y+3$ Rearrange terms.

$= y+3$ Combine like terms.

95. $8x-(x-3) = 8x-x+3$ Use the distributive property.

$= 7x+3$ Combine like terms.

97. $2(x-3)-(x+3) = 2x-6-x-3$ Use the distributive property.

$= 2x-x-6-3$ Rearrange terms.

$= x-9$ Combine like terms.

99. $4(x-3)+2(x-2)+4 = 4x-12+2x-4+4$ Use the distributive property.

$= 4x+2x-12-4+4$ Rearrange terms.

$= 6x-12$ Combine like terms.

101. $2(x-4)-3x+6 = 2x-8-3x+6$ Use the distributive property.

$= 2x-3x-8+6$ Rearrange terms.

$= x-2$ Combine like terms.

103. $-3(x-4)+2x-6=-3x+12+2x-6$ — Use the distributive property.
$=-3x+2x+12-6$ — Rearrange terms.
$=-x+6$ — Combine like terms.

105. $4(x-3)+4x-7=4x-12+4x-7$ — Use the distributive property.
$=4x+4x-12-7$ — Rearrange terms.
$=8x-19$ — Combine like terms.

107. $0.4+(x+5)-0.6+2=0.4+x+5-0.6+2$ — Use the distributive property.
$=x+0.4+5-0.6+2$ — Rearrange terms.
$=x+6.8$ — Combine like terms.

109. $9-(-3x+4)-5=9+3x-4-5$ — Use the distributive property.
$=3x+9-4-5$ — Rearrange terms.
$=3x$ — Combine like terms.

111. $4(x+2)-3(x-4)-5=4x+8-3x+12-5$ — Use the distributive property.
$=4x-3x+8+12-5$ — Rearrange terms.
$=x+15$ — Combine like terms.

113. $-0.2(2-x)+4(y+0.2)=-0.4+0.2x+4y+0.8$ — Use the distributive property
$=0.2x+4y-0.4+0.8$ — Rearrange terms.
$=0.2x+4y+0.4$ — Combine like terms.

115. $-6x+3y-(6+x)+(x+3)=-6x+3y-6-x+x+3$ — Use the distributive property.
$=-6x-x+x+3y-6+3$ — Rearrange terms.
$=-6x+3y-3$ — Combine like terms.

117. $-(x+3)+(2x+4)-6=-x-3+2x+4-6$ — Use the distributive property.
$=-x+2x-3+4-6$ — Rearrange terms.
$=x-5$ — Combine like terms.

119. $\frac{2}{3}(x-2)-\frac{1}{2}(x+4)=\frac{2}{3}x-\frac{4}{3}-\frac{1}{2}x-2$ Use the distributive property.

$=\frac{2}{3}x-\frac{1}{2}x-\frac{4}{3}-2$ Rearrange terms.

$=\frac{1}{6}x-\frac{10}{3}$ Combine like terms.

121. The signs of all the terms inside the parentheses are changed when the parentheses are removed.

123. **(a)** $2x^2, 3x, -5$; Terms are added or subtracted.

(b) $1, 2, x, 2x, x^2, 2x^2$; Expressions that are multiplied are factors.

Cumulative Review Exercises: 2.1

124. $|-7| = 7$

125. $-|-16| = -(16) = -16$

126. Individual answer.

127. Substitute -1 for each x in the expression.

$-x^2 + 5x - 6 = -(-1)^2 + 5(-1) - 6 = -1 + (-5) - 6 = -6 - 6 = -12$

Group Activity/Challenge Problems

1. $4x + 5y + 6(3x - 5y) - 4x + 3 = 4x + 5y + 18x - 30y - 4x + 3$

$= 4x + 18x - 4x + 5y - 30y + 3 = 18x - 25y + 3$

3. $x^2 + 2y - y^2 + 3x + 5x^2 + 6y^2 + 5y$

$= x^2 + 5x^2 - y^2 + 6y^2 + 3x + 2y + 5y$

$= 6x^2 + 5y^2 + 3x + 7y$

5. $3x^2 - 10x + 8$

(a) $3x^2$, $-10x$, 8; Expressions added are terms.

(b) 1, 3, x, $3x$, x^2, $3x^2$; Expressions multiplied are factors.

(c) 1, 2, 4, 8, −1, −2, −4, −8

Exercise Set 2.2

1. Substitute 4 for x, $x = 4$

$$2x - 3 = 5$$
$$2(4) - 3 = 5$$
$$8 - 3 = 5$$
$$5 = 5$$

Since we obtain a true statement, 4 is a solution.

3. Substitute −3 for x, $x = -3$

$$2x - 5 = 5(x + 2)$$
$$2(-3) - 5 = 5[(-3) + 2]$$
$$-6 - 5 = 5(-1)$$
$$-11 = -5$$

Since we obtain a false statement, −3 is not a solution.

5. Substitute 0 for x, $x = 0$

$$3x - 5 = 2(x + 3) - 11$$
$$3(0) - 5 = 2(0 + 3) - 11$$
$$0 - 5 = 2(3) - 11$$
$$-5 = 6 - 11$$
$$-5 = -5$$

Since we obtain a true statement, 0 is a solution.

7. Substitute 2.3 for x, $x = 2.3$

$$5(x + 2) - 3(x - 1) = 4$$
$$5(2.3 + 2) - 3(2.3 - 1) = 4$$
$$5(4.3) - 3(1.3) = 4$$
$$21.5 - 3.9 = 4$$
$$17.6 = 4$$

Since we obtain a false statement, 2.3 is not a solution.

9. Substitute $\frac{1}{2}$ for x, $x = \frac{1}{2}$

$$4x - 4 = 2x - 3$$
$$4\left(\frac{1}{2}\right) - 4 = 2\left(\frac{1}{2}\right) - 3$$
$$2 - 4 = 1 - 3$$
$$-2 = -2$$

Since we obtain a true statement, $\frac{1}{2}$ is a solution.

11. Substitute $\frac{11}{2}$ for x, $x = \frac{11}{2}$

$$3(x + 2) = 5(x - 1)$$
$$3\left(\frac{11}{2} + 2\right) = 5\left(\frac{11}{2} - 1\right)$$
$$3\left(\frac{15}{2}\right) = 5\left(\frac{9}{2}\right)$$
$$\frac{45}{2} = \frac{45}{2}$$

Since we obtain a true statement, $\frac{11}{2}$ is a solution.

13. **(a)** $x+5=8$

(b) $x+5=8$

$x+5-5=8-5$

Subtract 5 from both sides of the equation.

$x+0=3$

$x=3$

15. **(a)** $12=x+3$

(b) $12=x+3$

Subtract 3 from both sides of the equation.

$12-3=x+3-3$

$9=x+0$

$9=x$

17. **(a)** $10=x+7$

(b) $10=x+7$

Subtract 7 from both sides of the equation.

$10-7=x+7-7$

$3=x+0$

$3=x$

19. **(a)** $x+6=4+11$

(b) $x+6=4+11$

Subtract 6 from both sides of the equation.

$x+6=15$

$x+6-6=15-6$

$x=9$

21. $x+2=6$

Subtract 2 from both sides of the equation.

$x+2-2=6-2$

$x+0=4$

$x=4$

Check: $x+2=6$

$4+2=6$

$6=6$ True

23. $x+7=-3$

Subtract 7 from both sides of the equation.

$x+7-7=-3-7$

$x+0=-10$

$x=-10$

Check: $x+7=-3$

$-10+7=-3$

$-3=-3$ True

25. $x+4=-5$

Subtract 4 from both sides of the equation.

$x+4-4=-5-4$

$x+0=-9$

$x=-9$

Check: $x+4=-5$

$-9+4=-5$

$-5=-5$ True

27. $x+43=-18$

Subtract 43 from both sides of the equation.

$x+43-43=-18-43$

$x+0=-61$

$x=-61$

Check: $x + 43 = -18$

$-61 + 43 = -18$

$-18 = -18$ True

29. $-8 + x = 14$

Add 8 to both sides of the equation.

$-8 + 8 + x = 14 + 8$

$0 + x = 22$

$x = 22$

Check: $-8 + x = 14$

$-8 + 22 = 14$

$14 = 14$ True

31. $27 = x - 16$

Add 16 to both sides of the equation.

$27 + 16 = x - 16 + 16$

$43 = x + 0$

$43 = x$

Check: $27 = x - 16$

$27 = 43 - 16$

$27 = 27$ True

33. $-13 = x - 1$

Add 1 to both sides of the equation.

$-13 + 1 = x - 1 + 1$

$-12 = x + 0$

$-12 = x$

Check: $-13 = x - 1$

$-13 = -12 - 1$

$-13 = -13$ True

35. $29 = -43 + x$

Add 43 to both sides of the equation.

$29 + 43 = -43 + 43 + x$

$72 = 0 + x$

$72 = x$

Check: $29 = -43 + x$

$29 = -43 + 72$

$29 = 29$ True

37. $7 + x = -19$

Subtract 7 from both sides of the equation.

$7 - 7 + x = -19 - 7$

$0 + x = -26$

$x = -26$

Check: $7 + x = -19$

$7 + (-26) = -19$

$-19 = -19$ True

39. $x + 29 = -29$

Subtract 29 from both sides of the equation.

$x + 29 - 29 = -29 - 29$

$x + 0 = -58$

$x = -58$

Check: $x + 29 = -29$

$-58 + 29 = -29$

$-29 = -29$ True

41. $9 = x - 3$

Add 3 to both sides of the equation.

$9 + 3 = x - 3 + 3$

$12 = x + 0$

$12 = x$

Check: $9 = x - 3$

$9 = 12 - 3$

$9 = 9$ True

43. $x + 7 = -5$

Subtract 7 from both sides of the equation.

$x + 7 - 7 = -5 - 7$

$x + 0 = -12$

$x = -12$

Check: $x + 7 = -5$

$-12 + 7 = -5$

$-5 = -5$ True

45. $9 + x = 12$

Subtract 9 from both sides of the equation.

$9 - 9 + x = 12 - 9$

$0 + x = 3$

$x = 3$

Check: $9 + x = 12$

$9 + 3 = 12$

$12 = 12$ True

47. $-5 = 4 + x$

Subtract 4 from both sides of the equation.

$-5 - 4 = 4 - 4 + x$

$-9 = 0 + x$

$-9 = x$

Check: $-5 = 4 + x$

$-5 = 4 + 9$

$-5 = -5$ True

49. $40 = x - 13$

Add 13 to both sides of the equation.

$40 + 13 = x - 13 + 13$

$53 = x + 0$

$53 = x$

Check: $40 = x - 13$

$40 = 53 - 13$

$40 = 40$ True

51. $x - 12 = -9$

Add 12 to both sides of the equation.

$x - 12 + 12 = -9 + 12$

$x + 0 = 3$

$x = 3$

Check: $x - 12 = -9$

$3 - 12 = -9$

$-9 = -9$ True

53. $4 + x = 9$

Subtract 4 from both sides of the equation.

$4 - 4 + x = 9 - 4$

$0 + x = 5$

$x = 5$

Check: $4 + x = 9$

$4 + 5 = 9$

$9 = 9$ True

55. $-8 = -9 + x$

Add 9 to both sides of the equation.

$-8 + 9 = -9 + 9 + x$

$1 = 0 + x$

$1 = x$

Check: $-8 = -9 + x$

$-8 = -9 + 1$

$-8 = -8$ True

57. $5 = x - 12$

Add 12 to both sides of the equation.

$5 + 12 = x - 12 + 12$

$17 = x + 0$

$17 = x$

Check: $5 = x - 12$

$5 = 17 - 12$

$5 = 5$ True

59. $-50 = x - 24$

Add 24 to both sides of the equation.

$-50 + 24 = x - 24 + 24$

$-26 = x + 0$

$-26 = x$

Check: $-50 = x - 24$

$-50 = -26 - 24$

$-50 = -50$ True

61. $16 + x = -20$

Subtract 16 from both sides of the equation.

$16 - 16 + x = -20 - 16$

$0 + x = -36$

$x = -36$

Check: $16 + x = -20$

$16 - 36 = -20$

$-20 = -20$ True

63. $40.2 + x = -7.3$

Subtract 40.2 from both sides of the equation.

$40.2 - 40.2 + x = -7.3 - 40.2$

$0 + x = -47.5$

$x = -47.5$

Check: $40.2 + x = -7.3$

$40.2 + (-47.5) = -7.3$

$-7.3 = -7.3$ True

65. $-37 + x = 9.5$

Add 37 to both sides of the equation.

$-37 + 37 + x = 9.5 + 37$

$0 + x = 46.5$

$x = 46.5$

Check: $-37 + x = 9.5$

$-37 + 46.5 = 9.5$

$9.9 = 9.9$ True

67. $x - 8.42 = -30$

Add 8.42 to both sides of the equation.

$x - 8.42 + 8.42 = -30 + 8.42$

$x + 0 = -21.58$

$x = -21.58$

Check: $x - 8.42 = -30$

$-21.58 - 8.42 = -30$

$-30 = -30$ True

69. $9.75 = x + 9.75$

Subtract 9.75 from both sides of the equation.

$9.75 - 9.75 = x + 9.75 - 9.75$

$0 = x + 0$

$0 = x$

Check: $9.75 = x + 9.75$

$9.75 = 0 + 9.75$

$9.75 = 9.75$ True

71. $600 = x - 120$

Add 120 to both sides of the equation.

$600 + 120 = x - 120 + 120$

$720 = x + 0$

$720 = x$

Check: $600 = x - 120$

$600 = 720 - 120$

$600 = 600$ True

73. **(a)** The number of numbers that make the equation a true statement

(b) To find the solutions to an equation

75. Individual answer

77. **(a)** Get the variable by itself on one side of the equation.

(b) Individual answer

79. Subtract 3 from both sides of the equation.

Cumulative Review Exercises: 2.1

124. $|-7| = 7$

125. $-|-16| = -(16) = -16$

126. Individual answer.

127. Substitute −1 for each x in the expression.

$-x^2 + 5x - 6 = -(-1)^2 + 5(-1) - 6$

$= -1 + (-5) - 6 = -6 - 6 = -12$

Group Activity/Challenge Problems

1. $4x + 5y + 6(3x - 5y) - 4x + 3$

$= 4x + 5y + 18x - 30y - 4x + 3$

$= 4x + 18x - 4x + 5y - 30y + 3$

$= 18x - 25y + 3$

3. $x^2 + 2y - y^2 + 3x + 5x^2 + 6y^2 + 5y$

$= x^2 + 5x^2 - y^2 + 6y^2 + 3x + 2y + 5y$

$= 6x^2 + 5y^2 + 3x + 7y$

5. $3x^2 - 10x + 8$

(a) $3x^2, -10x, 8$; Expressions added are terms.

(b) $1, 3, x, 3x, x^2, 3x^2$; Expressions multiplied are factors.

(c) $1, 2, 4, 8, -1, -2, -4, -8$

Exercise Set 2.3

1. **(a)** $2x = 10$

(b) $2x = 10$

Divide both sides of the equation by 2.

$\frac{2x}{2} = \frac{10}{2}$

$x = 5$

3. **(a)** $6 = 3x$

(b) $6 = 3x$

Divide both sides of the equation by 3.

$$\frac{6}{3} = \frac{3x}{3}$$
$$2 = x$$

5. **(a)** $2x = 5$

(b) $2x = 5$

Divide both sides of the equation by 2.

$$\frac{2x}{2} = \frac{5}{2}$$
$$x = \frac{5}{2}$$

7. **(a)** $4 = 3x$

(b) $4 = 3x$

Divide both sides of the equation by 3.

$$\frac{4}{3} = \frac{3x}{3}$$
$$\frac{4}{3} = x$$

9. **(a)** $2x = 6$

Divide both sides of the equation by 2.

$$\frac{2x}{2} = \frac{6}{2}$$
$$x = 3$$

Check: $2x = 6$

$2(3) = 6$

$6 = 6$ True

11. $\frac{x}{2} = 4$

Multiply both sides of the equation by 2.

$$2\left(\frac{x}{2}\right) = 2(4)$$
$$x = 2(4)$$
$$x = 8$$

Check: $\frac{x}{2} = 4$

$\frac{8}{2} = 4$

$4 = 4$ True

13. $-4x = 8$

Divide both sides of the equation by -4.

$$\frac{-4x}{-4} = \frac{8}{-4}$$
$$x = -2$$

Check: $-4x = 8$

$-4(-2) = 8$

$8 = 8$ True

15. $\frac{x}{6} = -2$

Multiply both sides of the equation by 6.

$$6\left(\frac{x}{6}\right) = 6(-2)$$
$$x = 6(-2)$$
$$x = -12$$

Check: $\frac{x}{6} = -2$

$\frac{-12}{6} = -2$

$-2 = -2$ True

17. $\frac{x}{5}=1$

Multiply both sides of the equation by 5.

$5\left(\frac{x}{5}\right)=5(1)$

$x=5(1)$

$x=5$

Check: $\frac{x}{5}=1$

$\frac{5}{5}=1$

$1=1$ True

19. $-32x=-96$

Divide both sides of the equation by –32.

$\frac{-32x}{-32}=\frac{-96}{-32}$

$x=3$

Check: $-32x=-96$

$-32(3)=-96$

$-96=-96$ True

21. $-6=4z$

Divide both sides of the equation by 4

$\frac{-6}{4}=\frac{4z}{4}$

$-\frac{3}{2}=z$

Check: $-6=4z$

$-6=4\left(-\frac{3}{2}\right)$

$-6=-6$ True

23. $-x=-6$

$-1x=-6$

Multiply both sides of the equation by –1.

$(-1)(-1x)=(-1)(-6)$

$1x=6$

$x=6$

Check: $-x=-6$

$-6=-6$ True

25. $-2=-y$

$-2=-1y$

Multiply both sides of the equation by –1.

$(-1)(-2)=(-1)(-1y)$

$2=1y$

$2=y$

Check: $-2=-y$

$-2=-2$ True

27. $-\frac{x}{7}=-7$

$\frac{x}{-7}=-7$

Multiply both sides of the equation by –7.

$(-7)\left(\frac{x}{-7}\right)=(-7)(-7)$

$x=49$

Check: $-\frac{x}{7}=-7$

$-\frac{49}{7}=-7$

$-7=-7$ True

29. $4 = -12x$

Divide both sides of the equation by –12.

$$\frac{4}{-12} = \frac{-12x}{-12}$$

$$-\frac{1}{3} = x$$

Check: $4 = -12x$

$$4 = -12\left(-\frac{1}{3}\right)$$

$$4 = 4 \qquad \text{True}$$

31. $-\frac{x}{3} = -2$

$$\frac{x}{-3} = -2$$

Multiply both sides of the equation by –3.

$$(-3)\left(\frac{x}{-3}\right) = (-3)(-2)$$

$$x = 6$$

Check: $-\frac{x}{3} = -2$

$$-\frac{6}{3} = -2$$

$$-2 = -2 \qquad \text{True}$$

33. $13x = 10$

Divide both sides of the equation by 13.

$$\frac{13x}{13} = \frac{10}{13}$$

$$x = \frac{10}{13}$$

Check: $13x = 10$

$$13\left(\frac{10}{13}\right) = 10$$

$$10 = 10 \qquad \text{True}$$

35. $-4.2x = -8.4$

Divide both sides of the equation by –4.2.

$$\frac{-4.2x}{-4.2} = \frac{-8.4}{-4.2}$$

$$x = 2$$

Check: $-4.2x = -8.4$

$$-4.2(2) = -8.4$$

$$-8.4 = -8.4 \qquad \text{True}$$

37. $7x = -7$

Divide both sides of the equation by 7.

$$\frac{7x}{7} = \frac{-7}{7}$$

$$x = -1$$

Check: $7x = -7$

$$7(-1) = -7$$

$$-7 = -7 \qquad \text{True}$$

39. $5x = -\frac{3}{8}$

Multiply both sides of the equation by $\frac{1}{5}$.

$$\frac{1}{5} \cdot 5x = \left(\frac{1}{5}\right) \cdot \left(-\frac{3}{8}\right)$$

$$x = \frac{(1) \cdot (-3)}{(5) \cdot (8)}$$

$$x = -\frac{3}{40}$$

Check: $5x = -\frac{3}{8}$

$$5\left(-\frac{3}{40}\right) = -\frac{3}{8}$$

$$-\frac{3}{8} = -\frac{3}{8} \qquad \text{True}$$

41. $15 = -\frac{x}{5}$

$15 = \frac{x}{-5}$

Multiply both sides of the equation by –5

$(-5)(15) = -5 \cdot \left(\frac{x}{-5}\right)$

$-75 = x$

Check: $15 = -\frac{x}{5}$

$15 = -\frac{(-75)}{5}$

$15 = 15$ True

43. $-\frac{x}{5} = -25$

$\frac{x}{-5} = -25$

Multiply both sides of the equation by –5

$(-5)\left(\frac{x}{-5}\right) = (-5)(-25)$

$x = 125$

Check: $-\frac{x}{5} = -25$

$-\frac{125}{5} = -25$

$-25 = -25$ True

45. $\frac{x}{5} = -7$

Multiply both sides of the equation by 5.

$5 \cdot \left(\frac{x}{5}\right) = 5(-7)$

$x = -35$

Check: $\frac{x}{5} = -7$

$\frac{-35}{5} = -7$

$-7 = -7$ True

47. $5 = \frac{x}{4}$

Multiply both sides of the equation by 4.

$4 \cdot 5 = 4\left(\frac{x}{4}\right)$

$20 = x$

Check: $5 = \frac{x}{4}$

$5 = \frac{20}{4}$

$5 = 5$ True

49. $6d = -30$

Divide both sides of the equation by 6.

$\frac{6d}{6} = \frac{-30}{6}$

$d = -5$

Check: $6d = -30$

$6(-5) = -30$

$-30 = -30$ True

51. $\frac{y}{-2} = -6$

Multiply both sides of the equation by –2.

$(-2)\left(\frac{y}{-2}\right) = (-2)(-6)$

$y = 12$

Check: $\frac{y}{-2} = -6$

$\frac{12}{-2} = -6$

$-6 = -6$ True

53. $\frac{-3}{8}w = 6$

Mulitply both sides of the equation by $-\frac{8}{3}$.

$\left(\frac{-8}{3}\right)\left(\frac{-3}{8}\right)w = \left(\frac{-8}{3}\right)\cdot 6$

$w = -16$

Check: $\frac{-3}{8}w = 6$

$\frac{-3}{8}(-16) = 6$

$6 = 6$ True

55. $\frac{1}{3}x = -12$

Multiply both sides of the equation by 3.

$3\left(\frac{1}{3}\right)x = 3(-12)$

$x = -36$

Check: $\frac{1}{3}x = -12$

$\frac{1}{3}(-36) = -12$

$-12 = -12$ True

57. $-4 = -\frac{2}{3}z$

Multiply both sides of the equation by $-\frac{3}{2}$

$\left(-\frac{3}{2}\right)(-4) = \left(-\frac{3}{2}\right)\left(-\frac{2}{3}\right)z$

$6 = z$

Check: $-4 = -\frac{2}{3}z$

$-4 = -\frac{2}{3}\cdot 6$

$-4 = -4$ True

59. $-1.4x = 28.28$

Divide both sides of the equation by -1.4.

$\frac{-1.4x}{-1.4} = \frac{28.28}{-1.4}$

$x = -20.2$

Check: $-1.4x = 28.28$

$-1.4(-20.2) = 28.28$

$28.28 = 28.28$ True

61. $2x = -\frac{5}{2}$

Multiply both sides of the equation by $\frac{1}{2}$.

$\frac{1}{2}\cdot 2x = \frac{1}{2}\cdot\left(-\frac{5}{2}\right)$

$x = -\frac{5}{4}$

Check: $2x = -\frac{5}{2}$

$2\left(-\frac{5}{4}\right) = -\frac{5}{2}$

$-\frac{5}{2} = -\frac{5}{2}$ True

63. $\frac{2}{3}x = 6$

$\frac{3}{2} \cdot \frac{2}{3}x = \frac{3}{2} \cdot 6$

Multiply both sides of the equation by $\frac{3}{2}$.

$x = 9$

Check: $\frac{2}{3}x = 6$

$\frac{2}{3}(9) = 6$

$6 = 6$ True

65. **(a)** $-x = a$

Multiply both sides of the equation by −1.

$-1x = a$

$(-1)(-1x) = (-1)a$

$1x = -a$

$x = -a$

(b) $-x = 5$

Multiply both sides of the equation by −1.

$-1x = 5$

$(-1)(-1x) = (-1)5$

$1x = -5$

$x = -5$

(c) $-x = -5$

Multiply both sides of the equation by −1

$-1x = -5$

$(-1)(-1x) = (-1)(-5)$

$1x = 5$

$x = 5$

67. Divide by −2.

69. Multiply by $\frac{1}{4}$.

$4x = \frac{3}{5}$

$\left(\frac{1}{4}\right)4x = \frac{3}{5}\left(\frac{1}{4}\right)$

$x = \frac{3}{20}$

Cumulative Review Exercises: 2.3

71. $-8 - (-4) = -8 + 4 = -4$

72. $6 - (-3) - 5 - 4 = 6 + 3 - 5 - 4$
$= 9 - 5 - 4 = 4 - 4 = 0$

73. $-(x + 3) - 5(2x - 7) + 6$
Use the distributive property.
$= -x - 3 - 10x + 35 + 6$
Rearrange terms.
$= -x - 10x - 3 + 35 + 6$
Combine like terms.
$= -11x + 38$

74. $-48 = x + 9$
Subtract 9 from both sides of the equation.
$-48 - 9 = x + 9 - 9$
$-57 = x + 0$
$-57 = x$

Group Activity/Challenge Problems

1. **(a)** 4
(b) $2x + 6 = 14$

(c) $2x + 6 = 14$
$2x + 6 - 6 = 14 - 6$
$2x = 8$
$\frac{2x}{2} = \frac{8}{2}$
$x = 4$

3. (a) 1
(b) $6 = 2x + 4$
(c) $6 = 2x + 4$
$6 - 4 = 2x + 4 - 4$
$2 = 2x$
$\frac{2}{2} = \frac{2x}{2}$
$1 = x$ or $x = 1$

Exercise Set 2.4

1. (a) $2x + 4 = 16$

(b) $2x + 4 = 16$

Subtract 4 from both sides of the equation.

$2x + 4 - 4 = 16 - 4$
$2x = 12$

Divide both sides of the equation by 2.

$\frac{2x}{2} = \frac{12}{2}$
$x = 6$

3. (a) $30 = 2x + 12$

(b) $30 = 2x + 12$

Subtract 12 from both sides of the equation.

$30 - 12 = 2x + 12 - 12$
$18 = 2x$

Divide both sides of the equation by 2.

$\frac{18}{2} = \frac{2x}{2}$
$9 = x$

5. (a) $3x + 10 = 4$

(b) $3x + 10 = 4$

Subtract 10 from both sides of the equation.

$3x + 10 - 10 = 4 - 10$
$3x = -6$

Divide both sides of the equation by 3.

$\frac{3x}{3} = \frac{-6}{3}$
$x = -2$

7. (a) $5 + 3x = 12$

(b) $5 + 3x = 12$

Subtract 5 from both sides of the equation.

$5 - 5 + 3x = 12 - 5$
$3x = 7$

Divide both sides of the equation by 3.

$\frac{3x}{3} = \frac{7}{3}$
$x = \frac{7}{3}$

9. $2x + 4 = 10$

Subtract 4 from both sides of the equation.

$2x + 4 - 4 = 10 - 4$
$2x = 6$

Divide both sides of the equation by 2.

$$\frac{2x}{2}=\frac{6}{2}$$
$$x=3$$

11. $-2x-5=7$

Add 5 to both sides of the equation.

$$-2x-5+5=7+5$$
$$-2x=12$$

Divide both sides of the equation by –2.

$$\frac{-2x}{-2}=\frac{12}{-2}$$
$$x=-6$$

13. $5x-6=19$

Add 6 to both sides of the equation.

$$5x-6+6=19+6$$
$$5x=25$$

Divide both sides of the equation by 5.

$$\frac{5x}{5}=\frac{25}{5}$$
$$x=5$$

15. $5x-2=10$

Add 2 to both sides of the equation.

$$5x-2+2=10+2$$
$$5x=12$$

Divide both sides of the equation by 5.

$$\frac{5x}{5}=\frac{12}{5}$$
$$x=\frac{12}{5}$$

17. $-x-4=8$

Add 4 to both sides of the equation.

$$-x-4+4=8+4$$
$$-x=12$$

Multiply both sides of the equation by –1.

$$(-1)(-x)=(-1)(12)$$
$$x=-12$$

19. $12-x=9$

Subtract 12 from both sides of the equation.

$$12-12-x=9-12$$
$$-x=-3$$

Multiply both sides of the equation by –1.

$$(-1)(-x)=(-1)(-3)$$
$$x=3$$

21. $8+3x=19$

Subtract 8 from both sides of the equation.

$$8-8+3x=19-8$$
$$3x=11$$

Divide both sides of the equation by 3.

$$\frac{3x}{3}=\frac{11}{3}$$
$$x=\frac{11}{3}$$

23. $16x+5=-14$

Subtract 5 from both sides of the equation.

$$16x+5-5=-14-5$$
$$16x=-19$$

Divide both sides of the equation by 16.

$$\frac{16x}{16}=\frac{-19}{16}$$
$$x=-\frac{19}{16}$$

25. $-4.2 = 2x + 1.6$

Subtract 1.6 from both sides of the equation.

$$-4.2 - 1.6 = 2x + 1.6 - 1.6$$
$$-5.8 = 2x$$

Divide both sides of the equation by 2.

$$\frac{-5.8}{2} = \frac{2x}{2}$$
$$-2.9 = x$$

27. $6x - 9 = 21$

Add 9 to both sides of the equation.

$$6x - 9 + 9 = 21 + 9$$
$$6x = 30$$

Divide both sides of the equation by 6.

$$\frac{6x}{6} = \frac{30}{6}$$
$$x = 5$$

29. $12 = -6x + 5$

Subtract 5 from both sides of the equation.

$$12 - 5 = -6x + 5 - 5$$
$$7 = -6x$$

Divide both sides of the equation by –6.

$$\frac{7}{-6} = \frac{-6x}{-6}$$
$$-\frac{7}{6} = x$$

31. $-2x - 7 = -13$

Add 7 to both sides of the equation.

$$-2x - 7 + 7 = -13 + 7$$
$$-2x = -6$$

Divide both sides of the equation by –2.

$$\frac{-2x}{-2} = \frac{-6}{-2}$$
$$x = 3$$

33. $x + 0.05x = 21$

Combine like terms.

$$1.05x = 21$$

Divide both sides of the equation by 1.05.

$$\frac{1.05x}{1.05} = \frac{21}{1.05}$$
$$x = 20$$

35. $2 \cdot 3x - 9.34 = 6.3$

Add 9.34 to both sides of the equation.

$$2 \cdot 3x - 9.34 + 9.34 = 6.3 + 9.34$$
$$2.3x = 15.64$$

Divide both sides of the equation by 2.3.

$$\frac{2.3x}{2.3} = \frac{15.64}{2.3}$$
$$x = 6.8$$

37. $28.8 = x - 0.10x$

Combine like terms.

$$28.8 = 0.90x$$

Divide both sides of the equation by 0.90.

$$\frac{28.8}{0.90} = \frac{0.90x}{0.90}$$
$$32 = x$$

39. $3(x + 2) = 6$

Use the distributive property.

$$3x + 6 = 6$$

Subtract 6 from both sides of the equation.

$3x + 6 - 6 = 6 - 6$

$3x = 0$

Divide both sides of the equation by 3.

$\frac{3x}{3} = \frac{0}{3}$

$x = 0$

Subtract 12 from both sides of the equation.

$12 - 12 = 4x + 12 - 12$

$0 = 4x$

Divide both sides of the equation by 4.

$\frac{0}{4} = \frac{4x}{4}$

$0 = x$

41. $4(3 - x) = 12$

Use the distributive property.

$12 - 4x = 12$

Subtract 12 from both sides of the equation.

$12 - 12 - 4x = 12 - 12$

$-4x = 0$

Divide both sides of the equation by –4.

$\frac{-4x}{-4} = \frac{0}{-4}$

$x = 0$

43. $-4 = -(x + 5)$

Use the distributive property.

$-4 = -x - 5$

Add 5 to both sides of the equation.

$-4 + 5 = -x - 5 + 5$

$1 = -x$

Multiply both sides of the equation by –1.

$(-1)(1) = (-1)(-x)$

$-1 = x$

45. $12 = 4(x + 3)$

Use the distributive property.

$12 = 4x + 12$

47. $5 = 2(3x + 6)$

Use the distributive property.

$5 = 6x + 12$

Subtract 12 from both sides of the equation.

$5 - 12 = 6x + 12 - 12$

$-7 = 6x$

Divide both sides of the equation by 6.

$\frac{-7}{6} = \frac{6x}{6}$

$-\frac{7}{6} = x$

49. $2x + 3(x + 2) = 11$

Use the distributive property.

$2x + 3x + 6 = 11$

Combine like terms.

$5x + 6 = 11$

Subtract 6 from both sides of the equation.

$5x + 6 - 6 = 11 - 6$

$5x = 5$

Divide both sides of the equation by 5.

$\frac{5x}{5} = \frac{5}{5}$

$x = 1$

51. $x-3(2x+3)=11$

Use the distributive property.

$x-6x-9=11$

Combine like terms.

$-5x-9=11$

Add 9 to both sides of the equation.

$-5x-9+9=11+9$

$-5x=20$

Divide both sides of the equation by –5.

$\frac{-5x}{-5}=\frac{20}{-5}$

$x=-4$

53. $5x+3x-4x-7=9$

Combine like terms.

$4x-7=9$

Add 7 to both sides of the equation.

$4x-7+7=9+7$

$4x=16$

Divide both sides of the equation by 4.

$\frac{4x}{4}=\frac{16}{4}$

$x=4$

55. $0.7(x+3)=4.2$

Use the distributive property.

$0.7x+2.1=4.2$

Subtract 2.1 from both sides of the equation.

$0.7x+2.1-2.1=4.2-2.1$

$0.7x=2.1$

Divide both sides of the equation by 0.7.

$\frac{0.7x}{0.7}=\frac{2.1}{0.7}$

$x=3$

57. $1.4(5x-4)=-1.4$

Use the distributive property.

$7x-5.6=-1.4$

Add 5.6 to both sides of the equation.

$7x-5.6+5.6=-1.4+5.6$

$7x=4.2$

Divide both sides of the equation by 7.

$\frac{7x}{7}=\frac{4.2}{7}$

$x=0.6$

59. $3-2(x+3)+2=1$

Use the distributive property.

$3-2x-6+2=1$

Combine like terms.

$-2x-1=1$

Add 1 to both sides of the equation.

$-2x-1+1=1+1$

$-2x=2$

Divide both sides of the equation by –2.

$\frac{-2x}{-2}=\frac{2}{-2}$

$x=-1$

61. $1-(x+3)+2x=4$

Use the distributive property.

$1-x-3+2x=4$

Combine like terms.

$x-2=4$

Add 2 to both sides of the equation.

$x-2+2=4+2$

$x=6$

63. $4.22 - 6.4x + 9.60 = 0.38$

Combine like terms.

$-6.4x + 13.82 = 0.38$

Subtract 13.82 from both sides of the equation.

$$-6.4x + 13.82 - 13.82 = 0.38 - 13.82$$
$$-6.4x = -13.44$$

Divide both sides of the equation by –6.4.

$$\frac{-6.4x}{-6.4} = \frac{-13.44}{-6.4}$$
$$x = 2.1$$

65. $5.76 - 4.24x - 1.9x = 27.864$

Combine like terms.

$5.76 - 6.14x = 27.864$

Subtract 5.76 from both sides of the equation.

$5.76 - 5.76 - 6.14x = 27.864 - 5.76$

$-6.14x = 22.104$

Divide both sides of the equation by –6.14.

$$\frac{-6.14x}{-6.14} = \frac{22.104}{-6.14}$$
$$x = -3.6$$

67. Addition Property

69. $4x - 2(x + 3) = 4$
Use the distributive property.
$4x - 2x - 6 = 4$
Combine like terms.
$2x - 6 = 4$
Add 6 to both sides of the equation.
$2x - 6 + 6 = 4 + 6$
$2x = 10$
Divide both sides of the equation by 2.
$\frac{2x}{2} = 10$
$x = 5$

Cumulative Review Exercises: 2.4

70. $\frac{5}{8} = \frac{5}{8} \cdot \frac{5}{5} = \frac{25}{40}$ and $\frac{3}{5} = \frac{3}{5} \cdot \frac{8}{8} = \frac{24}{40}$

$\frac{5}{8} + \frac{3}{5} = \frac{25}{40} + \frac{24}{40} = \frac{49}{40}$ or $1\frac{9}{40}$

71. $[5(2-6) + 3(8 \div 4)^2]^2$

$= [5(-4) + 3(2)^2]^2$

$= [-20 + 3(4)]^2$

$= [-20 + 12]^2$

$= [-8]^2$

$= 64$

72. Isolate the variable on one side of the equation.

73. Divide both sides of the equation by –4.

Group Activity/Challenge Problems

1. $3(x-2) - (x+5) - 2(3-2x) = 18$

$3x - 6 - x - 5 - 6 + 4x = 18$

$6x - 17 = 18$

$6x = 35$

$x = \frac{35}{6}$

2. $-6 = -(x-5) - 3(5+2x) - 4(2x-4)$

$-6 = -x + 5 - 15 - 6x - 8x + 16$

$-6 = -15x + 6$

$-12 = -15x$

$\frac{-12}{-15} = x$

$\frac{4}{5} = x$ or $x = \frac{4}{5}$

3. $4[3 - 2(x + 4)] - (x + 3) = 13$

$4(3 - 2x - 8) - x - 3 = 13$

$4(-2x - 5) - x - 3 = 13$

$-8x - 20 - x - 3 = 13$

$-9x - 23 = 13$

$-9x = 36$

$x = -4$

4. $2x + 2 = 8$

$2x = 6$

$x = 3$

A single kiss is \$3.

5. $3x + 6 = 42$

$3x = 36$

$x = 12$

The box of stationary is \$12.

7. **(a)** $2x = x + 3$

(b) $2x = x + 3$

$x = 3$

9. **(a)** $2x + 3 = 4x + 2$

$3 - 2 = 4x - 2x$

$1 = 2x$

$\frac{1}{2} = x$ or $x = \frac{1}{2}$

Exercise Set 2.5

1. **(a)** $2x = x + 6$

(b) $2x = x + 6$

Subtract x from both sides of the equation.

$2x - x = x - x + 6$

$x = 6$

3. **(a)** $5 + 2x = x + 19$

(b) $5 + 2x = x + 19$

Subtract x from both sides of the equation.

$5 + 2x - x = x - x + 19$

$5 + x = 19$

Subtract 5 from both sides of the equation.

$5 - 5 + x = 19 - 5$

$x = 14$

5. **(a)** $5 + x = 2x + 5$

(b) $5 + x = 2x + 5$

Subtract x from both sides of the equation.

$5 + x - x = 2x - x + 5$

$5 = x + 5$

Subtract 5 from both sides of the equation.

$5 - 5 = x + 5 - 5$

$0 = x$

7. **(a)** $2x + 8 = x + 4$

(b) $2x + 8 = x + 4$

Subtract x from both sides of the equation.

$2x - x + 8 = x - x + 4$

$x + 8 = 4$

Subtract 8 from both sides of the equation.

$x + 8 - 8 = 4 - 8$

$x = -4$

9. $4x = 3x + 5$

Subtract $3x$ from both sides of the equation.

$4x - 3x = 3x - 3x + 5$

$x = 5$

11. $-4x + 10 = 6x$

Add $4x$ to both sides of the equation.

$-4x + 4x + 10 = 6x + 4x$

$10 = 10x$

Divide both sides of the equation by 10.

$\frac{10}{10} = \frac{10x}{10}$

$1 = x$

13. $5x + 3 = 6$

Subtract 3 from both sides of the equation.

$5x + 3 - 3 = 6 - 3$

$5x = 3$

Divide both sides of the equation by 5.

$\frac{5x}{5} = \frac{3}{5}$

$x = \frac{3}{5}$

15. $15 - 3x = 4x - 2x$

Combine like terms.

$15 - 3x = 2x$

Add $3x$ to both sides of the equation.

$15 - 3x + 3x = 2x + 3x$

$15 = 5x$

Divide both sides of the equation by 5.

$\frac{15}{2} = \frac{5x}{5}$

$3 = x$

17. $2x - 4 = 3x - 6$

Subtract $2x$ from both sides of the equation.

$2x - 2x - 4 = 3x - 2x - 6$

$-4 = x - 6$

Add 6 to both sides of the equation.

$-4 + 6 = x - 6 + 6$

$2 = x$

19. $3 - 2y = 3 - 8y$

$3 + 6y = 9$

$8y$ was added to both sides of the equation.

$6y = 6$

3 was subtracted from both sides of the equation.

$y = 1$

Both sides of the equation were divided by 6.

21. $4 - 0.6x = 2.4x - 8.48$

$4 = 3x - 8.48$

$0.6x$ was added to both sides of the equation.

$12.48 = 3x$

8.48 was added to both sides of the equation.

$4.16 = x$

Both sides of the equation were divided by 3.

23. $5x = 2(x+6)$

$5x = 2x + 12$

Distributive property was used.

$3x = 12$

$2x$ was subtracted from both sides of the equation.

$x = 4$

Both sides of the equation were divided by 3.

25. $x - 25 = 12x + 9 + 3x$

$x - 25 = 15x + 9$

Like terms were combined.

$-25 = 14x + 9$

x was subtracted from both sides of the equation.

$-34 = 14x$

9 was subtracted from both sides of the equation.

$-\frac{17}{7} = x$

Both sides of the equation were divided by 14.

27. $2(x+2) = 4x + 1 - 2x$

$2x + 4 = 4x + 1 - 2x$

Distributive property was used.

$2x + 4 = 2x + 1$

Like terms were combined.

$2x - 2x + 4 = 2x - 2x + 1$

$2x$ was subtracted from both sides of the equation.

$4 = 1$ False

Since a false statement is obtained, there is no solution.

29. $-(w+2) = -6w + 32$

$-w - 2 = -6w + 32$

Distributive property was used.

$-2 = -5w$

w was added to both sides of the equation.

$-34 = -5w\ \ 32$

32 was subtracted from both sides of the equation.

$\frac{34}{5} = w$

Both sides of the equation were divided by -5.

31. $4 - (2x+5) = 6x + 31$

$4 - 2x - 5 = 6x + 31$

Distributive property was used.

$-2x - 1 = 6x + 31$

Like terms were combined.

$-1 = 8x + 31$

$2x$ was added to both sides of the equation.

$-32 = 8x$

31 was subtracted from both sides of the equation.

$-4 = x$

Both sides of the equation were divided by 8.

33. $0.1(x + 10) = 0.3x - 4$

$0.1x + 1 = 0.3x - 4$

Distributive property was used.

$1 = 0.2x - 4$

$0.1x$ was subtracted from both sides of the equation.

$5 = 0.2x$

4 was added to both sides of the equation.

$25 = x$

Boths sides of the equation were divided by 0.2.

35. $2(x+4) = 4x + 3 - 2x + 5$

$2x + 8 = 4x + 3 - 2x + 5$

Distributive property was used.

$2x + 8 = 2x + 8$

Like terms were combined

The equation is true for all values of x. Thus the solution is all real numbers.

37. $9(-y+3) = -6y + 15 - 3y + 12$

$-9y + 27 = -6y + 15 - 3y + 12$

Distributive property was used.

$-9y + 27 = -9y + 27$

Like terms were combined.

The equation is true for all values of x. Thus the solution is all real numbers.

39. $-(3-p) = -(2p+3)$

$-3 + p = -2p - 3$

Distributive property was used.

$-3 + 3p = -3$

$2p$ was added to both sides of the equation.

$3p = 0$

3 was added to both sides of the equation.

$p = 0$

Both sides of the equation were divided by 3.

41. $-(x+4) + 5 = 4x + 1 - 5x$

$-x - 4 + 5 = 4x + 1 - 5x$

Distributive property was used.

$-x + 1 = -x + 1$

Like terms were combined.

The equation is true for all values of x. Thus the solution is all real numbers.

43. $35(2x+12) = 7(x-4) + 3x$

$70x + 420 = 7x - 28 + 3x$

Distributive property was used.

$70x + 420 = 10x - 28$

Like terms were combined.

$60x + 420 = -28$

$10x$ was subtracted from both sides of the equation.

$60x = -448$

420 was subtracted from both sides of the equation.

$x = -\frac{112}{15}$

Both sides of the equation were divided by 60.

45. $0.4(x+0.7) = 0.6(x-4.2)$

$0.4x + 0.28 = 0.6x - 2.52$

Distributive property was used.

$0.28 = 0.2x - 2.52$

$0.4x$ was subtracted from both sides of the equation.

$2.8 = 0.2x$

2.52 was added to both sides of the equation.

$14 = x$

Both sides of the equation were divided by 0.2.

47. $-(x-5)+2=3(4-x)+5x$

$-x+5+2=12-3x+5x$

Distributive property was used.

$-x+7=12+2x$

Like terms were combined.

$7=12+3x$

x was added to both sides of the equation.

$-5=3x$

12 was subtracted from both sides of the equation.

$-\frac{5}{3}=x$

Both sides of the equation were divided by 3.

49. $2(x-6)+3(x+1)=4x+3$

$2x-12+3x+3=4x+3$

Distributive property was used.

$5x-9=4x+3$

Like terms were combined.

$x-9=3$

$4x$ was subtracted from both sides of the equation.

$x=12$

9 was added to both sides of the equation.

51. $5+2x=6(x+1)-5(x-3)$

$5+2x=6x+6-5x+15$

Distributive property was used.

$5+2x=x+21$

Like terms were combined.

$5+x=21$

x was subtracted from both sides of the equation.

$x=16$

5 was subtracted from both sides of the equation.

53. $5-(x-5)=2(x+3)-6(x+1)$

$5-x+5=2x+6-6x-6$

Distributive property was used.

$-x+10=-4x$

Like terms were combined.

$10=-3x$

x was added to both sides of the equation.

$-\frac{10}{3}=x$

Both sides of the equation were divided by −3.

55. Individual answer

57. Will obtain a false statement

59. $4x+3(x+2)=5x-10$
Use the distributive property.
$4x+3x+6=5x-10$
Combine like terms.
$7x+6=5x-10$
Subtract $5x$ from both sides of the equation.
$2x+6=-10$
Subtract 6 from both sides of the equation.
$2x=-16$
Divide both sides of the equation by 2.
$x=-8$

Cumulative Review Exercises: 2.5

60. $\left(\frac{2}{3}\right)^5=0.131687243$

61. Numbers or letters multiplied together are factors; numbers or letters added or subtracted are terms.

62. $2(x-3)+4x-(4-x)$

$=2x-6+4x-4+x$

Use the distributive property.

$=7x-10$

Combine like terms.

63. $2(x-3)+4x-(4-x)=0$

$2x-6+4x-4+x=0$

Distributive property was used.

$7x-10=0$

Like terms were combined.

$7x=10$

10 was added to both sides of the equation.

$x=\frac{10}{7}$

Both sides of the equation were divided by 7.

64. $(x+4)-(4x-3)=16$

$x+4-4x+3=16$

Distributive property was used.

$-3x+7=16$

Like terms were combined.

$-3x=9$

7 was subtracted from both sides of the equation.

$x=-3$

Both sides of the equation were divided by −3.

Group Activity/Challenge Problems

1. $-2(x+3)+5x=-3(5-2x)+3(x+2)+6x$

$-2x-6+5x=-15+6x+3x+6+6x$

$3x-6=15x-9$

$-6=12x-9$

$3=12x$

$\frac{3}{12}=x$

$\frac{1}{4}=x$ or $x=\frac{1}{4}$

3. $4-[5-3(x+2)]=x-3$

$4-(5-3x-6)=x-3$

$4-5+3x+6=x-3$

$3x+5=x-3$

$2x+5=-3$

$2x=-8$

$x=-4$

5. $3x=x+20$

$2x=20$

$x=10$

The box weighs 10 lb.

Exercise Set 2.6

1. 5: 8

3. $\frac{4}{2}=\frac{2}{1}$

Ratio of D's to F's is 2:1

5. Total grades = 5 + 6 + 8 + 4 + 2 = 25

Ratio of total grades to D's = 25:4

7. 5:3

9. $\frac{20}{60} = \frac{1}{3}$, ratio is 1:3

11. 4 hours = $4 \times 60 = 240$ minutes

Ratio is $\frac{240}{40} = \frac{6}{1}$, or 6:1

13. 4 pounds is $4 \times 16 = 64$ ounces

Ratio is $\frac{24}{64} = \frac{13}{32}$ or 13:32

15. Gear ratio

$= \frac{\text{number of teeth on driving gear}}{\text{number of teeth on driven gear}}$

$= \frac{40}{5} = \frac{8}{1}$

Gear ratio is 8:1

17. **(a)** 5.3:3.6

(b) 5.3:3.6

Divide both parts of the ratio by 3.6.

$\frac{5.3}{3.6} : \frac{3.6}{3.6}$

About 1.47 : 1

19. **(a)** 20,000:165

(b) 20,000:165

Divide both parts of the ratio by 165.

$\frac{20{,}000}{165} : \frac{165}{165}$

About 121:1

21. $\frac{4}{x} = \frac{5}{20}$

$4 \cdot 20 = x \cdot 5$

$80 = 5x$

$\frac{80}{5} = x$

$16 = x$

23. $\frac{5}{3} = \frac{75}{x}$

$5 \cdot x = 3 \cdot 75$

$5x = 225$

$x = \frac{225}{5}$

$x = 45$

25. $\frac{90}{x} = \frac{-9}{10}$

$90 \cdot 10 = x(-9)$

$900 = -9x$

$\frac{900}{-9} = x$

$-100 = x$

27. $\frac{1}{9} = \frac{x}{45}$

$1 \cdot 45 = 9 \cdot x$

$45 = 9x$

$\frac{45}{9} = x$

$5 = x$

29. $\frac{3}{z} = \frac{2}{-20}$

$3(-20) = 2 \cdot z$

$-60 = 2z$

$\frac{-60}{2} = z$

$-30 = z$

31. $\frac{15}{20} = \frac{x}{8}$

$15 \cdot 8 = 20 \cdot x$

$120 = 20x$

$\frac{120}{20} = x$

$6 = x$

33. Let x = number of miles car can travel on 12 gallons.

$\frac{32 \text{ miles}}{1 \text{ gallon}} = \frac{x \text{ miles}}{12 \text{ gallons}}$

$\frac{32}{1} = \frac{x}{12}$

$32 \cdot 12 = 1 \cdot x$

$384 = x$

Thus the car can travel 384 mi on 12 gallons.

35. Let x = gallons of paint needed.

$\frac{1 \text{ gallon}}{825 \text{ sq feet}} = \frac{x \text{ gallons}}{5775 \text{ sq feet}}$

$\frac{1}{825} = \frac{x}{5775}$

$1 \cdot 5775 = 825 \cdot x$

$5775 = 825x$

$\frac{5775}{825} = x = 7$

Thus 7 gallons of paint are needed for the house.

37. Let x = length of mall on the blueprint.

$\frac{1 \text{ foot (on print)}}{510 \text{ feet (actual)}} = \frac{x \text{ feet (on print)}}{190 \text{ feet (actual)}}$

$\frac{1}{150} = \frac{x}{190}$

$1 \cdot 190 = 150 \cdot x$

$190 = 150x$

$\frac{190}{150} = x = 1.27$

Thus the mall will appear 1.27 ft long on the blueprint.

39. Let x = actual height of cactus.

$\frac{0.6 \text{ in. (in photo)}}{48 \text{ in. (actual)}} = \frac{3.25 \text{ in. (in photo)}}{x \text{ in. (actual)}}$

$\frac{0.6}{48} = \frac{3.25}{x}$

$0.6x = (48)(3.25)$

$0.6x = 156$

$x = \frac{156}{0.6} = 260$

Thus the actual height of cactus is 260 in. or 21 ft 8 in.

41. Let x = number of teaspoons needed for sprayer.

$\frac{3 \text{ teaspoons}}{1 \text{ gallon water}} = \frac{x \text{ teaspoons}}{8 \text{ gallons water}}$

$\frac{3}{1} = \frac{x}{8}$

$3 \cdot 8 = 1 \cdot x$

$24 = x$

Thus 24 teaspoons are needed for the sprayer.

43. Let x = number of trees saved.

$$\frac{17 \text{ trees}}{1 \text{ ton recycled paper}}$$

$$= \frac{x \text{ trees}}{20 \text{ tons recycled paper}}$$

$$\frac{17}{1} = \frac{x}{20}$$

$$17 \cdot 20 = 1 \cdot x$$

$$340 = x$$

Thus 340 trees have been saved.

45. Let x = number of milliliters to be given.

$$\frac{1 \text{ milliliter}}{400 \text{ micrograms}} = \frac{x \text{ milliliter}}{220 \text{ micrograms}}$$

$$\frac{1}{400} = \frac{x}{220}$$

$$1 \cdot 220 = 400 \cdot x$$

$$220 = 400x$$

$$\frac{220}{400} = x = 0.55$$

Thus 0.55 ml should be given.

47. Let x = number of seconds to go from 0 to 800.

$$\frac{30 \text{ seconds}}{250 \text{ counts on VCR}}$$

$$= \frac{x \text{ seconds}}{800 \text{ counts on VCR}}$$

$$\frac{30}{250} = \frac{x}{800}$$

$$30 \cdot 800 = x \cdot 250$$

$$24{,}000 = 250x$$

$$\frac{24{,}000}{250} = x = 96$$

Thus you should fast forward for 96 sec or 1 min 36 sec.

49. Let x = population in 1990.

$$\frac{16 \text{ births}}{1000 \text{ people}} = \frac{4{,}179{,}200 \text{ births}}{x \text{ people}}$$

$$\frac{16}{1000} = \frac{4{,}179{,}200}{x}$$

$$16 \cdot x = (1000)(4{,}179{,}200)$$

$$16x = 4{,}179{,}200{,}000$$

$$x = \frac{4{,}179{,}200{,}000}{16} = 261{,}200{,}000$$

Thus the population in 1990 was about 261,000 thousand people or 261,200,000 people.

51. $$\frac{12 \text{ inches}}{1 \text{ foot}} = \frac{57 \text{ inches}}{x \text{ feet}}$$

$$\frac{12}{1} = \frac{57}{x}$$

$$12x = 57$$

$$x = \frac{57}{12} = 4.75$$

Thus 57 inches equals 4.75 feet.

53. $$\frac{9 \text{ square feet}}{1 \text{ square yard}} = \frac{26.1 \text{ square feet}}{x \text{ square yards}}$$

$$\frac{9}{1} = \frac{26.1}{x}$$

$$9x = 26.1$$

$$x = \frac{26.1}{9} = 2.9$$

Thus 26.1 square feet equals 2.9 square yards.

55. $\frac{2.54 \text{ cm}}{1 \text{ inch}} = \frac{26.67 \text{ cm}}{x \text{ inches}}$

$\frac{2.54}{1} = \frac{26.67}{x}$

$2.54x = 26.67$

$x = \frac{26.67}{2.54} = 10.5$

Thus the length of the book is 10.5 inches.

57. $\frac{1.6 \text{ kilometers}}{1 \text{ mile}} = \frac{25 \text{ kilometers}}{x \text{ miles}}$

$\frac{1.6}{1} = \frac{25}{x}$

$1.6x = 25$

$x = \frac{25}{1.6} = 15.63$

Thus the distance of the crossing is 15.63 miles.

59. $\frac{480 \text{ grains}}{400 \text{ dollars}} = \frac{1 \text{ grain}}{x \text{ dollars}}$

$\frac{480}{400} = \frac{1}{x}$

$480x = 400$

$x = \frac{400}{480} = 0.83$

Thus the cost per grain is \$0.83.

61. $\frac{3.2 \text{ standard deviations}}{16 \text{ points}}$

$= \frac{1 \text{ standard deviation}}{x \text{ points}}$

$\frac{3.2}{16} = \frac{1}{x}$

$3.2x = 16$

$x = \frac{16}{3.2} = 5$

Thus 1 standard deviation equals 5 points.

63. $\frac{0.00059 \text{ dollars}}{1 \text{ lira}} = \frac{1200 \text{ dollars}}{x \text{ lira}}$

$\frac{0.00059}{1} = \frac{1200}{x}$

$0.00059x = 1200$

$x = \frac{1200}{0.00059}$

$= 2{,}033{,}898.3$

Thus 2,033,898.3 lira are needed to obtain 1200 U.S. dollars.

65. $\frac{3}{12} = \frac{8}{x}$

$3x = (8)(12)$

$3x = 96$

$x = \frac{96}{3} = 32$

Thus the side is 32 in. in length.

67. $\frac{5}{7} = \frac{8}{x}$

$5x = (7)(8)$

$5x = 56$

$x = \frac{56}{5} = 11.2$

Thus the side is 11.2 ft in length.

69. $\frac{20}{8} = \frac{14}{x}$

$20x = (8)(14)$

$20x = 112$

$x = \frac{112}{20} = 5.6$

Thus the side is 5.6 in. in length.

71. The ratio of low density to high density cholesterol is $\frac{127}{60}$.

If we divide 127 by 60 we obtain 2.12.

Thus Mrs. Sanchez' ratio is equivalent to 2.12:1.

Therefore her ratio is less than the desired 4:1 ratio.

73. Individual answer

75. Need a given ratio and one of the two parts of a second ratio

Cumulative Review Exercises: 2.6

76. Commutative property of addition

77. Associative property of multiplication

78. Distributive property

79. $-(2x+6)=2(3x-6)$

$-2x-6=6x-12$

The distributive property was used.

$-6=8x-12$

$2x$ was added to both sides of the equation.

$6=8x$

12 was added to both sides of the equation.

$\frac{3}{4}=x$

Both sides of the equation were divided by 8.

Group Activity/Challenge Problems

1. **(a)** 750: 10,000

(b) $\frac{750}{10{,}000}:\frac{10{,}000}{10{,}000}$

0.075: 1

2. flour $\frac{12}{8}=\frac{\frac{1}{2}}{x}$, $12x=\left(\frac{1}{2}\right)8$,

$12x=4$, $x=\frac{1}{3}$ cup

nutmeg $\frac{12}{8}=\frac{1}{x}$, $12x=8$, $x=\frac{2}{3}$ tsp

cinnamon $\frac{12}{8}=\frac{1}{x}$, $12x=8$, $x=\frac{2}{3}$ tsp

salt $\frac{12}{8}=\frac{\frac{1}{4}}{x}$, $12x=\left(\frac{1}{4}\right)8$,

$12x=2$, $x=\frac{1}{6}$ tsp

butter $\frac{12}{8}=\frac{2}{x}$, $12x=8(2)$,

$12x=16$, $x=\frac{4}{3}$ or $1\frac{1}{3}$ tbsp

sugar $\frac{12}{8}=\frac{\frac{3}{2}}{x}$, $12x=8\left(\frac{3}{2}\right)$,

$12x=12$, $x=1$ cup

3. $\frac{1}{40}=\frac{x}{25}$

$40x=25$

$x=0.625$

0.625 cubic centimeters

5. **(a)** $\frac{26}{1000}=\frac{x}{5420}$

$26(5420)=1000x$

$140920=1000x$

$140.92=x$

About 140,920,000 people

(b) $\frac{9}{1000}=\frac{x}{5420}$

$9(5420)=1000x$

$48780=1000x$

$48.78=x$

About 48,780,000 people

(c) $140.92 - 48.78 = 92.14$
About 92,140,000 people

Exercise Set 2.7

1. $x + 3 > 7$

Subtract 3 from both sides of the inequality.

$x + 3 - 3 > 7 - 3$

$x > 4$

3. $x + 5 \geq 3$

Subtract 5 from both sides of the inequality.

$x + 5 - 5 \geq 3 - 5$

$x \geq -2$

5. $-x + 3 < 8$

Subtract 3 from both sides of the inequality.

$-x + 3 - 3 < 8 - 3$

$-x < 5$

Multiply both sides of the inequality by –1 and change the direction of the inequality symbol.

$(-1)(-x) > (-1)(5)$

$x > -5$

7. $6 > x - 4$

Add 4 to both sides of the inequality.

$6 + 4 > x - 4 + 4$

$10 > x$

$x < 10$

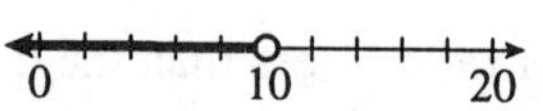

9. $8 \leq 4 - x$

Add x to both sides of the inequality.

$8 + x \leq 4 - x + x$

$8 + x \leq 4$

Subtract 8 from both sides of the inequality

$8 - 8 + x \leq 4 - 8$

$x \leq -4$

11. $-2x < 3$

Divide both sides of the inequality by –2 and change the direction of the inequality symbol.

$$\frac{-2x}{-2} > \frac{3}{-2}$$

$$x > -\frac{3}{2}$$

13. $2x + 3 \leq 5$

Subtract 3 from both sides of the inequality.

$2x + 3 - 3 \leq 5 - 3$

$2x \leq 2$

Divide both sides of the inequality by 2.

$\frac{2x}{2} \le \frac{2}{2}$

$x \le 1$

−5 0 5

15. $12x + 24 < -12$

Subtract 24 from both sides of the inequality.

$12x + 24 - 24 < -12 - 24$

$12x < -36$

Divide both sides of the inequality by 12.

$\frac{12x}{12} < \frac{-36}{12}$

$x < -3$

−5 −3 0 5

17. $4 - 6x > -5$

Subtract 4 from both sides of the inequality.

$4 - 4 - 6x > -5 - 4$

$-6x > -9$

Divide both sides of the inequality by −6 and change the direction of the inequality symbol.

$\frac{-6x}{-6} < \frac{-9}{-6}$

$x < \frac{3}{2}$

−5 0 $\frac{3}{2}$ 5

19. $15 > -9x + 50$

Add $9x$ to both sides of the inequality.

$9x + 15 > -9x + 9x + 50$

$9x + 15 > 50$

Subtract 15 from both sides of the inequality.

$9x + 15 - 15 > 50 - 15$

$9x > 35$

Divide both sides of the inequality by 9.

$\frac{9x}{9} > \frac{35}{9}$

$x > \frac{35}{9}$

0 $\frac{35}{9}$ 5 10

21. $4 < 3x + 12$

Subtract 12 from both sides of the inequality.

$4 - 12 < 3x + 12 - 12$

$-8 < 3x$

Divide both sides of the inequality by 3.

$\frac{-8}{3} < \frac{3x}{3}$

$\frac{-8}{3} < x$

$x > -\frac{8}{3}$

−5 $-\frac{8}{3}$ 0 5

23. $6x + 2 \le 3x - 9$

$6x \le 3x - 11$

2 was subtracted from both sides of the inequality.

$3x \le -11$

$3x$ was subtracted from both sides of the inequality.

$x \le -\frac{11}{3}$

Both sides of the inequality were divided by 3

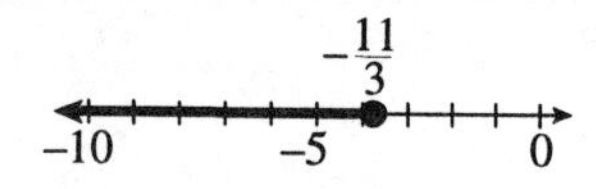

25. $x-4\le 3x+8$

$x-12\le 3x$

8 was subtracted from both sides of the inequality.

$-12\le 2x$

x was subtracted from both sides of the inequality.

$-6\le x$

Both sides of the equation were divided by 2.

$x\ge -6$

27. $-x+4<-3x+6$

$-x<-3x+2$

4 was subtracted from both sides of the inequality.

$2x<2$

$3x$ was added to both sides of the inequality.

$x<1$

Both sides of the inequality were divided by 2.

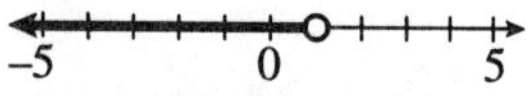

29. $-3(2x-4)>2(6x-12)$

$-6x+12>12x-24$

Distributive property was used.

$-6x+36>12x$

24 was added to both sides of the inequality.

$36>18x$

$6x$ was added to both sides of the inequality.

$2>x$

Both sides of the inequality were divided by 18.

$x<2$

31. $x+3<x+4$

Subtract x from both sides of the inequality.

$$x-x+3<x-x+4$$
$$3<4$$

Since 3 is always less than 4, the solution is all real numbers.

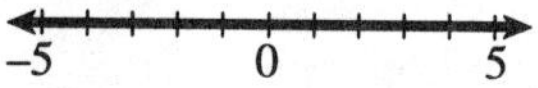

33. $6(3-x)<2x+12$
$18-6x<2x+12$

Distributive property was used.
$6-6x<2x$
12 was subtracted from both sides of the inequality.

$6<8x$

$6x$ was added to both sides of the inequality.

$\frac{3}{4}<x$

Both sides of the inequality. were divided by 8.

$x>\frac{3}{4}$

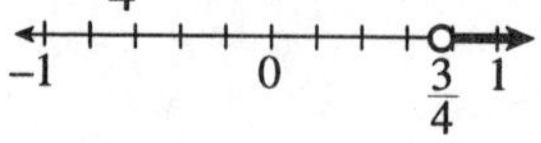

35. $-21(2-x)+3x > 4x+4$

$-42+21x+3x > 4x+4$

Distributive property was used.

$-42+24x > 4x+4$

Like terms were combined.

$24x > 4x+46$

42 was added to both sides of the inequality.

$20x > 46$

$4x$ was subtracted from both sides of the inequality.

$x > \frac{23}{10}$

Both sides of the inequality were divided by 20.

–5 0 $\frac{23}{10}$ 5

37. $4x-4 < 4(x-5)$

$4x-4 < 4x-20$

Distributive property was used.

$-4 < -20$

$4x$ was subtracted from both sides of the inequality.

Since –4 is never less than –20, the answer is no solution.

–5 0 5

39. $5(2x+3) \ge 6+(x+2)-2x$

$10x+15 \ge 6+x+2-2x$

Distributive property was used.

$10x+15 \ge 8-x$

Like terms were combined.

$10x \ge -7-x$

15 was subtracted from both sides of the inequality.

$11x \ge -7$

x was added to both sides of the inequality.

$x \ge -\frac{7}{11}$

Both sides of the inequality were divided by 11.

–1 $-\frac{7}{11}$ 0 1

41. Since 3 is always less than 5, the solution is all real numbers.

43. Since 5 is never less than 2, the answer is no solution.

45. The direction of the inequality symbol is changed when multiplying or dividing by a negative number.

Cumulative Review Exercises: 2.7

47. Substitute 3 for x.

$-x^2 = -(3)^2 = -(3)(3) = -9$

48. Substitute –5 for x.

$-x^2 = -(-5)^2 = -(-5)(-5) = -25$

49. $4-3(2x-4) = 5-(x+3)$

$4-6x+12 = 5-x-3$

The distributive property was used.

$-6x+16 = 2-x$

Like terms were combined.

$16 = 2+5x$

$6x$ was added to both sides of the equation.

$14 = 5x$

2 was subtracted from both sides of the equation.

$\frac{14}{5} = x$

Both sides of the equation were divided by 5.

$x = \frac{14}{5}$ or $2\frac{4}{5}$

50. Let x = number of kilowatt hours of electricity used.

$\frac{\$0.174}{1 \text{ kilowatt hour}} = \frac{\$87}{x \text{ kilowatt hours}}$

$\frac{0.174}{1} = \frac{87}{x}$

$0.174x = 87$

$x = \frac{87}{0.174} = 500$

Thus the Cisneros used 500 kilowatt hours of electricity in July.

Group Activity/Challenge Problems

1. $3(2 - x) - 4(2x - 3) \le 6 + 2x - 6(x - 5) + 2x$

$6 - 3x - 8x + 12 \le 6 + 2x - 6x + 30 + 2x$

$-11x + 18 \le -2x + 36$

$18 \le 9x + 36$

$-18 \le 9x$

$-2 \le x$ or $x \ge -2$

4. **(a)** $\$30,000 \times 0.001 - \$20.00 = \$10.00$

(b) $\$200,000 \times 0.001 - \$40.00 = \$160.00$

Chapter 2 Review Exercises

1. $2(x+4) = 2x + 2(4) = 2x + 8$

2. $3(x-2) = 3[x + (-2)] = 3x + 3(-2)$
$= 3x + (-6) = 3x - 6$

3. $2(4x-3) = 2[4x + (-3)] = 2(4x) + 2(-3)$
$= 8x + (-6) = 8x - 6$

4. $-2(x+4) = -2x + (-2)(4) = -2x + (-8)$
$= -2x - 8$

5. $-(x+2) = -1(x+2) = (-1)(x) + (-1)(2)$
$= -x + (-2) = -x - 2$

6. $-(x-2) = -1(x-2) = -1[x + (-2)]$
$= (-1)(x) + (-1)(-2) = -x + 2$

7. $-4(4-x) = -4[4 + (-x)]$
$= (-4)(4) + (-4)(-x) = -16 + 4x$

8. $3(6-2x) = 3[6 + (-2x)] = 3(6) + 3(-2x)$
$= 18 + (-6x) = 18 - 6x$

9. $4(5x-6) = 4[5x + (-6)] = 4(5x) + 4(-6)$
$= 20x + (-24) = 20x - 24$

10. $-3(2x-5) = -3[2x + (-5)]$
$= -3(2x) + (-3)(-5) = -6x + 15$

11. $6(6x-6) = 6[6x + (-6)]$
$= 6(6x) + 6(-6) = 36x + (-36)$
$= 36x - 36$

12. $4(-x+3) = 4(-x) + 4(3) = -4x + 12$

13. $-3(x+y) = -3(x) + -3(y) = -3x - 3y$

14. $-2(3x-2) = -2[3x + (-2)]$
$= -2(3x) + (-2)(-2) = -6x + 4$

15. $-(3+2y) = -1(3+2y)$
$= (-1)(3) + (-1)(2y) = -3 + (-2y) = -3 - 2y$

16. $-(x+2y-z) = -1(x+2y-z)$
$= -1(x) + (-1)(2y) + (-1)(-z)$
$= -x + (-2y) + z = -x - 2y + z$

17. $3(x+3y-2z) = 3[x+3y+(-2z)]$
$= 3x + 3(3y) + 3(-2z)$
$= 3x + 9y + (-6z) = 3x + 9y - 6z$

18. $-2(2x-3y+7) = -2[2x + (-3y) + 7]$
$= -2(2x) + (-2)(-3y) + (-2)(7)$
$= -4x + 6y + (-14) = -4x + 6y - 14$

19. Since $2 + 3 = 5$, then $2x + 3x = 5x$.

20. The $4y$ and $3y$ are like terms.
Combining like terms.
$4y + 3y + 2 = 7y + 2$

21. The 4 and 3 are like terms.
$4 - 2y + 3$
Rearrange terms.
$= -2y + 4 + 3$
Combine like terms.
$= -2y + 7$

22. The $3x$ and $2x$ are like terms.
Combining like terms.
$1 + 3x + 2x = 1 + 5x$
Rearranging terms.
$= 5x + 1$

23. The $2y$ and y are like terms.
Combining like terms.
$6x + 2y + y = 6x + 3y$

24. The $-2x$ and $-x$ are like terms.
Combining like terms.
$-2x - x + 3y = -3x + 3y$

25. The $2x$ and $4x$ are like terms.
The $3y$ and $5y$ are like terms.
Rearranging terms.
$3x + 3y + 4x + 5y = 2x + 4x + 3y + 5y$
Combining like terms.
$= 6x + 8y$

26. There are no like terms.
$6x + 3y + 2$ cannot be further simplified.

27. The $2x$ and $-3x$ are like terms.
Combining like terms.
$2x - 3x - 1 = -x - 1$

28. The $5x$ and $-2x$ are like terms.
Combining like terms.
$5x - 2x + 3y + 6 = 3x + 3y + 6$

29. The x, $8x$, and $-9x$ are like terms.
Combining like terms.
$x + 8x - 9x + 3 = 9x - 9x + 3 = 3$

30. The $-4x$ and $-8x$ are like terms:
Combining like terms.
$-4x - 8x + 3 = -12x + 3$

31. Use the distributive property.
$3(x + 2) + 2x = 3x + 6 + 2x$
Combine like terms.
$= 5x + 6$

32. Use the distributive property.
$-2(x + 3) + 6 = -2x - 6 + 6$
Combine like terms.
$= -2x$

33. Use the distributive property.
$2x + 3(x + 4) - 5 = 2x + 3x + 12 - 5$
Combine like terms.
$= 5x + 7$

34. Use the distributive property.
$4(3 - 2x) - 2x = 12 - 8x - 2x$
Rearrange terms.
$= -8x - 2x + 12$
Combine like terms.
$= -10x + 12$

35. Use the distributive property.
$6 - (-x + 3) + 4x = 6 + x - 3 + 4x$
Rearrange terms.
$= x + 4x + 6 - 3$
Combine like terms.
$= 5x + 3$

36. Use the distributive property.
$2(2x + 5) - 10 - 4 = 4x + 10 - 10 - 4$
Combine like terms.
$= 4x - 4$

37. $-6(4 - 3x) - 18 + 4x$
Use the distributive property.
$= -24 + 18x - 18 + 4x$
Rearrange terms.
$= 18x + 4x - 24 - 18$
Combine like terms.
$= 22x - 42$

38. Use the distributive property.
$6 - 3(x + y) + 6x = 6 - 3x - 3y + 6x$
Rearrange terms.
$= 6x - 3x - 3y + 6$
Combine like terms.
$= 3x - 3y + 6$

39. Use the distributive property.
$3(x + y) - 2(2x - y) = 3x + 3y - 4x + 2y$
Rearrange terms.
$= 3x - 4x + 3y + 2y$
Combine like terms.
$= -x + 5y$

40. Use the distributive property.
$3x - 6y + 2(4y + 8) = 3x - 6y + 8y + 16$
Combine like terms.
$= 3x + 2y + 16$

41. Use the distributive property.

$3-(x-y)+(x-y)=3-x+y+x-y$

Rearrange terms.

$=-x+x+y-y+3$

Combine like terms.

$=3$

42. $(x+y)-(2x+3y)+4$

Use the distributive property.

$=x+y-2x-3y+4$

Rearrange terms.

$=x-2x+y-3y+4$

Combine like terms.

$=-x-2y+4$

43. $2x=4$

Divide both sides of the equation by 2.

$$\frac{2x}{2}=\frac{4}{2}$$

$x=2$

44. $x+3=-5$

Subtract 3 from both sides of the equation.

$x+3-3=-5-3$

$x=-8$

45. $x-4=7$

Add 4 to both sides of the equation.

$x-4+4=7+4$

$x=11$

46. $\frac{x}{3}=-9$

Multiply both sides of the equation by 3.

$$3\left(\frac{x}{3}\right)=3(-9)$$

$x=-27$

47. $2x+4=8$

Subtract 4 from both sides of the equation.

$2x+4-4=8-4$

$2x=4$

Divide both sides of the equation by 2.

$$\frac{2x}{2}=\frac{4}{2}$$

$x=2$

48. $14=3+2x$

Subtract 3 from both sides of the equation.

$14-3=3-3+2x$

$11=2x$

Divide both sides of the equation by 2.

$$\frac{11}{2}=\frac{2x}{2}$$

$$\frac{11}{2}=x$$

49. $8x-3=-19$

Add 3 to both sides of the equation.

$8x-3+3=-19+3$

$8x=-16$

Divide both sides of the equation by 8.

$$\frac{8x}{8}=\frac{-16}{8}$$

$x=-2$

50. $6 - x = 9$

Add x to both sides of the equation.

$6 - x + x = 9 + x$

$6 = 9 + x$

Subtract 9 from both sides of the equation.

$6 - 9 = 9 - 9 + x$

$-3 = x$

51. $-x = -12$

$-1x = -12$

Multiply both sides of the equation by -1.

$(-1)(-1x) = (-1)(-12)$

$1x = 12$

$x = 12$

52. $2(x + 2) = 6$

$2x + 4 = 6$

The distributive property was used.

$2x = 2$

4 was subtracted from both sides of the equation.

$x = 1$

Both sides of the equation were divided by 2.

53. $-3(2x - 8) = -12$

$-6x + 24 = -12$

The distributive property was used.

$-6x = -36$

24 was subtracted from both sides of the equation.

$x = 6$

Both side of the equation were divided by -6.

54. $4(6 + 2x) = 0$

$24 + 8x = 0$

The distributive property was used.

$8x = -24$

24 was subtracted from both sides of the equation.

$x = -3$

Both sides of the equation were divided by 8.

55. $3x + 2x + 6 = -15$

$5x + 6 = -15$

Like terms were combined.

$5x = -21$

6 was subtracted from both sides of the equation.

$x = -\frac{21}{5}$

Both sides of the equation were divided by 5.

56. $4 = -2(x + 3)$

$4 = -2x - 6$

The distributive property was used.

$10 = -2x$

6 was added to both sides of the equation.

$-5 = x$

Both sides of the equation were divided by -2.

57. $27 = 46 + 2x - x$

$27 = 46 + x$

Like terms were combined.

$-19 = x$

46 was subtracted from both sides of the equation.

58. $4x+6-7x+9=18$

$-3x+15=18$

Like terms were combined.

$-3x=3$

15 was subtracted from both sides of the equation.

$x=-1$

Both sides of the equation were divided by -3.

59. $4+3(x+2)=12$

$4+3x+6=12$

The distributive property was used.

$3x+10=12$

Like terms were combined.

$3x=2$

10 was subtracted from both sides of the equation.

$x=\frac{2}{3}$

Both sides of the equation were divided by 3.

60. $-3+3x=-2(x+1)$

$-3+3x=-2x-2$

The distributive property was used.

$3x=-2x+1$

3 was added to both sides of the equation.

$5x=1$

$2x$ was added to both sides of the equation.

$x=\frac{1}{5}$

Both sides of the equation were divided by 5.

61. $3x-6=-5x+30$

$3x=-5x+36$

6 was added to both sides of the equation.

$8x=36$

$5x$ was added to both sides of the equation.

$x=\frac{9}{2}$

Both sides of the equation were divided by 8.

62. $-(x+2)=2(3x-6)$

$-x-2=6x-12$

The distributive property was used.

$-x+10=6x$

12 was added to both sides of the equation.

$10=7x$

x was added to both sides of the equation.

$\frac{10}{7}=x$

Both sides of the equation were divided by 7.

63. $2x+6=3x+9$

$2x=3x+3$

6 was subtracted from both sides of the equation.

$-x=3$

$3x$ was subtracted from both sides of the equation.

$x=-3$

Both sides of the equation were multiplied by -1.

64. $-5x+3=2x+10$

$-5x-7=2x$

10 was subtracted from both sides of the equation.

$-7 = 7x$

$5x$ was added to both sides of the equation.

$-1 = x$

Both sides of the equation were divided by 7.

65. $3x - 12x = 24 - 6x$

$-9x = 24 - 6x$

Like terms were combined.

$-3x = 24$

$6x$ was added to both sides of the equation.

$x = -8$

Both sides of the equation were divided by −3.

66. $2(x + 4) = -3(x + 5)$

$2x + 8 = -3x - 15$

The distributive property was used.

$5x + 8 = -15$

$3x$ was added to both sides of the equation.

$5x = -23$

8 was subtracted from both sides of the equation.

$x = -\frac{23}{5}$

Both sides of the equation were divided by 5.

67. $4(2x - 3) + 4 = 9x + 2$

$8x - 12 + 4 = 9x + 2$

The distributive property was used.

$8x - 8 = 9x + 2$

Like terms were combined.

$8x - 10 = 9x$

2 was subtracted from both sides of the equation.

$-10 = x$

$8x$ was subtracted from both sides of the equation.

68. $6x + 11 = -(6x + 5)$

$6x + 11 = -6x - 5$

The distributive property was used.

$6x = -6x - 16$

11 was subtracted from both sides of the equation.

$12x = -16$

$6x$ was added to both sides of the equation.

$x = -\frac{4}{3}$

Both sides of the equation were divided by 12.

69. $2(x + 7) = 6x + 9 - 4x$

$2x + 14 = 6x + 9 - 4x$

The distributive property was used.

$2x + 14 = 2x + 9$

Like terms were combined.

$2x - 2x + 14 = 2x - 2x + 9$

Subtract $2x$ from both sides of the equation.

$14 = 9$ false

Since a false statement was obtained, there is no solution.

70. $-5(3 - 4x) = -6 + 20x - 9$

$-15 + 20x = -6 + 20x - 9$

The distributive property was used.

$-15 + 20x = -15 + 20x$

Like terms were combined.

The statement is true for all values of x, thus the solution is all real numbers.

71. $4(x-3)-(x+5)=0$

$4x-12-x-5=0$

The distributive property was used.

$3x-17=0$

Like terms were combined.

$3x=17$

17 was added to both sides of the equation.

$x=\frac{17}{3}$

Both sides of the equation were divided by 3.

72. $-2(4-x)=6(x+2)+3x$

$-8+2x=6x+12+3x$

The distributive property was used.

$-8+2x=9x+12$

Like terms were combined.

$-20+2x=9x$

12 was subtracted from both sides of the equation.

$-20=7x$

$2x$ was subtracted from both sides of the equation.

$-\frac{20}{7}=x$

Both sides of the equation were divided by 7.

73. $\frac{15}{20}=\frac{3}{4}$ so the ratio is 3:4

74. 80 ounces $=\frac{80}{16}=5$ pounds

The ratio of 80 ounces to 12 pounds is thus 5:12.

75. 32 ounces $=\frac{32}{16}=2$ pounds

The ratio of 32 ounces to 2 pounds is $\frac{2}{2}=\frac{1}{1}$

The ratio is 1:1.

76. $\frac{x}{9}=\frac{6}{18}$

$x\cdot 18=9\cdot 6$

$18x=54$

$x=\frac{54}{18}=3$

77. $\frac{15}{100}=\frac{x}{20}$

$(15)(20)=100\cdot x$

$300=100x$

$\frac{300}{100}=x$

$3=x$

78. $\frac{3}{x}=\frac{15}{45}$

$3\cdot 45=15\cdot x$

$135=15x$

$\frac{135}{15}=x$

$9=x$

79. $\frac{20}{45}=\frac{15}{x}$

$20\cdot x=15\cdot 45$

$20x=675$

$x=\frac{675}{20}=\frac{135}{4}$

80. $\frac{6}{2}=\frac{-12}{x}$

$6\cdot x=2(-12)$

$6x=-24$

$x=\frac{-24}{6}=-4$

81. $\frac{x}{9}=\frac{8}{-3}$

$-3\cdot x=9\cdot 8$

$-3x=72$

$x=\frac{72}{-3}=-24$

82. $\frac{-4}{9}=\frac{-16}{x}$

$-4\cdot x=9(-16)$

$-4x=-144$

$x=\frac{-144}{-4}=36$

83. $\frac{x}{-15}=\frac{30}{-5}$

$x(-5)=(-15)(30)$

$-5x=-450$

$x=\frac{-450}{-5}=90$

84. $\frac{6}{8}=\frac{30}{x}$

$6\cdot x=8\cdot 30$

$6x=240$

$x=\frac{240}{6}=40$

The length of the side is thus 40 in.

85. $\frac{7}{3.5}=\frac{2}{x}$

$7\cdot x=2(3.5)$

$7x=7$

$x=\frac{7}{7}=1$

The length of the side is thus 1 ft.

86. $2x+4\geq 8$

$2x\geq 4$

4 was subtracted from both sides of the inequality.

$x\geq 2$

Both sides of the inequality were divided by 2.

−5 0 5

87. $6-2x>4x-12$

$18-2x>4x$

12 was added to both sides of the inequality.

$18>6x$

$2x$ was added to both sides of the inequality.

$3>x$

Both sides of the inequality were divided by 6.

$x < 3$

–5 0 5

88. $6 - 3x \le 2x + 18$

$-12 - 3x \le 2x$

18 was subtracted from both sides of the inequality.

$-12 \le 5x$

$3x$ was added to both sides of the inequality.

$\frac{-12}{5} \le x$

Both sides of the inequality were divided by 5.

$x \ge -\frac{12}{5}$

–5 $-\frac{12}{5}$ 0 5

89. $2(x + 4) \le 2x - 5$

$2x + 8 \le 2x - 5$

Distributive property was used.

$2x - 2x + 8 \le 2x - 2x - 5$

$2x$ was subtracted from both sides of the inequality.

$8 \le -5$

Since 8 is never less than or equal to –5, the answer is no solution.

–5 0 5

90. $2(x + 3) > 6x - 4x + 4$

$2x + 6 > 6x - 4x + 4$

Distributive property was used.

$2x + 6 > 2x + 4$

Like terms were combined.

$2x - 2x + 6 > 2x - 2x + 4$

$2x$ was subtracted from both sides of the inequality.

$6 > 4$

Since 6 is always greater than 4, the answer is all real numbers.

–5 0 5

91. $x + 6 > 9x + 30$

$x - 24 > 9x$

30 was subtracted from both sides of the inequality.

$-24 > 8x$

x was subtracted from both sides of the inequality.

$-3 > x$

Both sides of the inequality were divided by 8.

$x < -3$

–5 0 5

92. $x - 2 \le -4x + 7$

$x \le -4x + 9$

2 was added to both sides of the inequality.

$5x \le 9$

$4x$ was added to both sides of the inequality.

$x \le \frac{9}{5}$

Both sides of the inequality were divided by 5.

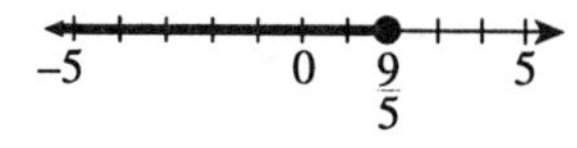

93. $-(x+2) < -2(-2x+5)$

$-x-2 < 4x-10$

Distributive property was used.

$-x+8 < 4x$

10 was added to both sides of the inequality.

$8 < 5x$

x was added to both sides of the inequality.

$\frac{8}{5} < x$

Both sides of the inequality were divided by 5

$x > \frac{8}{5}$

94. $2(x+3) < -(x+3)+4$

$2x+6 < -x-3+4$

Distributive property was used.

$2x+6 < -x+1$

Like terms were combined.

$2x < -x-5$

6 was subtracted from both sides of the inequality.

$3x < -5$

x was added to both sides of the inequality.

$x < -\frac{5}{3}$

Both sides of the inequality were divided by 3.

95. $-6x-3 \ge 2(x-4)+3x$

$-6x-3 \ge 2x-8+3x$

Distributive property was used.

$-6x-3 \ge 5x-8$

Like terms were combined.

$-6x+5 \ge 5x$

8 was added to both sides of the inequality.

$5 \ge 11x$

$6x$ was added to both sides of the inequality.

$\frac{5}{11} \ge x$

Both sides of the inequality were divided by 11.

$x \le \frac{5}{11}$

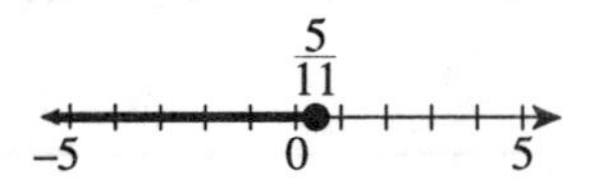

96. $-2(x-4) \le 3x+6-5x$

$-2x+8 \le 3x+6-5x$

Distributive property was used.

$-2x+8 \le -2x+6$

Like terms were combined.

$-2x+2x+8 \le -2x+2x+6$

$2x$ was added to both sides of the inequality.

$8 \le 6$

Since 8 is never less than or equal to 6, the answer is no solution.

97. $2(2x+4) > 4(x+2)-6$

$4x+8 > 4x+8-6$

Distributive property was used.

$4x+8 > 4x+2$

Like terms were combined.

$4x - 4x + 8 > 4x - 4x + 2$

$4x$ was subtracted from both sides of the inequality.

$8 > 2$

Since 8 is always greater than 2, the solution is all real numbers.

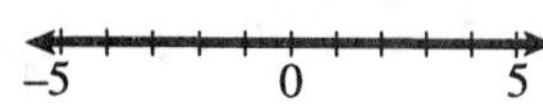

98. Let x = number of calories in 6 ounce piece of cake.

$$\frac{4\text{ ounces}}{160\text{ calories}} = \frac{6\text{ ounces}}{x\text{ calories}}$$

$$\frac{4}{160} = \frac{6}{x}$$

$$4 \cdot x = 160 \cdot 6$$

$$4x = 960$$

$$x = \frac{960}{4} = 240$$

Thus, a 6 ounce piece of cake has 240 calories.

99. Let x = number of pages that can be copied in 22 minutes.

$$\frac{1\text{ minute}}{5\text{ pages}} = \frac{22\text{ minutes}}{x\text{ pages}}$$

$$\frac{1}{5} = \frac{22}{x}$$

$$x = 22 \cdot 5$$

$$x = 110$$

110 pages can be copied in 22 minutes.

100. Let x = distance representing 380 miles.

$$\frac{60\text{ miles}}{1\text{ inch}} = \frac{380\text{ miles}}{x\text{ inches}}$$

$$\frac{60}{1} = \frac{380}{x}$$

$$60x = 380$$

$$x = \frac{380}{60} = 6\frac{1}{3}$$

$6\frac{1}{3}$ inches on the map represent 380 miles.

101. Let x = size of actual car

$$\frac{1\text{ inch}}{0.9\text{ feet}} = \frac{10.5\text{ inches}}{x\text{ feet}}$$

$$\frac{1}{0.9} = \frac{10.5}{x}$$

$$x = (0.9)(10.5)$$

$$x = 9.45$$

The size of the actual car is 9.45 ft.

102. $$\frac{3.1165\text{ pesos}}{1\text{ dollar}} = \frac{1\text{ peso}}{x\text{ dollars}}$$

$$\frac{3.1165}{1} = \frac{1}{x}$$

$$3.1165x = 1$$

$$x = \frac{1}{3.1165} = 0.3209$$

The value of 1 peso is about \$0.3209.

103. $$\frac{3\text{ radians}}{171.9\text{ degrees}} = \frac{1\text{ radian}}{x\text{ degrees}}$$

$$\frac{3}{171.9} = \frac{1}{x}$$

$$3x = 171.9$$

$$x = \frac{171.9}{3} = 57.3$$

There are 57.3 degrees in 1 radian.

104. Let x = number of bottles the machine can fill and cap in 2 minutes.

$$\frac{50 \text{ seconds}}{80 \text{ bottles}} = \frac{120 \text{ seconds}}{x \text{ bottles}}$$

$$\frac{50}{80} = \frac{120}{x}$$

$50x = (80)(120)$

$50x = 9600$

$x = \frac{9600}{50} = 192$

The machine can fill and cap 192 bottles in 2 minutes.

Chapter 2 Practice Test

1. $-2(4-2x) = -2[4+(-2x)]$

$= -2(4) + (-2)(-2x)$

$= -8 + 4x$

$= 4x - 8$

2. $-(x+3y-4) = -[x+3y+(-4)]$

$= -1[x+3y+(-4)]$

$= (-1)(x) + (-1)(3y) + (-1)(-4)$

$= -x + (-3y) + 4$

$= -x - 3y + 4$

3. The $3x$ and $-x$ are like terms.

Combining like terms: $3x - x + 4 = 2x + 4$

4. The $2x$ and $-3x$ are like terms.

The 4 and 6 are like terms.

Rearranging terms: $2x - 3x + 4 + 6$

Combining like terms: $-x + 10$

5. The $-2x$ and $-4x$ are like terms.

Rearranging terms: $-2x - 4x + y - 6$

Combining like terms: $-6x + y - 6$

6. The x and $6x$ are like terms.

The $-4y$ and $-y$ are like terms.

Rearranging terms: $x + 6x - 4y - y + 3$

Combining like terms: $7x - 5y + 3$

7. Use the distributive property.

$2x + 3 + 2(3x-2) = 2x + 3 + 6x - 4$

Rearrange terms.

$= 2x + 6x + 3 - 4$

Combine like terms.

$= 8x - 1$

8. $2x + 4 = 12$

Subtract 4 from both sides of the equation.

$2x + 4 - 4 = 12 - 4$

$2x = 8$

Divide both sides of the equation by 2.

$\frac{2x}{2} = \frac{8}{2}$

$x = 4$

9. $-x - 3x + 4 = 12$

$-4x + 4 = 12$

Like terms were combined.

$-4x = 8$

4 was subtracted from both sides of the equation.

$x = -2$

Both sides of the equation were divided by -4.

10. $4x - 2 = x + 4$

$4x = x + 6$

2 was added to both sides of the equation.

$3x = 6$

x was subtracted from both sides of the equation.

$x = 2$

Both sides of the equation were divided by 3.

11. $3(x-2) = -(5-4x)$

$3x - 6 = -5 + 4x$

The distributive property was used.

$3x - 1 = 4x$

5 was added to both sides of the equation.

$-1 = x$

$3x$ was subtracted from both sides of the equation.

12. $2x - 3(-2x+4) = -13 + x$

$2x + 6x - 12 = -13 + x$

The distributive property was used.

$8x - 12 = -13 + x$

Like terms were combined.

$8x = -1 + x$

12 was added to both sides of the equation.

$7x = -1$

x was subtracted from both sides of the equation.

$x = -\frac{1}{7}$

Both sides of the equation were divided by 7.

13. $3x - 4 - x = 2(x+5)$

Use the distributive property.

$3x - 4 - x = 2x + 10$

Combine like terms.

$2x - 4 = 2x + 10$

Subtract $2x$ from both sides of the equation.

$2x - 2x - 4 = 2x - 2x + 10$

$-4 = 10$ False

Since a false statement is obtained, there is no solution.

14. $-3(2x+3) = -2(3x+1) - 7$

Use the distributive property.

$-6x - 9 = -6x - 2 - 7$

Combine like terms.

$-6x - 9 = -6x - 9$

Since the equation is true for all values of x, the solution is all real numbers.

15. $\frac{9}{x} = \frac{3}{-15}$

$9(-15) = 3x$

$-135 = 3x$

$\frac{-135}{3} = x$

$-45 = x$

16. $2x - 4 < 4x + 10$

$2x - 14 < 4x$

10 was subtracted from both sides of the inequality.

$-14 < 2x$

$2x$ was subtracted from both sides of the inequality.

$-7 < x$

Both sides of the inequality were divided by 2.

$x > -7$

17. $3(x+4) \geq 5x - 12$

$3x + 12 \geq 5x - 12$

The distributive property was used.

$3x + 24 \geq 5x$

12 was added to both sides of the inequality.

$24 \geq 2x$

$3x$ was subtracted from both sides of the inequality.

$12 \geq x$

Both sides of the inequality were divided by 2.

$x \leq 12$

0 10 20

18. $4(x+3) + 2x < 6x - 3$

Use the distributive property.

$4x + 12 + 2x < 6x - 3$

Combine like terms.

$6x + 12 < 6x - 3$

Subtract $6x$ from both sides of the inequality.

$6x - 6x + 12 < 6x - 6x - 3$

$12 < -3$

Since 12 is never less than –3, the answer is no solution.

–5 0 5

19. $\frac{3}{4} = \frac{8}{x}$

$3x = 4 \cdot 8$

$3x = 32$

$x = \frac{32}{3}$

The length of side x is $\frac{32}{3}$ or $10\frac{2}{3}$ feet.

20. Let x = number of gallons needed.

$\frac{3 \text{ acres}}{6 \text{ gallons}} = \frac{75 \text{ acres}}{x \text{ gallons}}$

$\frac{3}{6} = \frac{75}{x}$

$3x = 6 \cdot 75$

$3x = 450$

$x = \frac{450}{3} = 150$

150 gallons are needed to treat 75 acres.

Chapter 2 Cumulative Review Test

1. $\frac{16}{20} \cdot \frac{4}{5} = \frac{16}{5} \cdot \frac{1}{5} = \frac{16 \cdot 1}{5 \cdot 5} = \frac{16}{25}$

2. $\frac{8}{24} \div \frac{2}{3} = \frac{8}{24} \cdot \frac{3}{2} = \frac{4}{8} \cdot \frac{1}{1} = \frac{4 \cdot 1}{8 \cdot 1} = \frac{4}{8} = \frac{1}{2}$

3. $|-2| > 1$ since $|-2| = 2$

4. $-6 - (-3) + 5 - 8 = -6 + 3 + 5 - 8$
$= -3 + 5 - 8 = 2 - 8 = -6$

5. $-12 - (-4) = -12 + 4 = -8$

6. $16 - 6 \div 2 \cdot 3 = 16 - 3 \cdot 3 = 16 - 9 = 7$

7. $3[6 - (4 - 3^2)] - 30 = 3[6 - (4 - 9)] - 30$
$= 3[6 - (-5)] - 30 = 3[6 + 5] - 30$
$= 3(11) - 30 = 33 - 30 = 3$

8. Substitute –2 for each x.

$-3x^2 - 4x + 5 = -3(-2)^2 - 4(-2) + 5$

$= -3(4) - (-8) + 5 = -12 + 8 + 5$

$= -4 + 5 = 1$

9. Associative property of addition

10. $6x$ and $4x$ are like terms.

$2y$ and $-y$ are like terms.

Rearranging terms: $6x + 4x + 2y - y$

Combining like terms: $10x + y$

11. $3x$ and $2x$ and $-2x$ are like terms.

Rearranging terms: $3x - 2x + 2x + 16$

Combining like terms: $3x + 16$

12. $4x - 2 = 10$

Add 2 to both sides of the equation.

$4x - 2 + 2 = 10 + 2$

$4x = 12$

Divide both sides of the equation by 4.

$$\frac{4x}{4} = \frac{12}{4}$$

$x = 3$

13. $\frac{1}{4}x = -10$

Multiply both sides of the equation by 4.

$$4\left(\frac{1}{4}x\right) = 4(-10)$$

$x = -40$

14. $6x + 5x + 6 = 28$

Combine like terms.

$11x + 6 = 28$

Subtract 6 from both sides of the equation.

$11x = 22$

Divide both sides of the equation by 11.

$x = 2$

15. $3(x - 2) = 5(x - 1) + 3x + 4$

$3x - 6 = 5x - 5 + 3x + 4$

The distributive property was used.

$3x - 6 = 8x - 1$

Like terms were combined.

$3x - 5 = 8x$

1 was added to both sides of the equation.

$-5 = 5x$

$3x$ was subtracted from both sides of the equation.

$-1 = x$

Both sides of the equation were divided by 5.

16. $\frac{15}{30} = \frac{3}{x}$

$15 \cdot x = 3 \cdot 30$

$15x = 90$

$$x = \frac{90}{15} = 6$$

17. $x - 4 > 6$

Add 4 to both sides of the inequality.

$x - 4 + 4 > 6 + 4$

$x > 10$

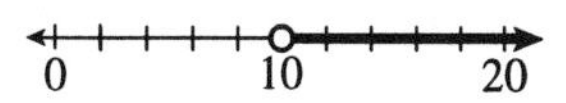

18. $2x-7\le 3x+5$

Subtract 5 from both sides of the inequality.

$2x-7-5\le 3x+5-5$

$2x-12\le 3x$

Subtract $2x$ from both sides of the inequality.

$2x-2x-12\le 3x-2x$

$-12\le x$

$x\ge -12$

−20 −10 0

19. Let x = number of pounds of fertilizer needed.

$$\frac{5000 \text{ square feet}}{36 \text{ pounds}}=\frac{22{,}000 \text{ square feet}}{x \text{ pounds}}$$

$$\frac{5000}{36}=\frac{22{,}000}{x}$$

$5000x=(36)(22{,}000)$

$5000x=792{,}000$

$$x=\frac{792{,}000}{5000}=158.4$$

158.4 pounds are needed to fertilize a 22,000-square-foot lawn.

20. Let x = amount he earns after 8 hours.

$$\frac{2 \text{ hours}}{\$10.50}=\frac{8 \text{ hours}}{x \text{ dollars}}$$

$$\frac{2}{10.5}=\frac{8}{x}$$

$2x=(10.5)(8)$

$2x=84$

$$x=\frac{84}{2}=42$$

He earns $42 after 8 hours.

CHAPTER 3

Exercise Set 3.1

1. Substitute 5 for s.

$A = s^2$

$A = (5)^2 = 25$

3. Substitute 6 for l and 5 for w.

$P = 2l + 2w$

$P = 2(6) + 2(5)$

$P = 12 + 10$

$P = 22$

5. Substitute 6 for h, 18 for b and 24 for d.

$A = \frac{1}{2}h(b + d)$

$A = \frac{1}{2}(6)(18 + 24)$

$A = \frac{1}{2}(6)(42) = 126$

7. Substitute 2 for r and 3.14 for π.

$C = 2\pi r$

$C = 2(3.14)(2) = 12.56$

9. Substitute 30 for A and 6 for b

$A = \frac{1}{2}bh$

$30 = \frac{1}{2}(6)h$

$30 = 3h$

$\frac{30}{3} = \frac{3h}{3}$

$10 = h$

11. Substitute 18 for V, 1 for w, and 3 for h.

$V = lwh$

$18 = l(1)(3)$

$18 = 3l$

$\frac{18}{3} = \frac{3l}{3}$

$6 = l$

13. Substitute 1000 for P, 0.08 for r, and 1 for t.

$A = P(1 + rt)$

$A = 1000[1 + (0.08)(1)]$

$A = 1000[1 + 0.08]$

$A = 1000(1.08) = 1080$

15. Substitute 36 for M and 16 for a.

$M = \frac{a+b}{2}$

$36 = \frac{16+b}{2}$

$2(36) = 2\left(\frac{16+b}{2}\right)$

$72 = 16 + b$

$72 - 16 = b$

$56 = b$

17. Substitute 41 for F.

$\text{C} = \frac{5}{9}(\text{F} - 32)$

$\text{C} = \frac{5}{9}(41 - 32)$

$\text{C} = \frac{5}{9}(9) = 5$

19. Substitute 2 for z, 50 for m, and 5 for s.

$$z = \frac{x - m}{s}$$

$$2 = \frac{x - 50}{5}$$

$$5(2) = 5\left(\frac{x - 50}{5}\right)$$

$$10 = x - 50$$

$$60 = x$$

21. Substitute 288 for K and 6 for v.

$$K = \frac{1}{2}mv^2$$

$$288 = \frac{1}{2}m(6)^2$$

$$288 = \frac{1}{2}m(36)$$

$$288 = 18m$$

$$\frac{288}{18} = \frac{18m}{18}$$

$$16 = m$$

23. Substitute 678.24 for v, 3.14 for π, and 6 for r.

$$v = \pi r^2 h$$

$$678.24 = (3.14)(6)^2 h$$

$$678.24 = (3.14)(36)h$$

$$678.24 = 113.04h$$

$$\frac{678.24}{113.04} = \frac{113.04h}{113.04}$$

$$6 = h$$

25. $2x + y = 8$

Subtract $2x$ from both sides of the equation.

$$2x - 2x + y = -2x + 8$$

$$y = -2x + 8$$

Substitute 2 for x.

$$y = -2(2) + 8$$

$$y = -4 + 8 = 4$$

27. $2x = 6y - 4$

Add 4 to both sides of the equation.

$$2x + 4 = 6y - 4 + 4$$

$$2x + 4 = 6y$$

Divide both sides of the equation by 6.

$$\frac{2x + 4}{6} = \frac{6y}{6}$$

$$\frac{2x + 4}{6} = y$$

Substitute 10 for x.

$$y = \frac{2(10) + 4}{6}$$

$$y = \frac{20 + 4}{6} = \frac{24}{6} = 4$$

29. $2y = 6 - 3x$

$$2y = -3x + 6$$

Divide both sides of the equation by 2.

$$\frac{2y}{2} = \frac{-3x + 6}{2}$$

$$y = \frac{-3x + 6}{2}$$

Substitute 2 for x.

$$y = \frac{-3(2) + 6}{2}$$

$$y = \frac{-6 + 6}{2} = \frac{0}{2} = 0$$

31. $-4x+5y=-20$

Add $4x$ to both sides of the equation.

$-4x+4x+5y=4x-20$

$5y=4x-20$

Divide both sides of the equation by 5.

$$\frac{5y}{5}=\frac{4x-20}{5}$$

$$y=\frac{4x-20}{5}$$

Substitute 4 for x.

$$y=\frac{4(4)-20}{5}=\frac{16-20}{5}=-\frac{4}{5}$$

33. $-3x=18-6y$

Subtract 18 from both sides of the equation.

$-3x-18=18-18-6y$

$-3x-18=-6y$

Divide both sides of the equation by -6.

$$\frac{-3x-18}{-6}=\frac{-6y}{-6}$$

$$\frac{-3x-18}{-6}=y$$

$$\frac{(-1)(3x+18)}{(-1)(6)}=y$$

$$\frac{3x+18}{6}=y$$

Substitute 0 for x.

$$y=\frac{3(0)+18}{6}$$

$$y=\frac{0+18}{6}=\frac{18}{6}=3$$

35. $-8=-x-2y$

Add x to both sides of the equation.

$x-8=x-x-2y$

$x-8=-2y$

Divide both sides of the equation by -2.

$$\frac{x-8}{-2}=\frac{-2y}{-2}$$

$$\frac{-x+8}{2}=y$$

Substitute -4 for x.

$$y=\frac{-(-4)+8}{2}=\frac{4+8}{2}=\frac{12}{2}=6$$

37. $d=rt$

Divide both sides of the equation by r.

$$\frac{d}{r}=\frac{rt}{r}$$

$$\frac{d}{r}=t$$

39. $i=prt$

Divide both sides of the equation by rt.

$$\frac{i}{rt}=\frac{prt}{rt}$$

$$\frac{i}{rt}=p$$

41. $C=\pi d$

Divide both sides of the equation by π.

$$\frac{C}{\pi}=\frac{\pi d}{\pi}$$

$$\frac{C}{\pi}=d$$

43. $A = \frac{1}{2}bh$

Multiply both sides of the equation by 2.

$2A = 2(\frac{1}{2}bh)$

$2A = bh$

Divide both sides of the equation by h.

$\frac{2A}{h} = \frac{bh}{h}$

$\frac{2A}{h} = b$

45. $P = 2l + 2w$

Subtract $2l$ from both sides of the equation.

$P - 2l = 2l - 2l + 2w$

$P - 2l = 2w$

Divide both sides of the equation by 2.

$\frac{P - 2l}{2} = \frac{2w}{2}$

$\frac{P - 2l}{2} = w$

47. $4n + 3 = m$

Subtract 3 from both sides of the equation.

$4n + 3 - 3 = m - 3$

$4n = m - 3$

Divide both sides of the equation by 4.

$\frac{4n}{4} = \frac{m - 3}{4}$

$n = \frac{m - 3}{4}$

49. $y = mx + b$

Subtract mx from both sides of the equation.

$y - mx = mx - mx + b$

$y - mx = b$

51. $I = P + Prt$

Subtract P from both sides of the equation.

$I - P = P - P + Prt$

$I - P = Prt$

Divide both sides of the equation by Pt.

$\frac{I - P}{Pt} = \frac{Prt}{Pt}$

$\frac{I - P}{Pt} = r$

53. $A = \frac{m + 2d}{3}$

Multiply both sides of the equation by 3.

$3A = 3\left(\frac{m + 2d}{3}\right)$

$3A = m + 2d$

Subtract m from both sides of the equation.

$3A - m = m - m + 2d$

$3A - m = 2d$

Divide both sides of the equation by 2.

$\frac{3A - m}{2} = \frac{2d}{2}$

$\frac{3A - m}{2} = d$

55. $d = a + b + c$

Subtract a from both sides of the equation.

$d - a = a - a + b + c$

$d - a = b + c$

Subtract c from both sides of the equation.

$d - a - c = b + c - c$

$d - a - c = b$

57. $ax + by = c$

Subtract ax from both sides of the equation.

$ax - ax + by = -ax + c$

$by = -ax + c$

Divide both sides of the equation by b.

$$\frac{by}{b} = \frac{-ax + c}{b}$$

$$y = \frac{-ax + c}{b}$$

59. $V = \pi r^2 h$

Divide both sides of the equation by πr^2.

$$\frac{V}{\pi r^2} = \frac{\pi r^2 h}{\pi r^2}$$

$$\frac{V}{\pi r^2} = h$$

61. Substitute 10 for n.

$$d = \frac{1}{2}n^2 - \frac{3}{2}n$$

$$d = \frac{1}{2}(10)^2 - \frac{3}{2}(10)$$

$$d = \frac{1}{2}(100) - 15 = 50 - 15 = 35$$

63. Substitute 50 for F.

$$\text{C} = \frac{5}{9}(\text{F}-32)$$

$$\text{C} = \frac{5}{9}(50 - 32)$$

$$\text{C} = \frac{5}{9}(18)$$

$\text{C} = 10$

The equivalent temperature is 10°C.

65. Substitute 35 for C.

$$\text{F} = \frac{9}{5}\text{C} + 32$$

$$\text{F} = \frac{9}{5}(35) + 32$$

$\text{F} = 63 + 32 = 95$

The equivalent temperature is 95°F.

67. $P = \dfrac{KT}{V}$

$$P = \frac{(1)(10)}{1} = \frac{10}{1} = 10$$

69. $P = \dfrac{KT}{V}$

$$80 = \frac{(K)(100)}{5}$$

$80 = 20K$

$$\frac{80}{20} = \frac{20K}{20}$$

$4 = K$

71. Substitute 5 for n.

$S = n^2 + n$

$S = (5)^2 + 5$

$S = 25 + 5 = 30$

73. $i = prt$

$i = (4000)(0.12)(3)$

$i = 1440$

He paid $1440 interest.

75. $i = prt$

$1050 = p(0.07)(3)$

$1050 = 0.21p$

$\frac{1050}{0.21} = \frac{0.21p}{0.21}$

$5000 = p$

Ms. Levy placed $5000 in the savings account.

77. $P = a + b + c$

$P = 5 + 12 + 13$

$P = 30$

The perimeter of the triangle is 30 inches.

79. $A = \frac{1}{2}bh$

$A = \frac{1}{2}(6)(8)$

$A = 24$

The area of the triangle is 24 square centimeters.

81. $A = \pi r^2$

$A = (3.14)(4)^2$

$A = 3.14(16)$

$A = 50.24$

The area of the circle is 50.24 square inches.

83. The radius is half the diameter, so $r = \frac{8}{2} = 4$ inches.

$C = 2\pi r$

$C = 2(3.14)(4)$

$C = 25.12$

The circumference of the circle is 25.12 inches.

85. $A = lw$

$48 = 6w$

$\frac{48}{6} = \frac{6w}{6}$

$8 = w$

The width of the post office is 8 feet.

87. **(a)** $C = 2\pi r$

$390 = 2(3.14)(r)$

$390 = 6.28r$

$\frac{390}{6.28} = \frac{6.28r}{6.28}$

$62.1 = r$

The radius of the roots is 62.1 feet.

(b) The diameter is twice the radius, so $d = 2(62.1) = 124.2$

The diameter is 124.2 feet.

89. The radius is half the diameter, so $r = \frac{22}{2} = 11$ inches.

Height = 4 feet = $12 \times 4 = 48$ inches

$V = \pi r^2 h$

$V = (3.14)(11)^2(48)$

$V = 18{,}237.12$

The volume of the drum is 18,237.12 cubic inches.

91. When you multiply a unit by the same unit, you get a square unit.

93. **(a)** $C = 2\pi r$. $\pi = \dfrac{C}{2r}$ or $\pi = \dfrac{C}{d}$

(b) π or about 3.14.

(c) Individual project

Cumulative Review Exercises: 3.1

94. $[4(12 \div 2^2 - 3)^2]^2$

$= [4(12 \div 4 - 3)^2]^2$

$= [4(3 - 3)^2]^2$

$= [4(0)^2]^2$

$= [0]^2$

$= 0$

95. $\dfrac{6}{4} = \dfrac{3}{2}$ so the ratio of Arabians to Morgans is 3:2.

96. Let x = number of minutes to siphon 13,500 gallons

$\dfrac{25 \text{ gallons}}{3 \text{ minutes}} = \dfrac{13{,}500 \text{ gallons}}{x \text{ minutes}}$

$\dfrac{25}{3} = \dfrac{13{,}500}{x}$

$25 \cdot x = 3(13{,}500)$

$25x = 40{,}500$

$x = \dfrac{40{,}500}{25} = 1620$

It will take 1620 minutes or 27 hours to empty the pool.

97. $2(x - 4) \ge 3x + 9$

Use the distributive property

$2x - 8 \ge 3x + 9$

9 was subtracted from both sides of the inequality

$2x - 17 \ge 3x$

$2x$ was subtracted from both sides of the inequality

$-17 \ge x$

$x \le -17$

Group Activity/Challenge Problems

1. **(a)** Shaded area =

Area of square – Area of circle

Area of square = d^2

Area of circle = $\pi r^2 = \pi\left(\dfrac{d}{2}\right)^2$

Shaded area = $d^2 - \pi\left(\dfrac{d}{2}\right)^2$

(b) Shaded area = $d^2 - \pi\left(\dfrac{d}{2}\right)^2$

$= (4)^2 - \pi\left(\dfrac{4}{2}\right)^2 = 16 - \pi(2)^2$

$= 16 - 4\pi = 16 - 4(3.14) = 3.44$

The shaded area is 3.44 square feet.

(c) Shaded area = $d^2 - \pi\left(\dfrac{d}{2}\right)^2$

$= (6)^2 - \pi\left(\dfrac{6}{2}\right)^2 = 36 - \pi(3)^2$

$= 36 - 9\pi = 36 - 9(3.14) = 7.74$

The shaded area is 7.74 square feet.

2. **(a)** $V = lwh$

$V = (3x)(x)(6x - 1)$

$V = 3x^2(6x - 1)$

$V = 18x^3 - 3x^2$

(b) $V = 18x^3 - 3x^2$

$V = 18(7)^3 - 3(7)^2$

$V = 6174 - 147$

$V = 6027$

Volume is 6027 cm^3.

(c) $S = 2lw + 2lh + 2wh$

$S = 2(3x)(x) + 2(3x)(6x - 1)$
$+ 2(x)(6x - 1)$

$S = 6x^2 + 36x^2 - 6x + 12x^2 - 2x$

$S = 54x^2 - 8x$

(d) $S = 54x^2 - 8x$

$S = 54(7)^2 - 8(7)$

$S = 2646 - 56$

$S = 2590$

Surface area is 2590 cm^2.

4. **(a)** $C = 2\pi r$

$446 = 2\pi r$

$\frac{446}{2\pi} = r$

$\frac{446}{2(3.14)} = r$

$71 = r$

The radius is 71 ft, so the diameter is $2(71) = 142$ ft.

(b) Inside diameter

= outside diameter – 2(4)

Inside diameter = 142 – 8 = 134

The inside diameter is 134 ft.

(c) $A = \pi r^2$

$A = (3.14)\left(\frac{134}{2}\right)^2$

$A = 14,095$

The surface area is 14,095 sq ft.

(d) $V = \pi r^2 h$

$V = (3.14)\left(\frac{134}{2}\right)^2(120)$

$V = 1,691,455$

The volume is 1,691,455 cu ft.

Exercise Set 3.2

1. $x + 5$

3. $4x$

5. $0.70x$

7. $0.10c$

9. $0.16p$

11. $6x - 3$

13. $\frac{3}{4}x + 7$

15. $2(x + 8)$

17. $4x$

19. $0.23x$

21. $8.20b$

23. $300n$

25. $25x$

27. $12x$

29. $16c$

31. $275x + 25y$

33. Six less than a number.

35. One more than four times a number.

37. Seven less than five times a number.

39. Four times a number, decreased by two.

41. Three times a number subtracted from two.

43. Twice the difference of a number and one.

45. Martin's salary is x; Eileen's salary is $x + 45$.

47. One number is x; the second number is $\frac{x}{3}$.

49. The first consecutive even integer is x; the second consecutive even integer is $x + 2$.

51. One number is x; the second number is $x + 12$ (or $x - 12$).

53. One number is x; the other number is

$\frac{x}{2} + 3$

55. One number is x; the second number is $3x - 4$.

57. The first consecutive odd integer is x; the second consecutive odd integer is $x + 2$.

59. One number is x; the second number is $x - 0.15x$.

61. $c, c - 0.10c$

63. $p, p - 0.50p$

65. $w, 2w$

67. $m, m + 0.15m$

69. Let x = one number, then $5x$ = second number.

First number + second number = 18

$x + 5x = 18$

71. Let x = smaller consecutive integer, then $x + 1$ = larger consecutive integer.

Smaller + larger = 47

$x + (x + 1) = 47$

73. Let x = the number.

Twice the number decreased by 8 is 12.

$2x - 8 = 12$

75. Let x = the number.

One-fifth of the sum of the number and 10 is 150.

$\frac{1}{5}(x+10)=150$

77. Let x = distance traveled by one train, then $2x-8$ = distance traveled by second train.

Distance of train 1 + distance of train 2 = total distance

$x+(2x-8)=1000$

79. Let x = the number

Then the number increased by 8% is

$x+0.08x$

$x+0.08x=92$

81. Let x = original cost of the jacket, then the discount is $0.25x$.

Original cost – discount = \$65

$x-0.25x=65$

83. Let x = original cost of the recorder, then the reduction is $0.20x$.

Original cost – reduction = \$215

$x-0.20x=215$

85. Let x = one number, then $2x-3$ = second number

First number + second number = 21

$x+(2x-3)=21$

87. $40t=180$

89. $15y=215$

91. $25q=150$

93. Three more than a number is six.

95. Three times a number, decreased by one, is four more than twice the number.

97. Four times the difference of a number and one is six.

99. Six more than five times a number is the difference of six times the number and one.

101. The sum of a number and the number increased by four is eight.

103. The sum of twice a number and the number increased by three is five.

105. Individual answer

Cumulative Review Exercises: 3.2

107. $\frac{1}{2}(6.7)=3.35$

3.35 teaspoons should be used.

108. $\dfrac{3 \text{ cups cat chow}}{1 \text{ cup water}} = \dfrac{\frac{1}{2} \text{ cup cat chow}}{x \text{ cups water}}$

$\dfrac{3}{1} = \dfrac{\frac{1}{2}}{x}$

$3x = \dfrac{1}{2}$

$\dfrac{1}{3}(3x) = \left(\dfrac{1}{3}\right)\left(\dfrac{1}{2}\right)$

$x = \dfrac{1}{6}$

Melinda should add $\frac{1}{6}$ cup water to $\frac{1}{2}$ cup cat chow.

109. Substitute 40 for P and 5 for w.

$P = 2l + 2w$

$40 = 2l + 2(5)$

$40 = 2l + 10$

$30 = 2l$

$15 = l$

110. $3x - 2y = 6$

Subtract $3x$ from both sides of the equation

$3x - 3x - 2y = -3x + 6$

$-2y = -3x + 6$

Divide both sides of the equation by -2

$\dfrac{-2y}{-2} = \dfrac{-3x+6}{-2}$

$y = \dfrac{3x-6}{2}$

Substitute 6 for x.

$y = \dfrac{3(6)-6}{2}$

$y = \dfrac{18-6}{2}$

$y = \dfrac{12}{2} = 6$

Group Activity/Challenge Problems

1. **(a)** 1 minute = 60 seconds

1 hour = 60 minutes = 3600 seconds

1 day = 24 hours = 1440 minutes

= 86,400 seconds

Answer: $86{,}400d + 3600h + 60m + s$

(b) $86{,}400d + 3600h + 60m + s$

$= 86{,}400(4) + 3600(6) + 60(15) + 25$

$= 368{,}125$ seconds

3. **(a)** Let t be the time of the shower.

(b) $30 = 6t$

5. **(a)** Let y be the number of years.

(b) $40{,}000y = 1{,}000{,}000$

7. **(a)** Let m be the number of miles.

(b) $20 + 0.60m = 30 + 0.45m$

Exercise Set 3.3

1. Let x = smaller consecutive integer, then $x + 1$ = larger consecutive integer.

Smaller number + larger number = 45

$x + (x + 1) = 45$;

$x + (x + 1) = 45$

$2x + 1 = 45$

$2x = 44$

$x = 22$

Smaller number = 22

Larger number = $x + 1 = 22 + 1 = 23$

3. Let x = smaller consecutive odd integer, then $x + 2$ = larger consecutive odd integer.

Smaller number + larger number = 68

$x + (x + 2) = 68;$

$x + (x + 2) = 68$

$2x + 2 = 68$

$2x = 66$

$x = 33$

Smaller number = 33

Larger number = $x + 2 = 33 + 2 = 35$

5. Let x = one number, then $3x - 5$ = second number

First number + second number = 43

$x + (3x - 5) = 43;$

$x + (3x - 5) = 43$

$4x - 5 = 43$

$4x = 48$

$x = 12$

First number = 12

Second number = $3x - 5$

$= 3(12) - 5 = 31$

7. Let x = smallest consecutive odd integer, then second and third consecutive odd integers are $x + 2$ and $x + 4$.

First number + second number

+ third number = 87

$x + (x + 2) + (x + 4) = 87;$

$x + (x + 2) + (x + 4) = 87$

$3x + 6 = 87$

$3x = 81$

$x = 27$

First number = 27

Second number = $x + 2 = 27 + 2 = 29$

Third number = $x + 4 = 27 + 4 = 31$

9. Let x = smaller integer, then larger integer $= 2x - 8$.

Larger number – smaller number = 17

$(2x - 8) - x = 17;$

$(2x - 8) - x = 17$

$x - 8 = 17$

$x = 25$

Smaller number = 25

Larger number = $2x - 8$

$= 2(25) - 8 = 42$

11. Let x = miles per gallon for Vector, then $5x + 3$ = miles per gallon for Geo.

Miles per gallon for Vector

+ miles per gallon for Geo = 69

$x + (5x + 3) = 69;$

$x + (5x + 3) = 69$

$6x + 3 = 69$

$6x = 66$

$x = 11$

Vector gets 11 miles per gallon.

Geo gets $5x + 3 = 5(11) + 3$

= 58 miles per gallon

13. Let x = number of reactors in Mexico, then $15x + 6$ = number of reactors in Canada.

Number reactors in Mexico

+ number reactors in Canada = 22

$x + (15x + 6) = 22;$

$x + (15x + 6) = 22$

$16x + 6 = 22$

$16x = 16$

$x = 1$

Mexico has 1 operating reactor; Canada has $15x + 6 = 15(1) + 6 = 21$ reactors.

15. Let x = selling price of house, then commission received by broker = $0.06x$.

Selling price – commission

= amount received by Mrs. Sanchez

$x - 0.06x = 65,800$;

$x - 0.06x = 65,800$

$0.94x = 65,800$

$x = \dfrac{65,800}{0.94} = 70,000$

The selling price was \$70,000.

17. Let x = percent of pure gold by weight.

Total weight x percent gold

= amount of gold

$20x = 15$;

$20x = 15$

$x = \dfrac{15}{20} = 0.75$

18 karat gold is 75% by weight pure gold.

19. Let x = number of people in attendance, then $1500 + 2x$ = total received.

$1500 + 2x = 3100$;

$1500 + 2x = 3100$

$2x = 1600$

$x = 800$

800 people were in attendance.

21. Let x = total amount collected at door; 2000 + 2% of admission fees = total amount received.

$2000 + 0.02x = 2400$;

$2000 + 0.02x = 2400$

$0.02x = 400$

$x = \dfrac{400}{0.02} = 20,000$

The total amount collected at the door was \$20,000.

23. Let x = amount Mary would need to spend for savings to equal fee, then amount saved = $0.08x$.

Amount saved = yearly fee

$0.08x = 60$;

$0.08x = 60$

$x = \dfrac{60}{0.08} = 750$

Mary would need to spend \$750 for savings to equal yearly fee.

25. Let x = Paul's resent salary, then his salary increase = $0.08x$

Present salary + increase in salary

= future salary

$x + 0.08x = 37,800$;

$x + 0.08x = 37,800$

$1.08x = 37,800$

$x = \dfrac{37,800}{1.08} = 35,000$

His present salary is \$35,000.

27. Let x = price before tax,

then sales tax = $0.07x$.

Price before tax + sales tax = 1.50

$x + 0.07x = 1.50;$

$x + 0.07x = 1.50$

$1.07x = 1.50$

$x = \frac{1.50}{1.07} = 1.40$

Price of a hot dog before tax is $1.40.

29. Let x = amount left to favorite charity, then $3x$ = amount left to each child.

Amount received by charity

+ amount received by children

= 210,000

$x + 3x + 3x = 210{,}000;$

$x + 3x + 3x = 210{,}000$

$7x = 210{,}000$

$x = 30{,}000$

Charity receives $30,000.

Each child receives $3x = 3(30{,}000)$

= $90,000.

31. Let x = number of years for number of landfills to drop to 2000, then $300x$ = decrease in number of landfills over x years.

Present number – decrease in number = future number

$4000 - 300x = 2000;$

$4000 - 300x = 2000$

$-300x = -2000$

$x = 6.67$

It will take about 6.67 years for the number of landfills to drop to 2000.

33. Let x = amount of water that would be used without the special shower head, then amount of water saved $= 0.60x$

$$\begin{pmatrix}\text{amount of water}\\\text{without special}\\\text{shower head}\end{pmatrix} - \begin{pmatrix}\text{amount of}\\\text{water saved}\end{pmatrix} = \begin{pmatrix}\text{amount of}\\\text{water used}\\\text{with special}\\\text{shower head}\end{pmatrix}$$

$x - 0.60x = 24;$

$x - 0.60x = 24$

$0.40x = 24$

$x = \frac{24}{0.40} = 60$

Without the special shower head, 60 gallons of water would be used.

35. Let x = number of miles per hour over speed limit she was traveling, then $15 + 5x$ = amount of fine.

$15 + 5x = 65;$

$15 + 5x = 65$

$5x = 50$

$x = 10$

She was traveling 10 miles per hour over the speed limit.

37. Let $x = 1980$ population.

The increase in population between 1980 and 1990 was $0.90x$.

1980 population + increase in population = 1990 population

$x + 0.90x = 110{,}000;$

$x + 0.90x = 110{,}000$

$1.90x = 110{,}000$

$x = \dfrac{110{,}000}{1.90} = 57{,}895$

The 1990 population was about 57,895.

39. Let x = maximum price of meal she can afford, then amount of tax $= 0.07x$ and amount of tip $= 0.15x$.

Price of meal + tax + tip = 20

$x + 0.07x + + 0.15x = 20;$

$x + 0.07x + 0.15x = 20$

$1.22x = 20$

$x = \dfrac{20}{1.22} = 16.39$

The maximum price she can afford for a meal is $16.39.

41. Let x = regular membership fee, then amount of reduction $= 0.10x$.

regular fee – reduction – 20 = new fee on a Monday

$x - 0.10x - 20 = 250;$

$x - 0.10x - 20 = 250$

$x - 0.10x = 270$

$0.90x = 270$

$x = \dfrac{270}{0.90} = 300$

Regular fee is $300.

43. Let x = amount of oil, then $15x$ = amount of gasoline.

Amount of oil + amount of gasoline = 4

$x + 15x = 4;$

$x + 15x = 4$

$16x = 4$

$x = \dfrac{1}{4}$

Amount of oil needed is $\frac{1}{4}$ or 0.25 gallons.

Amount of gas needed $= 15\left(\dfrac{1}{4}\right) = 3\frac{3}{4}$ or 3.75 gallons.

45. Let x = number of marriages performed in 1991, then decrease in marriages from 1991 to 1992 $= 0.01x$.

$$\begin{pmatrix}\text{Number of}\\ \text{marriages}\\ \text{in 1991}\end{pmatrix} - \begin{pmatrix}\text{decrease in}\\ \text{marriages from}\\ \text{1991 to 1992}\end{pmatrix} = \begin{pmatrix}\text{number of}\\ \text{marriages}\\ \text{in 1992}\end{pmatrix}$$

$x - 0.01x = 2.362;$

$x - 0.01x = 2.362$

$0.99x = 2.362$

$x = \dfrac{2.362}{0.99} = 2.386$

In 1991, about 2.386 million marriages were performed.

47. **(a)** With the 9.5% mortgage, the monthly payments are
$50(9.33) = \$466.50$.
With the 8.00% mortgage, the monthly payments are $50(8.37) = \$418.50$. In addition the Yearstons must pay $0.04(50{,}000) = \$2000$ because of the 4 points.
Let x = number of months when total payments from both mortgages are equal.
Monthly payments for 9.5% mortgage = monthly payments for 8.00% mortgage + points.
$466.50x = 418.50x + 2000$
$48x = 2000$
$x = \frac{2000}{48} = 41.7$
The total payments are the same after about 42 months or 3.5 years.

(b) After about 3.5 years the Citibank mortgage is cheaper because of the lower monthly payments. If they plan to live in the house for 20 years, the Citibank mortgage is less expensive.

49. **(a)** With the 9.00% mortgage, the monthly payments are
$100(9.00) = \$900$.
With the 8.875% mortgage, the monthly payments are $100(8.92) = \$892$.
Let y = cost of credit check and application fee at Key Mortgage, then $y + 150$ = cost of credit check and application fee at Countrywide.
Let x = number of months when total payments from the two mortgages are equal.
Total payments from Key Mortgage = total payments from Countrywide
$900x + y = 892x + (y + 150)$
$900x = 892x + 150$
$8x = 150$
$x = \frac{150}{8} = 18.75$
The total costs from the two mortgages are equal after 18.75 months or about 1.56 years.

(b) After about 1.56 years the Countrywide mortgage is cheaper because of the lower monthly payments. If John plans to live in the house for 10 years, the Countrywide Mortgage is less expensive.

51. **(a)** If they refinance, the monthly payments will be $50(8.29) = \$414.50$.

Their monthly savings will be $\$740 - \$414.50 = \$325.50$.

Let x = number of months when money saved = closing cost.

Money saved = closing cost

$325.50x = 3000$

$x = \frac{3000}{325.50} = 9.2$

It will take about 9.2 months for the money saved to equal the closing cost.

(b) Their monthly payments will be $\$740 - \$414.50 = \$325.50$ lower

53. Individual answer

Cumulative Review Exercises: 3.3

54. $\frac{1}{4}+\frac{3}{4}\div\frac{1}{2}-\frac{1}{3}=\frac{3}{12}+\frac{9}{12}\div\frac{6}{12}-\frac{4}{12}$

$=\frac{3}{12}+\frac{9}{12}\cdot\frac{12}{6}-\frac{4}{12}=\frac{3}{12}+\frac{9}{6}-\frac{4}{12}$

$=\frac{3}{12}+\frac{18}{12}-\frac{4}{12}=\frac{21}{12}-\frac{4}{12}=\frac{17}{12}$

55. Associative property of addition

56. Commutative property of multiplication

57. Distributive property

58. Let x = number of pounds of coleslaw needed.

$$\frac{5 \text{ people}}{\frac{1}{2} \text{ pound coleslaw}} = \frac{560 \text{ people}}{x \text{ pounds coleslaw}}$$

$$\frac{5}{\frac{1}{2}} = \frac{560}{x}$$

$$5x = \left(\frac{1}{2}\right)(560)$$

$$5x = 280$$

$$x = \frac{280}{5} = 56$$

He will need 56 pounds of coleslaw.

59. $M = \dfrac{a+b}{2}$

$$2M = 2\left(\frac{a+b}{2}\right)$$

$$2M = a+b$$

$$2M - a = a - a + b$$

$$2M - a = b$$

Group Activity/Challenge Problems

3. n

$4n$

$4n + 6$

$$\frac{4n+6}{2} = 2n+3$$

$$2n + 3 - 3 = 2n$$

4. $14{,}000 + 14{,}000x = 40{,}000$

$14{,}000x = 26{,}000$

$x = 1.857$

About 185.7% increase.

6. $x + 0.008x = 253{,}488$

$1.008x = 253{,}488$

$x = 251{,}476.2$

About 251,476.2 people per square mile.

Exercises Set 3.4

1. Let x = length of each side of the triangle, then $P = x + x + x = 3x$.

Perimeter = 28.5

$$x = \frac{28.5}{3} = 9.5$$

The length of each side is 9.5 inches.

3. Let x = measure of angle A, then $3x - 8$ = measure of angle B.

Sum of the 2 angles = 180

$x + (3x - 8) = 180$

$4x - 8 = 180$

$4x = 188$

$$x = \frac{188}{4} = 47$$

Measure of angle $A = 47°$

Measure of angle $B = 3(47) - 8$

$= 141 - 8 = 133°$

5. Let x = smallest angle

Then second angle = $x + 10$

and third angle = $2x - 30$

Sum of the 3 angles = 180

$x + (x + 10) + (2x - 30) = 180$

$4x - 20 = 180$

$4x = 200$

$x = \frac{200}{4} = 50$

The first angle is 50°.

The second angle is 50 + 10 = 60°.

The third angle is 2(50) – 30 = 70°.

7. Let x = length of each of the two equal sides, then $x - 2$ = length of third side,

Sum of the three sides = perimeter

$x + x + (x - 2) = 10$

$3x - 2 = 10$

$3x = 12$

$x = 4$

The length of the two equal sides is 4 meters.

The length of the third side is 4 – 2 = 2 meters.

9. Let w = width, then $2w - 24$ = length.

$P = 2l + 2w$

$240 = 2(2w - 24) + 2w$

$240 = 4w - 48 + 2w$

$288 = 6w$

$48 = w$

The width is 48 feet.

The length is 2(48) – 24 = 72 feet.

11. Let x = measure of each smaller angle, then $2x - 27$ = measure of each larger angle.

$$\begin{pmatrix}\text{measure of}\\ \text{the two}\\ \text{smaller angles}\end{pmatrix} + \begin{pmatrix}\text{measure of}\\ \text{the two}\\ \text{larger angles}\end{pmatrix}$$

$= 360°$

$x + x + (2x - 27) + (2x - 27) = 360°$

$6x - 54 = 360°$

$6x = 414°$

$x = 69°$

Each smaller angle is 69°.

Each larger angle is 2(69) – 27 = 111°.

13. Let x = length of a shelf, then $x + 2$ = height of bookcase

4 shelves + 2 sides = total lumber available

$4x + 2(x + 2) = 20$

$4x + 2x + 4 = 20$

$6x + 4 = 20$

$6x = 16$

$$x = \frac{16}{6} = 2\frac{2}{3}$$

The width of the bookcase is $2\frac{2}{3}$ feet.

The height of the bookcase is $2\frac{2}{3} + 2$ $= 4\frac{2}{3}$ feet.

15. Let x = height of shelves, then
$3x$ = length of shelves.

4 shelves + 3 sides = total lumber available.

$4(3x) + 3x = 45$

$12x + 3x = 45$

$15x = 45$

$x = 3$

The height of the unit is 3 feet.

The length of the unit is 3(3) = 9 feet.

17. $A = (2l) \cdot \left(\frac{w}{2}\right) = l \cdot w$

The area remains the same.

19. $V = 2l \cdot 2w \cdot 2h = 8l \cdot w \cdot h$

The volume becomes eight times as great.

20. Individual answer

Cumulative Review Exercises: 3.4

22. $-|-6| < |-4|$ since $-|-6| = -6$ and $|-4| = 4$

23. $|-3| > -|3|$ since $|-3| = 3$ and $-|3| = -3$

24. $-6 - (-2) + (-4) = -6 + 2 + (-4)$
$= -4 + (-4) = -8$

25. $-6y + x - 3(x - 2) + 2y$

Use the distributive property

$= -6y + x - 3x + 6 + 2y$

Rearrange terms

$= x - 3x - 6y + 2y + 6$

Combine like terms

$= -2x - 4y + 6$

26. $2x + 3y = 9$

$2x - 2x + 3y = -2x + 9$

$3y = -2x + 9$

$\frac{3y}{3} = \frac{-2x + 9}{3}$

$y = \frac{-2x + 9}{3}$ or $y = -\frac{2}{3}x + 3$

Substitute 3 for x.

$y = \frac{-2(3) + 9}{3}$

$= \frac{-6 + 9}{3} = \frac{3}{3} = 1$

Group Activity/Challenge Problems

2. $ac + ad + bc + bd$

3. **(a)** $P = 14.70 + 0.43x$

$162 = 14.70 + 0.43x$

$147.3 = 0.43x$

$342.56 = x$

The submarine can go 342.56 ft deep.

(b) $P = 14.70 + 0.43x$

$97.26 = 14.70 + 0.43x$

$82.56 = 0.43x$

$192 = x$

The submarine is 192 ft deep.

Chapter 3 Review Exercises

1. Substitute 4 for d and 3.14 for π.

$C = \pi d$

$C = (3.14)(4)$

$C = 12.56$

2. Substitute 12 for b and 8 for h.

$A = \frac{1}{2}bh$

$A = \frac{1}{2}(12)(8)$

$A = 48$

3. Substitute 6 for l and 4 for w.

$P = 2l + 2w$

$P = 2(6) + 2(4)$

$P = 12 + 8 = 20$

4. Substitute 1000 for p, 0.15 for r, and 2 for t.

$i = prt$

$i = (1000)(0.15)(2)$

$i = 300$

5. $E = IR$

$E = (0.12)(2000)$

$E = 240$

6. $A = \pi r^2$

$A = (3.14)(3)^2$

$A = (3.14)(9)$

$A = 28.26$

7. $V = \frac{4}{3}\pi r^3$

$V = \frac{4}{3}(3.14)(3)^3$

$V = \frac{4}{3}(3.14)(27) = 113.04$

8. $Fd^2 = km$

$(60)(2)^2 = k(12)$

$(60)(4) = 12k$

$240 = 12k$

$\frac{240}{12} = \frac{12k}{12}$

$20 = k$

9. $y = mx + b$

$15 = (3)(-2) + b$

$15 = -6 + b$

$15 + 6 = -6 + 6 + b$

$21 = b$

10. $2x + 3y = -9$

$2(12) + 3y = -9$

$24 + 3y = -9$

$24 - 24 + 3y = -9 - 24$

$3y = -33$

$\frac{3y}{3} = \frac{-33}{3}$

$y = -11$

11. $4x - 3y = 15 + x$

$4(-3) - 3y = 15 + (-3)$

$-12 - 3y = 12$

$-12 + 12 - 3y = 12 + 12$

$-3y = 24$

$\frac{-3y}{-3} = \frac{24}{-3}$

$y = -8$

12. $2x = y + 3z + 4$

$2(5) = y + 3(-3) + 4$

$10 = y + (-9) + 4$

$10 = y - 5$

$10 + 5 = y - 5 + 5$

$15 = y$

13. $IR = E + Rr$

$(5)(200) = 100 + (200)r$

$1000 = 100 + 200r$

$900 = 200r$

$\frac{900}{200} = \frac{200r}{200}$

$4.5 = r$

14. $2x - y = 12$

$2x - y + y = 12 + y$

$2x = 12 + y$

$2x - 12 = 12 - 12 + y$

$2x - 12 = y$

Substitute 10 for x.

$y = 2(10) - 12$

$y = 20 - 12 = 8$

15. $3x - 2y = -4$

$3x - 3x - 2y = -3x - 4$

$-2y = -3x - 4$

$\frac{-2y}{-2} = \frac{-3x - 4}{-2}$

$y = \frac{3x + 4}{2}$

Substitute 2 for x.

$y = \frac{3(2) + 4}{2}$

$y = \frac{6 + 4}{2}$

$y = \frac{10}{2} = 5$

16. $3x = 5 + 2y$

$3x - 5 = 5 - 5 + 2y$

$3x - 5 = 2y$

$\frac{3x - 5}{2} = \frac{2y}{2}$

$\frac{3x - 5}{2} = y$

Substitute −3 for x.

$y = \frac{3(-3) - 5}{2}$

$y = \frac{-9 - 5}{2}$

$y = \frac{-14}{2} = -7$

17. $-6x - 2y = 20$

$-6x + 6x - 2y = 6x + 20$

$-2y = 6x + 20$

$\frac{-2y}{-2} = \frac{6x + 20}{-2}$

$y = -3x - 10$

Substitute 0 for x.

$y = -3(0) - 10$

$y = 0 - 10 = -10$

18. $6 = -3x - 2y$

$3x + 6 = 3x - 3x - 2y$

$3x + 6 = -2y$

$\frac{3x + 6}{-2} = \frac{-2y}{-2}$

$\frac{-3x - 6}{2} = y$

Substitute –6 for x.

$y = \frac{-3(-6) - 6}{2}$

$y = \frac{18 - 6}{2}$

$y = \frac{12}{2} = 6$

19. $3y - 4x = -3$

$3y - 4x + 4x = 4x - 3$

$3y = 4x - 3$

$\frac{3y}{3} = \frac{4x - 3}{3}$

$y = \frac{4x - 3}{3}$

Substitute 2 for x.

$y = \frac{4(2) - 3}{3}$

$y = \frac{8 - 3}{3}$

$y = \frac{5}{3}$

20. $F = ma$

$\frac{F}{a} = \frac{ma}{a}$

$\frac{F}{a} = m$

21. $A = \frac{1}{2}bh$

$2A = 2\left(\frac{1}{2}bh\right)$

$2A = bh$

$\frac{2A}{b} = \frac{bh}{b}$

$\frac{2A}{b} = h$

22. $i = prt$

$\frac{i}{pr} = \frac{prt}{pr}$

$\frac{i}{pr} = t$

23. $P = 2l + 2w$

$P - 2l = 2l - 2l + 2w$

$P - 2l = 2w$

$\frac{P - 2l}{2} = \frac{2w}{2}$

$\frac{P - 2l}{2} = w$

24. $2x - 3y = 6$

$2x - 2x - 3y = -2x + 6$

$-3y = -2x + 6$

$\frac{-3y}{-3} = \frac{-2x + 6}{-3}$

$y = \frac{2x - 6}{3}$

25. $A = \frac{B + C}{2}$

$2A = 2\left(\frac{B + C}{2}\right)$

$2A = B + C$

$2A - C = B + C - C$

$2A - C = B$

26. $V = \pi r^2 h$

$\frac{V}{\pi r^2} = \frac{\pi r^2 h}{\pi r^2}$

$\frac{V}{\pi r^2} = h$

27. Substitute 600 for p, 0.15 for r, and 2 for t.

$i = prt$

$i = (600)(0.15)(2)$

$i = 180$

Karen will pay $180 interest.

28. $P = 2l + 2w$

$16 = 2l + 2(2)$

$16 = 2l + 4$

$12 = 2l$

$6 = l$

The length of the rectangle is 6 inches.

29. Let x = the smaller number.

Then $x + 4$ = the larger number.

Smaller number + larger number = 62

$x + (x + 4) = 62$

$2x + 4 = 62$

$2x = 58$

$x = 29$

The smaller number is 29 and the larger number is 29 + 4 = 33.

30. Let x = smaller consecutive integer.

Then $x + 1$ = larger consecutive integer

Smaller number + larger number = 255

$x + (x + 1) = 255$

$2x + 1 = 255$

$2x = 254$

$x = 127$

The smaller number is 127 and the larger number is 127 + 1 = 128

31. Let x = the smaller integer

Then $5x + 3$ = the larger integer

Larger number – smaller number = 31

$(5x + 3) - x = 31$

$4x + 3 = 31$

$4x = 28$

$x = 7$

The smaller number is 7 and the larger number is 5(7) + 3 = 38

32. Let x = cost of car before tax.
Then $0.05x$ = amount of tax.
Cost of car before tax + tax on car
= cost of car after tax

$$x + 0.05x = 8400$$
$$1.05x = 8400$$
$$x = \frac{8400}{1.05} = 8000$$

The cost of the car before tax is \$8000.

33. Let x = weekly dollar sales that would make total salary of both companies the same.
The commission at present company = $0.03x$ and commission at new company = $0.08x$
Salary + commission for present company = salary + commission for new company

$$500 + 0.03x = 400 + 0.08x$$
$$100 + 0.03x = 0.08x$$
$$100 = 0.05x$$
$$\frac{100}{0.05} = x$$
$$2000 = x$$

Paul's weekly sales would have to be \$2000 for the total salary of both companies to be the same.

34. Let x = original price of camcorder.
Then $0.20x$ = reduction during first week
Original price – first reduction – second reduction = price during second week

$$x - 0.20x - 25 = 495$$
$$0.8x - 25 = 495$$
$$0.8x = 520$$
$$x = \frac{520}{0.8} = 650$$

The original price of the camcorder was \$650

35. **(a)** The monthly payments with the 8.875% mortgage are 60(7.96) = \$477.60. With the 8.625% mortgage, the monthly payments are 60(7.78) = \$466.80. In addition the Johnsons must pay 0.03(60,000) = \$1800 because of the three points.
Let x = number of months when total payments from the two mortgages are the same.
Monthly payments for Comerica = monthly payments + points for Mellon

$$477.60x = 466.80x + 1800$$
$$10.8x = 1800$$
$$x = \frac{1800}{10.8} = 166.7$$

Total payments from the two mortgages are the same after about 166.7 months or 13.9 years.

(b) After about 13.9 years the Mellon mortgage is cheaper because of the lower monthly payments. If the Johnsons plan to keep the house for 20 years, they should apply to the Mellon Bank.

36. **(a)** If she refinances, her monthly payments will be 70(8.68) = \$607.60.
Her monthly savings will be
\$750 – \$607.60 = \$142.40
The cost of the point will be
0.01(70,000) = \$700
Let x = number of months when savings equals cost of the point and the closing cost.

$$142.40x = 700 + 3200$$
$$142.40x = 3900$$
$$x = \frac{3900}{142.40} = 27.4$$

The money she saves will equal the cost of the point and the closing cost after about 27.4 months or 2.3 years.

(b) After about 2.3 years the money she saves on the monthly payments exceeds the cost of the point and the closing cost. If she plans to live in the house for 10 more years it does pay her to refinance.

37. Let x = measure of the smallest angle
Then $x + 10$ = measure of second angle and $2x - 10$ = measure of third angle.
Sum of the three angles = 180

$x + (x + 10) + (2x - 10) = 180$

$4x = 180$

$x = 45$

The angles are 45°, 45 + 10 = 55°, and 2(45) – 10 = 80°.

38. Let x = measure of the smallest angle

Then $x + 10$ = measure of second angle,

$5x$ = measure of third angle,

$4x + 20$ = measure of the fourth angle.

Sum of the four angles = 360

$x + (x + 10) + 5x + (4x + 20) = 360$

$11x + 30 = 360$

$11x = 330$

$x = \dfrac{330}{11} = 30$

The angles are 30°, 30 + 10 = 40°,

5(30) = 150°, and 4(30) + 20 = 140°.

39. Let w = width of garden.

Then $w + 4$ = length of garden

$P = 2l + 2w$

$70 = 2(w + 4) + 2w$

$70 = 4w + 8$

$62 = 4w$

$15.5 = w$

The width will be 15.5 feet and the length will be 15.5 + 4 = 19.5 feet.

40. Let x = smaller consecutive odd integer.

Then $x + 2$ = larger consecutive odd integer.

Smaller number + larger number = 208

$x + (x + 2) = 208$

$2x + 2 = 208$

$2x = 206$

$x = 103$

The smaller number is 103 and the larger number is 103 + 2 = 105

41. Let x = cost of television before tax.

Then amount of tax = $0.06x$

Cost of television before tax + tax on television = cost of television after tax

$x + 0.06x = 477$

$1.06x = 477$

$x = \dfrac{477}{1.06} = 450$

The cost of the television before tax is $450.

42. Let x = his dollar sales

Then $0.05x$ = amount of commission

Salary + commission = 900

$300 + 0.05x = 900$

$0.05x = 600$

$x = \frac{600}{0.05} = 12{,}000$

His dollar sales last week were $12,000.

43. Let x = measure of the smallest angle

Then $x + 8$ = measure of second angle

and $2x + 4$ = measure of third angle

Sum of the three angles = 180

$x + (x + 8) + (2x + 4) = 180$

$4x + 12 = 180$

$4x = 168$

$x = 42$

The angles are 42°, 42 + 8 = 50°, and 2(42) + 4 = 88°.

44. Let t = number of years

Then $25t$ = increase in employees over t years

Present number of employees + increase in employees = future number of employees

$427 + 25t = 627$

$25t = 200$

$t = \frac{200}{25} = 8$

It will take 8 years before they reach 627 employees.

45. Let x = measure of each smaller angle

Then $x + 40$ = measure of each larger angle

(measure of the two smaller angles) + (measure of the two larger angles) = 360°

$x + x + (x + 40) + (x + 40) = 360$

$4x + 80 = 360$

$4x = 280$

$x = 70$

Each of the smaller angles is 70° and each of the two larger angles is 70 + 40 = 110°.

46. **(a)** Let x = number of copies that would result in both centers charging the same.

Then charge for copies at Copy King = $0.04x$ and charge for copies at King Kopie = $0.03x$

Monthly fee + charge for copies at Copy King = monthly fee + charge for copies at King Kopie.

$20 + 0.04x = 25 + 0.03x$

$0.04x = 5 + 0.03x$

$0.01x = 5$

$x = \frac{5}{0.01} = 500$

500 copies would result in both centers charging the same.

(b) Charge for 1000 copies at Copy King = 20 + (0.04)(1000) = 20 + 40 = $60.

Charge for copies at King Kopie = 25 + (0.03)(1000) = 25 + 30 = $55

King Kopie would cost less for 1000 copies by 60 – 55 = $5

Chapter 3 Practice Test

1. $P = 2l + 2w$

$P = 2(6) + 2(3)$

$P = 12 + 6 = 18$

Perimeter = 18 feet

2. $A = P + Prt$

$A = 100 + (100)(0.15)(3)$

$A = 100 + 45 = 145$

3. $V = \frac{1}{3}\pi r^2 h$

$V = \frac{1}{3}(3.14)(4)^2(6)$

$V = \frac{1}{3}(3.14)(16)(6)$

$V = 100.48$

4. $P = IR$

$\frac{P}{I} = \frac{IR}{I}$

$\frac{P}{I} = R$

5. $3x - 2y = 6$

$3x - 3x - 2y = -3x + 6$

$-2y = -3x + 6$

$\frac{-2y}{-2} = \frac{-3x+6}{-2}$

$y = \frac{3x-6}{2}$

6. $A = \frac{a+b}{3}$

$3A = 3\left(\frac{a+b}{3}\right)$

$3A = a + b$

$3A - b = a + b - b$

$3A - b = a$

7. $D = R(c + a)$

$\frac{D}{R} = \frac{R(c+a)}{R}$

$\frac{D}{R} = c + a$

$\frac{D}{R} - a = c + a - a$

$\frac{D}{R} - a = c$

$c = \frac{D}{R} - a$ or $\frac{D - Ra}{R}$

8. Let x = smaller integer

Then $2x - 10$ = larger integer

Smaller number + larger number = 158

$x + (2x - 10) = 158$

$3x - 10 = 158$

$3x = 168$

$x = 56$

The smaller number is 56 and the larger number is $2(56) - 10 = 102$.

9. Let x = smallest consecutive integer

Then $x + 1$ and $x + 2$ are the other two consecutive integers

Sum of the three integers = 42

$x + (x + 1) + (x + 2) = 42$

$3x + 3 = 42$

$3x = 39$

$x = 13$

The integers are 13, $13 + 1 = 14$, and $13 + 2 = 15$.

10. Let x = price of most expensive meal he can order

Then $0.15x$ = tip and $0.07x$ = tax

Price of meal + tip + tax = 20

$x + 0.15x + 0.07x = 20$

$1.22x = 20$

$x = \frac{20}{1.22} = 16.39$

The price of the most expensive meal he can order is $16.39.

11. Let x = length of smallest side

Then $x + 15$ = length of second side and

$2x$ = length of third side

Sum of the three sides = perimeter

$x + (x + 15) + 2x = 75$

$4x + 15 = 75$

$4x = 60$

$x = 15$

The three sides are 15 inches, 15 + 15 = 30 inches, and 2(15) = 30 inches.

12. Let x = measure of each smaller angle

Then $2x + 30$ = measure of each larger angle

(measure of the two smaller angles) + (measure of the two larger angles) = 360°

$x + x + (2x + 30) + (2x + 30) = 360°$

$6x + 60 = 360$

$6x = 300$

$x = 50$

Each of the smaller angles is 50° and each of the larger angles is 2(50) + 30 = 130°.

CHAPTER 4

Exercise Set 4.1

1. $x^2 \cdot x^4 = x^{2+4} = x^6$

3. $y \cdot y^2 = y^1 \cdot y^2 = y^{1+2} = y^3$

5. $3^2 \cdot 3^3 = 3^{2+3} = 3^5$ or 243

7. $y^2 \cdot y^3 = y^{2+3} = y^5$

9. $y^4 \cdot y = y^4 \cdot y^1 = y^{4+1} = y^5$

11. $\frac{x^{10}}{x^3} = x^{10-3} = x^7$

13. $\frac{5^4}{5^2} = 5^{4-2} = 5^2$ or 25

15. $\frac{y^2}{y} = y^{2-1} = y$

17. $\frac{x^2}{x^2} = x^{2-2} = x^0 = 1$

19. $\frac{y^{12}}{y^{11}} = y^{12-11} = y^1 = y$

21. $x^0 = 1$

23. $3x^0 = 3\left(x^0\right) = 3 \cdot 1 = 3$

25. $(3x)^0 = 1$

27. $(-4x)^0 = 1$

29. $\left(x^5\right)^2 = x^{5\cdot 2} = x^{10}$

31. $\left(x^5\right)^5 = x^{5\cdot 5} = x^{25}$

33. $\left(x^3\right)^1 = x^{3\cdot 1} = x^3$

35. $\left(x^4\right)^3 = x^{4\cdot 3} = x^{12}$

37. $\left(x^5\right)^3 = x^{5\cdot 3} = x^{15}$

39. $(1.3x)^2 = (1.3)^2 \cdot x^2 = 1.69x^2$

41. $(-x)^2 = (-1 \cdot x)^2 = (-1)^2 \cdot x^2 = x^2$

43. $\left(4x^2\right)^3 = 4^3 x^{2\cdot 3} = 64x^6$

45. $\left(-3x^3\right)^3 = (-3)^3 x^{3\cdot 3} = -27x^9$

47. $\left(2x^2y\right)^3 = 2^3 x^{2\cdot 3} y^3 = 8x^6y^3$

49. $\left(8.6x^2y^5\right)^2 = (8.6)^2 x^{2\cdot2}\, y^{5\cdot2}$

$= 73.96x^4y^{10}$

51. $\left(-6\,x^3y^2\right)^3 = (-6)^3 x^{3\cdot3}\, y^{2\cdot3}$

$= -216\,x^9\,y^6$

53. $\left(-x^4y^5z^6\right)^3 = (-1)^3 x^{4\cdot3}\, y^{5\cdot3}z^{6\cdot3}$

$= -x^{12}y^{15}z^{18}$

55. $\left(\frac{x}{y}\right)^2 = \frac{x^2}{y^2}$

57. $\left(\frac{x}{4}\right)^3 = \frac{x^3}{4^3} = \frac{x^3}{64}$

59. $\left(\frac{y}{x}\right)^5 = \frac{y^5}{x^5}$

61. $\left(\frac{6}{x}\right)^3 = \frac{6^3}{x^3} = \frac{216}{x^3}$

63. $\left(\frac{3x}{y}\right)^3 = \frac{3^3x^3}{y^3} = \frac{27x^3}{y^3}$

65. $\left(\frac{3x}{5}\right)^2 = \frac{3^2x^2}{5^2} = \frac{9x^2}{25}$

67. $\left(\frac{4y^3}{x}\right)^3 = \frac{4^3y^{3\cdot3}}{x^3} = \frac{64y^9}{x^3}$

69. $\left(\frac{-3x^3}{4}\right)^3 = \frac{(-3)^3x^{3\cdot3}}{4^3} = \frac{-27x^9}{64}$

71. $\frac{x^3y^2}{xy^6} = \frac{x\cdot x^2\cdot y^2}{x\cdot y^2\cdot y^4} = \frac{x^2}{y^4}$

73. $\frac{x^5y^7}{x^{12}y^3} = \frac{x^5\cdot y^3\cdot y^4}{x^5\cdot x^7\cdot y^3} = \frac{y^4}{x^7}$

75. $\frac{10x^3y^8}{2xy^{10}} = \frac{2\cdot5\cdot x\cdot x^2\cdot y^8}{2\cdot x\cdot y^8\cdot y^2} = \frac{5x^2}{y^2}$

77. $\frac{4xy}{16x^3y^2} = \frac{4xy}{4\cdot4\cdot x\cdot x^2\cdot y\cdot y} = \frac{1}{4x^2y}$

79. $\frac{35x^4y^7}{10x^9y^{12}} = \frac{5\cdot7\cdot x^4\cdot y^7}{2\cdot5\cdot x^4\cdot x^5\cdot y^7\cdot y^5} = \frac{7}{2x^5y^5}$

81. $\frac{-36xy^9z}{12x^4y^5z^2} = \frac{-3\cdot12\cdot x\cdot y^5\cdot y^4\cdot z}{12\cdot x\cdot x^3\cdot y^5\cdot z\cdot z} = \frac{-3y^4}{x^3z}$

83. $\frac{-6x^2y^7z^5}{2x^5y^9z^6} = \frac{-3\cdot2\cdot x^2\cdot y^7\cdot z^5}{2\cdot x^2\cdot x^3\cdot y^7\cdot y^2\cdot z^5\cdot z}$

$= \frac{-3}{x^3y^2z}$

85. $\left(\frac{4x^4}{2x^6}\right)^3 = \left(\frac{4}{2}\cdot\frac{x^4}{x^6}\right)^3 = \left(\frac{2}{x^2}\right)^3$

$= \frac{2^3}{x^{2\cdot3}} = \frac{8}{x^6}$

87. $\left(\frac{6y^7}{2y^3}\right)^3 = \left(\frac{6}{2}\cdot\frac{y^7}{y^3}\right)^3 = \left(3y^4\right)^3$

$= 3^3 y^{4\cdot 3} = 27y^{12}$

89. $\left(\frac{27x^9}{30x^5}\right)^0 = 1$

91. $\left(\frac{x^4y^3}{x^2y^5}\right)^2 = \left(\frac{x^4}{x^2}\cdot\frac{y^3}{y^5}\right)^2 = \left(\frac{x^2}{y^2}\right)^2$

$= \frac{x^{2\cdot 2}}{y^{2\cdot 2}} = \frac{x^4}{y^4}$

93. $\left(\frac{9y^2z^7}{18y^7z}\right)^4 = \left(\frac{9}{18}\cdot\frac{y^2}{y^7}\cdot\frac{z^7}{z}\right)^4 = \left(\frac{z^6}{2y^5}\right)^4$

$= \frac{z^{6\cdot 4}}{2^4 y^{5\cdot 4}} = \frac{z^{24}}{16y^{20}}$

95. $\left(\frac{4x^2y^5}{y^2}\right)^3 = \left(4x^2\cdot\frac{y^5}{y^2}\right)^3 = \left(4x^2y^3\right)^3$

$= 4^3 x^{2\cdot 3} y^{3\cdot 3} = 64x^6y^9$

97. $\left(\frac{-x^4y^6}{x^2}\right)^2 = \left(\frac{-x^4}{x^2}\cdot\frac{y^6}{1}\right)^2 = \left(-x^2y^6\right)^2$

$= (-1)^2 x^{2\cdot 2} y^{6\cdot 2} = x^4y^{12}$

99. $\left(\frac{-12x}{16x^7y^2}\right)^2 = \left(\frac{-12}{16}\cdot\frac{x}{x^7}\cdot\frac{1}{y^2}\right)^2$

$= \left(\frac{-3}{4x^6y^2}\right)^2 = \frac{(-3)^2}{4^2 x^{6\cdot 2} y^{2\cdot 2}} = \frac{9}{16x^{12}y^4}$

101. $\left(3xy^4\right)^2 = 3^2 x^2 y^{4\cdot 2} = 9x^2y^8$

103. $\left(-6xy^5\right)\left(3x^2y^4\right) = -6\cdot 3\cdot x\cdot x^2\cdot y^5\cdot y^4$

$= -18x^3y^9$

105. $\left(2x^4y^2\right)\left(4xy^6\right) = 2\cdot 4\cdot x^4\cdot x\cdot y^2\cdot y^6$

$= 8x^5y^8$

107. $(5xy)\left(2xy^6\right) = 5\cdot 2\cdot x\cdot x\cdot y\cdot y^6 = 10x^2y^7$

109. $(2xy)^2\left(3xy^2\right)^0$: $(2xy)^2 = 2^2x^2y^2$

$= 4x^2y^2$ and $\left(3xy^2\right)^0 = 1$,

so $(2xy)^2\left(3xy^2\right)^0$

$= \left(4x^2y^2\right)(1) = 4x^2y^2$

111. $\left(x^4y^6\right)^3\left(3x^2y^5\right)$: $\left(x^4y^6\right)^3 = x^{4\cdot 3}y^{6\cdot 3}$

$= x^{12}y^{18}$, so $\left(x^4y^6\right)^3\left(3x^2y^5\right)$

$= \left(x^{12}y^{18}\right)\left(3x^2y^5\right)$

$= 3\cdot x^{12}\cdot x^2\cdot y^{18}\cdot y^5 = 3x^{14}y^{23}$

113. $\left(2x^2y^5\right)\left(3x^5y^4\right)^3$: $\left(3x^5y^4\right)^3$

$= 3^3x^{5\cdot 3}y^{4\cdot 3}$

$= 27x^{15}y^{12}$, so $\left(2x^2y^5\right)\left(3x^5y^4\right)^3$

$= \left(2x^2y^5\right)\left(27x^{15}y^{12}\right)$

$= 2\cdot 27\cdot x^2\cdot x^{15}\cdot y^5\cdot y^{12}$

$= 54x^{17}y^{17}$

115. $\left(x^7y^5\right)\left(xy^2\right)^4$: $\left(xy^2\right)^4 = x^4y^{2\cdot 4} = x^4y^8$,

so $\left(x^7y^5\right)\left(xy^2\right)^4 = \left(x^7y^5\right)\left(x^4y^8\right)$

$= x^7\cdot x^4\cdot y^5\cdot y^8$

$= x^{11}y^{13}$

117. $\left(3x^4y^{10}\right)^2\left(2x^2y^8\right)$:

$\left(3x^4y^{10}\right)^2 3^2x^{4\cdot 2}y^{10\cdot 2}$

$= 9x^8y^{20}$, so $\left(3x^4y^{10}\right)^2\left(2x^2y^8\right)$

$= \left(9x^8y^{20}\right)\left(2x^2y^8\right)$

$= 9\cdot 2\cdot x^8\cdot x^2\cdot y^{20}\cdot y^8 = 18x^{10}y^{28}$

119. Cannot be simplified.

121. Cannot be simplified.

123. Cannot be simplified.

125. $\dfrac{x^2y^2}{x^2} = y^2$

127. Cannot be simplified.

129. $\dfrac{x^4}{x^2y} = \dfrac{x^2x^2}{x^2y} = \dfrac{x^2}{y}$

131. $x = 0$

133. The sign will be negative due to the odd exponent.

135. The sign will be positive due to the even exponent.

Cumulative Review Exercises: 4.1

137. A linear equation is an equation of the form $ax + b = c$.

138. A conditional linear equation is a linear equation that has only one solution.

139. An identity is an equation which is true for all real numbers.

140. $C = 2\pi r = 2\cdot\pi\cdot 3 \approx 18.84$ in.

$A = \pi r^2 = \pi\cdot 3^2 \approx 28.26 \text{ in.}^2$

141. $2x - 5y = 6,\ -5y = -2x + 6$

$y = \dfrac{-2x+6}{-5} = \dfrac{2x+(-6)}{5} = \dfrac{2}{5}x - \dfrac{6}{5}$

Group Activity/Challenge Problems

1. $\left(\frac{3x^4y^5}{6x^6y^8}\right)^3\left(\frac{9x^7y^8}{3x^3y^5}\right)^2$

$=\left(\frac{1}{2x^2y^3}\right)^3\left(\frac{3x^4y^3}{1}\right)^2$

$=\frac{1^3}{2^3x^6y^9}\cdot\frac{3^2x^8y^6}{1^2}$

$\frac{9x^8y^6}{8x^6y^9}=\frac{9x^2}{8y^3}$

3. 15; $x^{15}\cdot x^5=x^{15+5}=x^{20}$

5. 3; $(2x^3)^4=2^4x^{3\cdot4}=16x^{12}$

7. 2, 5; $\frac{x^2y^8}{x^3y^5}=\frac{y^{8-5}}{x^{3-2}}=\frac{y^3}{x}$

9. 4, 11; $\left(\frac{x^6y^{11}}{x^4y^8}\right)^3=(x^2y^3)^3=x^6y^9$

11. 3, 2, 4;
$(3x^2y^4)^2(5x^3y^5)=(3^2x^4y^8)(5x^3y^5)$
$=9\cdot5x^{4+3}y^{8+5}=45x^7y^{13}$

Exercise Set 4.2

1. $x^{-2}=\frac{1}{x^2}$

3. $4^{-1}=\frac{1}{4}$

5. $\frac{1}{x^{-3}}=x^3$

7. $\frac{1}{x^{-1}}=x^1=x$

9. $\frac{1}{4^{-2}}=4^2=16$

11. $\left(x^{-2}\right)^3=x^{-2\cdot3}=x^{-6}=\frac{1}{x^6}$

13. $\left(y^{-7}\right)^3=y^{-7\cdot3}=y^{-21}=\frac{1}{y^{21}}$

15. $\left(x^4\right)^{-2}=x^{4(-2)}=x^{-8}=\frac{1}{x^8}$

17. $\left(2^{-3}\right)^{-2}=2^{(-3)(-2)}=2^6=64$

19. $x^4\cdot x^{-1}=x^{4-1}=x^3$

21. $x^7\cdot x^{-5}=x^{7-5}=x^2$

23. $3^{-2}\cdot3^4=3^{-2+4}=3^2=9$

25. $\frac{x^8}{x^{10}}=x^{8-10}=x^{-2}=\frac{1}{x^2}$

27. $\frac{y^0}{y^{-3}} = y^{0-(-3)} = y^3$

29. $\frac{x^{-7}}{x^{-3}} = x^{-7-(-3)} = x^{-4} = \frac{1}{x^4}$

31. $\frac{3^2}{3^{-1}} = 3^{2-(-1)} = 3^3 = 27$

33. $3^{-3} = \frac{1}{3^3} = \frac{1}{27}$

35. $\frac{1}{z^{-9}} = z^9$

37. $\left(x^5\right)^{-5} = x^{5(-5)} = x^{-25} = \frac{1}{x^{25}}$

39. $\left(y^{-2}\right)^{-3} = y^{(-2)(-3)} = y^6$

41. $x^3 \cdot x^{-7} = x^{3-7} = x^{-4} = \frac{1}{x^4}$

43. $x^{-8} \cdot x^{-7} = x^{-8-7} = x^{-15} = \frac{1}{x^{15}}$

45. $\frac{x^{-3}}{x^5} = x^{-3-5} = x^{-8} = \frac{1}{x^8}$

47. $\frac{y^9}{y^{-1}} = y^{9-(-1)} = y^{10}$

49. $\frac{2^{-3}}{2^{-3}} = 2^{-3-(-3)} = 2^0 = 1$

51. $\left(2^{-1} + 3^{-1}\right)^0 = 1$

53. $\frac{1}{1^{-5}} = 1^{1-(-5)} = 1^6 = 1$

55. $\left(x^{-4}\right)^{-1} = x^{(-4)(-1)} = x^4$

57. $\left(x^0\right)^{-3} = x^{0(-3)} = x^0 = 1$

59. $2^{-3} \cdot 2 = 2^{-3+1} = 2^{-2} = \frac{1}{2^2} = \frac{1}{4}$

61. $6^{-4} \cdot 6^2 = 6^{-4+2} = 6^{-2} = \frac{1}{6^2} = \frac{1}{36}$

63. $\frac{x^{-1}}{x^{-4}} = x^{-1-(-4)} = x^3$

65. $\left(3^2\right)^{-1} = 3^{2(-1)} = 3^{-2} = \frac{1}{3^2} = \frac{1}{9}$

67. $\frac{5}{5^{-2}} = 5^{1+2} = 5^3 = 125$

69. $\frac{2^{-4}}{2^{-2}} = 2^{-4-(-2)} = 2^{-2} = \frac{1}{2^2} = \frac{1}{4}$

71. $\frac{7^{-1}}{7^{-1}} = 7^{-1-(-1)} = 7^0 = 1$

73. $2x^{-1}y = 2 \cdot \frac{1}{x} \cdot y = \frac{2y}{x}$

75. $\left(3x^3\right)^{-1} = 3^{-1}x^{3(-1)} = 3^{-1}x^{-3} = \frac{1}{3x^3}$

77. $5x^4y^{-1} = 5 \cdot x^4 \cdot \frac{1}{y} = \frac{5x^4}{y}$

79. $\left(3x^2y^3\right)^{-2} = 3^{-2}x^{2(-2)}y^{3(-2)}$

$= 3^{-2}x^{-4}y^{-6} = \frac{1}{3^2x^4y^6} = \frac{1}{9x^4y^6}$

81. $\left(x^5y^{-3}\right)^{-3} = x^{5(-3)}y^{(-3)(-3)}$

$= x^{-15}y^9 = \frac{y^9}{x^{15}}$

83. $3x\left(5x^{-4}\right) = 3 \cdot 5 \cdot x \cdot x^{-4} = 15x^{-3} = \frac{15}{x^3}$

85. $2x^5\left(3x^{-6}\right) = 2 \cdot 3 \cdot x^5 \cdot x^{-6} = 6x^{-1} = \frac{6}{x}$

87. $9x^5\left(-3x^{-7}\right) = -9 \cdot 3 \cdot x^5 \cdot x^{-7}$

$= -27x^{-2} = \frac{-27}{x^2}$

89. $\left(2x^{-3}y^{-2}\right)\left(x^4y^0\right)$

$= 2 \cdot x^{-3} \cdot x^4 \cdot y^{-2} \cdot y^0 = 2xy^{-2} = \frac{2x}{y^2}$

91. $\left(3y^{-2}\right)\left(5x^{-1}y^3\right) = 3 \cdot 5 \cdot x^{-1} \cdot y^{-2} \cdot y^3$

$= 15x^{-1}y = \frac{15y}{x}$

93. $\frac{8x^4}{4x^{-1}} = \frac{8}{4} \cdot \frac{x^4}{x^{-1}} = 2x^5$

95. $\frac{12x^{-2}}{3x^5} = \frac{12}{3} \cdot \frac{x^{-2}}{x^5} = 4 \cdot \frac{1}{x^7} = \frac{4}{x^7}$

97. $\frac{5x^{-2}}{25x^{-5}} = \frac{5}{25} \cdot \frac{x^{-2}}{x^{-5}} = \frac{1}{5} \cdot x^3 = \frac{x^3}{5}$

99. $\frac{12x^{-2}y^0}{2x^3y^2} = \frac{12}{2} \cdot \frac{x^{-2}}{x^3} \cdot \frac{y^0}{y^2}$

$= 6 \cdot \frac{1}{x^5} \cdot \frac{1}{y^2} = \frac{6}{x^5y^2}$

101. $\frac{16x^{-7}y^{-2}}{4x^5y^2} = \frac{16}{4} \cdot \frac{x^{-7}}{x^5} \cdot \frac{y^{-2}}{y^2}$

$= 4 \cdot \frac{1}{x^{12}} \cdot \frac{1}{y^4} = \frac{4}{x^{12}y^4}$

103. $\dfrac{9x^4y^{-7}}{18x^{-1}y^{-1}}=\dfrac{9}{18}\cdot\dfrac{x^4}{x^{-1}}\cdot\dfrac{y^{-7}}{y^{-1}}$

$=\dfrac{1}{2}\cdot x^5\cdot\dfrac{1}{y^6}=\dfrac{x^5}{2y^6}$

105. **(a)** Yes. $a^{-1}b^{-1}=\dfrac{1}{a}\cdot\dfrac{1}{b}=\dfrac{1}{ab}$

(b) No. $a^{-1}+b^{-1}=\dfrac{1}{a}+\dfrac{1}{b}\neq\dfrac{1}{a+b}$

107. Answers will vary.

Cumulative Review Exercises: 4.2

108. $2[6-(4-5)]\div 2-5^2$

$=2[6-(-1)]\div 2-5^2$

$=2[7]\div 2-5^2=2\cdot 7\div 2-25$

$=7-25=-18$

109. $\dfrac{-3^2\cdot 4\div 2}{\sqrt{9}-2^2}=\dfrac{-9\cdot 4\div 2}{3-4}=\dfrac{-18}{-1}=18$

110. Let x equal the amount of cleaner which should be used. Then $\dfrac{x}{2.5}=\dfrac{8}{3}$ or $x=\dfrac{(2.5)8}{3}\approx 6.67$. So, we should use 6.67 oz. of cleaner.

111. Let x = the smaller of the two numbers. Then $3x+1$ is the larger of the two numbers. Then the smaller plus the larger equals 37.

$x+3x+1=37$

$4x=36$

$x=9$

$3x+1=3(9)+1=28$

So, the two numbers are 9 and 28.

Group Activity/Challenge Problems

1. **(a)** $\left(\dfrac{3x^2y^3}{z}\right)^{-2}=\dfrac{3^{-2}x^{2(-2)}y^{3(-2)}}{z^{-2}}$

$=\dfrac{3^{-2}x^{-4}y^{-6}}{z^{-2}}=\dfrac{\frac{1}{9}\cdot\frac{1}{x^4}\cdot\frac{1}{y^6}}{\frac{1}{z^2}}$

$=\dfrac{z^2}{9x^4y^6}$

(b) $\left(\dfrac{3x^2y^3}{z}\right)^{-2}=\left(\dfrac{z}{3x^2y^3}\right)^2$

$=\dfrac{z^2}{3^2x^{2\cdot 2}y^{3\cdot 2}}=\dfrac{z^2}{9x^4y^6}$

3. -2; $5^{-2}=\dfrac{1}{5^2}=\dfrac{1}{25}$

5. -2; $(x^{-2}y^3)^{-2}=x^{(-2)(-2)}y^{3(-2)}$

$=x^4y^{-6}=\dfrac{x^4}{y^6}$

7. -3; $(x^4y^{-3})^{-3}=x^{4(-3)}y^{(-3)(-3)}$

$=x^{-12}y^9=\dfrac{y^9}{x^{12}}$

9. $3^0 - 3^{-1} = 1 - \frac{1}{3} = \frac{3}{3} - \frac{1}{3} = \frac{2}{3}$

11. $2 \cdot 4^{-1} + 4 \cdot 3^{-1} = 2 \cdot \frac{1}{4} + 4 \cdot \frac{1}{3} = \frac{2}{4} + \frac{4}{3}$

$= \frac{1}{2} + \frac{4}{3} = \frac{3}{6} + \frac{8}{6} = \frac{11}{6}$

Exercise Set 4.3

1. $42{,}000 = 4.2 \times 10^4$

3. $900 = 9 \times 10^2$

5. $0.053 = 5.3 \times 10^{-2}$

7. $19{,}000 = 1.9 \times 10^4$

9. $0.00000186 = 1.86 \times 10^{-6}$

11. $0.00000914 = 9.14 \times 10^{-6}$

13. $107 = 1.07 \times 10^2$

15. $0.153 = 1.53 \times 10^{-1}$

17. $4.2 \times 10^3 = 4200$

19. $4 \times 10^7 = 40{,}000{,}000$

21. $2.13 \times 10^{-5} = 0.0000213$

23. $3.12 \times 10^{-1} = 0.312$

25. $9 \times 10^6 = 9{,}000{,}000$

27. $5.35 \times 10^2 = 535$

29. $3.5 \times 10^4 = 35{,}000$

31. $1 \times 10^4 = 10{,}000$

33. $\left(4 \times 10^2\right)\left(3 \times 10^5\right) = (4 \times 3)\left(10^2 \times 10^5\right)$

$= 12 \times 10^7 = 120{,}000{,}000$

35. $\left(5.1 \times 10^1\right)\left(3 \times 10^{-4}\right)$

$= (5.1 \times 3)\left(10^1 \times 10^{-4}\right)$

$= 15.3 \times 10^{-3} = 0.0153$

37. $\left(6.2 \times 10^4\right)\left(1.5 \times 10^{-2}\right)$

$= (6.2 \times 1.5)\left(10^4 \times 10^{-2}\right)$

$= 9.3 \times 10^2 = 930$

39. $\frac{6.4 \times 10^5}{2 \times 10^3} = \left(\frac{6.4}{2}\right)\left(\frac{10^5}{10^3}\right)$

$= 3.2 \times 10^2 = 320$

41. $\dfrac{8.4\times10^{-6}}{4\times10^{-3}}=\left(\dfrac{8.4}{4}\right)\left(\dfrac{10^{-6}}{10^{-3}}\right)$

$=2.1\times10^{-3}=0.0021$

43. $\dfrac{4\times10^{5}}{2\times10^{4}}=\left(\dfrac{4}{2}\right)\left(\dfrac{10^{5}}{10^{4}}\right)=2\times10^{1}=20$

45. $(700{,}000)(6{,}000{,}000)$

$=\left(7\times10^{5}\right)\left(6\times10^{6}\right)$

$=(7\times6)\left(10^{5}\times10^{6}\right)$

$=42\times10^{11}=4.2\times10^{12}$

47. $(0.003)(0.00015)$

$=\left(3\times10^{-3}\right)\left(1.5\times10^{-4}\right)$

$=(3\times1.5)\left(10^{-3}\times10^{-4}\right)=4.5\times10^{-7}$

49. $\dfrac{1{,}400{,}000}{700}=\dfrac{1.4\times10^{6}}{7\times10^{2}}=\left(\dfrac{1.4}{7}\right)\left(\dfrac{10^{6}}{10^{2}}\right)$

$=0.2\times10^{4}=2\times10^{3}$

51. $\dfrac{0.00004}{200}=\dfrac{4\times10^{-5}}{2\times10^{2}}=\left(\dfrac{4}{2}\right)\left(\dfrac{10^{-5}}{10^{2}}\right)$

$=2\times10^{-7}$

53. $\dfrac{150{,}000}{0.0005}=\dfrac{1.5\times10^{5}}{5\times10^{-4}}=\left(\dfrac{1.5}{5}\right)\left(\dfrac{10^{5}}{10^{-4}}\right)$

$=0.3\times10^{9}=3\times10^{8}$

55. 9.2×10^{-5}, 1.3×10^{-1}, 8.4×10^{3}, 6.2×10^{4}

57. $(8{,}000{,}000{,}000{,}000)(0.000004)$

$=\left(8\times10^{12}\right)\left(4\times10^{-6}\right)$

$=(8\times4)\left(10^{12}\times10^{-6}\right)=32\times10^{6}$

$=3.2\times10^{7}$ It would take the computer 3.2×10^{7} seconds.

59. Minimum volume

$=\left(100{,}000\ \text{ft}^3/\text{sec}\right)(60\ \text{sec/min})$

$\cdot(60\ \text{min/hr})(24\ \text{hrs})$

$=8{,}640{,}000{,}000\ \text{ft}^3$

61. $(0.05)\left(4\times10^{9}\right)=\left(5\times10^{-2}\right)\left(4\times10^{9}\right)$

$=(5\times4)\left(10^{-2}\times10^{9}\right)=20\times10^{7}$

$=2\times10^{8}$ 200,000,000 pounds of plastic are recycled.

63. **(a)** 18 billion $=18{,}000{,}000{,}000$

$=1.8\times10^{10}$

(b) length $=(14)\left(2.38\times10^{5}\right)$

$=33.32\times10^{5}=3.332\times10^{6}$ miles or 3,332,000 mi

65. Let x be the unknown exponent.

$$\frac{(4\times 10^3)(6\times 10^x)}{24\times 10^{-5}}=1,$$

$$\frac{(4\times 6)(10^3\times 10^x)}{24\times 10^{-5}}=1,$$

$$\frac{24\times 10^{3+x}}{24\times 10^{-5}}=1$$

$$10^{8+x}=1$$

so $x=-8$ since $8+x=0$

Cumulative Review Exercises: 4.3

67. $4x^2+3x+\frac{x}{2}$

$4\cdot 0^2+3\cdot 0+\frac{0}{2}=0+0+0=0$

68. **(a)** $-x=-\frac{3}{2},\ x=\frac{3}{2}$

(b) $5x=0,\ x=\frac{0}{5}=0$

69. $2x-3(x-2)=x+2$

$2x-3x+6=x+2$

$-x+6=x+2$

$-2x+6=2$

$-2x=-4$

$x=2$

70. $\left(\frac{-2x^5y^7}{8x^8y^3}\right)^3=\left(\frac{-2}{8}\cdot\frac{x^5}{x^8}\cdot\frac{y^7}{y^3}\right)^3$

$=\left(\frac{-1}{4}\cdot\frac{1}{x^3}\cdot y^4\right)^3=\left(\frac{-y^4}{4x^3}\right)^3$

$=\frac{(-1)^3y^{4\cdot 3}}{4^3x^{3\cdot 3}}=\frac{-y^{12}}{64x^9}$

Group Activity/Challenge Problems

1. **(a)** million = 1,000,000 = 1.0×10^6

billion = 1,000,000,000 = 1.0×10^9

trillion =

1,000,000,000 = 1.0×10^{12}

(b) $\frac{1{,}000{,}000}{1{,}000}=\frac{1.0\times 10^6}{1.0\times 10^3}$

$=1.0\times 10^3=1000$

1,000 days or approximately 2.74 years

(c) $\frac{1.0\times 10^9}{1.0\times 10^3}=1.0\times 10^6$

1,000,000,000 days or approximately

2740 years

(d) $\frac{1.0\times 10^{12}}{1.0\times 10^3}=1.0\times 10^9$

1,000,000,000 days or approximately

2,740,000 years

(e) $\frac{1.0\times 10^9}{1.0\times 10^6}=1.0\times 10^3=1000$

2. **(a)** $2032 - 1992 = 40$

$2(5.4 \times 10^9) = 1.08 \times 10^{10}$

(b) $\dfrac{5.4 \times 10^9}{(40)(365)} = 3.7 \times 10^5$ or

approximately 370,000 people

Exercise Set 4.4

1. Monomial

3. Monomial

5. Binomial

7. Trinomial

9. Not a polynomial

11. Binomial

13. Monomial

15. Polynomial

17. Trinomial

19. Not a polynomial

21. Already in descending order, first degree.

23. $2x^2 + x - 6$, second degree.

25. $-x^2 - 4x - 8$, second degree.

27. Already in descending order, third degree.

29. Already in descending order, second degree.

31. $-6x^3 + x^2 - 3x + 4$, third degree.

33. $5x^2 - 2x - 4$, second degree.

35. $-2x^3 + 3x^2 + 5x - 6$, third degree.

37. $(2x+3)+(4x-2) = 2x+3+4x-2$

$= 2x+4x+3-2 = 6x+1$

39. $(-4x+8)+(2x+3) = -4x+8+2x+3$

$= -4x+2x+8+3 = -2x+11$

41. $(5x+8)+(-6x-10) = 5x+8-6x-10$

$= 5x-6x+8-10 = -x-2$

43. $(9x-12)+(12x-9) = 9x-12+12x-9$

$= 9x+12x-12-9 = 21x-21$

45. $\left(x^2+2.6x-3\right)+(4x+3.8)$

$x^2+2.6x-3+4x+3.8$

$= x^2+2.6x+4x-3+3.8$

$= x^2+6.6x+0.8$

47. $(5x-7)+(2x^2+3x+12)$

$=5x-7+2x^2+3x+12$

$=2x^2+5x+3x-7+12$

$=2x^2+8x+5$

49. $(3x^2-4x+8)+(2x^2+5x+12)$

$=3x^2-4x+8+2x^2+5x+12$

$=3x^2+2x^2-4x+5x+8+12$

$=5x^2+x+20$

51. $(-x^2-4x+8)+\left(5x-2x^2+\frac{1}{2}\right)$

$=-x^2-4x+8+5x-2x^2+\frac{1}{2}$

$=-x^2-2x^2-4x+5x+8+\frac{1}{2}$

$=-3x^2+x+\frac{17}{2}$

53. $(8x^2+4)+(-2.6x^2-5x)$

$=8x^2+4-2.6x^2-5x$

$=8x^2-2.6x^2-5x+4$

$=5.4x^2-5x+4$

55. $(-7x^3-3x^2+4)+(4x+5x^3-7)$

$=-7x^3-3x^2+4+4x+5x^3-7$

$=-7x^3+5x^3-3x^2+4x+4-7$

$=-2x^3-3x^2+4x-3$

57. $(x^2+xy-y^2)+(2x^2-3xy+y^2)$

$=x^2+xy-y^2+2x^2-3xy+y^2$

$=x^2+2x^2+xy-3xy-y^2+y^2$

$=3x^2-2xy$

59. $(2x^2y+2x-3)+(3x^2y-5x+5)$

$=2x^2y+2x-3+3x^2y-5x+5$

$=2x^2y+3x^2y+2x-5x-3+5$

$=5x^2y-3x+2$

61.
$$\begin{array}{r} 3x-6 \\ \underline{4x+5} \\ 7x-1 \end{array}$$

63.
$$\begin{array}{r} x^2-2x+4 \\ \underline{\quad 3x+12} \\ x^2+x+16 \end{array}$$

65.
$$\begin{array}{r} -2x^2+4x-12 \\ \underline{-x^2-2x\quad\quad} \\ -3x^2+2x-12 \end{array}$$

67.
$$\begin{array}{r} 3x^2+4x-5 \\ \underline{4x^2+3x-8} \\ 7x^2+7x-13 \end{array}$$

69.
$$\begin{array}{r} 2x^3+3x^2+6x-9 \\ \underline{-4x^2\qquad+7} \\ 2x^3-x^2+6x-2 \end{array}$$

71. $6x^3 - 4x^2 + x - 9$

$\underline{-x^3 - 3x^2 - x + 7}$

$5x^3 - 7x^2 \quad -2$

73. $xy + 6x + 4$

$\underline{2xy - 3x - 1}$

$3xy + 3x + 3$

75. $(3x-4)-(2x+2) = 3x-4-2x-2$

$= 3x - 2x - 4 - 2 = x - 6$

77. $(-2x-3)-(-5x-7) = -2x-3+5x+7$

$= -2x + 5x - 3 + 7 = 3x + 4$

79. $(-x+4)-(-x+9) = -x+4+x-9$

$= -x + x + 4 - 9 = -5$

81. $(6-12x)-(3-5x) = 6-12x-3+5x$

$= -12x + 5x + 6 - 3 = -7x + 3$

83. $\left(9x^2+7x-5\right)-\left(3x^2+3.5\right)$

$= 9x^2 + 7x - 5 - 3x^2 - 3.5$

$= 9x^2 - 3x^2 + 7x - 5 - 3.5$

$= 6x^2 + 7x - 8.5$

85. $\left(5x^2-x-1\right)-\left(-3x^2-2x-5\right)$

$= 5x^2 - x - 1 + 3x^2 + 2x + 5$

$= 5x^2 + 3x^2 - x + 2x - 1 + 5$

$= 8x^2 + x + 4$

87. $\left(5x^2-x+12\right)-(5+x)$

$= 5x^2 - x + 12 - 5 - x$

$= 5x^2 - x - x + 12 - 5$

$= 5x^2 - 2x + 7$

89. $(7x-0.6)-\left(-2x^2+4x-8\right)$

$= 7x - 0.6 + 2x^2 - 4x + 8$

$= 2x^2 + 7x - 4x - 0.6 + 8$

$= 2x^2 + 3x + 7.4$

91. $\left(2x^3-4x^2+5x-7\right)-\left(3x+\frac{3}{5}x^2-5\right)$

$= 2x^3 - 4x^2 + 5x - 7 - 3x - \frac{3}{5}x^2 + 5$

$= 2x^3 - 4x^2 - \frac{3}{5}x^2 + 5x - 3x - 7 + 5$

$= 2x^3 - \frac{23}{5}x^2 + 2x - 2$

93. $\left(9x^3-\frac{1}{5}\right)-\left(x^2+5x\right)$

$= 9x^3 - \frac{1}{5} - x^2 - 5x = 9x^3 - x^2 - 5x - \frac{1}{5}$

95. $(3x+5)-(4x-6) = 3x+5-4x+6$

$= 3x - 4x + 5 + 6 = -x + 11$

97. $\left(2x^2-4x+8\right)-(5x-6)$

$=2x^2-4x+8-5x+6$

$=2x^2-4x-5x+8+6$

$=2x^2-9x+14$

99. $\left(3x^3+5x^2+9x-7\right)-\left(4x^3-6x^2\right)$

$=3x^3+5x^2+9x-7-4x^3+6x^2$

$=3x^3-4x^3+5x^2+6x^2+9x-7$

$=-x^3+11x^2+9x-7$

101. $5x+10$ $\quad$ $5x+10$

$\underline{-(2x-7)}$ or $\underline{-2x+7}$

$3x+17$

103. $-5x+3$ $\quad$ $-5x+3$

$\underline{-(-9x-4)}$ or $\underline{9x+4}$

$4x+7$

105. $9x^2+7x-9$ $\quad$ $9x^2+7x-9$

$\underline{-\left(4x^2 \quad -7\right)}$ or $\underline{-4x^2 \quad +7}$

$5x^2+7x-2$

107. $x-6$ $\quad$ $x-6$

$\underline{-\left(-4x^2+6x \quad \right)}$ or $\underline{4x^2-6x}$

$4x^2-5x-6$

109. $4x^3-6x^2+7x-9$

$\underline{-\left(x^2+6x-7\right)}$ or

$4x^3-6x^2+7x-9$

$\underline{-x^2-6x+7}$

$4x^3-7x^2+x-2$

111. A polynomial is a sum of a finite number of terms of the form ax^n where a is a real number and n is a whole number.

113. **(a)** The degree of a term in one variable is the exponent on the variable.

(b) The degree of a polynomial in one variable is equal to the degree of the highest term in the polynomial.

115. Write the polynomial with the exponents on the variables descending from left to right.

117. Answers will vary. One possible solution is: $\left(x^2+2x-3\right)+\left(x^2+3x-3\right)$.

119. Answers will vary. One possible solution is: $\left(x^2+2x-1\right)-\left(2x^2-2x+4\right)$.

Cumulative Review Exercises

120. $|-4|<|-6|$

121. False

122. True.

123. False

124. False.

125. $\left(\dfrac{3x^4y^5}{6x^7y^4}\right)^3 = \left(\dfrac{3}{6}\cdot\dfrac{x^4}{x^7}\cdot\dfrac{y^5}{y^4}\right)^3$

$= \left(\dfrac{1}{2}\cdot\dfrac{1}{x^3}\cdot y\right)^3 = \left(\dfrac{y}{2x^3}\right)^3 = \dfrac{y^3}{2^3x^{3\cdot 3}}$

$= \dfrac{y^3}{8x^9}$

Group Activity/Challenge Problems

1. $(3x^2-6x+3)-(2x^2-x-6)$

$-(x^2+7x-9)$
$= 3x^2-6x+3-2x^2+x+6-x^2-7x+9$

$= (3x^2-2x^2-x^2)+(-6x+x-7x)$

$+(3+6+9)$
$= -12x+18$

2. $3x^2y-6xy-2xy+9xy^2-5xy+3x$
$= 3x^2y+(-6xy-2xy-5xy)+9xy^2+3x$
$= 3x^2y-13xy+9xy^2+3x$

3. $4(x^2+2x-3)-6(2-4x-x^2)$

$-2x(x+2)$
$= 4x^2+8x-12-(12-24x-6x^2)$

$-(2x^2+4x)$
$= 4x^2+8x-12-12+24x+6x^2$

$-2x^2-4x$
$= (4x^2+6x^2-2x^2)+(8x+24x-4x)$

$+(-12-12)$
$= 8x^2+28x-24$

Exercise Set 4.5

1. $x^2\cdot 3xy = 3x^{2+1}y = 3x^3y$

3. $5x^4y^5(6xy^2) = 5\cdot 6x^{4+1}y^{5+2} = 30x^5y^7$

5. $4x^4y^6(-7x^2y^9) = -4\cdot 7x^{4+2}y^{6+9}$

$= -28x^6y^{15}$

7. $9xy^6\cdot 6x^5y^8 = 9\cdot 6x^{1+5}y^{6+8} = 54x^6y^{14}$

9. $(6x^2y)\left(\dfrac{1}{2}x^4\right) = 6\cdot\dfrac{1}{2}x^{2+4}y = 3x^6y$

11. $(1.5x^4y^3)(6xy) = 1.5\cdot 6x^{4+1}y^{3+1}$

$= 9x^5y^4$

13. $3(x+4) = 3\cdot x+3\cdot 4 = 3x+12$

15. $2x(x-3) = 2x\cdot x+2x(-3) = 2x^2-6x$

17. $-4x(-2x+6)=(-4x)(-2x)+(-4x)(6)=8x^2-24x$

19. $2x\left(x^2+3x-1\right)=2x\cdot x^2+2x(3x)+2x(-1)=2x^3+6x^2-2x$

21. $-2x\left(x^2-2x+5\right)=(-2x)\left(x^2\right)+(-2x)(-2x)+(-2x)(5)=-2x^3+4x^2-10x$

23. $5x\left(-4x^2+6x-4\right)=5x\left(-4x^2\right)+5x(6x)+5x(-4)=-20x^3+30x^2-20x$

25. $\left(3x^2+4x-5\right)8x=3x^2\cdot 8x+4x\cdot 8x+(-5)(8x)=24x^3+32x^2-40x$

27. $0.3x(2xy+5x-6y)=(0.3x)(2xy)+(0.3x)(5x)+(0.3x)(-6y)=0.6x^2y+1.5x^2-1.8xy$

29. $(x-y-3)y=x\cdot y+(-y)y+(-3)y=xy-y^2-3y$

31. $(x+3)(x+4)=x\cdot x+x\cdot 4+3\cdot x+3\cdot 4=x^2+4x+3x+12=x^2+7x+12$

33. $(2x+5)(3x-6)=(2x)(3x)+2x(-6)+5\cdot 3x+5(-6)=6x^2-12x+15x-30=6x^2+3x-30$

35. $(2x-4)(2x+4)=(2x)(2x)+(2x)(4)+(-4)(2x)+(-4)(4)=4x^2+8x-8x-16=4x^2-16$

37. $(5-3x)(6+2x)=5\cdot 6+5(2x)+(-3x)(6)+(-3x)(2x)=30+10x-18x-6x^2=30-8x-6x^2$

$=-6x^2-8x+30$

39. $(-x+3)(2x+5)=(-x)(2x)+(-x)(5)+3\cdot 2x+3\cdot 5=-2x^2-5x+6x+15=-2x^2+x+15$

41. $(x+4)(x+3)=x\cdot x+x\cdot 3+4\cdot x+4\cdot 3=x^2+3x+4x+12=x^2+7x+12$

43. $(x+4)(x-2)=x\cdot x+x(-2)+4\cdot x+4(-2)=x^2-2x+4x-8=x^2+2x-8$

45. $(3x+4)(2x+5)=(3x)(2x)+(3x)5+4(2x)+4\cdot 5=6x^2+15x+8x+20=6x^2+23x+20$

47. $(3x+4)(2x-3)=(3x)(2x)+(3x)(-3)+4(2x)+4(-3)=6x^2-9x+8x-12=6x^2-x-12$

49. $(x-1)(x+1)=x\cdot x+x\cdot 1+(-1)\cdot x+(-1)\cdot 1=x^2+x-x-1=x^2-1$

51. $(2x-3)(2x-3)=(2x)(2x)+(2x)(-3)+(-3)(2x)+(-3)(-3)=4x^2-6x-6x+9$

$=4x^2-12x+9$

53. $(4-x)(3+2x)=4\cdot 3+4(2x)+(-x)3+(-x)(2x)=12+8x-3x-2x^2=12+5x-2x^2$

$=-2x^2+5x+12$

55. $(2x+3)(4-2x)=(2x)4+(2x)(-2x)+3\cdot 4+3(-2x)=8x-4x^2+12-6x=-4x^2+2x+12$

57. $(x+y)(x-y)=x\cdot x+x(-y)+y\cdot x+y(-y)=x^2-xy+xy-y^2=x^2-y^2$

59. $(2x-3y)(3x+2y)=(2x)(3x)+(2x)(2y)+(-3y)(3x)+(-3y)(2y)$

$=6x^2+4xy-9xy-6y^2=6x^2-5xy-6y^2$

61. $(4x-3y)(2y-3)=(4x)(2y)+(4x)(-3)+(-3y)(2y)+(-3y)(-3)=8xy-12x-6y^2+9y$

63. $(x+0.6)(x+0.3)=x\cdot x+x(0.3)+(0.6)x+(0.6)(0.3)=x^2+0.3x+0.6x+0.18$

$=x^2+0.9x+0.18$

65. $(2x+4)\left(x+\frac{1}{2}\right)=(2x)x+(2x)\left(\frac{1}{2}\right)+4\cdot x+4\left(\frac{1}{2}\right)=2x^2+x+4x+2=2x^2+5x+2$

67. $(x+4)(x-4)=(x)^2-(4)^2=x^2-16$

69. $(2x-1)(2x+1)=(2x)^2-(1)^2=4x^2-1$

71. $(x+y)^2=(x)^2+2(x)(y)+(y)^2=x^2+2xy+y^2$

73. $(x-0.2)^2=(x)^2-2(x)(0.2)+(0.2)^2=x^2-0.4x+0.04$

75. $(3x+5)(3x-5)=(3x)^2-(5)^2=9x^2-25$

77. $(0.4x+y)^2=(0.4x)^2+2(0.4x)(y)+(y)^2=0.16x^2+0.8xy+y^2$

79. $(x+3)\left(2x^2+4x-1\right)=x\left(2x^2+4x-1\right)+3\left(2x^2+4x-1\right)=2x^3+4x^2-x+6x^2+12x-3$

$=2x^3+10x^2+11x-3$

81. $(5x+4)\left(x^2-x+4\right)=5x\left(x^2-x+4\right)+4\left(x^2-x+4\right)=5x^3-5x^2+20x+4x^2-4x+16$

$=5x^3-x^2+16x+16$

83. $\left(-2x^2-4x+1\right)(7x-3)=-2x^2(7x-3)-4x(7x-3)+1(7x-3)$

$=-14x^3+6x^2-28x^2+12x+7x-3=-14x^3-22x^2+19x-3$

85. $(-3x+9)\left(-6x^2+5x-3\right)=-3x\left(-6x^2+5x-3\right)+9\left(-6x^2+5x-3\right)$

$=18x^3-15x^2+9x-54x^2+45x-27=18x^3-69x^2+54x-27$

87. $\left(3x^2-2x+4\right)\left(2x^2+3x+1\right)=3x^2\left(2x^2+3x+1\right)-2x\left(2x^2+3x+1\right)+4\left(2x^2+3x+1\right)$

$=6x^4+9x^3+3x^2-4x^3-6x^2-2x+8x^2+12x+4=6x^4+5x^3+5x^2+10x+4$

89. $\left(x^2-x+3\right)\left(x^2-2x\right)=x^2\left(x^2-2x\right)-x\left(x^2-2x\right)+3\left(x^2-2x\right)$

$=x^4-2x^3-x^3+2x^2+3x^2-6x=x^4-3x^3+5x^2-6x$

91. $(3x^3+2x^2-x)(x-3)=3x^3(x-3)+2x^2(x-3)-x(x-3)$

$=3x^4-9x^3+2x^3-6x^2-x^2+3x=3x^4-7x^3-7x^2+3x$

93. $(a+b)(a^2-ab+b^2)=a(a^2-ab+b^2)+b(a^2-ab+b^2)$

$=a^3-a^2b+ab^2+a^2b-ab^2+b^3=a^3+b^3$

95. Yes.

97. No. For example, $(x-y)(x+y)=x^2-y^2$.

99. **(a)** $A=(x+2)(2x+1)=x(2x)+x\cdot 1+2\cdot 2x+2\cdot 1=2x^2+x+4x+2=2x^2+5x+2$

(b) If $x=4$, $A=2\cdot 4^2+5\cdot 4+2=54$ square feet

(c) For the rectangle to be a square, all sides must have the same length. So,

$x+2=2x+1$

$x+2=2x+1$

$2=x+1$

$1=x$

So, $x=1$ ft

Cumulative Review Exercises: 4.5

100. Let x equal the maximum distance. Then

$2+1.5(x-1)=20$

$2+1.5x-1.5=20$

$1.5x+0.5=20$

$1.5x=19.5$

$x=\dfrac{19.5}{1.5}=13$ miles

101. $\left(\dfrac{4x^8y^5}{8x^8y^6}\right)^4=\left(\dfrac{4}{8}\cdot\dfrac{x^8}{x^8}\cdot\dfrac{y^5}{y^6}\right)^4=\left(\dfrac{1}{2}\cdot\dfrac{1}{y}\right)^4=\left(\dfrac{1}{2y}\right)^4=\dfrac{1}{2^4y^4}=\dfrac{1}{16y^4}$

102. **(a)** $-6^3 = -(6^3) = -216$

(b) $6^{-3} = \frac{1}{6^3} = \frac{1}{216}$

103. $\left(-x^2 - 6x + 5\right) - \left(5x^2 - 4x - 3\right) = -x^2 - 6x + 5 - 5x^2 + 4x + 3 = -6x^2 - 2x + 8$

Group Activity/Challenge Problems

1. $\sqrt{5}x\left(2x^2 + \sqrt{5}x - \frac{1}{2}\right) = 2x^3\sqrt{5} + 5x^2 - \frac{x\sqrt{5}}{2}$

3. $(2x^3 - 6x^2 + 5x - 3)(3x^3 - 6x + 4)$

$= 2 \cdot 3x^{3+3} - 2 \cdot 6x^{3+1} + 2 \cdot 4x^3 - 6 \cdot 3x^{2+3} - 6(-6)x^{2+1}$
$\quad -6 \cdot 4x^2 + 5 \cdot 3x^{1+3} - 5 \cdot 6x^{1+1} + 5 \cdot 4x - 3 \cdot 3x^3 - 3(-6)x - 3 \cdot 4$

$= 6x^6 - 12x^4 + 8x^3 - 18x^5 + 36x^3 - 24x^2 + 15x^4 - 30x^2 + 20x - 9x^3 + 18x - 12$

$= 6x^6 - 18x^5 + 3x^4 + 35x^3 - 54x^2 + 38x - 12$

5. $3x^2;\ 3x^2(4x^2 - 2x + 5) = 12x^4 - 6x^3 + 15x^2$

6. $-4xy^3;\ -4xy^3(3x^2 - 6x + 8) = -12x^3y^3 + 24x^2y^3 - 32xy^3$

7. $x + 2;\ (x + 2)(x + 3) = x^2 + 3x + 2x + 6 = x^2 + 5x + 6$

11. $x + 3,\ x - 2;\ x^2 + x - 6 = x^2 - 2x + 3x - 6 = (x + 3)(x - 2)$

Exercise Set 4.6

1. $\frac{x^2 - 2x - 15}{x + 3} = x - 5$ or $\frac{x^2 - 2x - 15}{x - 5} = x + 3$

3. $\dfrac{2x^2+5x+3}{2x+3}=x+1$ or $\dfrac{2x^2+5x+3}{x+1}=2x+3$

5. $\dfrac{4x^2-9}{2x+3}=2x-3$ or $\dfrac{4x^2-9}{2x-3}=2x+3$

7. $\dfrac{2x+4}{2}=\dfrac{2x}{2}+\dfrac{4}{2}=x+2$

9. $\dfrac{2x+6}{2}=\dfrac{2x}{2}+\dfrac{6}{2}=x+3$

11. $\dfrac{3x+8}{2}=\dfrac{3x}{2}+\dfrac{8}{2}=\dfrac{3}{2}x+4$

13. $\dfrac{-6x+4}{2}=\dfrac{-6x}{2}+\dfrac{4}{2}=-3x+2$

15. $\dfrac{-9x-3}{-3}=\dfrac{(-1)(-9x-3)}{(-1)(-3)}=\dfrac{9x+3}{3}=\dfrac{9x}{3}+\dfrac{3}{3}=3x+1$

17. $\dfrac{3x+6}{x}=\dfrac{3x}{x}+\dfrac{6}{x}=3+\dfrac{6}{x}$

19. $\dfrac{9-3x}{-3x}=\dfrac{(-1)(9-3x)}{(-1)(-3x)}=\dfrac{-9+3x}{3x}=\dfrac{-9}{3x}+\dfrac{3x}{3x}=-\dfrac{3}{x}+1$

21. $\dfrac{3x^2+6x-9}{3x^2}=\dfrac{3x^2}{3x^2}+\dfrac{6x}{3x^2}+\dfrac{-9}{3x^2}=1+\dfrac{2}{x}-\dfrac{3}{x^2}$

23. $\dfrac{-4x^5+6x+8}{2x^2}=\dfrac{-4x^5}{2x^2}+\dfrac{6x}{2x^2}+\dfrac{8}{2x^2}=-2x^3+\dfrac{3}{x}+\dfrac{4}{x^2}$

25. $\dfrac{x^6+4x^4-3}{x^3}=\dfrac{x^6}{x^3}+\dfrac{4x^4}{x^3}+\dfrac{-3}{x^3}=x^3+4x-\dfrac{3}{x^3}$

27. $\dfrac{6x^5-4x^4+12x^3-5x^2}{2x^3}=\dfrac{6x^5}{2x^3}+\dfrac{-4x^4}{2x^3}+\dfrac{12x^3}{2x^3}+\dfrac{-5x^2}{2x^3}=3x^2-2x+6-\dfrac{5}{2x}$

29. $\dfrac{4x^3+6x^2-8}{-4x}=\dfrac{(-1)(4x^3+6x^2-8)}{(-1)(-4x)}=\dfrac{-4x^3-6x^2+8}{4x}=\dfrac{-4x^3}{4x}+\dfrac{-6x^2}{4x}+\dfrac{8}{4x}=-x^2-\dfrac{3}{2}x+\dfrac{2}{x}$

31. $\dfrac{9x^6+3x^4-10x^2-9}{3x^2}=\dfrac{9x^6}{3x^2}+\dfrac{3x^4}{3x^2}+\dfrac{-10x^2}{3x^2}+\dfrac{-9}{3x^2}=3x^4+x^2-\dfrac{10}{3}-\dfrac{3}{x^2}$

33.

$$\begin{array}{r} x+3 \\ x+1\overline{)\,x^2+4x+3} \\ \underline{-x^2-x} \\ 3x+3 \\ \underline{-3x-3} \\ 0 \end{array}$$

Thus $\dfrac{x^2+4x+3}{x+1}=x+3$

35.

$$\begin{array}{r} 2x+3 \\ x+5\overline{)\,2x^2+13x+15} \\ \underline{-2x^2-10x} \\ 3x+15 \\ \underline{-3x-15} \\ 0 \end{array}$$

Thus $\dfrac{2x^2+13x+15}{x+5}=2x+3$

37.

$$\begin{array}{r} 2x+4 \\ 3x+2\overline{)6x^2+16x+8} \\ \underline{-6x^2-4x} \\ 12x+8 \\ \underline{-12x-8} \\ 0 \end{array}$$

Thus $\dfrac{6x^2+16x+8}{3x+2}=2x+4$

39.

$$\begin{array}{r} x-2 \\ 2x+5\overline{)2x^2+x-10} \\ \underline{-2x^2-5x} \\ -4x-10 \\ \underline{+4x+10} \\ 0 \end{array}$$

Thus $\dfrac{2x^2+x-10}{2x+5}=x-2$

41.

$$\begin{array}{r} x+5 \\ 2x-3\overline{)2x^2+7x-18} \\ \underline{-2x^2+3x} \\ 10x-18 \\ \underline{-10x+15} \\ -3 \end{array}$$

Thus $\dfrac{2x^2+7x-18}{2x-3}=x+5-\dfrac{3}{2x-3}$

43.

$$\begin{array}{r} 2x+3 \\ 2x-3\overline{)4x^2\qquad-9} \\ \underline{-4x^2+6x} \\ 6x-9 \\ \underline{-6x+9} \\ 0 \end{array}$$

Thus $\dfrac{4x^2-9}{2x-3}=2x+3$

45.

$$\require{enclose}\begin{array}{r} 2x-3 \\ 4x+9\enclose{longdiv}{8x^2+6x-25} \\ \underline{-8x^2-18x} \\ -12x-25 \\ \underline{12x+27} \\ 2 \end{array}$$

Thus $\dfrac{8x+6x^2-25}{4x+9}=2x-3+\dfrac{2}{4x+9}$

47.

$$\begin{array}{r} 4x-3 \\ 2x+3\enclose{longdiv}{8x^2+6x-12} \\ \underline{-8x^2-12x} \\ -6x-12 \\ \underline{6x+9} \\ -3 \end{array}$$

Thus $\dfrac{6x+8x^2-12}{2x+3}=4x-3-\dfrac{3}{2x+3}$

49.

$$\begin{array}{r} 2x^2+3x-1 \\ 2x+3\enclose{longdiv}{4x^3+12x^2+7x-3} \\ \underline{-4x^3-6x^2} \\ 6x^2+7x \\ \underline{-6x^2-9x} \\ -2x-3 \\ \underline{2x+3} \\ 0 \end{array}$$

Thus $\dfrac{4x^3+12x^2+7x-3}{2x+3}=2x^2+3x-1$

51.

$$\begin{array}{r|l} & 3x^2-3x-1 \\ 3x+2 & 9x^3-3x^2-9x+4 \end{array}$$

$$\underline{-9x^3-6x^2}$$
$$-9x^2-9x$$
$$\underline{9x^2+6x}$$
$$-3x+4$$
$$\underline{3x+2}$$
$$6$$

Thus $\dfrac{9x^3-3x^2-9x+4}{3x+2}=3x^2-3x-1+\dfrac{6}{3x+2}$

53.

$$\begin{array}{r|l} & x^2+3x+9 \\ x-3 & x^3 \qquad -8 \end{array}$$

$$\underline{-x^3+3x^2}$$
$$3x^2$$
$$\underline{-3x^2+9x}$$
$$9x-8$$
$$\underline{-9x+27}$$
$$19$$

Thus $\dfrac{x^3-8}{x-3}=x^2+3x+9+\dfrac{19}{x-3}$

55.

$$\begin{array}{r|l} & x^2+3x+9 \\ x-3 & x^3 \qquad -27 \end{array}$$

$$\underline{-x^3+3x^2}$$
$$3x^2$$
$$\underline{-3x^2+9x}$$
$$9x-27$$
$$\underline{-9x+27}$$
$$0$$

Thus $\dfrac{x^3-27}{x-3}=x^2+3x+9$

57.

$$\begin{array}{r} 2x^2+x-2 \\ 2x-1\overline{)4x^3 \qquad\quad -5x} \\ \underline{-4x^3+2x^2} \\ 2x^2-5x \\ \underline{-2x^2+x} \\ -4x \\ \underline{4x-2} \\ -2 \end{array}$$

Thus $\frac{4x^3-5x}{2x-1}=2x^2+x-2-\frac{2}{2x-1}$

59.

$$\begin{array}{r} -x^2-7x-5 \\ x-1\overline{)-x^3-6x^2+2x-3} \\ \underline{x^3-x^2} \\ -7x^2+2x \\ \underline{7x^2-7x} \\ -5x-3 \\ \underline{5x-5} \\ -8 \end{array}$$

Thus $\frac{-x^3-6x^2+2x-3}{x-1}=-x^2-7x-5-\frac{8}{x-1}$

61. $3x^2$ is what each term must be divided by to obtain the result.

63. Since the shaded areas minus 2 must equal 5, 4, 2, and 0 respectively, the shaded areas are 7, 6, 4, and 2 respectively.

Cumulative Review Exercises: 4.6

64. **(a)** 2

(b) 2, 0

(c) $2,\ -5,\ 0,\ \frac{2}{5},\ -6.3,\ -\frac{23}{34}$

(d) $\sqrt{7},\ \sqrt{3}$

(e) $2,\ -5,\ 0,\ \sqrt{7},\ \frac{2}{5},\ -6.3,\ \sqrt{3},\ -\frac{23}{34}$

65. (a) $\frac{0}{1}=0,$

(b) $\frac{1}{0}$ is undefined

66. Evaluate expressions in parentheses first, then exponents, followed by multiplications and divisions from left to right, and finally additions and subtractions from left to right

67.
$$\begin{aligned} 2(x+3)+2x &= x+4 \\ 2x+6+2x &= x+4 \\ 4x+6 &= x+4 \\ 4x &= x-2 \\ 3x &= -2 \\ x &= -\tfrac{2}{3} \end{aligned}$$

Group Activity/Challenge Problems

1.
$$\begin{array}{r} 2x^2-3x+\frac{5}{2} \\ 2x+3\overline{)4x^3+0x^2-4x+6} \\ \underline{4x^3+6x^2} \qquad\qquad \\ -6x^2-4x \qquad \\ \underline{-6x^2-9x} \qquad \\ 5x+6 \\ \underline{5x+\frac{15}{2}} \\ -\frac{3}{2} \end{array}$$

Answer: $2x^2-3x+\frac{5}{2}-\frac{3}{2(2x+3)}$

3.

$$
\begin{array}{r|l}
 & -3x+3 \\
-x-3 & 3x^2+6x-10 \\
 & \underline{3x^2+9x} \\
 & -3x-10 \\
 & \underline{-3x-9} \\
 & -1
\end{array}
$$

Answer: $-3x+3+\dfrac{1}{x+3}$

5. $(x+3)(x+1)-1=x^2+x+3x+3-1=x^2+4x+2$

Exercise Set 4.7

1. Rate $= \dfrac{\text{volume}}{\text{time}} = \dfrac{30}{6} = 5.$ Therefore, 5 gallons of water are used per minute.

3. Rate $= \dfrac{\text{number of copies}}{\text{time}} = \dfrac{100}{2.5} = 40.$ Therefore, the photocopying machine must make 40 copies per minute.

5. Rate $= \dfrac{\text{distance}}{\text{time}} = \dfrac{238{,}000}{87} = 2736$ mph, Therefore, the rate is about 2736 mph.

7. Rate $= \dfrac{\text{volume}}{\text{time}} = \dfrac{1500}{6} = 250$ cm^3/hr, Therefore, the flow rate should be 250 cc/hr.

9. $\dfrac{1}{3}$ hr $= 20$ min.Distance = rate $\cdot$time $= 3\cdot\dfrac{1}{3} = 1$, Therefore, he walks 1 mile.

11. 2 days $= 2\cdot 24$ hours $= 48$ hours. Volume = rate $\cdot$ time $= 7\cdot 48 = 336$
Therefore, the average American uses 336 gallons in two days.

13. Let t be the time it takes for the planes to be 4025 miles apart.

Plane	Rate	Time	Distance
Northbound	500	t	$500t$
Southbound	650	t	$650t$

$$500t + 650t = 4025$$
$$1150t = 4025$$
$$t = 3.5$$

Therefore, they will be 4025 miles apart in 3.5 hours.

15. Let r be the speed of the Santa Fe special. Then $r + 30$ is the speed of the Amtrak train.

Train	Rate	Time	Distance
Santa Fe	r	6	$6r$
Amtrak	$r + 30$	6	$6(r+30)$

$$6r + 6(r+30) = 804$$
$$6r + 6r + 180 = 804$$
$$12r = 624$$
$$r = 52$$

The speed of Santa Fe Special is 52 mph, and the speed of the Amtrak train is 82 mph.

17. **(a)** Let r be the speed of the cutter coming from the east. Then $r+5$ is the speed of the cutter coming from the west.

Cutter	Rate	Time	Distance
Eastbound	$r+5$	3	$3(r+5)$
Westbound	r	3	$3r$

$$3(r+5) + 3r = 225$$
$$3r + 15 + 3r = 225$$
$$6r = 210$$
$$r = 35$$

The speed of the westbound cutter is 35 mph, and the speed of the eastbound cutter is 40 mph.

(b) $35 \text{ mph} \cdot \dfrac{1 \text{ knot}}{1.15 \text{ mph}} \approx 30.43 \text{ knots}, \; 40 \text{ mph} \cdot \dfrac{1 \text{ knot}}{1.15 \text{ mph}} \approx 34.78 \text{ knots}$

The speed of the westbound cutter is about 30.43 knots and the speed of the eastbound cutter is about 34.78 knots.

19. Let r be the rate of the tape on SLP speed.

Speed	Rate	Time	Distance
SP	6.72	120	806.4
SLP	r	360	$360r$

$360r = 806.4$

$r = 2.24$

The rate of the tape of SLP speed is 2.24 ft/min.

21. Let r be the speed of Chestnut. Then the speed of Midnight is $r+3$.

Horse	Rate	Time	Distance
Chestnut	r	1.5	$1.5r$
Midnight	$r+3$	1.5	$1.5(r+3)$

$1.5r + 1.5(r+3) = 16.5$

$1.5r + 1.5r + 4.5 = 16.5$

$3r = 12$

$r = 4$

Chestnut is traveling at a rate of 4 mph, and Midnight is traveling at a rate of 7 mph.

23. Let t be the time it takes for Serge to catch up.

Climber	Rate	Time	Distance
Serge	20	t	$20t$
Francine	18	$t+30$	$18(t-30)$

$20t = 18(t+30)$

$20t = 18t + 540$

$2t = 540$

$t = 270$

Distance $= 20(270) = 5400$..

They will meet 5400 ft up the mountain.

25. Let r = speed of moving walkway (in ft/min)

	Rate	Time	Distance
Marguerita on foot	100 ft/min	2.75 min	275 ft
Marguerita on walkway	$100 + r$	1.25 min	$(100 + r)(1.25)$

(a) 275 ft

(b) distance "going" on foot = distance "returning" walking on moving walkway

$275 = (100 + r)(1.25)$

$275 = 125 + 1.25r$

$150 = 1.25r$

$120 = r$

The moving walkway moves at 120 ft/min.

27. Let r be the rate of clearing the bridge. Then $1.2 + r$ is the rate of clearing the road.

	Rate	Time	Distance
Road	$1.2 + r$	20	$20(1.2 + r)$
Bridge	r	60	$60r$

$$20(1.2 + r) + 60r = 124$$
$$24 + 20r + 60r = 124$$
$$80r = 100$$
$$r = 1.25$$

The crew will clear the bridge at a rate of 1.25 ft/day, and they will clear the road at a rate of 2.45 ft/day.

29. Let x be the amount in the 8% account. Then $8900 - x$ is the amount in the 11% account.

Principal	Rate	Time	Interest
x	8%	1	$0.08x$
$8900 - x$	11%	1	$0.11(8900 - x)$

$$0.08x + 0.11(8900 - x) = 874$$
$$0.08x + 979 - 0.11x = 874$$
$$-0.03x = -105$$
$$x = 3500$$

\$3500 is invested at 8% and \$5400 is invested at 11%.

31. Let x be the amount invested at 10%. Then $5000 - x$ is invested at 6%.

Principal	Rate	Time	Interest
x	10%	1	$0.10x$
$5000 - x$	6%	1	$0.06(5000 - x)$

$0.10x = 0.06(5000 - x)$

$0.1x = 300 — 0.06x$

$0.16x = 300$

$x = 1875$

\$1875 is invested at 10% and \$3125 is invested at 6%.

33. Let x be the number of months Charles paid \$13.20. Then $12 - x$ is the number of months he paid \$14.50.

Rate	Time	Amount
13.20	x	$13.20x$
14.50	$12 - x$	$14.50(12 - x)$

$13.20x + 14.50(12 - x) = 170.10$

$13.20x + 174 - 14.50x = 170.10$

$-1.30x = -3.9$

$x = 3$

The rate increased after three months, or in April.

35. Let x be the number of \$1 bills. Then Phil has $(12 - x)$ \$10 bills.

Bill	Number	Total
\$1	x	x
\$10	$12 - x$	$10(12 - x)$

$x + 10(12 - x) = 39$

$x + 120 - 10x = 39$

$-9x = -81$

$x = 9$

Phil has nine \$1 bills and three \$10 bills.

37. Let x be the number of hours worked at \$6.00 per hour. Then $18-x$ is the number of hours he worked at \$6.50 per hour.

Rate	Hours	Total
\$6.00	x	$6x$
\$6.50	$18-x$	$(6.5)(18-x)$

$$6x+6.5(18-x)=114$$
$$6x+117-6.5x=114$$
$$-0.5x=-3$$
$$x=6$$

Casey worked 6 hours at \$6.00 per hour and 12 hours at \$6.50 per hour.

39. Let x be the number of pounds of coffee costing \$6.20 per pound.

Type	Cost	Pounds	Total
\$6.20	\$6.20	x	$6.2x$
\$5.60	\$5.60	18	\$100.80
Mixture	\$5.80	$x+18$	$5.8(x+18)$

$$6.2x+100.80=5.8(x+18)$$
$$6.2x+100.80=5.8x+104.40$$
$$0.4x=3.60$$
$$x=9$$

9 pounds of coffee costing \$6.20 should be mixed.

41. Let x be the amount of 12% sulfuric acid solution.

Solution	Strength	Liters	Amount
20%	0.20	1	0.20
12%	0.12	x	$0.12x$
Mixture	0.15	$x+1$	$0.15(x+1)$

$$0.20+0.12x=0.15(x+1)$$
$$0.20+0.12x=0.15x+0.15$$
$$-0.03x=-0.05$$
$$x=\frac{5}{3}$$

$1\frac{2}{3}$ liters of 12% sulfuric acid should be mixed.

43. Let x be the percent of orange juice in the new mixture.

Percent	Quarts	Orange Juice
12%	6	0.72
x	$6\frac{1}{2}$	$\frac{13}{2}x$

$$0.72=\frac{13}{2}x$$
$$x=0.111$$

The new mixture consists of 11.1% orange juice.

45. Let x be the amount of Family grass seed. Then $10-x$ is the amount of Spot Filler grass seed.

Seed	Price	Amount	Total
Family	\$2.25	x	$2.25x$
Filler	\$1.90	$10-x$	$1.90(10-x)$
Mixture	\$2.00	10	20

$$2.25x+1.90(10-x)=20$$
$$2.25x+19-1.90x=20$$
$$0.35x=1$$
$$x=2.86$$

2.86 pounds of Family grass seed and 7.14 pounds of Spot Filler grass seed should be mixed together.

47. **(a)** Let x be the number of shares of Mattel. Then $4x$ is the number of shares of United Airlines.

Company	Shares	Price	Total
Mattel	x	\$22	$22x$
United	$4x$	\$140	$560x$

$22x + 560x = 8000$

$582x = 8000$

$x = 13.74$

The investor should buy 13 shares of Mattel and 52 shares of United Airlines.

(b) The investor spent $(13)(22) + (52)(140) = \$7566$. She has \$434 left over.

Cumulative Review Exercises: 4.7

48. **(a)** $2\frac{3}{4} \div 1\frac{5}{8} = \frac{11}{4} \div \frac{13}{8} = \frac{11/4}{13/8}$

$= \frac{11}{4} \cdot \frac{8}{13} = \frac{22}{13}$ or $1\frac{9}{13}$

(b) $2\frac{3}{4} + 1\frac{5}{8} = \frac{11}{4} + \frac{13}{8}$

$= \frac{22}{8} + \frac{13}{8} = \frac{35}{8}$ or $4\frac{3}{8}$

49. $6(x-3) = 4x - 18 + 2x$

$6x - 18 = 6x - 18$

All real numbers are solutions.

50. $\frac{6}{x} = \frac{72}{9}$

$6 \cdot 9 = 72x$

$54 = 72x$

$x = \frac{54}{72} = \frac{3}{4}$ or 0.75

51. $3x - 4 \le -4x + 3(x-1)$

$3x - 4 \le -4x + 3x - 3$

$3x - 4 \le -x - 3$

$4x \le 1$

$x \le \frac{1}{4}$

Group Activity/Challenge Problems

1. **(a)** distance = rate × time

$(5280)(10) = (1472)(3)t$

$52{,}800 = 4416t$

(b) $\frac{52{,}800}{4416} = t$

$t = 11.96$; approximately 11.96 hours

3. $d = rt$

$r = \frac{d}{t}$

$r = \frac{6\text{ ft}}{100\text{ ft}/4(1.47\text{ ft}/\text{sec})}$

$= 0.3528\text{ ft}/\text{sec}$

Approximately 0.35 ft/sec.

Chapter 4 Review Exercises

1. $x^4 \cdot x^2 = x^{4+2} = x^6$

2. $x^3 \cdot x^5 = x^{3+5} = x^8$

3. $3^2 \cdot 3^3 = 3^{2+3} = 3^5 = 243$

4. $2^4 \cdot 2 = 2^{4+1} = 2^5 = 32$

5. $\frac{x^4}{x} = x^{4-1} = x^3$

6. $\frac{x^6}{x^6} = x^{6-6} = x^0 = 1$

7. $\frac{3^5}{3^3} = 3^{5-3} = 3^2 = 9$

8. $\frac{4^5}{4^3} = 4^{5-3} = 4^2 = 16$

9. $\frac{x^6}{x^8} = \frac{1}{x^{8-6}} = \frac{1}{x^2}$

10. $\frac{x^7}{x^2} = x^{7-2} = x^5$

11. $x^0 = 1$

12. $3x^0 = 3(x^0) = 3 \cdot 1 = 3$

13. $(3x)^0 = 1$

14. $4^0 = 1$

15. $(2x)^2 = 2^2 x^2 = 4x^2$

16. $(3x)^3 = 3^3 x^3 = 27x^3$

17. $(-2x)^2 = (-2)^2 x^2 = 4x^2$

18. $(-3x)^3 = (-3)^3 x^3 = -27x^3$

19. $(2x^2)^4 = 2^4 x^{2 \cdot 4} = 16x^8$

20. $(-x^4)^3 = (-1)^3 x^{4 \cdot 3} = -x^{12}$

21. $(-x^3)^4 = (-1)^4 x^{3 \cdot 4} = x^{12}$

22. $\left(\frac{2x^3}{y}\right)^2 = \frac{2^2 x^{3 \cdot 2}}{y^2} = \frac{4x^6}{y^2}$

23. $\left(\frac{3x^4}{2y}\right)^3 = \frac{3^3 x^{4 \cdot 3}}{2^3 y^3} = \frac{27x^{12}}{8y^3}$

24. $6x^2 \cdot 4x^3 = 6 \cdot 4x^{2+3} = 24x^5$

25. $\frac{16x^2y}{4xy^2} = \frac{16}{4} \cdot \frac{x^2}{x} \cdot \frac{y}{y^2} = 4x\frac{1}{y} = \frac{4x}{y}$

26. $(2x^2y)^2 \cdot 3x$:

$(2x^2y)^2 = 2^2 x^{2 \cdot 2} y^2 = 4x^4y^2$. So,

$(2x^2y)^2 \cdot 3x = (4x^4y^2) \cdot 3x = 4 \cdot 3x^{4+1}y^2$

$= 12x^5y^2$

27. $\left(\frac{9x^2y}{3xy}\right)^2 = \left(\frac{9}{3} \cdot \frac{x^2}{x} \cdot \frac{y}{y}\right)^2 = (3x)^2$

$= 3^2x^2 = 9x^2$

28. Since $(2x^2y)^3 = 2^3x^{2\cdot3}y^3 = 8x^6y^3$

$(2x^2y)^3(3xy^4) = (8x^6y^3)(3xy^4)$

$= 8\cdot3x^{6+1}y^{3+4} = 24x^7y^7$

29. Since $(2x^3y^4)^2 = 2^2x^{3\cdot2}y^{4\cdot2} = 4x^6y^8$,

then $(4x^2y^3)(2x^3y^4)^2 = (4x^2y^3)(4x^6y^8)$

$= 4\cdot4x^{2+6}y^{3+8} = 16x^8y^{11}$

30. Since $(2x^3y^4)^2 = 2^2x^{3\cdot2}y^{4\cdot2} = 4x^6y^8$,

then $6x^4(2x^3y^4)^2 = 6x^4(4x^6y^8)$

$= 6\cdot4x^{4+6}y^8 = 24x^{10}y^8$

31. $\left(\frac{8x^4y^3}{2xy^5}\right)^2 = \left(\frac{8}{2}\cdot\frac{x^4}{x}\cdot\frac{y^3}{y^5}\right)^2 = \left(4x^3\cdot\frac{1}{y^2}\right)^2$

$= \left(\frac{4x^3}{y^2}\right)^2 = \frac{4^2x^{3\cdot2}}{y^{2\cdot2}} = \frac{16x^6}{y^4}$

32. $\left(\frac{5x^4y^7}{10xy^{10}}\right)^3 = \left(\frac{5}{10}\cdot\frac{x^4}{x}\cdot\frac{y^7}{y^{10}}\right)^3$

$= \left(\frac{1}{2}\cdot x^3\cdot\frac{1}{y^3}\right)^3 = \left(\frac{x^3}{2y^3}\right)^3$

$= \frac{x^{3\cdot3}}{2^3y^{3\cdot3}} = \frac{x^9}{8y^9}$

33. $x^{-3} = \frac{1}{x^3}$

34. $y^{-7} = \frac{1}{y^7}$

35. $5^{-2} = \frac{1}{5^2} = \frac{1}{25}$

36. $\frac{1}{x^{-3}} = x^3$

37. $\frac{1}{x^{-7}} = x^7$

38. $\frac{1}{3^{-2}} = 3^2 = 9$

39. $x^3\cdot x^{-5} = x^{3-5} = x^{-2} = \frac{1}{x^2}$

40. $x^{-2}\cdot x^{-3} = x^{-2-3} = x^{-5} = \frac{1}{x^5}$

41. $x^4\cdot x^{-7} = x^{4-7} = x^{-3} = \frac{1}{x^3}$

42. $x^{13}\cdot x^{-5} = x^{13-5} = x^8$

43. $\frac{x^2}{x^{-3}} = x^{2-(-3)} = x^5$

44. $\frac{x^5}{x^{-2}} = x^{5-(-2)} = x^7$

45. $\frac{x^{-3}}{x^3} = \frac{1}{x^{3+3}} = \frac{1}{x^6}$

46. $(3x^4)^{-2} = 3^{-2}x^{4(-2)} = 3^{-2}x^{-8}$

$= \dfrac{1}{3^2x^8} = \dfrac{1}{9x^8}$

47. $(4x^{-3}y)^{-3} = 4^{-3}x^{(-3)(-3)}y^{-3} = 4^{-3}x^9y^{-3}$

$= \dfrac{x^9}{4^3y^3} = \dfrac{x^9}{64y^3}$

48. $(-5x^{-2})^3 = (-5)^3x^{(-2)3} = -125x^{-6}$

$= \dfrac{-125}{x^6}$

49. $-2x^3 \cdot 4x^5 = -2 \cdot 4x^{3+5} = -8x^8$

50. $(3x^{-2}y)^3 = 3^3x^{(-2)3}y^3 = 27x^{-6}y^3 = \dfrac{27y^3}{x^6}$

51. $(4x^{-2}y^3)^{-2} = 4^{-2}x^{(-2)(-2)}y^{3(-2)}$

$= 4^{-2}x^4y^{-6} = \dfrac{x^4}{4^2y^6} = \dfrac{x^4}{16y^6}$

52. $2x(3x^{-2}) = 2 \cdot 3x^{1-2} = 6x^{-1} = \dfrac{6}{x}$

53. $(5x^{-2}y)(2x^4y) = 5 \cdot 2x^{-2+4}y^{1+1} = 10x^2y^2$

54. $4x^5(6x^{-7}y^2) = 4 \cdot 6x^{5-7}y^2 = 24x^{-2}y^2$

$= \dfrac{24y^2}{x^2}$

55. $2x^{-4}(3x^{-2}y^{-1}) = 2 \cdot 3x^{-4-2}y^{-1}$

$= 6x^{-6}y^{-1} = \dfrac{6}{x^6y}$

56. $\dfrac{6xy^4}{2xy^{-1}} = \dfrac{6}{2} \cdot \dfrac{x}{x} \cdot \dfrac{y^4}{y^{-1}} = 3y^5$

57. $\dfrac{9x^{-2}y^3}{3xy^2} = \dfrac{9}{3} \cdot \dfrac{x^{-2}}{x} \cdot \dfrac{y^3}{y^2} = 3 \cdot \dfrac{1}{x^3} \cdot y = \dfrac{3y}{x^3}$

58. $\dfrac{25xy^{-6}}{5y^{-2}} = \dfrac{25}{5} \cdot x \cdot \dfrac{y^{-6}}{y^{-2}} = 5 \cdot x \cdot \dfrac{1}{y^4} = \dfrac{5x}{y^4}$

59. $\dfrac{36x^4y^7}{9x^5y^{-3}} = \dfrac{36}{9} \cdot \dfrac{x^4}{x^5} \cdot \dfrac{y^7}{y^{-3}} = 4 \cdot \dfrac{1}{x} \cdot y^{10}$

$= \dfrac{4y^{10}}{x}$

60. $\dfrac{4x^5y^{-2}}{8x^7y^3} = \dfrac{4}{8} \cdot \dfrac{x^5}{x^7} \cdot \dfrac{y^{-2}}{y^3} = \dfrac{1}{2} \cdot \dfrac{1}{x^2} \cdot \dfrac{1}{y^5}$

$= \dfrac{1}{2x^2y^5}$

61. $364{,}000 = 3.64 \times 10^5$

62. $1{,}640{,}000 = 1.64 \times 10^6$

63. $0.00763 = 7.63 \times 10^{-3}$

64. $0.176 = 1.76 \times 10^{-1}$

65. $2080 = 2.08 \times 10^3$

66. $0.000314 = 3.14 \times 10^{-4}$

67. $4.2 \times 10^{-3} = 0.0042$

68. $1.65 \times 10^4 = 16,500$

69. $9.7 \times 10^5 = 970,000$

70. $4.38 \times 10^{-6} = 0.00000438$

71. $9.14 \times 10^{-1} = 0.914$

72. $5.36 \times 10^2 = 536$

73. $(2.3 \times 10^2)(2 \times 10^4) = (2.3 \times 2)(10^2 \times 10^4) = 4.6 \times 10^6 = 4,600,000$

74. $(4.2 \times 10^{-3})(3 \times 10^5) = (4.2 \times 3)(10^{-3} \times 10^5) = 12.6 \times 10^2 = 1260$

75. $(6.4 \times 10^{-3})(3.1 \times 10^3) = (6.4 \times 3.1)(10^{-3} \times 10^3) = 19.84 \times 10^0 = 19.84$

76. $\dfrac{6.8 \times 10^3}{2 \times 10^{-2}} = \left(\dfrac{6.8}{2}\right)\left(\dfrac{10^3}{10^{-2}}\right) = 3.4 \times 10^5 = 340,000$

77. $\dfrac{36 \times 10^4}{4 \times 10^6} = \left(\dfrac{36}{4}\right)\left(\dfrac{10^4}{10^6}\right) = 9 \times 10^{-2} = 0.09$

78. $\dfrac{15 \times 10^{-3}}{5 \times 10^2} = \left(\dfrac{15}{5}\right)\left(\dfrac{10^{-3}}{10^2}\right) = 3 \times 10^{-5} = 0.00003$

79. $(60,000)(20,000) = (6 \times 10^4)(2 \times 10^4) = (6 \times 2)(10^4 \times 10^4) = 12 \times 10^8 = 1.2 \times 10^9$

80. $(0.00004)(600,000) = (4 \times 10^{-5})(6 \times 10^5) = (4 \times 6)(10^{-5} \times 10^5) = 24 \times 10^0 = 2.4 \times 10^1$

81. $(0.00023)(40,000) = (2.3 \times 10^{-4})(4 \times 10^4) = (2.3 \times 4)(10^{-4} \times 10^4) = 9.2 \times 10^0 = 9.2$

82. $\dfrac{40{,}000}{0.0002}=\dfrac{4\times10^4}{2\times10^{-4}}=\left(\dfrac{4}{2}\right)\left(\dfrac{10^4}{10^{-4}}\right)=2\times10^8$

83. $\dfrac{0.000068}{0.02}=\dfrac{6.8\times10^{-5}}{2\times10^{-2}}=\left(\dfrac{6.8}{2}\right)\left(\dfrac{10^{-5}}{10^{-2}}\right)=3.4\times10^{-3}$

84. $\dfrac{1{,}500{,}000}{0.003}=\dfrac{1.5\times10^6}{3\times10^{-3}}=\left(\dfrac{1.5}{3}\right)\left(\dfrac{10^6}{10^{-3}}\right)=0.5\times10^9=5\times10^8$

85. **(a)** Times greater $=\dfrac{2.29\times10^7}{8.17\times10^5}=\left(\dfrac{2.29}{8.17}\right)\left(\dfrac{10^7}{10^5}\right)=0.28\times10^2=28$; 28 times greater

(b) $2.29\times10^7-8.17\times10^5=22{,}900{,}000-817{,}000=22{,}083{,}000$ more copies

86. Distance $=(520)(5.87\times10^{12})=(5.2\times10^2)(5.87\times10^{12})=(5.2\times5.87)(10^2\times10^{12})$

$=30.524\times10^{14}=3.0524\times10^{15}$ miles

87. Binomial, first degree.

88. Monomial, zero degree.

89. x^2+3x-4, trinomial, second degree.

90. $4x^2-x-3$, trinomial, second degree.

91. Binomial, second degree.

92. Not a polynomial.

93. $-4x^2+x$, binomial, second degree.

94. Not a polynomial.

95. $x^3 + 4x^2 - 2x - 6$, third degree polynomial.

96. $(x+3)+(2x+4) = x+3+2x+4 = x+2x+3+4 = 3x+7$

97. $(5x-5)+(4x+6) = 5x-5+4x+6 = 5x+4x-5+6 = 9x+1$

98. $(-3x+4)+(5x-9) = -3x+4+5x-9 = -3x+5x+4-9 = 2x-5$

99. $(-x^2+6x-7)+(-2x^2+4x-8) = -x^2+6x-7-2x^2+4x-8$

$= -x^2-2x^2+6x+4x-7-8 = -3x^2+10x-15$

100. $(12x^2+4x-8)+(-x^2-6x+5) = 12x^2+4x-8-x^2-6x+5$

$= 12x^2-x^2+4x-6x-8+5 = 11x^2-2x-3$

101. $(2x-4.3)-(x+2.4) = 2x-4.3-x-2.4 = 2x-x-4.3-2.4 = x-6.7$

102. $(-4x+8)-(-2x+6) = -4x+8+2x-6 = -4x+2x+8-6 = -2x+2$

103. $\left(9x^2 - \frac{3}{4}x\right) - \left(\frac{1}{2}x - 4\right) = 9x^2 - \frac{3}{4}x - \frac{1}{2}x + 4 = 9x^2 - \frac{5}{4}x + 4$

104. $(6x^2-6x+1)-(12x+5) = 6x^2-6x+1-12x-5 = 6x^2-6x-12x+1-5 = 6x^2-18x-4$

105. $(-2x^2+8x-7)-(3x^2+12) = -2x^2+8x-7-3x^2-12 = -2x^2-3x^2+8x-7-12$

$= -5x^2+8x-19$

106. $(x^2+7x-3)-(x^2+3x-5) = x^2+7x-3-x^2-3x+5 = x^2-x^2+7x-3x-3+5 = 4x+2$

107. $x(2x-4) = x(2x)+x(-4) = 2x^2-4x$

108. $4.5x(x^2-3x) = 4.5x(x^2)+4.5x(-3x) = 4.5x^3-13.5x^2$

109. $3x(2x^2-4x+7)=3x(2x^2)+3x(-4x)+3x(7)=6x^3-12x^2+21x$

110. $-x(3x^2-6x-1)=(-x)(3x^2)+(-x)(-6x)+(-x)(-1)=-3x^3+6x^2+x$

111. $-4x(-6x^2+4x-2)=(-4x)(-6x^2)+(-4x)(4x)+(-4x)(-2)=24x^3-16x^2+8x$

112. $(x+4)(x+5)=x\cdot x+x\cdot 5+4\cdot x+4\cdot 5=x^2+5x+4x+20=x^2+9x+20$

113. $(2x+4)(x-3)=2x\cdot x+2x(-3)+4\cdot x+4(-3)=2x^2-6x+4x-12=2x^2-2x-12$

114. $(4x+6)^2=(4x)^2+2(4x)(6)+(6)^2=16x^2+48x+36$

115. $(6-2x)(2+3x)=6\cdot 2+6\cdot 3x+(-2x)(2)+(-2x)(3x)=12+18x-4x-6x^2$

$=12+14x-6x^2=-6x^2+14x+12$

116. $(x+4)(x-4)=(x)^2-(4)^2=x^2-16$

117. $(3x+1)(x^2+2x+4)=3x(x^2+2x+4)+1(x^2+2x+4)=3x^3+6x^2+12x+x^2+2x+4$

$=3x^3+7x^2+14x+4$

118. $(x-1)(3x^2+4x-6)=x(3x^2+4x-6)-1(3x^2+4x-6)=3x^3+4x^2-6x-3x^2-4x+6$

$=3x^3+x^2-10x+6$

119. $(-5x+2)(-2x^2+3x-6)=-5x(-2x^2+3x-6)+2(-2x^2+3x-6)$

$=10x^3-15x^2+30x-4x^2+6x-12=10x^3-19x^2+36x-12$

120. $\dfrac{2x+4}{2}=\dfrac{2x}{2}+\dfrac{4}{2}=x+2$

121. $\frac{4x-8}{4}=\frac{4x}{4}-\frac{8}{4}=x-2$

122. $\frac{8x^2+4x}{x}=\frac{8x^2}{x}+\frac{4x}{x}=8x+4$

123. $\frac{6x^2+9x-4}{3}=\frac{6x^2}{3}+\frac{9x}{3}-\frac{4}{3}=2x^2+3x-\frac{4}{3}$

124. $\frac{8x^2+6x-4}{x}=\frac{8x^2}{x}+\frac{6x}{x}-\frac{4}{x}=8x+6-\frac{4}{x}$

125. $\frac{8x^5-4x^4+3x^2-2}{2x}=\frac{8x^5}{2x}-\frac{4x^4}{2x}+\frac{3x^2}{2x}-\frac{2}{2x}=4x^4-2x^3+\frac{3}{2}x-\frac{1}{x}$

126. $\frac{16x-4}{-2}=\frac{(-1)(16x-4)}{(-1)(-2)}=\frac{-16x+4}{2}=\frac{-16x}{2}+\frac{4}{2}=-8x+2$

127. $\frac{12+6x^2+3x}{-3x}=\frac{(-1)(6x^2+3x+12)}{(-1)(-3x)}=\frac{-6x^2-3x-12}{3x}=-\frac{6x^2}{3x}-\frac{3x}{3x}-\frac{12}{3x}=-2x-1-\frac{4}{x}$

128. $\frac{5x^3+10x+2}{2x^2}=\frac{5x^3}{2x^2}+\frac{10x}{2x^2}+\frac{2}{2x^2}=\frac{5}{2}x+\frac{5}{x}+\frac{1}{x^2}$

129.

$$\begin{array}{r} x+4 \\ x-3\overline{)x^2+\ x-12} \\ \underline{-x^2+3x} \\ 4x-12 \\ \underline{-4x+12} \\ 0 \end{array}$$

Thus, $\frac{x^2+x-12}{x-3}=x+4$

130.

$$\begin{array}{r} 2x-3 \\ 3x-1\overline{)6x^2-11x+3} \\ \underline{-6x^2+2x} \\ -9x+3 \\ \underline{+9x-3} \\ 0 \end{array}$$

Thus, $\dfrac{6x^2-11x+3}{3x-1}=2x-3$

131.

$$\begin{array}{r} 5x-2 \\ x+6\overline{)5x^2+28x-10} \\ \underline{-5x^2-30x} \\ -2x-10 \\ \underline{+2x+12} \\ 2 \end{array}$$

Thus, $\dfrac{5x^2+28x-10}{x+6}=5x-2+\dfrac{2}{x+6}$

132.

$$\begin{array}{r} 2x^2+3x-4 \\ 2x+3\overline{)4x^3+12x^2+x-12} \\ \underline{-4x^3-6x^2} \\ 6x^2+x \\ \underline{-6x^2-9x} \\ -8x-12 \\ \underline{+8x+12} \\ 0 \end{array}$$

Thus, $\dfrac{4x^3+12x^2+x-12}{2x+3}=2x^2+3x-4$

133.

$$\begin{array}{r} 2x^2 + x - 2 \\ 2x-1\overline{)4x^3 \qquad\quad - 5x + 4} \\ \underline{-4x^3 + 2x^2} \qquad\qquad \\ 2x^2 - 5x \qquad \\ \underline{-2x^2 + x} \qquad \\ -4x + 4 \\ \underline{+4x - 2} \\ 2 \end{array}$$

Thus, $\dfrac{4x^3 - 5x + 4}{2x - 1} = 2x^2 + x - 2 + \dfrac{2}{2x - 1}$

134. $\text{Time} = \dfrac{\text{distance}}{\text{rate}} = \dfrac{280}{70} = 4 \text{ hr}$

135. $\text{Rate} = \dfrac{\text{distance}}{\text{time}} = \dfrac{3500}{6.5} = 538.46 \text{ mph}$

136. Let t be the time it takes for the joggers to be 4 kilometers apart.

Jogger	Rate	Time	Distance
Marty	8	t	$8t$
Nick	6	t	$6t$

$8t - 6t = 4$

$2t = 4$

$t = 2$

So, it takes the joggers 2 hours to be 4 kilometers apart.

137. Let x be the amount invested at 8%. Then 12,000 - x is the amount invested at $7\frac{1}{4}\%$.

Principal	Rate	Time	Interest
x	8%	1	$0.08x$
12,000 - x	$7\frac{1}{4}\%$	1	0.0725(12,000 - x)

$0.08x + 0.0725(12{,}000 - x) = 900$

$0.08x + 870 - 0.0725x = 900$

$0.0075x = 30$

$x = 4000$

Kathy should invest \$4000 at 8% and \$8000 at $7\frac{1}{4}\%$.

138. Let x be the number of liters of the 10% solution. Then $2 - x$ is the number of liters of the 5% acid solution.

Solution	Strength	Liters	Amount
10%	0.10	x	$0.10x$
5%	0.05	$2 - x$	$0.05(2 - x)$
Mixture	0.08	2	0.16

$0.10x + 0.05(2 - x) = 0.16$

$0.10x + 0.10 - 0.05x = 0.16$

$0.05x = 0.06$

$x = 1.2$

The chemist should mix 1.2 liters of 10% solution with 0.8 liters of 5% solution.

139. $\text{Speed} = \dfrac{\text{distance}}{\text{time}} = \dfrac{26}{4} = 6.5 \text{ mph}$

140. Let t be the amount of time it takes for the trains to be 440 miles apart.

Train	Rate	Time	Distance
First	50	t	$50t$
Second	60	t	$60t$

$50t + 60t = 440$

$110t = 440$

$t = 4$

After 4 hours, the two trains will be 440 miles apart.

141. Let x be the amount of \$3.50 per pound hamburger. Then 80 - x is the amount of \$4.10 per pound hamburger.

Hamburger	Price	Amount	Total
\$3.50	3.50	x	$3.50x$
\$4.10	4.10	$80 - x$	$4.10(80 - x)$
Mixture	3.65	80	292

$3.50x + 4.10(80 - x) = 292$

$3.50x + 328 - 4.10x = 292$

$-0.60x = -36$

$x = 60$

The butcher should mix 60 lbs of \$3.50 per pound hamburger with 20 lbs of \$4.10 per pound hamburger.

142. Let x be the number of 32 cent stamps. Then 40 - x is the number of 22 cent stamps.

Stamp	Cost	Number	Total
32 cent	0.32	x	$0.32x$
22 cent	0.22	$40 - x$	$0.22(40 - x)$

$0.32x + 0.22(40 - x) = 12.40$

$0.32x + 8.80 - 0.22x = 12.40$

$0.10x = 3.60$

$x = 36$

Joan bought thirty-six 32 cent stamps and four 22 cent stamps.

143. Let r be the speed of the older brother. Then $r + 5$ is the speed of the younger brother.

Brother	Speed	Time	Distance
Older	r	2	$2r$
Younger	$r + 5$	2	$2(r + 5)$

$2r + 2(r + 5) = 230$

$2r + 2r + 10 = 230$

$4r = 220$

$r = 55$

The speed of the older brother is 55 mph and the speed of the younger brother is 60 mph.

144. Let x be the number of liters of the 30% solution.

Solution	Strength	Liters	Amount
30%	0.30	x	$0.30x$
12%	0.12	2	0.24
Mixture	0.15	$x+2$	$0.15(x+2)$

$0.30x + 0.24 = 0.15(x+2)$

$0.30x + 0.24 = 0.15x + 0.30$

$0.15x = 0.06$

$x = 0.4$

Hence, 0.4 liters of the 30% solution should be mixed with 2 liters of the 12% solution.

Chapter 4 Practice Test

1. $2x^2 \cdot 3x^4 = 2 \cdot 3x^{2+4} = 6x^6$

2. $(3x^2)^3 = 3^3 x^{2 \cdot 3} = 27x^6$

3. $\dfrac{8x^4}{2x} = \dfrac{8}{2}x^{4-1} = 4x^3$

4. $\left(\dfrac{3x^2y}{6xy^3}\right)^3 = \left(\dfrac{3}{6} \cdot \dfrac{x^2}{x} \cdot \dfrac{y}{y^3}\right)^3 = \left(\dfrac{1}{2} \cdot x \cdot \dfrac{1}{y^2}\right)^3 = \left(\dfrac{x}{2y^2}\right)^3 = \dfrac{x^3}{2^3 y^{2 \cdot 3}} = \dfrac{x^3}{8y^6}$

5. $(2x^3y^{-2})^{-2} = 2^{-2}x^{3(-2)}y^{(-2)(-2)} = 2^{-2}x^{-6}y^4 = \dfrac{y^4}{2^2x^6} = \dfrac{y^4}{4x^6}$

6. $\dfrac{2x^4y^{-2}}{10x^7y^4} = \dfrac{2}{10} \cdot \dfrac{x^4}{x^7} \cdot \dfrac{y^{-2}}{y^4} = \dfrac{1}{5} \cdot \dfrac{1}{x^3} \cdot \dfrac{1}{y^6} = \dfrac{1}{5x^3y^6}$

7. Trinomial

8. Monomial

9. Not a polynomial

10. $6x^3 - 2x^2 + 5x - 5$, third degree.

11. $(2x+4)+(3x^2-5x-3) = 2x+4+3x^2-5x-3 = 3x^2+2x-5x+4-3 = 3x^2-3x+1$

12. $(x^2-4x+7)-(3x^2-8x+7) = x^2-4x+7-3x^2+8x-7$

$= x^2-3x^2-4x+8x+7-7 = -2x^2+4x$

13. $(4x^2-5)-(x^2+x-8) = 4x^2-5-x^2-x+8 = 4x^2-x^2-x-5+8 = 3x^2-x+3$

14. $3x(4x^2-2x+5) = 3x(4x^2)+3x(-2x)+3x(5) = 12x^3-6x^2+15x$

15. $(4x+7)(2x-3) = (4x)(2x)+(4x)(-3)+7(2x)+7(-3) = 8x^2-12x+14x-21$

$= 8x^2+2x-21$

16. $(6-4x)(5+3x) = 6\cdot 5+6(3x)+(-4x)\cdot 5+(-4x)(3x) = 30+18x-20x-12x^2$

$= -12x^2-2x+30$

17. $(2x-4)(3x^2+4x-6) = 2x(3x^2+4x-6)-4(3x^2+4x-6)$

$= 6x^3+8x^2-12x-12x^2-16x+24 = 6x^3-4x^2-28x+24$

18. $\dfrac{16x^2+8x-4}{4} = \dfrac{16x^2}{4}+\dfrac{8x}{4}-\dfrac{4}{4} = 4x^2+2x-1$

19. $\dfrac{3x^2-6x+5}{-3x} = \dfrac{(-1)(3x^2-6x+5)}{(-1)(-3x)} = \dfrac{-3x^2+6x-5}{3x} = -\dfrac{3x^2}{3x}+\dfrac{6x}{3x}-\dfrac{5}{3x} = -x+2-\dfrac{5}{3x}$

20.

$$\begin{array}{r} 4x+5 \\ 2x-3\overline{)8x^2-2x-15} \\ \underline{-8x^2+12x} \\ 10x-15 \\ \underline{-10x+15} \\ 0 \end{array}$$

Thus, $\dfrac{8x^2-2x-15}{2x-3} = 4x+5$

21. Time $= \dfrac{\text{amount}}{\text{rate}} = \dfrac{40}{0.7} = 57.14$. It will take Madison 57.14 hrs.

22. Let x be the speed of train B.

Train	Rate	Time	Distance
A	60	4	240
B	x	3	$3x$

$3x = 240$

$x = 80$

Train B is moving at a rate of 80 mph.

23. Let x be the number of liters of the 20% salt solution.

Solution	Strength	Liters	Amount
20%	0.20	x	$0.20x$
40%	0.40	60	24
Mixture	0.35	$60 + x$	$0.35(x+60)$

$0.20x + 24 = 0.35(x+60)$

$0.20x + 24 = 0.35x + 21$

$-0.15x = -3$

$x = 20$

20 liters of the 20% salt solution must be added.

24. $(42{,}000)(30{,}000) = (4.2\times10^4)(3\times10^4) = (4.2\times3)(10^4\times10^4) = 12.6\times10^8 = 1.26\times10^9$

25. $\dfrac{0.0008}{4000}=\dfrac{8\times10^{-4}}{4\times10^{3}}=\dfrac{8}{4}\cdot\dfrac{10^{-4}}{10^{3}}=2\times10^{-7}$

Chapter 4 Cumulative Review Test

1. $16\div(4-6)\cdot5=16\div(-2)\cdot5=-40$

2. $2x+5=3(x-5)$

$2x+5=3x-15$

$-x=-20$

$x=20$

3. $3(x-2)-(x+4)=2x-10$

$3x-6-x-4=2x-10$

$2x-10=2x-10$

all real numbers

4. $2x-14>5x+1$

$-3x>15$

$x<-5$

-5

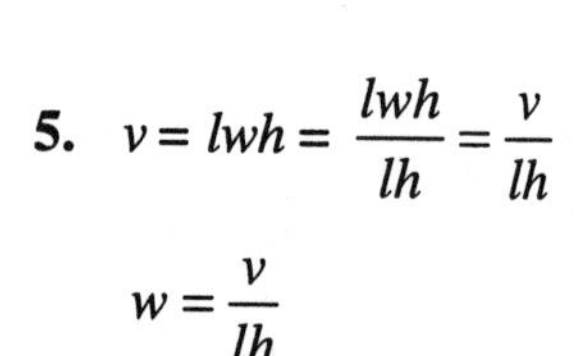

5. $v=lwh=\dfrac{lwh}{lh}=\dfrac{v}{lh}$

$w=\dfrac{v}{lh}$

6. $4x-3y=6$

$4x-6=3y$

$y=\dfrac{4x-6}{3}$ or $y=\dfrac{4}{3}x-2$

7. $(3x^4)(2x^5)=3\cdot2x^{4+5}=6x^9$

8. $(3x^2y^4)^3(5x^2y)$: $(3x^2y^4)^3 = 3^3x^{2\cdot3}y^{4\cdot3} = 27x^6y^{12}$

So, $(3x^2y^4)^3(5x^2y) = (27x^6y^{12})(5x^2y) = 27\cdot5x^{6+2}y^{12+1} = 135x^8y^{13}$

9. $3x^2 - 2x - 5$, second degree

10. $(2x^2 - 9x - 7) - (6x^2 - 3x + 4) = 2x^2 - 9x - 7 - 6x^2 + 3x - 4 = 2x^2 - 6x^2 - 9x + 3x - 7 - 4$

$= -4x^2 - 6x - 11$

11. $(2x^2 + 4x - 3) + (6x^2 - 7x + 12) = 2x^2 + 4x - 3 + 6x^2 - 7x + 12 = 2x^2 + 6x^2 + 4x - 7x - 3 + 12$

$= 8x^2 - 3x + 9$

12. $(4x^2 - 5x - 2) - (3x^2 - 2x + 5) = 4x^2 - 5x - 2 - 3x^2 + 2x - 5 = 4x^2 - 3x^2 - 5x + 2x - 2 - 5$

$= x^2 - 3x - 7$

13. $(2x - 3)(3x - 5) = (2x)(3x) + (2x)(-5) + (-3)(3x) + (-3)(-5) = 6x^2 - 10x - 9x + 15$

$= 6x^2 - 19x + 15$

14. $(2x^2 + 4x + 8)(x - 5) = 2x^2(x - 5) + 4x(x - 5) + 8(x - 5) = 2x^3 - 10x^2 + 4x^2 - 20x + 8x - 40$

$= 2x^3 - 6x^2 - 12x - 40$

15. $\dfrac{9x^2 - 6x + 8}{3x} = \dfrac{9x^2}{3x} - \dfrac{6x}{3x} + \dfrac{8}{3x} = 3x - 2 + \dfrac{8}{3x}$

16.

$$\begin{array}{r}
2x - 3 \\
x + 2\overline{)2x^2 + x - 6} \\
\underline{-2x^2 - 4x} \\
-3x - 6 \\
\underline{+3x + 6} \\
0
\end{array}$$

Thus, $\dfrac{2x^2 + x - 6}{x + 2} = 2x - 3$

17. $\frac{x}{8} = \frac{1.25}{3}$

$x = \frac{8(1.25)}{3} = 3.33$; Eight cans of soup cost \$3.33.

18. $11 + 2x = 19$

$2x = 8$

$x = 4$

19. Let x = width of rectangle. Then the length = $2x + 4$

Perimeter = 2 × length of rectangle + 2 × width of of rectangle

So $2(2x + 4) + 2x = 26$

$4x + 8 + 2x = 26$

$6x = 18$

$x = 3$

Length = $2x + 4 = 2(3) + 4 = 10$

Therefore, the width is 3 ft and the length is ft

20. Let t be the number of hours.

Runner	Rate	Time	Distance
First	6	t	$6t$
Second	8	t	$8t$

$6t + 8t = 28$

$14t = 28$

$t = 2$

The runners will be 28 miles apart after 2 hours.

CHAPTER 5

Exercise Set 5.1

1. $36 = 6 \cdot 6 = 2 \cdot 3 \cdot 2 \cdot 3 = 2^2 \cdot 3^2$

3. $90 = 9 \cdot 10 = 3 \cdot 3 \cdot 2 \cdot 5 = 2 \cdot 3^2 \cdot 5$

5. $200 = 20 \cdot 10 = 2 \cdot 10 \cdot 2 \cdot 5$
$= 2 \cdot 2 \cdot 5 \cdot 2 \cdot 5 = 2^3 \cdot 5^2$

7. $20 = 2^2 \cdot 5$, $24 = 2^3 \cdot 3$, so the greatest common factor is 2^2 or 4.

9. $60 = 2^2 \cdot 3 \cdot 5$, $84 = 2^2 \cdot 3 \cdot 7$, so the greatest common factor is $2^2 \cdot 3$ or 12.

11. $72 = 2^3 \cdot 3^2$, $90 = 2 \cdot 3^2 \cdot 5$, so the greatest common factor is $2 \cdot 3^2$ or 18.

13. The greatest common factor is x.

15. The greatest common factor is $3x$.

17. The greatest common factor is 1.

19. The greatest common factor is xy.

21. The greatest common factor is x^3y^5.

23. The greatest common factor is 5.

25. The greatest common factor is x^2y^2.

27. The greatest common factor is x.

29. The greatest common factor is $x + 3$.

31. The greatest common factor is $2x - 3$.

33. The greatest common factor is $3x - 4$.

35. The greatest common factor is $x + 4$.

37. The greatest common factor is 3.
$3x + 6 = 3 \cdot x + 3 \cdot 2 = 3(x + 2)$

39. The greatest common factor is 5.
$15x - 5 = 5 \cdot 3x - 5 \cdot 1 = 5(3x - 1)$

41. $13x + 5$ cannot be factored.

43. The greatest common factor is $3x$.
$9x^2 - 12x = 3x \cdot 3x - 3x \cdot 4 = 3x(3x - 4)$

45. The greatest common factor is $2p$.
$26p^2 - 8p = 2p \cdot 13p - 2p \cdot 4$
$= 2p(13p - 4)$

47. The greatest common factor is $2x$.
$4x^3 - 6x = 2x \cdot 2x^2 - 2x \cdot 3 = 2x(2x^2 - 3)$

49. The greatest common factor is $12x^8$.
$36x^{12} - 24x^8 = 12x^8 \cdot 3x^4 - 12x^8 \cdot 2$
$= 12x^8(3x^4 - 2)$

51. The greatest common factor is $3y^3$.
$24y^{15} - 9y^3 = 3y^3 \cdot 8y^{12} - 3y^3 \cdot 3$
$= 3y^3(8y^{12} - 3)$

53. The greatest common factor is x.

$x + 3xy^2 = x \cdot 1 + x \cdot 3y^2 = x(1 + 3y^2)$

55. $6x + 5y$ cannot be factored.

57. The greatest common factor is $4xy$.

$16xy^2z + 4x^3y = 4xy \cdot 4yz + 4xy \cdot x^2 =$
$4xy(4yz + x^2)$

59. The greatest common factor is $2xy^2$.

$34x^2y^2 + 16xy^4 = 2xy^2 \cdot 17x + 2xy^2 \cdot 8y^2$

$= 2xy^2(17x + 8y^2)$

61. The greatest common factor is $36xy^2z$.

$36xy^2z^3 + 36x^3y^2z$

$= 36xy^2z \cdot z^2 + 36xy^2z \cdot x^2$

$= 36xy^2z(z^2 + x^2)$

63. The greatest common factor is y^3z^5.

$14y^3z^5 - 9xy^3z^5 = y^3z^5 \cdot 14 - y^3z^5 \cdot 9x =$
$y^3z^5(14 - 9x)$

65. The greatest common factor is 3.

$3x^2 + 6x + 9 = 3 \cdot x^2 + 3 \cdot 2x + 3 \cdot 3$

$= 3(x^2 + 2x + 3)$

67. The greatest common factor is 3.

$9x^2 + 18x + 3 = 3 \cdot 3x^2 + 3 \cdot 6x + 3 \cdot 1$

$= 3(3x^2 + 6x + 1)$

69. The greatest common factor is $4x$.

$4x^3 - 8x^2 + 12x$

$= 4x \cdot x^2 - 4x \cdot 2x + 4x \cdot 3$

$= 4x(x^2 - 2x + 3)$

71. $35x^2 - 16x + 10$ cannot be factored.

73. The greatest common factor is 3.

$15p^2 - 6p + 9 = 3 \cdot 5p^2 - 3 \cdot 2p + 3 \cdot 3$

$= 3(5p^2 - 2p + 3)$

75. The greatest common factor is $4x^3$.

$24x^6 + 8x^4 - 4x^3$

$= 4x^3 \cdot 6x^3 + 4x^3 \cdot 2x - 4x^3 \cdot 1$

$= 4x^3(6x^3 + 2x - 1)$

77. The greatest common factor is xy.

$8x^2y + 12xy^2 + 9xy$

$= xy \cdot 8x + xy \cdot 12y + xy \cdot 9$

$= xy(8x + 12y + 9)$

79. The greatest common factor is $x + 4$.

$x(x + 4) + 3(x + 4) = (x + 3)(x + 4)$

81. The greatest common factor is $4x - 3$.

$7x(4x - 3) - 4(4x - 3) = (7x - 4)(4x - 3)$

83. The greatest common factor is $2x + 1$.

$4x(2x + 1) + 1(2x + 1)$

$= (4x + 1)(2x + 1)$

85. The greatest common factor is $2x + 1$.

$4x(2x + 1) + 2x + 1$

$= 4x(2x + 1) + 1(2x + 1)$

$= (4x + 1)(2x + 1)$

87. A factored expression is an expression written as a product of factors.

89. Answers may vary.

Cumulative Review Problems: 5.1

91. $3x-(x-6)+4(3-x)$

$=3x-x+6+12-4x=-2x+18$

92. $2(x+3)-x=5x+2$

$2x+6-x=5x+2$

$x+6=5x+2$

$-4x+6=2$

$-4x=-4$

$x=1$

93. $1000=500(1+r\cdot 2)$

$1000=500+1000r$

$500=1000r$

$r=\frac{500}{1000}=\frac{1}{2}$

94. $A=\frac{1}{2}bh$

$2A=bh$

$\frac{2A}{b}=h$ or $h=\frac{2A}{b}$

Group Activity/Challenge Problems

1. The greatest common factor is $2(x-3)$.

$4x^2(x-3)^3-6x(x-3)^2+4(x-3)$

$=2(x-3)\cdot 2x^2(x-3)^2+2(x-3)$

$\quad\cdot 3x(x-3)+2(x-3)\cdot 2$

$=2(x-3)[2x^2(x-3)^2+3x(x-3)+2]$

2. The greatest common factor is $2x^2(2x-7)$.

$6x^5(2x+7)+4x^3(2x+7)-2x^2(2x+7)$

$=2x^2(2x-7)\cdot 3x^3+2x^2(2x-7)\cdot 2x$

$\quad+2x^2(2x-7)\cdot(-1)$

$=2x^2(2x-7)(3x^3+2x-1)$

3. First factor $x^{1/3}$ from terms.

$x^{7/3}+5x^{4/3}+6x^{1/3}=x^{1/3}(x^2+5x+6)$

$=x^{1/3}(x+2)(x+3)$

4. The greatest common factor is $5x^{-1/2}$.

$15x^{1/2}+5x^{-1/2}=5x^{-1/2}\cdot 3x+5x^{-1/2}\cdot 1$

$=5x^{-1/2}(3x+1)$

5. $x^2+2x+3x+6=x(x+2)+3(x+2)$

$=(x+3)(x+2)$

6. **(a)** $1+2-3+4+5-6+7+8-9+10$
$\quad+11-12+13+14-15$

$=0+3+6+9+12$

$=3\cdot 0+3\cdot 1+3\cdot 2+3\cdot 3+3\cdot 4$

(b) $3\cdot 0+3\cdot 1+3\cdot 2+3\cdot 3+3\cdot 4$

$=3(0+1+2+3+4)$

(c) $3(0+1+2+3+4)=3(10)=30$

(d) $1+2-3+\ldots+31+32-33$

$=0+3+6+\ldots+30$

$=3\cdot 0+3\cdot 1+3\cdot 2+\ldots+3\cdot 10$

$=3(0+1+2+\ldots+10)$

$=3(55)=165$

Exercise Set 5.2

1. $x^2+4x+3x+12=x(x+4)+3(x+4)$
$=(x+3)(x+4)$

3. $x^2+2x+4x+8=x(x+2)+4(x+2)$
$=(x+4)(x+2)$

5. $x^2+2x+5x+10=x(x+2)+5(x+2)$
$=(x+5)(x+2)$

7. $x^2+3x-5x-15=x(x+3)-5(x+3)$
$=(x-5)(x+3)$

9. $4x^2+6x-6x-9$
$=2x(2x+3)-3(2x+3)$
$=(2x-3)(2x+3)$

11. $3x^2+9x+x+3=3x(x+3)+1(x+3)$
$=(3x+1)(x+3)$

13. $4x^2-2x-2x+1$
$=2x(2x-1)-1(2x-1)$
$=(2x-1)(2x-1)=(2x-1)^2$

15. $8x^2+32x+x+4$
$=8x(x+4)+1(x+4)=(8x+1)(x+4)$

17. $3x^2-2x+3x-2=x(3x-2)+1(3x-2)$
$=(x+1)(3x-2)$

19. $2x^2-4x-3x+6=2x(x-2)-3(x-2)$
$=(2x-3)(x-2)$

21. $15x^2-9x+25x-15$
$=3x(5x-3)+5(5x-3)$
$=(3x+5)(5x-3)$

23. $x^2+2xy-3xy-6y^2$
$=x(x+2y)-3y(x+2y)$
$=(x-3y)(x+2y)$

25. $6x^2-9xy+2xy-3y^2$
$=3x(2x-3y)+y(2x-3y)$
$=(3x+y)(2x-3y)$

27. $10x^2-12xy-25xy+30y^2$
$=2x(5x-6y)-5y(5x-6y)$
$=(2x-5y)(5x-6y)$

29. $x^2+bx+ax+ab=x(x+b)+a(x+b)$
$=(x+a)(x+b)$

31. $xy+4x-2y-8$
$=x(y+4)-2(y+4)=(x-2)(y+4)$

33. $a^2+2a+ab+2b=a(a+2)+b(a+2)$
$=(a+b)(a+2)$

35. $xy-x+5y-5=x(y-1)+5(y-1)$
$=(x+5)(y-1)$

37. $12+8y-3x-2xy$

$=4(3+2y)-x(3+2y)=(4-x)(3+2y)$

39. $a^3+2a^2+a+2=a^2(a+2)+1(a+2)$

$=(a^2+1)(a+2)$

41. $x^3+4x^2-3x-12$

$=x^2(x+4)-3(x+4)=(x^2-3)(x+4)$

43. $2x^2-12x+8x-48$

$=2\cdot x^2-2\cdot 6x+2\cdot 4x-2\cdot 24$

$=2(x^2-6x+4x-24)$

$=2[x(x-6)+4(x-6)]$

$=2(x+4)(x-6)$

45. $4x^2+8x+8x+16$

$=4\cdot x^2+4\cdot 2x+4\cdot 2x+4\cdot 4$

$=4(x^2+2x+2x+4)$

$=4[x(x+2)+2(x+2)]$

$=4(x+2)(x+2)=4(x+2)^2$

47. $6x^3+9x^2-2x^2-3x$

$=x\cdot 6x^2+x\cdot 9x-x\cdot 2x-x\cdot 3$

$=x(6x^2+9x-2x-3)$

$=x[3x(2x+3)-1(2x+3)]$

$=x(3x-1)(2x+3)$

49. $2x^2-4xy+8xy-16y^2$

$=2\cdot x^2-2\cdot 2xy+2\cdot 4xy-2\cdot 8y^2$

$=2(x^2-2xy+4xy-8y^2)$

$=2[x(x-2y)+4y(x-2y)]$

$=2(x+4y)(x-2y)$

51. $3x+2y+6+xy=3x+xy+6+2y$

$=x(3+y)+2(3+y)=(x+2)(3+y)$

53. $6x+5y+xy+30=6x+xy+30+5y$

$=x(6+y)+5(6+y)=(x+5)(6+y)$

55. $ax+by+ay+bx=ax+bx+ay+by$

$=x(a+b)+y(a+b)=(x+y)(a+b)$

57. $cd-12-4d+3c=cd+3c-4d-12$

$=c(d+3)-4(d+3)=(c-4)(d+3)$

59. $ac-bd-ad+bc=ac-ad+bc-bd$

$=a(c-d)+b(c-d)=(a+b)(c-d)$

61. The first step in any factoring by grouping problem is to factor out a common factor, if one exists.

63. If you multiply $(x-2)(x+4)$ by the FOIL method, you get the polynomial $x^2+4x-2x-8$.

Cumulative Review Exercises: 5.2

65. time $= \frac{\text{amount}}{\text{rate}} = \frac{25}{3.5}$, which is approximately equal to 7.14. The approximate age of the willow tree is 7.14 years.

66. Since one-way fare is \$1.25, a round trip costs \$2.50. The number of days Steve would have to travel to and from work in a month to make the monthly pass worthwhile is at least $\frac{52}{2.50}$ days. So, Steve would have to travel at least 21 days to and from work.

67. $\frac{15x^3 - 6x^2 - 9x + 5}{3x}$

$= \frac{15x^3}{3x} - \frac{6x^2}{3x} - \frac{9x}{3x} + \frac{5}{3x}$

$= 5x^2 - 2x - 3 + \frac{5}{3x}$

68.

$$\begin{array}{r|l} & x+3 \\ \hline x-3 & x^2 \qquad -9 \\ & \underline{-x^2 + 3x} \\ & \qquad 3x - 9 \\ & \qquad \underline{-3x + 9} \\ & \qquad\qquad 0 \end{array}$$

Thus, $\frac{x^2 - 9}{x - 3} = x + 3$.

Group Activity/Challenge Problems

1. $3x^5 - 15x^3 + 2x^3 - 10x$

$= x(3x^4 - 15x^2 + 2x^2 - 10)$

$= x[3x^2(x^2 - 5) + 2(x^2 - 5)]$

$= x(3x^2 + 2)(x^2 - 5)$

3. $18a^2 + 3ax^2 - 6ax - x^3$

$= 3a(6a + x^2) - x(6a + x^2)$

$= (3a - x)(6a + x^2)$

5. $3x^2 + 10x + 8$, $10x = 6x + 4x$

(a) $3x^2 + 6x + 4x + 8$

(b) $2x(x + 2) + 4(x + 2) = (3x + 4)(x + 2)$

7. $2x^2 - 11x + 15$, $-11x = -6x - 5x$

(a) $2x^2 - 6x - 5x + 15$

(b) $2x(x - 3) - 5(x - 3) = (2x - 5)(x - 3)$

9. $4x^2 - 17x - 15$, $-17x = -20x + 3x$

(a) $4x^2 - 20x + 3x - 15$

(b) $4x(x - 5) + 3(x - 5) = (4x + 3)(x - 5)$

Exercise Set 5.3

1. $x^2 + 3x + 2 = (x + 2)(x + 1)$

3. $x^2 + 6x + 8 = (x + 4)(x + 2)$

5. $x^2 + 7x + 12 = (x + 4)(x + 3)$

7. $x^2 - 7x + 9$ cannot be factored

9. $y^2 - 16y + 15 = (y - 1)(y - 15)$

11. $x^2 + x - 6 = (x + 3)(x - 2)$

13. $r^2 + 2r - 15 = (r + 5)(r - 3)$

15. $b^2 - 11b + 18 = (b-9)(b-2)$

17. $x^2 - 8x - 15$ cannot be factored

19. $a^2 + 12a + 11 = (a+1)(a+11)$

21. $x^2 + 13x - 30 = (x+15)(x-2)$

23. $x^2 + 4x + 4 = (x+2)(x+2) = (x+2)^2$

25. $k^2 + 6k + 9 = (k+3)(k+3) = (k+3)^2$

27. $x^2 + 10x + 25 = (x+5)(x+5) = (x+5)^2$

29. $w^2 - 18w + 45 = (w-15)(w-3)$

31. $x^2 + 22x - 48 = (x+24)(x-2)$

33. $x^2 - x - 20 = (x-5)(x+4)$

35. $y^2 - 9y + 14 = (y-7)(y-2)$

37. $x^2 + 12x - 64 = (x+16)(x-4)$

39. $x^2 - 14x + 24 = (x-12)(x-2)$

41. $x^2 - 2x - 80 = (x-10)(x+8)$

43. $x^2 - 17x + 60 = (x-5)(x-12)$

45. $x^2 + 30x + 56 = (x+2)(x+28)$

47. $x^2 - 4xy + 4y^2 = (x-2y)(x-2y)$
$= (x-2y)^2$

49. $x^2 + 8xy + 15y^2 = (x+3y)(x+5y)$

51. $2x^2 - 12x + 10 = 2(x^2 - 6x + 5)$
$= 2(x-5)(x-1)$

53. $5x^2 + 20x + 15 = 5(x^2 + 4x + 3)$
$= 5(x+1)(x+3)$

55. $2x^2 - 14x + 24 = 2(x^2 - 7x + 12)$
$= 2(x-4)(x-3)$

57. $x^3 - 3x^2 - 18x = x(x^2 - 3x - 18)$
$= x(x-6)(x+3)$

59. $2x^3 + 6x^2 - 56x = 2x(x^2 + 3x - 28)$
$= 2x(x+7)(x-4)$

61. $x^3 + 4x^2 + 4x = x(x^2 + 4x + 4)$
$= x(x+2)(x+2) = x(x+2)^2$

63. Since c is positive, both signs will be the same. Since b is positive, both signs will be positive.

65. Since c is negative, one sign will be positive, the other will be negative.

67. Since c is positive, both signs will be the same. Since b is negative, both signs will be negative.

69. The trinomial $x^2 - 11x + 24$ is obtained by using the FOIL method.

71. The trinomial $2x^2 - 8xy - 10y^2$ is obtained by multiplying all the factors and combining like terms.

73. The trinomial factoring problem may be checked by multiplying the factors using the FOIL method.

Cumulative Review Exercises: 5.3

75. $2(2x-3) = 2x+8$

$4x - 6 = 2x + 8$

$2x - 6 = 8$

$2x = 14$

$x = 7$

76. $(2x^2 + 5x - 6)(x-2) = 2x^2(x-2) + 5x(x-2) - 6(x-2) = 2x^3 - 4x^2 + 5x^2 - 10x - 6x + 12$

$= 2x^3 + x^2 - 16x + 12$

77.

$$\begin{array}{r} 3x + 2 \\ x-4\overline{)3x^2 - 10x - 10} \\ \underline{-3x^2 + 12x} \\ 2x - 10 \\ \underline{-2x + 8} \\ -2 \end{array}$$

Thus $\dfrac{3x^2 - 10x - 10}{x-4} = 3x + 2 - \dfrac{2}{x-4}$

78. Let x be the percent of acid in the mixture.

Solution	Strength	Liters	Amount
18%	0.18	4	0.72
26%	0.26	1	0.26
Mixture	$\frac{x}{100}$	5	$\frac{5x}{100}$

$$0.72 + 0.26 = \frac{5x}{100}$$

$$0.98 = \frac{x}{20}$$

$$19.6 = x$$

The mixture is a 19.6% acid solution.

79. $3x^2 + 5x - 6x - 10 = x(3x+5) - 2(3x+5) = (x-2)(3x+5)$

Group Activity/Challenge Problems

1. $x^2 + 0.6x + 0.08 = (x+0.4)(x+0.2)$

3. $x^2 + \frac{2}{5}x + \frac{1}{25} = \left(x + \frac{1}{5}\right)\left(x + \frac{1}{5}\right)$

$= \left(x + \frac{1}{5}\right)^2$

5. $-x^2 - 6x - 8 = -1(x^2 + 6x + 8)$

$= -(x+2)(x+4)$

7. $x^2 + 5x - 300 = (x+20)(x-15)$

9. $x^2 + 180x + 8000 = (x+100)(x+80)$

11. $x^2 + 20x - 8000 = (x+100)(x-80)$

13. $x^2 - 240x + 8000 = (x-200)(x-40)$

Exercise Set 5.4

1. $2x^2 + 7x + 6 = (2x+3)(x+2)$

3. $6x^2 + 13x + 6 = (2x+3)(3x+2)$

5. $3x^2 + 4x + 1 = (3x+1)(x+1)$

7. $2x^2 + 11x + 15 = (2x+5)(x+3)$

9. $4x^2 + 4x - 3 = (2x-1)(2x+3)$

11. $5y^2 - 8y + 3 = (5y-3)(y-1)$

13. $5a^2 - 12a + 6$ cannot be factored

15. $4x^2 + 13x + 3 = (4x+1)(x+3)$

17. $5x^2 + 11x + 4$ cannot be factored

19. $5y^2 - 16y + 3 = (5y-1)(y-3)$

21. $4x^2 + 4x - 15 = (2x+5)(2x-3)$

23. $7x^2 - 16x + 4 = (7x-2)(x-2)$

25. $3x^2 - 10x + 7 = (x-1)(3x-7)$

27. $5z^2 - 33z - 14 = (5z+2)(z-7)$

29. $8x^2 + 2x - 3 = (4x+3)(2x-1)$

31. $10x^2 - 27x + 5 = (5x-1)(2x-5)$

33. $8x^2 - 2x - 15 = (4x+5)(2x-3)$

35. $6x^2 + 33x + 15 = 3 \cdot 2x^2 + 3 \cdot 11x + 3 \cdot 5$
$= 3(2x^2 + 11x + 5) = 3(2x+1)(x+5)$

37. $6x^2 + 4x - 10 = 2 \cdot 3x^2 + 2 \cdot 2x - 2 \cdot 5$
$= 2(3x^2 + 2x - 5) = 2(3x+5)(x-1)$

39. $6x^3 + 5x^2 - 4x = x \cdot 6x^2 + x \cdot 5x - x \cdot 4$
$= x(6x^2 + 5x - 4) = x(2x-1)(3x+4)$

41. $4x^3 + 2x^2 - 6x$
$= 2x \cdot 2x^2 + 2x \cdot x - 2x \cdot 3$
$= 2x(2x^2 + x - 3) = 2x(2x+3)(x-1)$

43. $6x^3 + 4x^2 - 10x$
$= 2x \cdot 3x^2 + 2x \cdot 2x - 2x \cdot 5$
$= 2x(3x^2 + 2x - 5) = 2x(3x+5)(x-1)$

45. $60x^2 + 40x + 5 = 5 \cdot 12x^2 + 5 \cdot 8x + 5 \cdot 1$
$= 5(12x^2 + 8x + 1) = 5(6x+1)(2x+1)$

47. $2x^2 + 5xy + 2y^2 = (2x+y)(x+2y)$

49. $2x^2 - 7xy + 3y^2 = (2x - y)(x - 3y)$

51. $18x^2 + 18xy - 8y^2$

$= 2 \cdot 9x^2 + 2 \cdot 9xy - 2 \cdot 4y^2$

$= 2(9x^2 + 9xy - 4y^2)$

$= 2(3x - y)(3x + 4y)$

53. $6x^2 - 15xy - 36y^2$

$= 3 \cdot 2x^2 - 3 \cdot 5xy - 3 \cdot 12y^2$

$= 3(2x^2 - 5xy - 12y^2)$

$= 3(2x + 3y)(x - 4y)$

55. $6x^2 + x - 12$. This polynomial was obtained by multiplying the factors.

57. $6x^2 + 21x + 15$. This polynomial was obtained by multiplying the factors.

59. $2x^4 - x^3 - 3x^2$. This polynomial was obtained by multiplying the factors.

61. **(a)** The second factor can be found by dividing the polynomial by the binomial.

(b)
$$\begin{array}{r} 6x + \ \ 11 \\ 3x+10 \overline{\smash{)}18x^2 + 93x + 110} \\ \underline{-18x^2 - 60x} \\ 33x + 110 \\ \underline{-33x - 110} \\ 0 \end{array}$$

63. Any trinomial factoring problem may be checked by mulitplying the factors.

Cumulative Review Exercises: 5.4

65. $3x + 4 = -(x - 6)$

$3x + 4 = -x + 6$

$4x + 4 = 6$

$4x = 2$

$x = \frac{2}{4} = \frac{1}{2}$

66. Let x be the width. Then the length is $2x + 2$ and the perimeter is 2 times the length plus 2 times the width.

$2(2x + 2) + 2x = 22$

$4x + 4 + 2x = 22$

$6x = 18$

$x = 3$

Then the length is $2 \cdot 3 + 2 = 8$.

The length is 8 feet, and the width is 3 feet.

67. $36x^4y^3 - 12xy^2 + 24x^5y^6$

$= 12xy^2 \cdot 3x^3y - 12xy^2 \cdot 1$

$\quad + 12xy^2 \cdot 2x^4y^4$

$= 12xy^2(3x^3y - 1 + 2x^4y^4)$

68. $x^2 - 15x + 54 = (x - 9)(x - 6)$

Group Activity/Challenge Problems

1. $18x^2 + 9x - 20 = (6x - 5)(3x + 4)$

3. $15x^2 - 124x + 160 = (5x - 8)(3x - 20)$

5. $72x^2 - 180x - 200 = 4(18x^2 - 45x - 50)$
$= 4(6x + 5)(3x - 10)$

Exercise Set 5.5

1. $x^2 - 4 = x^2 - 2^2 = (x + 2)(x - 2)$

3. $y^2 - 25 = y^2 - 5^2 = (y + 5)(y - 5)$

5. $x^2 - 49 = x^2 - 7^2 = (x + 7)(x - 7)$

7. $x^2 - y^2 = (x + y)(x - y)$

9. $9y^2 - 16 = (3y)^2 - 4^2 = (3y + 4)(3y - 4)$

11. $64a^2 - 36b^2 = 4(16a^2 - 9b^2)$
$= 4[(4a)^2 - (3b)^2]$
$= 4(4a + 3b)(4a - 3b)$

13. $25x^2 - 16 = (5x)^2 - 4^2$
$= (5x + 4)(5x - 4)$

15. $z^4 - 81x^2 = (z^2)^2 - (9x)^2$
$= (z^2 + 9x)(z^2 - 9x)$

17. $9x^4 - 81y^2 = 9(x^4 - 9y^2)$
$= 9[(x^2)^2 - (3y)^2]$
$= 9(x^2 + 3y)(x^2 - 3y)$

19. $49m^4 - 16n^2 = (7m^2)^2 - (4n)^2$
$= (7m^2 + 4n)(7m^2 - 4n)$

21. $20x^2 - 180 = 20(x^2 - 9) = 20(x^2 - 3^2)$
$= 20(x + 3)(x - 3)$

23. $16x^2 - 100y^4 = 4(4x^2 - 25y^4)$
$= 4[(2x)^2 - (5y^2)^2]$
$= 4(2x - 5y^2)(2x + 5y^2)$

25. $x^3 + y^3 = (x + y)(x^2 - xy + y^2)$

27. $a^3 - b^3 = (a - b)(a^2 + ab + b^2)$

29. $x^3 + 8 = x^3 + 2^3 = (x + 2)(x^2 - 2x + 4)$

31. $x^3 - 27 = x^3 - 3^3 = (x - 3)(x^2 + 3x + 9)$

33. $a^3 + 1 = a^3 + 1^3 = (a + 1)(a^2 - a + 1)$

35. $8x^3 + 27 = (2x)^3 + 3^3$
$= (2x + 3)(4x^2 - 6x + 9)$

37. $27a^3 - 64 = (3a)^3 - 4^3$
$= (3a - 4)(9a^2 + 12a + 16)$

39. $27 - 8y^3 = 3^3 - (2y)^3$
$= (3 - 2y)(9 + 6y + 4y^2)$

41. $8x^3 - 27y^3 = (2x)^3 - (3y)^3$

$= (2x - 3y)(4x^2 + 6xy + 9y^2)$

43. $2x^2 - 2x - 12 = 2(x^2 - x - 6)$

$= 2(x - 3)(x + 2)$

45. $x^2y - 16y = y(x^2 - 16) = y(x^2 - 4^2)$

$= y(x + 4)(x - 4)$

47. $3x^2 + 6x + 3 = 3(x^2 + 2x + 1)$

$= 3(x + 1)(x + 1) = 3(x + 1)^2$

49. $5x^2 + 10x - 15 = 5(x^2 + 2x - 3)$

$= 5(x + 3)(x - 1)$

51. $3xy - 6x + 9y - 18 = 3(xy - 2x + 3y - 6)$

$= 3[x(y - 2) + 3(y - 2)] = 3(x + 3)(y - 2)$

53. $2x^2 - 72 = 2(x^2 - 36) = 2(x^2 - 6^2)$

$= 2(x + 6)(x - 6)$

55. $3x^2y - 27y = 3y(x^2 - 9) = 3y(x^2 - 3^2)$

$= 3y(x + 3)(x - 3)$

57. $3x^3y^2 + 3y^2 = 3y^2(x^3 + 1)$

$= 3y^2(x^3 + 1^3)$

$= 3y^2(x + 1)(x^2 - x + 1)$

59. $2x^3 - 16 = 2(x^3 - 8) = 2(x^3 - 2^3)$

$= 2(x - 2)(x^2 + 2x + 4)$

61. $6x^2 - 4x + 24x - 16$

$= 2(3x^2 - 2x + 12x - 8)$

$= 2[x(3x - 2) + 4(3x - 2)]$

$= 2(x + 4)(3x - 2)$

63. $3x^3 - 10x^2 - 8x = x(3x^2 - 10x - 8)$

$= x(3x + 2)(x - 4)$

65. $4x^2 + 5x - 6 = (x + 2)(4x - 3)$

67. $25b^2 - 100 = 25(b^2 - 4) = 25(b^2 - 2^2)$

$= 25(b + 2)(b - 2)$

69. $a^5b^2 - 4a^3b^4 = a^3b^2(a^2 - 4b^2)$

$= a^3b^2[a^2 - (2b)^2]$

$= a^3b^2(a + 2b)(a - 2b)$

71. $3x^4 - 18x^3 + 27x^2 = 3x^2(x^2 - 6x + 9)$

$= 3x^2(x - 3)(x - 3) = 3x^2(x - 3)^2$

73. $x^3 + 25x = x(x^2 + 25)$

75. $y^4 - 16 = (y^2)^2 - 4^2 = (y^2 + 4)(y^2 - 4)$

$= (y^2 + 4)(y^2 - 2^2)$

$= (y^2 + 4)(y + 2)(y - 2)$

77. $10a^2 + 25ab - 60b^2$
$= 5(2a^2 + 5ab - 12b^2)$
$= 5(2a - 3b)(a + 4b)$

79. $2ab + 4a - 3b - 6 = 2a(b+2) - 3(b+2)$
$= (2a - 3)(b + 2)$

81. $9 - 9y^4 = 9(1 - y^4) = 9[1^2 - (y^2)^2]$
$= 9(1 + y^2)(1 - y^2) = 9(1 + y^2)(1^2 - y^2)$
$= 9(1 + y^2)(1 + y)(1 - y)$

83. **(a)** $a^3 + b^3 = (a + b)(a^2 - ab + b^2)$
(b) Answers may vary.

Cumulative Review Problems: 5.5

85. $6(x - 2) < 4x - 3 + 2x$
$6x - 12 < 6x - 3$
$-12 < -3$
The solution is all real numbers.

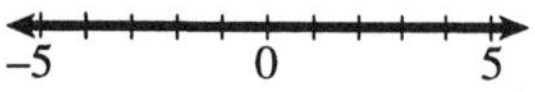

86. $2x - 5y = 6$
$-5y = -2x + 6$
$5y = 2x - 6$
$y = \frac{2x - 6}{5}$ or $y = \frac{2}{5}x - \frac{6}{5}$

87. $\left(\frac{4x^4y}{6xy^5}\right)^3 = \left(\frac{4}{6} \cdot \frac{x^4}{x} \cdot \frac{y}{y^5}\right)^3$
$= \left(\frac{2}{3} \cdot x^3 \cdot \frac{1}{y^4}\right)^3 = \left(\frac{2x^3}{3y^4}\right)^3$
$= \frac{2^3 x^{3\cdot 3}}{3^3 y^{4\cdot 3}} = \frac{8x^9}{27y^{12}}$

88. $x^{-2}x^{-3} = x^{-2-3} = x^{-5} = \frac{1}{x^5}$

Group Activity/Challenge Problems

2. $x^6 + 1 = (x^2)^3 + 1^3 = (x^2 + 1)(x^4 - x^2 + 1)$

3. $x^6 - 27y^9 = (x^2)^3 - (3y^3)^3$
$= (x^2 - 3y^2)(x^4 + 3x^2y^3 + 9y^6)$

4. $x^2 - 6x + 9 - 4y^2 = (x - 3)^2 - (2y)^2$
$= (x - 3 + 2y)(x - 3 - 2y)$

Exercise Set 5.6

1. $x(x + 2) = 0$
$x = 0$ or $x + 2 = 0$
$x = -2$

3. $4x(x - 9) = 0$
$4x = 0$ or $x - 9 = 0$
$x = \frac{0}{4}$ or $x = 9$
$x = 0$

5. $(2x+5)(x-3)=0$

$2x+5=0$ or $x-3=0$

$2x=-5 \qquad x=3$

$x=-\frac{5}{2}$

7. $x^2-16=0$

$(x+4)(x-4)=0$

$x+4=0$ or $x-4=0$

$x=-4$ or $x=4$

9. $x^2-12x=0$

$x(x-12)=0$

$x=0$ or $x-12=0$

$x=12$

11. $9x^2+18x=0$

$9x(x+2)=0$

$9x=0$ or $x+2=0$

$x=\frac{0}{9}$ or $x=-2$

$x=0$

13. $x^2+x-12=0$

$(x+4)(x-3)=0$

$x+4=0$ or $x-3=0$

$x=-4$ or $x=3$

15. $x^2-12x=-20$

$x^2-12x+20=0$

$(x-10)(x-2)=0$

$x-10=0$ or $x-2=0$

$x=10$ or $x=2$

17. $z^2+3z=18$

$z^2+3z-18=0$

$(z+6)(z-3)=0$

$z+6=0$ or $z-3=0$

$z=-6$ or $z=3$

19. $3x^2-6x-72=0$

$3(x^2-2x-24)=0$

$3(x-6)(x+4)=0$

$x-6=0$ or $x+4=0$

$x=6$ or $x=-4$

21. $x^2+19x=42$

$x^2+19x-42=0$

$(x-2)(x+21)=0$

$x-2=0$ or $x+21=0$

$x=2$ or $x=-21$

23. $2y^2+22y+60=0$

$2(y^2+11y+30)=0$

$2(y+6)(y+5)=0$

$y+6=0$ or $y+5=0$

$y=-6$ or $y=-5$

25. $-2x-8=-x^2$

$x^2-2x-8=0$

$(x-4)(x+2)=0$

$x-4=0$ or $x+2=0$

$x=4$ or $x=-2$

27. $-x^2+30x+64=0$

$x^2-30x-64=0$

$(x-32)(x+2)=0$

$x-32=0$ or $x+2=0$

$x=32$ or $x=-2$

29. $x^2-3x-18=0$

$(x-6)(x+3)=0$

$x-6=0$ or $x+3=0$

$x=6$ or $x=-3$

31. $3p^2=22p-7$

$3p^2-22p+7=0$

$(3p-1)(p-7)=0$

$3p-1=0$ or $p-7=0$

$3p=1$ or $p=7$

$p=\frac{1}{3}$

33. $3r^2+r=2$

$3r^2+r-2=0$

$(3r-2)(r+1)=0$

$3r-2=0$ or $r+1=0$

$3r=2$ or $r=-1$

$r=\frac{2}{3}$

35. $4x^2+4x-48=0$

$4(x^2+x-12)=0$

$4(x+4)(x-3)=0$

$x+4=0$ or $x-3=0$

$x=-4$ or $x=3$

37. $6x^2-5x=4$

$6x^2-5x-4=0$

$(3x-4)(2x+1)=0$

$3x-4=0$ or $2x+1=0$

$3x=4$ or $2x=-1$

$x=\frac{4}{3}$ or $x=-\frac{1}{2}$

39. $2x^2-10x=-12$

$2x^2-10x+12=0$

$2(x^2-5x+6)=0$

$2(x-2)(x-3)=0$

$x-2=0$ or $x-3=0$

$x=2$ or $x=3$

41. $2x^2=32x$

$2x^2-32x=0$

$2x(x-16)=0$

$2x=0$ or $x-16=0$

$x=\frac{0}{2}$ or $x=16$

$x=0$

43. $x^2=36$

$x^2-36=0$

$(x+6)(x-6)=0$

$x+6=0$ or $x-6=0$

$x=-6$ or $x=6$

45. $x^2 = 9$

$x^2 - 9 = 0$

$(x+3)(x-3) = 0$

$x + 3 = 0$ or $x - 3 = 0$

$x = -3$ or $x = 3$

47. $-t^2 = -81$

$t^2 = 81$

$t^2 - 81 = 0$

$(t+9)(t-9) = 0$

$t + 9 = 0$ or $t - 9 = 0$

$t = -9$ or $t = 9$

49. Let x be the first positive even number. Then $x + 2$ is the other positive even number.

$x(x+2) = 120$

$x^2 + 2x = 120$

$x^2 + 2x - 120 = 0$

$(x-10)(x+12) = 0$

$x - 10 = 0$ or $x + 12 = 0$

$x = 10$ or $x = -12$

Since x must be positive, $x = 10$. The two integers are 10 and 12.

51. Let x = first positive integer.

Then $4x$ is the second positive integer.

$x(4x) = 64$

$4x^2 = 64$

$4x^2 - 64 = 0$

$4(x^2 - 16) = 0$

$4(x+4)(x-4) = 0$

$x + 4 = 0$ or $x - 4 = 0$

$x = -4$ or $x = 4$

Since x must be positive, the integers are $x = 4$ and $4x = 4(4) = 16$.

53. Let w = width. Then length = $2w - 3$.

$A = l \cdot w$

$54 = (2w-3)(w)$

$54 = 2w^2 - 3w$

$0 = 2w^2 - 3w - 54$ or $2w^2 - 3w - 54 = 0$

$(2w+9)(w-6) = 0$

$2w = -9$ or $w = 6$

$w = -\frac{9}{2}$

Since dimensions must be positive, width = w = 6 inches and length = $2w - 3 = 2(6) - 3 = 9$ inches.

55. Let w = width. Then length = $w + 2$.

$A = l \cdot w$

$35 = (w+2)(w)$

$35 = w^2 + 2w$

$0 = w^2 + 2w - 35$ or $w^2 + 2w - 35 = 0$

$(w+7)(w-5) = 0$

$w + 7 = 0$ or $w - 5 = 0$

$w = -7$ or $w = 5$

Since dimensions must be positive, the width is $w = 5$ feet and the length is $w + 2 = 5 + 2 = 7$ feet.

57. Using the equation from Example 10:

$d = 16t^2$

$400 = 16t^2$

$\frac{400}{16} = t^2$

$25 = t^2$

$0 = t^2 - 25$ or $t^2 - 25 = 0$

$(t+5)(t-5) = 0$

$t+5 = 0$ or $t-5 = 0$

$t = -5$ or $t = 5$

Since time must be positive, it takes the diver 5 seconds to hit the water.

59. $C = x^2 - 27x - 20$

$70 = x^2 - 27x - 20$

$0 = x^2 - 27x - 90$

$0 = (x+3)(x-30)$

$x+3 = 0$ or $x-30 = 0$

$x = -3$ or $x = 30$

Since the number of water sprinklers manufactured must be greater than 0, 30 water sprinklers were manufactured at a cost of $70.

61. **(a)** $N = 26(26-1) = 26(25) = 650$. There are 650 possible first and second place finishers.

(b) $90 = x(x-1)$

$90 = x^2 - x$

$0 = x^2 - x - 90$ or $x^2 - x - 90 = 0$

$(x-10)(x+9) = 0$

$x - 10 = 0$ or $x + 9 = 0$

$x = 10$ or $x = -9$

There are 10 teams in the league, since the number of teams must be positive.

63. **(a)** $C = \frac{12(12-1)}{2} = \frac{12(11)}{2} = 66$.

A switchboard can make 66 simultaneous telephone connections.

(b) $55 = \frac{n(n-1)}{2}$

$110 = n(n-1)$

$110 = n^2 - n$

$0 = n^2 - n - 110$ or $n^2 - n - 110 = 0$

$(n-11)(n+10) = 0$

$n - 11 = 0$ or $n + 10 = 0$

$n = 11$ or $n = -10$

Since the number of lines must be positive, the switchboard has 11 lines.

65. You can eliminate the 3 by dividing both sides of the equation by 3 to get $(x-4)(x+5) = 0$.

Cumulative Review Exercises: 5.6

66. $16 - 2^2 \cdot 12 \div 3 - 3^2 = 16 - 4 \cdot 12 \div 3 - 9$

$= 16 - 48 \div 3 - 9 = 16 - 16 - 9 = -9$

67. $(3x+2) - (x^2 - 4x + 6)$

$= 3x + 2 - x^2 + 4x - 6$

$= -x^2 + 3x + 4x + 2 - 6 = -x^2 + 7x - 4$

68. $(3x^2 + 2x - 4)(2x - 1)$

$= 3x^2(2x-1) + 2x(2x-1) - 4(2x-1)$

$= 6x^3 - 3x^2 + 4x^2 - 2x - 8x + 4$

$= 6x^3 + x^2 - 10x + 4$

69.
$$\begin{array}{r} 2x-3 \\ 3x-5\overline{\smash{)}6x^2-19x+15} \\ \underline{-6x^2+10x} \\ -9x+15 \\ \underline{9x-15} \\ 0 \end{array}$$

Thus $\dfrac{6x^2-19x+15}{3x-5}=2x-3$

70. $\dfrac{6x^2-19x+15}{3x-5}=\dfrac{(3x-5)(2x-3)}{(3x-5)}$

$=2x-3$

Group Activity/Challenge Problems

1. Solutions are 6 and 3.

$(x-6)(x-3)=x^2-9x+18=0$

3. Solutions are –5 and –9.

$(x+5)(x+9)=x^2+14x+45=0$

5. Solutions are 0, 3, and 5.

$x(x-3)(x-5)=x(x^2-8x+15)$

$=x^3-8x^2+15x=0$

7. $C=2x^2-20x+600$, $R=x^2+50x-400$

$2x^2-20x+600=x^2+50x-400$

$x^2-70x+1000=0$

$(x-50)(x-20)=0$, $x=20$ and 50

9. $x^3+3x^2-10x=x(x+5)(x-2)=0$,

$x=0,-5,2$

Chapter 5 Review Exercises

1. The greatest common factor is x^2.

2. The greatest common factor is $3p$.

3. The greatest common factor is 6.

4. The greatest common factor is $4x^2y^2$.

5. The greatest common factor is 1.

6. The greatest common factor is $8x^2$.

7. The greatest common factor is $2x-7$.

8. The greatest common factor is $x+5$.

9. $4x-16=4(x-4)$

10. $10x+5=5(2x+1)$

11. $24y^2-4y=4y(6y-1)$

12. $55p^3-20p^2=5p^2(11p-4)$

13. $24x^2y+18x^3y^2=6x^2y(4+3xy)$

14. $6xy-12x^2y=6xy(1-2x)$

15. $2x^2+4x-8=2(x^2+2x-4)$

16. $60x^4y^4 + 6x^9y^3 - 18x^5y^2$
$= 6x^4y^2(10y^2 + x^5y - 3x)$

17. $24x^2 - 13y^2 + 6xy$ cannot be factored

18. $x(5x+3) - 2(5x+3) = (x-2)(5x+3)$

19. $3x(x-1) - 2(x-1) = (3x-2)(x-1)$

20. $2x(4x-3) + 4x - 3$
$= 2x(4x-3) + 1(4x-3)$
$= (2x+1)(4x-3)$

21. $x^2 + 4x + 2x + 8 = x(x+4) + 2(x+4)$
$= (x+2)(x+4)$

22. $x^2 - 3x + 4x - 12 = x(x-3) + 4(x-3)$
$= (x+4)(x-3)$

23. $x^2 - 7x + 7x - 49 = x(x-7) + 7(x-7)$
$= (x+7)(x-7)$

24. $2a^2 - 2ab - a + b = 2a(a-b) - 1(a-b)$
$= (2a-1)(a-b)$

25. $3xy + 3x + 2y + 2 = 3x(y+1) + 2(y+1)$
$= (3x+2)(y+1)$

26. $x^2 + 3x - 2xy - 6y = x(x+3) - 2y(x+3)$
$= (x-2y)(x+3)$

27. $5x^2 + 20x - x - 4 = 5x(x+4) - 1(x+4)$
$= (5x-1)(x+4)$

28. $5x^2 - xy + 20xy - 4y^2$
$= x(5x-y) + 4y(5x-y)$
$= (x+4y)(5x-y)$

29. $12x^2 - 8xy + 15xy - 10y^2$
$= 4x(3x-2y) + 5y(3x-2y)$
$= (4x+5y)(3x-2y)$

30. $12x^2 + 15xy - 8xy - 10y^2$
$= 3x(4x+5y) - 2y(4x+5y)$
$= (3x-2y)(4x+5y)$

31. $ab - a + b - 1 = a(b-1) + 1(b-1)$
$= (a+1)(b-1)$

32. $3x^2 - 9xy + 2xy - 6y^2$
$= 3x(x-3y) + 2y(x-3y)$
$= (3x+2y)(x-3y)$

33. $20x^2 - 12x + 15x - 9$
$= 4x(5x-3) + 3(5x-3)$
$= (4x+3)(5x-3)$

34. $6x^2 + 9x - 2x - 3$
$= 3x(2x+3) - 1(2x+3)$
$= (3x-1)(2x+3)$

35. $x^2+7x+10=(x+2)(x+5)$

36. $x^2-8x+15=(x-3)(x-5)$

37. $x^2-x-20=(x+4)(x-5)$

38. $x^2+x-20=(x-4)(x+5)$

39. $x^2-11x+30=(x-5)(x-6)$

40. $x^2-15x+56=(x-8)(x-7)$

41. $x^2-12x-44$ cannot be factored

42. $x^2+11x+24=(x+3)(x+8)$

43. $x^3+5x^2+4x=x(x^2+5x+4)$
$=x(x+1)(x+4)$

44. $x^3-3x^2-40x=x(x^2-3x-40)$
$=x(x-8)(x+5)$

45. $x^2-2xy-15y^2=(x-5y)(x+3y)$

46. $4x^3+32x^2y+60xy^2$
$=4x(x^2+8xy+15y^2)$
$=4x(x+3y)(x+5y)$

47. $2x^2+5x-3=(2x-1)(x+3)$

48. $3x^2+13x+4=(3x+1)(x+4)$

49. $4x^2-9x+5=(4x-5)(x-1)$

50. $5x^2-13x-6=(5x+2)(x-3)$

51. $4x^2+4x-15=(2x+5)(2x-3)$

52. $5x^2-32x+12=(5x-2)(x-6)$

53. $3x^2+13x+8$ cannot be factored

54. $6x^2+31x+5=(6x+1)(x+5)$

55. $4x^2+8x-21=(2x-3)(2x+7)$

56. $6x^2+11x-10=(3x-2)(2x+5)$

57. $8x^2-18x-35=(4x+5)(2x-7)$

58. $4x^2+20x+25=(2x+5)(2x+5)$
$=(2x+5)^2$

59. $9x^3-12x^2+4x=x(9x^2-12x+4)$
$=x(3x-2)(3x-2)=x(3x-2)^2$

60. $18x^3-24x^2-10x=2x(9x^2-12x-5)$
$=2x(3x+1)(3x-5)$

61. $4x^2-16xy+15y^2=(2x-3y)(2x-5y)$

62. $16x^2-22xy-3y^2=(8x+y)(2x-3y)$

63. $x^2-16=x^2-4^2=(x+4)(x-4)$

64. $x^2-64=x^2-8^2=(x+8)(x-8)$

65. $4x^2-16=4(x^2-4)=4(x^2-2^2)$
$=4(x+2)(x-2)$

66. $81x^2-9y^2=9(9x^2-y^2)$
$=9[(3x)^2-y^2]=9(3x+y)(3x-y)$

67. $64x^4-81y^4=(8x^2)^2-(9y^2)^2$
$=(8x^2+9y^2)(8x^2-9y^2)$

68. $16-25y^2=4^2-(5y)^2=(4+5y)(4-5y)$

69. $4x^4-9y^4=(2x^2)^2-(3y^2)^2$
$=(2x^2+3y^2)(2x^2-3y^2)$

70. $100x^4-121y^4=(10x^2)^2-(11y^2)^2$
$=(10x^2+11y^2)(10x^2-11y^2)$

71. $x^3-y^3=(x-y)(x^2+xy+y^2)$

72. $x^3+y^3=(x+y)(x^2-xy+y^2)$

73. $a^3+8=a^3+2^3=(a+2)(a^2-2a+4)$

74. $a^3-1=a^3-1^3=(a-1)(a^2+a+1)$

75. $a^3+27=a^3+3^3=(a+3)(a^2-3a+9)$

76. $x^3-8=x^3-2^3=(x-2)(x^2+2x+4)$

77. $8x^3-y^3=(2x)^3-y^3$
$=(2x-y)(4x^2+2xy+y^2)$

78. $27-8y^3=3^3-(2y)^3$
$=(3-2y)(9+6y+4y^2)$

79. $27x^4-75y^2=3(9x^4-25y^2)$
$=3[(3x^2)^2-(5y)^2]$
$=3(3x^2+5y)(3x^2-5y)$

80. $2x^3-128y^3=2(x^3-64y^3)$
$=2[x^3-(4y)^3]$
$=2(x-4y)(x^2+4xy+16y^2)$

81. $x^2-15x+50=(x-5)(x-10)$

82. $2x^2-16x+32=2(x^2-8x+16)$
$=2(x-4)(x-4)=2(x-4)^2$

83. $4x^2 - 36 = 4(x^2 - 9) = 4(x^2 - 3^2)$
$= 4(x+3)(x-3)$

84. $4y^2 - 64 = 4(y^2 - 16) = 4(y^2 - 4^2)$
$= 4(y+4)(y-4)$

85. $8x^2 + 16x - 24 = 8(x^2 + 2x - 3)$
$= 8(x+3)(x-1)$

86. $x^2 - 6x - 27 = (x-9)(x+3)$

87. $4x^2 - 4x - 15 = (2x+3)(2x-5)$

88. $6x^2 - 33x + 36 = 3(2x^2 - 11x + 12)$
$= 3(2x-3)(x-4)$

89. $8x^3 - 8 = 8(x^3 - 1) = 8(x^3 - 1^3)$
$= 8(x-1)(x^2 + x + 1)$

90. $x^3y - 27y = y(x^3 - 27) = y(x^3 - 3^3)$
$= y(x-3)(x^2 + 3x + 9)$

91. $x^2y - xy + 4xy - 4y = y(x^2 - x + 4x - 4)$
$= y[x(x-1) + 4(x-1)] = y(x+4)(x-1)$

92. $6x^3 + 30x^2 + 9x^2 + 45x$
$= 3x(2x^2 + 10x + 3x + 15)$
$= 3x[2x(x+5) + 3(x+5)]$
$= 3x(2x+3)(x+5)$

93. $x^2 + 5xy + 6y^2 = (x+2y)(x+3y)$

94. $2x^2 - xy - 10y^2 = (2x-5y)(x+2y)$

95. $4x^2 - 20xy + 25y^2 = (2x-5y)(2x-5y)$
$= (2x-5y)^2$

96. $16y^2 - 49z^2 = (4y)^2 - (7z)^2$
$= (4y+7z)(4y-7z)$

97. $ab + 7a + 6b + 42 = a(b+7) + 6(b+7)$
$= (a+6)(b+7)$

98. $16y^5 - 25y^7 = y^5(16 - 25y^2)$
$= y^5[4^2 - (5y)^2] = y^5(4+5y)(4-5y)$

99. $2x^3 + 12x^2y + 16xy^2$
$= 2x(x^2 + 6xy + 8y^2)$
$= 2x(x+2y)(x+4y)$

100. $6x^2 + 5xy - 21y^2 = (2x-3y)(3x+7y)$

101. $32x^3 + 32x^2 + 6x = 2x(16x^2 + 16x + 3)$
$= 2x(4x+1)(4x+3)$

102. $y^4 - 1 = (y^2)^2 - 1^2 = (y^2+1)(y^2-1)$
$= (y^2+1)(y^2 - 1^2) = (y^2+1)(y+1)(y-1)$

103. $x(x-5)=0$

$x=0$ or $x-5=0$

$x=5$

104. $(x+3)(x+4)=0$

$x+3=0$ or $x+4=0$

$x=-3 \qquad x=-4$

105. $(x-5)(3x+2)=0$

$x-5=0$ or $3x+2=0$

$x=5 \qquad 3x=-2$

$x=-\frac{2}{3}$

106. $x^2-3x=0$

$x(x-3)=0$

$x=0$ or $x-3=0$

$x=3$

107. $5x^2+20x=0$

$5x(x+4)=0$

$5x=0$ or $x+4=0$

$x=\frac{0}{5} \qquad x=-4$

$x=0$

108. $x^2-2x-24=0$

$(x-6)(x+4)=0$

$x-6=0$ or $x+4=0$

$x=6 \qquad x=-4$

109. $x^2+8x+15=0$

$(x+3)(x+5)=0$

$x+3=0$ or $x+5=0$

$x=-3 \qquad x=-5$

110. $x^2=-2x+8$

$x^2+2x-8=0$

$(x+4)(x-2)=0$

$x+4=0$ or $x-2=0$

$x=-4 \qquad x=2$

111. $x^2-12=-x$

$x^2+x-12=0$

$(x+4)(x-3)=0$

$x+4=0$ or $x-3=0$

$x=-4 \qquad x=3$

112. $3x^2+21x+30=0$

$3(x^2+7x+10)=0$

$3(x+2)(x+5)=0$

$x+2=0$ or $x+5=0$

$x=-2 \qquad x=-5$

113. $x^2-6x+8=0$

$(x-4)(x-2)=0$

$x-4=0$ or $x-2=0$

$x=4 \qquad x=2$

114. $6x^2+6x-12=0$

$6(x^2+x-2)=0$

$6(x+2)(x-1)=0$

$x+2=0$ or $x-1=0$

$x=-2 \qquad x=1$

115. $8x^2-3=-10x$

$8x^2+10x-3=0$

$(4x-1)(2x+3)=0$

$4x-1=0$ or $2x+3=0$

$4x=1 \qquad 2x=-3$

$x=\frac{1}{4}$ or $x=-\frac{3}{2}$

116. $2x^2+15x=8$

$2x^2+15x-8=0$

$(2x-1)(x+8)=0$

$2x-1=0$ or $x+8=0$

$2x=1 \qquad x=-8$

$x=\frac{1}{2}$

117. $4x^2-16=0$

$4(x^2-4)=0$

$4(x^2-2^2)=0$

$4(x+2)(x-2)=0$

$x+2=0$ or $x-2=0$

$x=-2 \qquad x=2$

118. $36x^2-49=0$

$(6x)^2-7^2=0$

$(6x+7)(6x-7)=0$

$6x+7=0$ or $6x-7=0$

$6x=-7 \qquad 6x=7$

$x=-\frac{7}{6}$ or $x=\frac{7}{6}$

119. $26x-48=-4x^2$

$4x^2+26x-48=0$

$2(2x^2+13x-24)=0$

$2(2x-3)(x+8)=0$

$2x-3=0$ or $x+8=0$

$2x=3 \qquad x=-8$

$x=\frac{3}{2}$

120. $-48x=-12x^2-45$

$12x^2-48x+45=0$

$3(4x^2-16x+15)=0$

$3(2x-3)(2x-5)=0$

$2x-3=0$ or $2x-5=0$

$2x=3 \qquad 2x=5$

$x=\frac{3}{2}$ or $x=\frac{5}{2}$

121. Let x be the smaller of the two integers. Then the other is $x + 1$.

$x(x + 1) = 110$

$x^2 + x = 110$

$x^2 + x - 110 = 0$

$(x + 11)(x - 10) = 0$

$x + 11 = 0$ or $x - 10 = 0$

$x = -11$ $\quad$ $x = 10$

Since the integers must be positive, the smaller is 10 and the larger is 11.

122. Let x be the smaller integer. The larger is $x + 2$.

$x(x + 2) = 48$

$x^2 + 2x = 48$

$x^2 + 2x - 48 = 0$

$(x + 8)(x - 6) = 0$

$x + 8 = 0$ or $x - 6 = 0$

$x = -8$ $\quad$ $x = 6$

Since the integers must be positive, they are 6 and 8.

123. Let x be the smaller integer. Then the larger is $2x - 2$.

$x(2x - 2) = 40$

$2x^2 - 2x = 40$

$2x^2 - 2x - 40 = 0$

$2(x^2 - x - 20) = 0$

$2(x - 5)(x + 4) = 0$

$x - 5 = 0$ or $x + 4 = 0$

$x = 5$ $\quad$ $x = -4$

Since the integers are positive, the smaller is 5 and the larger is 8.

124. Let w be the width of the rectangle. Then the length is $w + 2$.

$w(w + 2) = 63$

$w^2 + 2w = 63$

$w^2 + 2w - 63 = 0$

$(w + 9)(w - 7) = 0$

$w + 9 = 0$ or $w - 7 = 0$

$w = -9$ $\quad$ $w = 7$

Since the width must be positive, it is 7 feet, and the length is 9 feet.

125. Let x be the length of a side of the smaller square. Then $x + 4$ is the length of a side of the larger square.

$(x + 4)^2 = 81$

$x^2 + 8x + 16 = 81$

$x^2 + 8x - 65 = 0$

$(x + 13)(x - 5) = 0$

$x + 13 = 0$ or $x - 5 = 0$

$x = -13$ $\quad$ $x = 5$

Since lengths must be positive, the length of a side of the smaller square is 5 inches while the length of a side of the larger square is 9 inches.

126. $C = x^2 - 18x + 10$

$50 = x^2 - 18x + 10$

$0 = x^2 - 18x - 40$

$0 = (x - 20)(x + 2)$

$x - 20 = 0$ or $x + 2 = 0$

$x = 20$ $\quad$ $x = -2$

Since only a positive number of boxes may be manufactured. It costs \$50 to manufacture 20 boxes.

Chapter 5 Practice Test

1. The greatest common factor is $2x^2$

2. The greatest common factor is $3xy^2$

3. $4x^2y - 8xy = 4xy(x-2)$

4. $24x^2y - 6xy + 9x = 3x(8xy - 2y + 3)$

5. $x^2 - 3x + 2x - 6 = x(x-3) + 2(x-3)$
   $= (x+2)(x-3)$

6. $3x^2 - 12x + x - 4 = 3x(x-4) + 1(x-4)$
   $= (3x+1)(x-4)$

7. $5x^2 - 15xy - 3xy + 9y^2$
   $= 5x(x-3y) - 3y(x-3y)$
   $= (5x-3y)(x-3y)$

8. $x^2 + 12x + 32 = (x+4)(x+8)$

9. $x^2 + 5x - 24 = (x+8)(x-3)$

10. $x^2 - 9xy + 20y^2 = (x-4y)(x-5y)$

11. $2x^2 - 22x + 60 = 2(x^2 - 11x + 30)$
    $= 2(x-5)(x-6)$

12. $2x^3 - 3x^2 + x = x(2x^2 - 3x + 1)$
    $= x(2x-1)(x-1)$

13. $12x^2 - xy - 6y^2 = (3x+2y)(4x-3y)$

14. $x^2 - 9y^2 = x^2 - (3y)^2 = (x+3y)(x-3y)$

15. $x^3 + 27 = x^3 + 3^3 = (x+3)(x^2 - 3x + 9)$

16. $(x-2)(2x-5) = 0$
    $x - 2 = 0$ or $2x - 5 = 0$
    $x = 2 \qquad 2x = 5$
    $x = \frac{5}{2}$

17. $x^2 + 6 = -5x$
    $x^2 + 5x + 6 = 0$
    $(x+2)(x+3) = 0$
    $x + 2 = 0$ or $x + 3 = 0$
    $x = -2 \qquad x = -3$

18. $x^2 + 4x - 5 = 0$
    $(x+5)(x-1) = 0$
    $x + 5 = 0$ or $x - 1 = 0$
    $x = -5 \qquad x = 1$

19. Let x be the smaller of two integers. The larger of the two is $2x + 1$.

$x(2x+1) = 36$

$2x^2 + x = 36$

$2x^2 + x - 36 = 0$

$(2x+9)(x-4) = 0$

$2x+9 = 0$ or $x-4 = 0$

$2x = -9 \qquad x = 4$

$x = -\frac{9}{2}$

Since the two numbers are positive integers, the smaller is 4 and the larger is 9.

20. Let w be the width of the rectangle. Then the length is $w + 2$.

$w(w+2) = 24$

$w^2 + 2w = 24$

$w^2 + 2w - 24 = 0$

$(w+6)(w-4) = 0$

$w+6 = 0$ or $w-4 = 0$

$w = -6 \qquad w = 4$

Since the width is positive, it is 4 meters, and the length is 6 meters.

CHAPTER 6

Exercise Set 6.1

1. $x \neq 0$

3. $x - 6 = 0$

   $x \neq 6$

5. $x^2 - 4 = 0$

   $(x-2)(x+2) = 0$

   $x \neq 2,\ x \neq -2$

7. $x^2 + 6x - 16 = 0$

   $(x-2)(x+8) = 0$

   $x \neq 2,\ x \neq -8$

9. $\dfrac{x}{x+xy} = \dfrac{x}{x(1+y)} = \dfrac{1}{1+y}$

11. $\dfrac{4x+12}{x+3} = \dfrac{4(x+3)}{x+3} = 4$

13. $\dfrac{x^3+6x^2+3x}{2x} = \dfrac{x\left(x^2+6x+3\right)}{2x}$

    $= \dfrac{\left(x^2+6x+3\right)}{2}$

15. $\dfrac{x^2+2x+1}{x+1} = \dfrac{(x+1)(x+1)}{x+1} = x+1$

17. $\dfrac{x^2-2x}{x^2-4x+4} = \dfrac{x(x-2)}{(x-2)(x-2)} = \dfrac{x}{x-2}$

19. $\dfrac{x^2-x-6}{x^2-4} = \dfrac{(x-3)(x+2)}{(x-2)(x+2)} = \dfrac{x-3}{x-2}$

21. $\dfrac{2x^2-4x-6}{x-3} = \dfrac{2(x-3)(x+1)}{(x-3)}$

    $= 2(x+1)$

23. $\dfrac{2x-3}{3-2x} = \dfrac{2x-3}{-(2x-3)} = -1$

25. $\dfrac{x^2-2x-8}{4-x} = \dfrac{(x-4)(x+2)}{-(x-4)} = -(x+2)$

27. $\dfrac{x^2+3x-18}{-2x^2+6x} = \dfrac{(x+6)(x-3)}{-2x(x-3)} = \dfrac{-(x+6)}{2x}$

29. $\dfrac{2x^2+5x-3}{1-2x} = \dfrac{(2x-1)(x+3)}{-(2x-1)}$

    $= -(x+3)$

31. $\dfrac{6x^2+x-2}{2x-1} = \dfrac{(2x-1)(3x+2)}{2x-1} = 3x+2$

33. $\dfrac{6x^2+7x-20}{2x+5} = \dfrac{(2x+5)(3x-4)}{2x+5}$

    $= 3x-4$

35. $\dfrac{6x^2-13x+6}{3x-2} = \dfrac{(3x-2)(2x-3)}{3x-2}$

    $= 2x-3$

37. $\dfrac{x^2-3x+4x-12}{x-3}=\dfrac{x(x-3)+4(x-3)}{x-3}$

$=\dfrac{(x+4)(x-3)}{x-3}=x+4$

39. $\dfrac{2x^2-8x+3x-12}{2x^2+8x+3x+12}=\dfrac{2x(x-4)+3(x-4)}{2x(x+4)+3(x+4)}$

$=\dfrac{(x-4)(2x+3)}{(x+4)(2x+3)}=\dfrac{x-4}{x+4}$

41. $\dfrac{x^3-8}{x-2}=\dfrac{(x-2)\left(x^2+2x+4\right)}{x-2}$

$=x^2+2x+4$

43. x^2+1 is never zero if x is a real number.

45. $x-3$ cannot be zero. $x\neq 3$

47. $-\dfrac{x+4}{4-x}=-\dfrac{x+4}{-(x-4)}=\dfrac{x+4}{x-4}\neq -1$ No.

49. The numerator and denominator have no common factors other than 1.

51. $\dfrac{2+3x}{6}$ There is no factor common to both terms in the numerator.

53. $x^2-x-6=(x-3)(x+2)$

denominator $=x+2$

55. $(x+3)(x+4)=x^2+7x+12$

$=$ numerator

Cumulative Review Exercises: 6.1

57. $z=\dfrac{x-y}{2}$

$2z=x-y$

$2z-x=-y$

$y=x-2z$

58. Let x = smallest angle. Then the second angle = $x+30$ and third angle = $3x+10$.

angle 1 + angle 2 + angle 3 = 180°

$x+(x+30)+(3x+10)=180°$

$5x+40=180$

$5x=140$

$x=28$

So $x+30=28+30=58$

and $3x+10=3(28)+10=84+10=94$.

The three angles are 28°, 58°, and 94°.

59. $\left(\dfrac{3x^6y^2}{9x^4y^3}\right)^2=\left(\dfrac{x^2}{3y}\right)^2=\dfrac{x^4}{9y^2}$

60. $6x^2-4x-8-(-3x^2+6x+9)$

$=6x^2-4x-8+3x^2-6x-9$

$=9x^2-10x-17$

Group Activity/Challenge Problems

1. (a) $\dfrac{x+3}{x^2-2x+3x-6}$

$=\dfrac{x+3}{x(x-2)+3(x-2)}$

$=\dfrac{x+3}{(x+3)(x-2)};\ x\neq -3,\ x\neq 2$

(b) $\dfrac{x+3}{(x+3)(x-2)}=\dfrac{1}{x-2}$

3. (a) $\dfrac{x+5}{2x^3+7x^2-15x}$

$=\dfrac{x+5}{x(2x^2+7x-15)}$

$=\dfrac{x+5}{x(2x-3)(x+5)}$;

$x\neq 0,\ x\neq \dfrac{3}{2},\ x\neq -5$

(b) $\dfrac{x+5}{x(2x-3)(x+5)}=\dfrac{1}{2(2x-3)}$

5. $\dfrac{\frac{1}{5}x^5-\frac{2}{3}x^4}{\frac{1}{5}x^5-\frac{2}{3}x^4}=1$

Exercise Set 6.2

1. $\dfrac{3x}{2y}\cdot\dfrac{y^2}{6}=\dfrac{3x}{2y}\cdot\dfrac{y\cdot y}{3\cdot 2}=\dfrac{xy}{4}$

3. $\dfrac{16x^2}{y^4}\cdot\dfrac{5x^2}{y^2}=\dfrac{80x^4}{y^6}$

5. $\dfrac{6x^5y^3}{5z^3}\cdot\dfrac{6x^4}{5yz^4}=\dfrac{36x^9y^2}{25z^7}$

7. $\dfrac{3x-2}{3x+2}\cdot\dfrac{4x-1}{1-4x}=\dfrac{3x-2}{3x+2}\cdot\dfrac{4x-1}{-(4x-1)}$

$=\dfrac{-3x+2}{3x+2}$

9. $\dfrac{x^2+7x+12}{x+4}\cdot\dfrac{1}{x+3}=\dfrac{(x+4)(x+3)}{(x+4)(x+3)}=1$

11. $\dfrac{a^2-b^2}{a}\cdot\dfrac{a^2+ab}{a+b}=\dfrac{\left(a^2-b^2\right)a(a+b)}{a(a+b)}$

$=a^2-b^2$

13. $\dfrac{6x^2-14x-12}{6x+4}\cdot\dfrac{x+3}{2x^2-2x-12}$

$=\dfrac{2\left(3x^2-7x-6\right)}{2(3x+2)}\cdot\dfrac{(x+3)}{2\left(x^2-x-6\right)}$

$=\dfrac{2(3x+2)(x-3)(x+3)}{2(3x+2)2(x-3)(x+2)}=\dfrac{(x+3)}{2(x+2)}$

15. $\dfrac{x+3}{x-3}\cdot\dfrac{x^3-27}{x^2+3x+9}$

$=\dfrac{(x+3)(x-3)\left(x^2+3x+9\right)}{(x-3)\left(x^2+3x+9\right)}=x+3$

17. $\dfrac{6x^3}{y}\div\dfrac{2x}{y^2}=\dfrac{6x^3}{y}\cdot\dfrac{y^2}{2x}=3x^2y$

19. $\dfrac{25xy^2}{7z}\div\dfrac{5x^2y^2}{14z^2}=\dfrac{25xy^2}{7z}\cdot\dfrac{14z^2}{5x^2y^2}$

$=\dfrac{10z}{x}$

21. $\dfrac{7a^2b}{xy}\div\dfrac{7}{6xy}=\dfrac{7a^2b}{xy}\cdot\dfrac{6xy}{7}=6a^2b$

23. $\frac{3x^2+6x}{x} \div \frac{2x+4}{x^2}$

$= \frac{3x(x+2)}{x} \cdot \frac{x^2}{2(x+2)} = \frac{3x^2}{2}$

25. $(x-3) \div \frac{x^2+3x-18}{x}$

$= (x-3) \cdot \frac{x}{(x-3)(x+6)} = \frac{x}{x+6}$

27. $\frac{x^2-12x+32}{x^2-6x-16} \div \frac{x^2-x-12}{x^2-5x-24}$

$= \frac{x^2-12x+32}{x^2-6x-16} \cdot \frac{x^2-5x-24}{x^2-x-12}$

$= \frac{(x-8)(x-4)}{(x-8)(x+2)} \cdot \frac{(x-8)(x+3)}{(x-4)(x+3)} = \frac{x-8}{x+2}$

29. $\frac{2x^2+9x+4}{x^2+7x+12} \div \frac{2x^2-x-1}{(x+3)^2}$

$= \frac{(2x+1)(x+4)}{(x+3)(x+4)} \cdot \frac{(x+3)^2}{(2x+1)(x-1)} = \frac{x+3}{x-1}$

31. $\frac{x^2-y^2}{x^2-2xy+y^2} \div \frac{x+y}{x-y}$

$= \frac{(x-y)(x+y)}{(x-y)^2} \cdot \frac{x-y}{x+y} = 1$

33. $\frac{12x^2}{6y^2} \cdot \frac{36xy^5}{12} = 6x^3y^3$

35. $\frac{45a^2b^3}{12c^3} \cdot \frac{4c}{9a^3b^5} = \frac{5}{3ab^2c^2}$

37. $\frac{-xy}{a} \div \frac{-2ax}{6y} = \frac{-xy}{a} \cdot \frac{6y}{-2ax} = \frac{3y^2}{a^2}$

39. $\frac{80m^4}{49x^5y^7} \cdot \frac{14x^{12}y^5}{25m^5} = \frac{32x^7}{35y^2m}$

41. $(2x+5) \cdot \frac{1}{4x+10} = \frac{(2x+5)}{2(2x+5)} = \frac{1}{2}$

43. $\frac{1}{7x^2y} \div \frac{1}{21x^3y} = \frac{1}{7x^2y} \cdot \frac{21x^3y}{1} = 3x$

45. $\frac{12a^2}{4bc} \div \frac{3a^2}{bc} = \frac{12a^2}{4bc} \cdot \frac{bc}{3a^2} = 1$

47. $\frac{5-2x}{x+8} \cdot \frac{-x-8}{2x-5} = \frac{-(2x-5)}{x+8} \cdot \frac{-(x+8)}{2x-5} = 1$

49. $\frac{2a+2b}{3} \div \frac{a^2-b^2}{a-b}$

$= \frac{2(a+b)}{3} \cdot \frac{a-b}{(a-b)(a+b)} = \frac{2}{3}$

51. $\frac{1}{-x-4} \div \frac{x^2-7x}{x^2-3x-28}$

$= \frac{1}{-(x+4)} \cdot \frac{(x+4)(x-7)}{x(x-7)} = -\frac{1}{x}$

53. $\frac{4x+4y}{xy^2}\cdot\frac{x^2y}{3x+3y}=\frac{4(x+y)}{y}\cdot\frac{x}{3(x+y)}$

$=\frac{4x}{3y}$

55. $\frac{x^2+10x+21}{x+7}\div(x+3)$

$=\frac{(x+3)(x+7)}{(x+7)}\cdot\frac{1}{x+3}=1$

57. $\frac{3x^2-x-2}{x+7}\div\frac{x-1}{4x^2+25x-21}$

$=\frac{(3x+2)(x-1)}{x+7}\cdot\frac{(4x-3)(x+7)}{x-1}$

$=(3x+2)(4x-3)=12x^2-x-6$

59. $\frac{5x^2+17x+6}{x+3}\cdot\frac{x-1}{5x^2+7x+2}$

$=\frac{(5x+2)(x+3)}{x+3}\cdot\frac{x-1}{(5x+2)(x+1)}$

$=\frac{x-1}{x+1}$

61. $\frac{x^2-y^2}{x^2+xy}\cdot\frac{3x^2+6x}{3x^2-2xy-y^2}$

$=\frac{(x-y)(x+y)}{x(x+y)}\cdot\frac{3x(x+2)}{(3x+y)(x-y)}$

$=\frac{3(x+2)}{3x+y}=\frac{3x+6}{3x+y}$

63. $\frac{2x^2+7x-15}{1-x}\div\frac{3x^2+13x-10}{x-1}$

$=\frac{(2x-3)(x+5)}{-(x-1)}\cdot\frac{x-1}{(3x-2)(x+5)}$

$=-\frac{2x-3}{3x-2}=\frac{-2x+3}{3x-2}$

65. $\frac{x^2-4y^2}{x^2+3xy+2y^2}\cdot\frac{x+y}{x^2-4xy+4y^2}$

$=\frac{(x-2y)(x+2y)}{(x+2y)(x+y)}\cdot\frac{(x+y)}{(x-2y)^2}=\frac{1}{x-2y}$

67. $\frac{x+3}{(x-4)}\cdot\frac{\square}{x+3}=x+2$

Numerator must be:

$=(x+2)(x-4)=x^2-2x-8$

69. $\frac{x-4}{x+5}\cdot\frac{x+5}{\square}=\frac{1}{x+3}$

Denominator must be:

$=(x+3)(x-4)$

$=x^2-x-12$

71. Multiply numerators, then multiply denominators and simplify.

Cumulative Review Exercises: 6.2

73. $\left(4x^3y^2z^4\right)\left(5\cdot xy^3z^7\right)$

$=4\cdot5\cdot x^3xy^2y^3z^4z^7=20x^4y^5z^{11}$

74. $\frac{4x^3-5x}{2x-1}=2x^2+x-2-\frac{2}{2x-1}$

$$\begin{array}{r} 2x^2+x-2 \\ 2x-1\overline{)\,4x^3+0x^2-5x+0} \\ \underline{-4x^3+2x^2} \\ +2x^2-5x \\ \underline{-2x^2+x} \\ -4x+0 \\ \underline{+4x-2} \\ -2 \end{array}$$

75. $3x^2-9x-30=3\left(x^2-3x-10\right)$

$=3(x-5)(x+2)$

76. $3x^2-9x-30=0$

$x^2-3x-10=0$

$(x-5)(x+2)=0$

$x=5,\ x=-2$

Group Activity/Challenge Problems

1. $\left(\frac{x+2}{x^2-4x-12}\cdot\frac{x^2-9x+18}{x-2}\right)$

$\div\frac{x^2+5x+6}{x^2-4}$

$=\frac{x+2}{(x-6)(x+2)}\cdot\frac{(x-3)(x-6)}{x-2}$

$\cdot\frac{(x+2)(x-2)}{(x+2)(x+3)}$

$=\frac{x-3}{x+3}$

2. $\left(\frac{x^2-x-6}{2x^2-9x+9}\div\frac{x^2+x-12}{x^2+3x-4}\right)$

$\cdot\frac{2x^2-5x+3}{x^2+x-2}$

$=\frac{(x-3)(x+2)}{(2x-3)(x-3)}\cdot\frac{(x+4)(x-1)}{(x+4)(x-3)}$

$\cdot\frac{(2x-3)(x-1)}{(x+2)(x-1)}$

$=\frac{x-1}{x-3}$

3. $\frac{\square}{\square}\cdot\frac{x^2+3x-4}{x^2-4x+3}=\frac{x-2}{x-5}$

$\frac{\square}{\square}\cdot\frac{(x+4)(x-1)}{(x-3)(x-1)}=\frac{x-2}{x-5}$

$\frac{(x-2)(x-3)}{(x-5)(x+4)}\cdot\frac{(x+4)(x-1)}{(x-3)(x-1)}=\frac{x-2}{x-5}$

Therefore, the numerator is $(x-2)(x-3)=x^2-5x+6$ and the denominator is $(x-5)(x+4)=x^2-x-20$.

Exercise Set 6.3

1. $\frac{x-1}{6}+\frac{x}{6}=\frac{x-1+x}{6}=\frac{2x-1}{6}$

3. $\frac{2x+3}{5}-\frac{x}{5}=\frac{2x+3-x}{5}=\frac{x+3}{5}$

5. $\frac{1}{x}+\frac{x+2}{x}=\frac{1+x+2}{x}=\frac{x+3}{x}$

7. $\frac{x-3}{x}+\frac{x+3}{x}=\frac{x-3+x+3}{x}=\frac{2x}{x}=2$

9. $\dfrac{x}{x-2}+\dfrac{2x+3}{x-2}=\dfrac{x+2x+3}{x-2}=\dfrac{3x+3}{x-2}$

11. $\dfrac{9x+7}{6x^2}-\dfrac{3x+4}{6x^2}=\dfrac{9x+7-(3x+4)}{6x^2}=\dfrac{9x+7-3x-4}{6x^2}=\dfrac{6x+3}{6x^2}=\dfrac{3(2x+1)}{3(2x^2)}=\dfrac{2x+1}{2x^2}$

13. $\dfrac{-2x+6}{x^2+x-6}+\dfrac{3x-3}{x^2+x-6}=\dfrac{-2x+6+3x-3}{x^2+x-6}=\dfrac{x+3}{(x+3)(x-2)}=\dfrac{1}{x-2}$

15. $\dfrac{x+4}{3x+2}-\dfrac{x+4}{3x+2}=\dfrac{x+4-(x+4)}{3x+2}=0$

17. $\dfrac{2x+4}{x-7}-\dfrac{6x+5}{x-7}=\dfrac{2x+4-(6x+5)}{x-7}=\dfrac{2x+4-6x-5}{x-7}=\dfrac{-4x-1}{x-7}$

19. $\dfrac{x^2+4x+3}{x+2}-\dfrac{5x+9}{x+2}=\dfrac{x^2+4x+3-(5x+9)}{x+2}=\dfrac{x^2+4x+3-5x-9}{x+2}$

$=\dfrac{x^2-x-6}{x+2}=\dfrac{(x-3)(x+2)}{x+2}=x-3$

21. $\dfrac{-2x+5}{5x-10}+\dfrac{2(x-5)}{5x-10}=\dfrac{-2x+5+2(x-5)}{5x-10}=\dfrac{-2x+5+2x-10}{5x-10}=\dfrac{-5}{5(x-2)}=\dfrac{-1}{x-2}$

23. $\dfrac{x^2-2x-3}{x^2-x-6}+\dfrac{x-3}{x^2-x-6}=\dfrac{x^2-2x-3+x-3}{x^2-x-6}=\dfrac{x^2-x-6}{x^2-x-6}=1$

25. $\dfrac{-x-7}{2x-9}-\dfrac{-3x-16}{2x-9}=\dfrac{-x-7-(-3x-16)}{2x-9}=\dfrac{-x-7+3x+16}{2x-9}=\dfrac{2x+9}{2x-9}$

27. $\dfrac{x^2+6x}{(x+9)(x+5)}-\dfrac{27}{(x+9)(x+5)}=\dfrac{x^2+6x-27}{(x+9)(x+5)}=\dfrac{(x+9)(x-3)}{(x+9)(x+5)}=\dfrac{x-3}{x+5}$

29. $\dfrac{3x^2-7x}{4x^2-8x}+\dfrac{x}{4x^2-8x}=\dfrac{3x^2-7x+x}{4x^2-8x}=\dfrac{3x^2-6x}{4x^2-8x}=\dfrac{3x(x-2)}{4x(x-2)}=\dfrac{3}{4}$

31. $\dfrac{2x^2-6x+5}{2x^2+18x+16}-\dfrac{8x+21}{2x^2+18x+16}=\dfrac{2x^2-6x+5-8x-21}{2x^2+18x+16}=\dfrac{2x^2-14x-16}{2x^2+18x+16}$

$=\dfrac{2\left(x^2-7x-8\right)}{2\left(x^2+9x+8\right)}=\dfrac{(x-8)(x+1)}{(x+8)(x+1)}=\dfrac{x-8}{x+8}$

33. $\dfrac{x^2+3x-6}{x^2-5x+4}-\dfrac{-2x^2+4x-4}{x^2-5x+4}=\dfrac{x^2+3x-6+2x^2-4x+4}{x^2-5x+4}=\dfrac{3x^2-x-2}{x^2-5x+4}$

$=\dfrac{(3x+2)(x-1)}{(x-4)(x-1)}=\dfrac{3x+2}{x-4}$

35. $\dfrac{5x^2+40x+8}{x^2-64}+\dfrac{x^2+9x}{x^2-64}=\dfrac{5x^2+40x+8+x^2+9x}{x^2-64}=\dfrac{6x^2+49x+8}{x^2-64}=\dfrac{(6x+1)(x+8)}{(x-8)(x+8)}=\dfrac{6x+1}{x-8}$

37. $\dfrac{4x-3}{5x+4}-\dfrac{2x-7}{5x+4}=\dfrac{4x-3-(2x-7)}{5x+4}=\dfrac{4x-3-2x+7}{5x+4}=\dfrac{4x-3-2x-7}{5x+4}$

The sign of the 7 is +. The negative in $-(2x-7)$ was not distributed.

39. $\dfrac{4x+5}{x^2-6x}-\dfrac{-x^2+3x+6}{x^2-6x}=\dfrac{4x+5+x^2-3x-6}{x^2-6x}\neq\dfrac{4x+5+x^2+3x+6}{x^2-6x}$

The negative sign in $-(-x^2+3x+6)$ was not distributed.

41. $\dfrac{x}{3}+\dfrac{x-1}{3}$

Least common denominator $=3$

43. $\dfrac{1}{2x}+\dfrac{1}{3}\qquad 2x\cdot3=6x$

45. $\frac{3}{5x}+\frac{7}{2}$ $\quad 5x \cdot 2 = 10x$

47. $\frac{2}{x^2}+\frac{3}{x}$

Least common denominator $= x^2$

49. $\frac{x+4}{2x+3}+\frac{x}{1}$

Least common denominator $= 2x+3$

51. $\frac{x}{x+1}+\frac{4}{x^2}$

Least common denominator $= x^2(x+1)$

53. $\frac{x+3}{16x^2y}-\frac{5}{9x^3}=\frac{x+3}{2^4x^2y}-\frac{5}{3^2x^3}$

Least common denominator

$= 2^4 \cdot 3^2 \cdot x^3 \cdot y = 144x^3y$

55. $\frac{x^2+3}{18x}-\frac{x-7}{12(x+5)}$

$18x = 2 \cdot 3^2 \cdot x$

$12(x+5) = 2^2 3(x+5)$

Least common denominator

$= 2^2 \cdot 3^2 x(x+5) = 36x(x+5)$

57. $\frac{2x-7}{x^2+x}-\frac{x^2}{x+1}$

$x^2+x = x(x+1)$

Least common denominator

$= x(x+1)$

59. $\frac{15}{36x^2y}+\frac{x+3}{15xy^3}$

$36x^2y = 2^2 \cdot 3^2 x^2 y$

$15xy^3 = 3 \cdot 5xy^3$

Least common denominator

$= 2^2 \cdot 3^2 \cdot 5x^2y^3 = 180x^2y^3$

61. $\frac{6}{2x+8}+\frac{6x+3}{3x-9}$

$2x+8 = 2(x+4)$

$3x-9 = 3(x-3)$

Least common denominator

$= 6(x+4)(x-3)$

63. $\frac{9x+4}{x+6}-\frac{3x-6}{x+5}$

Least common denominator

$= (x+6)(x+5)$

65. $\frac{x-2}{x^2-5x-24}+\frac{3}{x^2+11x+24}$

$x^2-5x-24 = (x-8)(x+3)$

$x^2+11x+24 = (x+8)(x+3)$

Least common denominator

$= (x-8)(x+3)(x+8)$

67. $\frac{6}{x+3}-\frac{x+5}{x^2-4x+3}$

$x^2-4x+3 = (x-3)(x-1)$

Least common denominator

$= (x+3)(x-3)(x-1)$

69. $\dfrac{2x}{x^2-x-2}-\dfrac{3}{x^2+4x+3}$

$x^2-x-2=(x-2)(x+1)$

$x^2+4x+3=(x+3)(x+1)$

Least common denominator
$=(x-2)(x+1)(x+3)$

71. $\dfrac{3x-5}{x^2+4x+4}+\dfrac{3}{x+2}$

Least common denominator

$=x^2+4x+4=(x+2)^2$

73. Least common denominator
$=x^2-9x+20=(x-5)(x-4)$

75. $3x^2+16x-12=(3x-2)(x+6)$

$3x^2+17x-6=(3x-1)(x+6)$

Least common denominator
$=(3x-2)(x+6)(3x-1)$

77. $4x^2+4x+1=(2x+1)^2$

$8x^2+10x+3=(2x+1)(4x+3)$

Least common denominator
$=(2x+1)^2(4x+3)$

79. Each term changes sign.

81. $-x^2-x+12=-(x^2+x-12)$

Least common denominator

$=x^2+x-12=(x+4)(x-3)$

83. $\dfrac{-x^2-4x+3}{2x+5}+\dfrac{\square}{2x+5}=\dfrac{5x-7}{2x+5}$

$-x^2-4x+3+\square=5x-7$

$\square=x^2+4x-3+5x-7$

$=x^2+9x-10$

85. $-3x^2-9-\square=x^2+3x$

$-3x^2-9-x^2-3x=\square$

$\square=-4x^2-3x-9$

Cumulative Review Exercises: 6.3

86. $4\dfrac{3}{5}-2\dfrac{5}{9}=(4-2)+\dfrac{3}{5}-\dfrac{5}{9}$

$=2+\dfrac{27}{45}-\dfrac{25}{45}=2+\dfrac{2}{45}=2\dfrac{2}{45}$

87. $6x+4=-(x+2)-3x+4$

$6x+4=-x-2-3x+4$

$6x+4=-4x+2$

$6x+4x=2-4$

$10x=-2$

$x=\dfrac{-2}{10}=-\dfrac{1}{5}$

88. $\dfrac{6}{128}=\dfrac{x}{24}$

$x=\dfrac{6\cdot 24}{128}=\dfrac{144}{128}=1\dfrac{16}{128}=1\dfrac{1}{8}$

$=1.125$ ounces

89. $C_1 =$ Cost of plan 1

$C_2 =$ Cost of plan 2

$h =$ Number of hours played

$C_1 = 100 + 2h \quad C_2 = 250$

(a) $100 + 2h = 250$

$2h = 150$

$h = 75$

(b) $4 \cdot 52 = 208 > 75$. Choose plan 2.

90. $\dfrac{x^2+x-6}{2x^2+7x+3} \div \dfrac{x^2+5x+6}{x^2-4}$

$= \dfrac{x^2+x-6}{2x^2+7x+3} \cdot \dfrac{x^2-4}{x^2+5x+6}$

$= \dfrac{(x-2)(x+3)}{(2x+1)(x+3)} \; \dfrac{(x-2)(x+2)}{(x+3)(x+2)}$

$= \dfrac{(x-2)^2}{(2x+1)(x+3)}$

Group Activity/Challenge Problems

1. $\dfrac{3x-2}{x^2-9} - \dfrac{4x^2-6}{x^2-9} + \dfrac{5x-1}{x^2-9}$

$= \dfrac{3x-2-4x^2+6+5x-1}{x^2-9}$

$= \dfrac{-4x^2+8x+3}{x^2-9}$

3. $\dfrac{3}{2x^3y^6} - \dfrac{5}{6x^5y^9} + \dfrac{1}{5x^{12}y^2}$

$= \dfrac{3}{2x^3y^6} - \dfrac{5}{2 \cdot 3x^5y^9} + \dfrac{1}{5x^{12}y^2}$

Least common denominator

$= 2 \cdot 3 \cdot 5x^{12}y^9 = 30x^{12}y^9$

5. $\dfrac{4}{x^2-x-12} + \dfrac{3}{x^2-6x+8} + \dfrac{5}{x^2+x-6}$

$= \dfrac{4}{(x-4)(x+3)} + \dfrac{3}{(x-4)(x-2)}$

$+ \dfrac{5}{(x+3)(x-2)}$

Least common denominator

$= (x-4)(x+3)(x-2)$

7. $\dfrac{7}{2x^2+x-10} + \dfrac{5}{2x^2+7x+5}$

$- \dfrac{4}{-x^2+x+2}$

$= \dfrac{7}{(2x+5)(x-2)} + \dfrac{5}{(2x+5)(x+1)}$

$+ \dfrac{4}{(x-2)(x+1)}$

Least common denominator

$= (2x+5)(x-2)(x+1)$

Exercise Set 6.4

1. $\dfrac{3}{x} + \dfrac{1}{2x} = \dfrac{6}{2x} + \dfrac{1}{2x} = \dfrac{7}{2x}$

3. $\dfrac{4}{x^2} + \dfrac{3}{2x} = \dfrac{4 \cdot 2}{2x^2} + \dfrac{3x}{2x^2} = \dfrac{3x+8}{2x^2}$

5. $2 - \dfrac{1}{x^2} = \dfrac{2x^2}{x^2} - \dfrac{1}{x^2} = \dfrac{2x^2-1}{x^2}$

7. $\dfrac{2}{x^2} + \dfrac{3}{5x} = \dfrac{5 \cdot 2}{5x^2} + \dfrac{3x}{5x^2} = \dfrac{3x+10}{5x^2}$

9. $\frac{3}{4x^2y}+\frac{7}{5xy^2}=\frac{3\cdot 5y}{20x^2y^2}+\frac{7\cdot 4x}{20x^2y^2}$

$=\frac{15y}{20x^2y^2}+\frac{28x}{20x^2y^2}=\frac{28x+15y}{20x^2y^2}$

11. $x+\frac{x}{y}=\frac{xy}{y}+\frac{x}{y}=\frac{xy+x}{y}$

13. $\frac{3x-1}{2x}+\frac{2}{3x}=\frac{(3x-1)3}{6x}+\frac{4}{6x}$

$=\frac{9x-3+4}{6x}=\frac{9x+1}{6x}$

15. $\frac{5x}{y}+\frac{y}{x}=\frac{5x^2}{xy}+\frac{y^2}{xy}=\frac{5x^2+y^2}{xy}$

17. $\frac{4}{5x^2}-\frac{6}{y}=\frac{4y}{5x^2y}-\frac{6(5x^2)}{5x^2y}=\frac{4y-30x^2}{5x^2y}$

19. $\frac{5}{x}+\frac{3}{x-2}=\frac{5(x-2)}{x(x-2)}+\frac{3x}{x(x-2)}$

$=\frac{8x-10}{x(x-2)}$

21. $\frac{9}{a+3}+\frac{2}{a}=\frac{9a}{a(a+3)}+\frac{2(a+3)}{a(a+3)}$

$=\frac{11a+6}{a(a+3)}$

23. $\frac{4}{3x}-\frac{2x}{3x+6}=\frac{4}{3x}-\frac{2x}{3(x+2)}$

$=\frac{4(x+2)}{3x(x+2)}-\frac{2x\cdot x}{3x(x+2)}=\frac{-2x^2+4x+8}{3x(x+2)}$

25. $\frac{4}{x-3}+\frac{2}{3-x}=\frac{4}{x-3}-\frac{2}{x-3}=\frac{2}{x-3}$

27. $\frac{7}{x+5}-\frac{4}{-x-5}=\frac{7}{x+5}+\frac{4}{x+5}=\frac{11}{x+5}$

29. $\frac{3}{x+1}+\frac{4}{x-1}$

$=\frac{3(x-1)}{(x+1)(x-1)}+\frac{4(x+1)}{(x+1)(x-1)}$

$=\frac{3x-3+4x+4}{(x+1)(x-1)}=\frac{7x+1}{(x+1)(x-1)}$

31. $\frac{x+5}{x-5}-\frac{x-5}{x+5}$

$=\frac{(x+5)^2}{(x-5)(x+5)}-\frac{(x-5)^2}{(x-5)(x+5)}$

$=\frac{x^2+10x+25-(x^2-10x+25)}{(x-5)(x+5)}$

$=\frac{20x}{(x-5)(x+5)}$

33. $\frac{x}{x^2-9}+\frac{4}{x+3}$

$=\frac{x}{(x+3)(x-3)}+\frac{4(x-3)}{(x+3)(x-3)}$

$=\frac{5x-12}{(x+3)(x-3)}$

35. $$\frac{x+3}{x^2-9}-\frac{3}{x+3}=\frac{x+3}{(x+3)(x-3)}-\frac{3(x-3)}{(x+3)(x-3)}=\frac{x+3-3x+9}{(x+3)(x-3)}=\frac{-2x+12}{(x+3)(x-3)}$$

37. $$\frac{2x+3}{x^2-7x+12}-\frac{2}{x-3}$$

$$\frac{2x+3}{(x-3)(x-4)}-\frac{2(x-4)}{(x-3)(x-4)}=\frac{2x+3-2x+8}{(x-3)(x-4)}=\frac{11}{(x-3)(x-4)}$$

39. $$\frac{x^2}{x^2+2x-8}-\frac{x-4}{x+4}=\frac{x^2}{(x+4)(x-2)}-\frac{(x-4)(x-2)}{(x+4)(x-2)}=\frac{x^2-(x^2-6x+8)}{(x+4)(x-2)}=\frac{6x-8}{(x+4)(x-2)}$$

41. $$\frac{x-1}{x^2+4x+4}+\frac{x-1}{x+2}=\frac{x-1}{(x+2)^2}+\frac{(x-1)(x+2)}{(x+2)^2}=\frac{x-1+x^2+x-2}{(x+2)^2}=\frac{x^2+2x-3}{(x+2)^2}$$

43. $$\frac{1}{x^2+3x-10}+\frac{3}{x^2+x-6}=\frac{1}{(x+5)(x-2)}+\frac{3}{(x-2)(x+3)}$$

$$=\frac{x+3}{(x+5)(x-2)(x+3)}+\frac{3(x+5)}{(x-2)(x+3)(x+5)}=\frac{4x+18}{(x+5)(x-2)(x+3)}$$

45. $$\frac{1}{x^2-4}+\frac{3}{x^2+5x+6}=\frac{1}{(x-2)(x+2)}+\frac{3}{(x+3)(x+2)}$$

$$=\frac{x+3}{(x-2)(x+2)(x+3)}+\frac{3(x-2)}{(x-2)(x+2)(x+3)}=\frac{4x-3}{(x-2)(x+2)(x+3)}$$

47. $$\frac{x}{3x^2+5x-2}-\frac{4}{2x^2+7x+6}=\frac{x}{(3x-1)(x+2)}-\frac{4}{(2x+3)(x+2)}$$

$$=\frac{x(2x+3)}{(3x-1)(x+2)(2x+3)}-\frac{4(3x-1)}{(2x+3)(x+2)(3x-1)}=\frac{2x^2+3x-12x+4}{(3x-1)(x+2)(2x+3)}$$

$$=\frac{2x^2-9x+4}{(3x-1)(x+2)(2x+3)}$$

49. $\dfrac{x}{4x^2+11x+6}-\dfrac{2}{8x^2+2x-3}=\dfrac{x}{(4x+3)(x+2)}-\dfrac{2}{(4x+3)(2x-1)}$

$=\dfrac{x(2x-1)-2(x+2)}{(4x+3)(x+2)(2x-1)}=\dfrac{2x^2-x-2x-4}{(4x+3)(x+2)(2x-1)}=\dfrac{2x^2-3x-4}{(4x+3)(x+2)(2x-1)}$

51. **(a)** Factor the denominators. Find a common denominator. Multiply the numerator and denominator of each fraction by the factors needed to build the common denominator. Once both fractions have a common denominator, add or subtract the numerators.

(b) $\dfrac{x}{x^2-x-6}+\dfrac{3}{x^2-4}=\dfrac{x}{(x-3)(x+2)}+\dfrac{3}{(x-2)(x+2)}$

$=\dfrac{x(x-2)}{(x-3)(x+2)(x-2)}+\dfrac{3(x-3)}{(x-2)(x+2)(x-3)}=\dfrac{x^2-2x+3x-9}{(x-3)(x+2)(x-2)}$

$=\dfrac{x^2+x-9}{(x-3)(x+2)(x-2)}$

Cumulative Review Exercises: 6.4

52. $\dfrac{18\text{ counts}}{2\text{ minutes}}\cdot\left(1\dfrac{1}{2}\text{ hours}\right)$

$\dfrac{18\text{ counts}}{2\text{ minutes}}(90\text{ minutes})=810\text{ counts}$

53. $3(x-2)+2<4(x+1)$

$3x-6+2<4x+4$

$3x-4<4x+4$

$-8<x$

$x>-8$

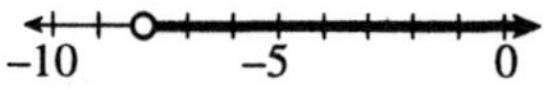

54.

$$\begin{array}{r} 4x-3 \\ 2x+3\overline{)\,8x^2+6x-13} \\ \underline{-8x^2-12x} \\ -6x-13 \\ \underline{+6x+9} \\ -4 \end{array}$$

Answer: $4x-3-\dfrac{4}{2x+3}$

55. $$\frac{x^2+xy-6y^2}{x^2-xy-2y^2}\cdot\frac{x^2-y^2}{x^2+2xy-3y^2}=\frac{(x+3y)(x-2y)}{(x+y)(x-2y)}\cdot\frac{(x-y)(x+y)}{(x+3y)(x-y)}=1$$

Group Activity/Challenge Problems

1. $$\frac{x}{x-2}+\frac{3}{x+2}+\frac{4}{x^2-4}=\frac{x(x+2)}{(x-2)(x+2)}+\frac{3(x-2)}{(x+2)(x-2)}+\frac{4}{(x+2)(x-2)}$$

$$=\frac{x^2+2x+3x-6+4}{(x+2)(x-2)}=\frac{x^2+5x-2}{(x+2)(x-2)}$$

3. $$\frac{x+6}{4-x^2}-\frac{x+3}{x+2}+\frac{x-3}{2-x}=\frac{x+6}{(2-x)(2+x)}-\frac{(x+3)(2-x)}{(2+x)(2-x)}+\frac{(x-3)(2+x)}{(2-x)(2+x)}$$

$$=\frac{x+6-2x+x^2-6+3x+2x+x^2-6-3x}{(2-x)(2+x)}=\frac{2x^2+x-6}{(2-x)(2+x)}=\frac{(2x-3)(x+2)}{(2-x)(2+x)}=\frac{2x-3}{2-x}$$

5. $$\frac{3x}{x^2-4}+\frac{4}{x^3+8}=\frac{3x(x^2-2x+4)}{(x+2)(x-2)(x^2-2x+4)}+\frac{4(x-2)}{(x+2)(x-2)(x^2-2x+4)}$$

$$=\frac{3x^3-6x^2+12x+4x-8}{(x+2)(x-2)(x^2-2x+4)}=\frac{3x^3-6x^2+16x-8}{(x+2)(x-2)(x^2-2x+4)}$$

Exercise Set 6.5

1. $$\frac{1+\frac{3}{5}}{2+\frac{1}{5}} = \frac{\left(1+\frac{3}{5}\right)5}{\left(2+\frac{1}{5}\right)5} = \frac{5+3}{10+1} = \frac{8}{11}$$

3. $$\frac{2+\frac{3}{8}}{1+\frac{1}{3}} = \frac{\left(2+\frac{3}{8}\right)8\cdot 3}{\left(1+\frac{1}{3}\right)8\cdot 3} = \frac{48+9}{24+8} = \frac{57}{32}$$

5. $$\frac{\frac{4}{9}-\frac{3}{8}}{4-\frac{3}{5}} = \frac{\left(\frac{4}{9}-\frac{3}{8}\right)9\cdot 8\cdot 5}{\left(4-\frac{3}{5}\right)9\cdot 8\cdot 5} = \frac{(32-27)5}{(20-3)9\cdot 8}$$
$$= \frac{5\cdot 5}{17\cdot 9\cdot 8} = \frac{25}{1224}$$

7. $$\frac{\frac{x^2y}{4}\cdot 4\cdot x}{\frac{2}{x}\cdot 4\cdot x} = \frac{x^3y}{8}$$

9. $$\frac{\frac{8x^2y}{3z^3}\cdot 9z^5}{\frac{4xy}{9z^5}\cdot 9z^5} = \frac{24x^2yz^2}{4xy} = 6xz^2$$

11. $$\frac{\left(x+\frac{1}{y}\right)y}{\left(\frac{x}{y}\right)y} = \frac{xy+1}{x}$$

13. $$\frac{\left(\frac{9}{x}+\frac{3}{x^2}\right)x^2}{\left(3+\frac{1}{x}\right)x^2} = \frac{9x+3}{3x^2+x} = \frac{3(3x+1)}{(3x+1)x} = \frac{3}{x}$$

15. $$\frac{\left(3-\frac{1}{y}\right)y}{\left(2-\frac{1}{y}\right)y} = \frac{3y-1}{2y-1}$$

17. $$\frac{\left(\frac{x}{y}-\frac{y}{x}\right)xy}{\left(\frac{x+y}{x}\right)xy} = \frac{x^2-y^2}{(x+y)y}$$
$$= \frac{(x+y)(x-y)}{(x+y)y} = \frac{x-y}{y}$$

19. $$\frac{\left(\frac{a^2}{b}-b\right)ab}{\left(\frac{b^2}{a}-a\right)ab} = \frac{(a^2-b^2)a}{(b^2-a^2)b} = -\frac{a}{b}$$

21. $$\frac{\left(\frac{a}{b}-2\right)b}{\left(\frac{-a}{b}+2\right)b} = \frac{a-2b}{-a+2b} = \frac{a-2b}{-(a-2b)} = -1$$

23. $$\frac{\left(\frac{4x+8}{3x^2}\right)6x^2}{\left(\frac{4x}{6}\right)6x^2} = \frac{(4x+8)2}{4x^3} = \frac{2(x+2)}{x^3}$$

25. $$\frac{\left(\frac{1}{a}+\frac{1}{b}\right)ab}{\left(\frac{1}{ab}\right)ab} = b+a$$

27. $$\frac{\left(\frac{a}{b}+\frac{1}{a}\right)ab}{\left(\frac{b}{a}+\frac{1}{a}\right)ab} = \frac{a^2+b}{b^2+b} = \frac{a^2+b}{b(b+1)}$$

29. $$\frac{\left(\frac{1}{x}-\frac{1}{y}\right)xy}{\left(\frac{1}{x}+\frac{1}{y}\right)xy}=\frac{y-x}{y+x}$$

31. $$\frac{\left(\frac{1}{x^2}+\frac{1}{x}\right)x^2y^2}{\left(\frac{1}{y}+\frac{1}{y^2}\right)x^2y^2}=\frac{y^2+xy^2}{x^2y+x^2}=\frac{y^2(x+1)}{x^2(y+1)}$$

33. A fraction whose numerator or denominator (or both) contains a fraction.

Cumulative Review Exercises: 6.5

35. $4x+3(x-2)=7x-6$

$4x+3x-6=7x-6$

$-6=-6$

True for all real numbers.

36. A polynomial is a sum of terms of the form ax^n where a is a real number and n is a whole number.

37. $(4x^2y^3)^2\cdot(3xy^4)=16x^4y^6 3xy^4$

$=48x^5y^{10}$

38. $$\frac{x}{3x^2+17x-6}-\frac{2}{x^2+3x-18}$$

$$=\frac{x}{(3x-1)(x+6)}-\frac{2}{(x+6)(x-3)}$$

$$=\frac{x(x-3)-2(3x-1)}{(3x-1)(x+6)(x-3)}$$

$$=\frac{x^2-3x-6x+2}{(3x-1)(x+6)(x-3)}$$

$$=\frac{x^2-9x+2}{(3x-1)(x+6)(x-3)}$$

Group Activity/Challenge Problems

1. (a) $$E=\frac{\frac{1}{2}\left(\frac{2}{3}\right)}{\frac{2}{3}+\frac{1}{2}}=\frac{\frac{2}{6}}{\frac{4+3}{6}}=\frac{2}{6}\cdot\frac{6}{7}=\frac{2}{7}$$

(b) $$E=\frac{\frac{1}{2}\left(\frac{4}{5}\right)}{\frac{4}{5}+\frac{1}{2}}=\frac{\frac{4}{10}}{\frac{8+5}{10}}=\frac{4}{10}\cdot\frac{10}{13}=\frac{4}{13}$$

3. $$\frac{\frac{a}{b}+b-\frac{1}{a}}{\frac{a}{b^2}-\frac{b}{a}+\frac{1}{a^2}}=\frac{\frac{a^2+b^2a-b}{ba}}{\frac{a^3-ab^3+b^2}{a^2b^2}}$$

$$=\frac{a^2+b^2a-b}{ba}\cdot\frac{a^2b^2}{a^3-ab^3+b^2}$$

$$=\frac{a^3b+a^2b^3-ab^2}{a^3-ab^3+b^2}$$

Exercise Set 6.6

1. $\frac{3}{5}=\frac{x}{10}$

Multiply both sides of the equation by the least common denominator, 10.

$10\left(\frac{3}{5}\right)=\left(\frac{x}{10}\right)10$

$2\left(\frac{3}{1}\right)=\left(\frac{x}{1}\right)1$

$6=x$

Check: $\frac{3}{5}=\frac{x}{10}$

$\frac{3}{5}=\frac{(6)}{10}$

$\frac{3}{5}=\frac{3}{5}$ True

3. $\frac{5}{12}=\frac{20}{x}$

Cross-multiply

$5x=20(12)$

$5x=240$

$x=48$

Check: $\frac{5}{12}=\frac{20}{x}$

$\frac{5}{12}=\frac{20}{(48)}$

$\frac{5}{12}=\frac{5}{12}$ True

5. $\frac{a}{25}=\frac{12}{10}$

Cross-multiply

$10a=12(25)$

$10a=300$

$a=30$

Check: $\frac{a}{25}=\frac{12}{10}$

$\frac{(30)}{25}=\frac{12}{10}$

$\frac{6}{5}=\frac{6}{5}$ True

7. $\frac{9}{3b}=\frac{-6}{2}$

Multiply both sides of the equation by the least common denominator, $6b$.

$6b\left(\frac{9}{3b}\right)=\left(\frac{-6}{2}\right)6b$

$2\left(\frac{9}{1}\right)=\left(\frac{-6}{1}\right)3b$

$18=-18b$

$-1=b$

Check: $\frac{9}{3b}=\frac{-6}{2}$

$\frac{9}{3(-1)}=\frac{-6}{2}$

$-3=-3$ True

9. $\frac{x+4}{9}=\frac{5}{9}$

Multiply both sides of the equation by the least common denominator, 9.

$9\left(\frac{x+4}{9}\right)=\left(\frac{5}{9}\right)9$

$1\left(\frac{x+4}{1}\right)=\left(\frac{5}{1}\right)1$

$x+4=5$

$x=1$

Check: $\frac{x+4}{9}=\frac{5}{9}$

$\frac{(1)+4}{9}=\frac{5}{9}$

$\frac{5}{9}=\frac{5}{9}$ True

11. $\frac{4x+5}{6}=\frac{7}{2}$

Cross-multiply

$2(4x+5)=7(6)$

Distributive property

$2(4x)+2(5)=7(6)$

$8x+10=42$

$8x=32$

$x=4$

Check: $\frac{4x+5}{6}=\frac{7}{2}$

$\frac{4(4)+5}{6}=\frac{7}{2}$

$\frac{21}{6}=\frac{7}{2}$

$\frac{7}{2}=\frac{7}{2}$ True

13. $\frac{6x+7}{10}=\frac{2x+9}{6}$

Cross-multiply

$6(6x+7)=10(2x+9)$

Distributive property

$6(6x)+6(7)=10(2x)+10(9)$

$36x+42=20x+90$

$16x+42=90$

$16x=48$

$x=3$

Check: $\frac{6x+7}{10}=\frac{2x+9}{6}$

$\frac{6(3)+7}{10}=\frac{2(3)+9}{6}$

$\frac{25}{10}=\frac{15}{6}$

$\frac{5}{2}=\frac{5}{2}$ True

15. $\frac{x}{3}-\frac{3x}{4}=\frac{1}{12}$

Multiply both sides of the equation by the least common denominator, 12.

$12\left(\frac{x}{3}-\frac{3x}{4}\right)=\left(\frac{1}{12}\right)12$

$12\left(\frac{x}{3}\right)-12\left(\frac{3x}{4}\right)=1$

$4x-9x=1$

$-5x=1$

$x=-\frac{1}{5}$

Check: $\frac{x}{3}-\frac{3x}{4}=\frac{1}{12}$

$$\frac{\left(-\frac{1}{5}\right)}{3}-\frac{3\left(-\frac{1}{5}\right)}{4}=\frac{1}{12}$$

$$\frac{5\left(-\frac{1}{5}\right)}{5\cdot 3}-\frac{3\left(-\frac{1}{5}\right)5}{4\cdot 5}=\frac{1}{12}$$

$$\frac{1\left(-\frac{1}{1}\right)}{5\cdot 3}-\frac{3\left(-\frac{1}{1}\right)1}{4\cdot 5}=\frac{1}{12}$$

$$\frac{-1}{5\cdot 3}-\frac{-3}{4\cdot 5}=\frac{1}{3\cdot 4}$$

Multiply both sides of the equation by the least common denominator, $3\cdot 4\cdot 5$.

$$3\cdot 4\cdot 5\left(\frac{-1}{5\cdot 3}-\frac{-3}{4\cdot 5}\right)$$

$$=\left(\frac{1}{3\cdot 4}\right)3\cdot 4\cdot 5$$

$$4\left(\frac{-1}{1}\right)-3\left(\frac{-3}{1}\right)=\left(\frac{1}{1}\right)5$$

$$-4+9=5$$

$$5=5 \qquad \text{True}$$

17. $\frac{3}{4}-x=2x$

Multiply both sides of the equation by the least common denominator, 4.

$$4\left(\frac{3}{4}-x\right)=(2x)4$$

$$1\left(\frac{3}{1}\right)-4(x)=(2x)4$$

$$3-4x=8x,\ 3=12x,\ \frac{1}{4}=x$$

A check will show that $\frac{1}{4}$ is a solution to the equation.

19. $\frac{5}{3x}+\frac{3}{x}=1$

Multiply both sides of the equation by the least common denominator, $3x$.

$$3x\left(\frac{5}{3x}+\frac{3}{x}\right)=(1)3x$$

$$1\left(\frac{5}{1}\right)+3\left(\frac{3}{1}\right)=3x(1)$$

$$5+9=3x$$

$$14=3x$$

$$\frac{14}{3}=x$$

A check will show that $\frac{14}{3}$ is a solution to the equation.

21. $\frac{x-1}{x-5}=\frac{4}{x-5}$

Multiply both sides of the equation by the least common denominator, $(x-5)$.

$$(x-5)\left(\frac{x-1}{x-5}\right)=\left(\frac{4}{x-5}\right)(x-5)$$

$$1\left(\frac{x-1}{1}\right)=\left(\frac{4}{1}\right)1$$

$$x-1=4$$

$$x=5$$

Check: $\frac{x-1}{x-5}=\frac{4}{x-5}$

$$\frac{5-1}{5-5}=\frac{4}{5-5}$$

$$\frac{4}{0}=\frac{4}{0}$$

$\frac{4}{0}$ is not a real number. Since $\frac{4}{0}$ is not a real number, 5 is an extraneous solution. Thus this equation has no solution.

23. $\dfrac{5y-3}{7}=\dfrac{15y-2}{28}$

Multiply both sides of the equation by the least common denominator, 28.

$$28\left(\frac{5y-3}{7}\right)=\left(\frac{15y-2}{28}\right)28$$

$$4\left(\frac{5y-3}{1}\right)=\left(\frac{15y-2}{1}\right)1$$

$$20y-12=15y-2$$

$$5y-12=-2$$

$$5y=10$$

$$y=2$$

A check will show that 2 is a solution to the equation.

25. $\dfrac{5}{-x-6}=\dfrac{2}{x}$

Cross-multiply

$$5(x)=2(-x-6)$$

$$5x=-2x-12$$

$$7x=-12$$

$$x=-\frac{12}{7}$$

A check will show that $-\dfrac{12}{7}$ is a solution to the equation.

27. $\dfrac{2x-3}{x-4}=\dfrac{5}{x-4}$

Multiply both sides of the equation by the least common denominator, $(x-4)$.

$$(x-4)\left(\frac{2x-3}{x-4}\right)=\left(\frac{5}{x-4}\right)(x-4)$$

$$1\left(\frac{2x-3}{1}\right)=\left(\frac{5}{1}\right)1$$

$$2x-3=5$$

$$2x=8,\ x=4$$

Check: $\dfrac{2x-3}{x-4}=\dfrac{5}{x-4}$

$$\frac{2(4)-3}{4-4}=\frac{5}{4-4}$$

$$\frac{5}{0}=\frac{5}{0}$$

$\dfrac{5}{0}$ is not a real number. Since $\dfrac{5}{0}$ is not a real number, 4 is an extraneous solution. Therefore, this equation has no solution.

29. $\dfrac{x-2}{x+4}=\dfrac{x+1}{x+10}$

Cross-multiply

$$(x-2)(x+10)=(x+1)(x+4)$$

$$x^2-2x+10x-20=x^2+x+4x+4$$

$$-2x+10x-20=x+4x+4$$

$$8x-20=5x+4$$

$$3x-20=4$$

$$3x=24$$

$$x=8$$

Check: $\dfrac{x-2}{x+4}=\dfrac{x+1}{x+10}$

$$\frac{8-2}{8+4}=\frac{8+1}{8+10}$$

$$\frac{6}{12}=\frac{9}{18}$$

$$\frac{1}{2}=\frac{1}{2} \qquad \text{True}$$

31. $\dfrac{2x-1}{3}-\dfrac{3x}{4}=\dfrac{5}{6}$

Multiply both sides of the equation by the least common denominator, 12.

$12\left(\dfrac{2x-1}{3}-\dfrac{3x}{4}\right)=\left(\dfrac{5}{6}\right)12$

$4\left(\dfrac{2x-1}{1}\right)-3\left(\dfrac{3x}{1}\right)=\left(\dfrac{5}{1}\right)2$

$8x-4-9x=10$

$-4-x=10$

$-4=10+x$

$-14=x$

A check will show that –14 is a solution to the equation.

33. $x+\dfrac{6}{x}=-5$

Multiply both sides of the equation by x.

$x\left(x+\dfrac{6}{x}\right)=(-5)x$

$x(x)+1\left(\dfrac{6}{1}\right)=(-5)x$

$x^2+6=-5x$

$x^2+5x+6=0$

$(x+2)(x+3)=0$

$x+2=0$ or $x+3=0$

$x=-2$ or $x=-3$

A check will show that –2 and –3 are the solutions to the equation.

35. $\dfrac{3y-2}{y+1} = 4 - \dfrac{y+2}{y-1}$

Multiply both sides of the equation by the least common denominator, $(y+1)(y-1)$.

$$(y+1)(y-1)\left(\frac{3y-2}{y+1}\right) = \left[4 - \frac{y+2}{y-1}\right](y+1)(y-1)$$

$$(y-1)\left(\frac{3y-2}{1}\right) = \left[4 - \frac{y+2}{y-1}\right](y+1)(y-1)$$

$$(y-1)(3y-2) = 4(y+1)(y-1) - \left(\frac{y+2}{y-1}\right)(y+1)(y-1)$$

$$3y^2 - 3y - 2y + 2 = 4(y^2 + y - y - 1) - (y+2)(y+1)$$

$$3y^2 - 3y - 2y + 2 = 4[y^2 + (-1)] - (y+2)(y+1)$$

$$3y^2 - 5y + 2 = 4y^2 - 4 - y^2 - 2y - y - 2$$

$$-5y + 2 = -3y - 6$$

$$2 = 2y - 6$$

$$8 = 2y$$

$$4 = y$$

A check will show that 4 is a solution to the equation.

37. $\dfrac{1}{x+3} + \dfrac{1}{x-3} = \dfrac{-5}{x^2-9}$

$\dfrac{1}{x+3} + \dfrac{1}{x-3} = \dfrac{-5}{(x-3)(x+3)}$ Factor $x^2 - 9$.

Multiply both sides of the equation by the least common denominator, $(x-3)(x+3)$.

$$(x-3)(x+3)\left[\frac{1}{x+3} + \frac{1}{x-3}\right] = \left[\frac{-5}{(x-3)(x+3)}\right](x-3)(x+3)$$

$$(x-3)\left[\frac{1}{1}\right] + (x+3)\left[\frac{1}{1}\right] = -5$$

$$x - 3 + x + 3 = -5$$

$$2x = -5$$

$$x = -\frac{5}{2}$$

A check will show that $-\dfrac{5}{2}$ is a solution to the equation.

39. $\frac{a}{a-3}+\frac{3}{2}=\frac{3}{a-3}$

Multiply both sides of the equation by the least common denominator, $2(a-3)$.

$$2(a-3)\left[\frac{a}{a-3}+\frac{3}{2}\right]=\left[\frac{3}{a-3}\right]2(a-3)$$

$$2(a-3)\left[\frac{a}{a-3}\right]+2(a-3)\left[\frac{3}{2}\right]=\left[\frac{3}{a-3}\right]2(a-3)$$

$$2\left[\frac{a}{1}\right]+(a-3)\left[\frac{3}{1}\right]=(3)2$$

$$2a+3a-9=6$$

$$5a-9=6$$

$$5a=15,\ a=3$$

Check: $\frac{a}{a-3}+\frac{3}{2}=\frac{3}{a-3}$

$$\frac{3}{3-3}+\frac{3}{2}=\frac{3}{3-3}$$

$$\frac{3}{0}+\frac{3}{2}=\frac{3}{0}$$

$\frac{3}{0}$ is not a real number. Since $\frac{3}{0}$ is not a real number, 3 is an extraneous solution. Therefore, the equation has no solution.

41. $\frac{2}{x-3}-\frac{4}{x+3}=\frac{8}{x^2-9}$ First factor x^2-9.

$$\frac{2}{x-3}-\frac{4}{x+3}=\frac{8}{(x-3)(x+3)}$$

Multiply both sides of the equation by the least common denominator, $(x-3)(x+3)$.

$$(x-3)(x+3)\left[\frac{2}{x-3}-\frac{4}{x+3}\right]=\left[\frac{8}{(x-3)(x+3)}\right](x-3)(x+3)$$

$$(x-3)(x+3)\left[\frac{2}{x-3}\right]-(x-3)(x+3)\left[\frac{4}{x+3}\right]=\left[\frac{8}{(x-3)(x+3)}\right](x-3)(x+3)$$

$$(x+3)\left[\frac{2}{1}\right]-(x-3)\left[\frac{4}{1}\right]=\left[\frac{8}{1}\right]1$$

$$2x+6-4x+12=8$$

$$-2x+18=8$$

$$-2x=-10,\ x=5$$

Check: $\dfrac{2}{x-3}-\dfrac{4}{x+3}=\dfrac{8}{x^2-9}$

$$\frac{2}{5-3}-\frac{4}{5+3}=\frac{8}{(5)^2-9}$$

$$\frac{2}{2}-\frac{4}{8}=\frac{8}{16}$$

$$1-\frac{1}{2}=\frac{1}{2}$$

$$\frac{1}{2}=\frac{1}{2} \qquad \text{True}$$

43. $\dfrac{y}{2y+2}+\dfrac{2y-16}{4y+4}=\dfrac{y-3}{y+1}$

First factor $2y+2$ and $4y+4$.

$$\frac{y}{2(y+1)}+\frac{2y-16}{4(y+1)}=\frac{y-3}{y+1}$$

Multiply both sides of the equation by the least common denominator, $4(y+1)$.

$$4(y+1)\left[\frac{y}{2(y+1)}+\frac{2y-16}{4(y+1)}\right]=\left[\frac{y-3}{y+1}\right]4(y+1)$$

$$4(y+1)\left[\frac{y}{2(y+1)}\right]+4(y+1)\left[\frac{2y-16}{4(y+1)}\right]=\left[\frac{y-3}{y+1}\right]4(y+1)$$

$$2\left[\frac{y}{1}\right]+1\left[\frac{2y-16}{1}\right]=\left[\frac{y-3}{1}\right]4$$

$$2y+2y-16=4y-12$$

$$4y-16=4y-12$$

$$-16=-12 \qquad \text{False}$$

Since this is a false statement, the equation has no solution.

45. $\frac{1}{2}+\frac{1}{x-1}=\frac{2}{x^2-1}$ Factor x^2-1.

$\frac{1}{2}+\frac{1}{x-1}=\frac{2}{(x-1)(x+1)}$

Multiply both sides of the equation by the least common denominator, $2(x-1)(x+1)$.

$2(x-1)(x+1)\left[\frac{1}{2}+\frac{1}{x-1}\right]=\left(\frac{2}{(x-1)(x+1)}\right)2(x-1)(x+1)$

$2(x-1)(x+1)\left[\frac{1}{2}\right]+2(x-1)(x+1)\left[\frac{1}{x-1}\right]=\left(\frac{2}{1}\right)2$

$(x-1)(x+1)\left[\frac{1}{1}\right]+2(x+1)\left[\frac{1}{1}\right]=4$

$x^2-x+x-1+2x+2=4$

$x^2+2x+1=4$

$x^2+2x-3=0$

$(x+3)(x-1)=0$

$x+3=0$ or $x-1=0$

$x=-3$ or $x=1$

Check: $x=-3$

$\frac{1}{2}+\frac{1}{x-1}=\frac{2}{x^2-1}$

$\frac{1}{2}+\frac{1}{-3-1}=\frac{2}{(-3)^2-1}$

$\frac{1}{2}+\frac{1}{-4}=\frac{2}{8}$

$\frac{1}{2}-\frac{1}{4}=\frac{1}{4}$

$\frac{1}{4}=\frac{1}{4}$ True

Check: $x=1$

$\frac{1}{2}+\frac{1}{x-1}=\frac{2}{x^2-1}$

$\frac{1}{2}+\frac{1}{1-1}=\frac{2}{(1)^2-1}$

$\frac{1}{2}+\frac{1}{0}=\frac{2}{0}$

$\frac{2}{0}$ and $\frac{1}{0}$ are not real numbers.

Since $\frac{2}{0}$ and $\frac{1}{0}$ are not real numbers, 1 is an extraneous solution. Therefore, the only solution to the equation is -3.

47. $\dfrac{x+2}{x^2-x}=\dfrac{6}{x^2-1}$ Factor (x^2-x) and (x^2-1).

$\dfrac{x+2}{x(x-1)}=\dfrac{6}{(x-1)(x+1)}$

Multiply both sides of the equation by the least common denominator, $x(x-1)(x+1)$.

$x(x-1)(x+1)\left(\dfrac{x+2}{x(x-1)}\right)=\left(\dfrac{6}{(x-1)(x+1)}\right)x(x-1)(x+1)$

$(x+1)\left(\dfrac{x+2}{1}\right)=\left(\dfrac{6}{1}\right)x$

$(x+1)(x+2)=6x$

$x^2+x+2x+2=6x$

$x^2+3x+2=6x$

$x^2-3x+2=0$

$(x-2)(x-1)=0$

$x-2=0$ or $x-1=0$

$x=2$ or $x=1$

Check: $x=2$

$\dfrac{x+2}{x^2-x}=\dfrac{6}{x^2-1}$

$\dfrac{2+2}{(2)^2-2}=\dfrac{6}{(2)^2-1}$

$\dfrac{4}{2}=\dfrac{6}{3}$

$2=2$ True

Check: $x=1$

$\dfrac{x+2}{x^2-x}=\dfrac{6}{x^2-1}$

$\dfrac{1+2}{(1)^2-1}=\dfrac{6}{(1)^2-1}$

$\dfrac{3}{0}=\dfrac{6}{0}$

Neither $\dfrac{3}{0}$ nor $\dfrac{6}{0}$ is a real number.

Since $\dfrac{3}{0}$ and $\dfrac{6}{0}$ are not real numbers, 1 is an extraneous solution. Therefore, the only solution to the equation is 2.

49. **(a)** Factor all denominators completely.
Multiply both sides of the equation by the least common denominator of all the fractions in the equation.
Use the distributive property if necessary.
Simplify the equation.
Solve the equation which is linear, quadratic (or beyond the scope of this course).
Check the solutions to see if they are extraneous.

(b) $\frac{1}{x-1}-\frac{1}{x+1}=\frac{3x}{x^2-1}$ Factor x^2-1.

$\frac{1}{x-1}-\frac{1}{x+1}=\frac{3x}{(x-1)(x+1)}$

Multiply both sides of the equation by the least common denominator, $(x-1)(x+1)$.

$(x-1)(x+1)\left(\frac{1}{x-1}-\frac{1}{x+1}\right)=\left(\frac{3x}{(x-1)(x+1)}\right)(x-1)(x+1)$

$(x-1)(x+1)\left(\frac{1}{x-1}\right)-(x-1)(x+1)\left(\frac{1}{x+1}\right)=\left(\frac{3x}{1}\right)1$

$(x+1)\left(\frac{1}{1}\right)-(x-1)\left(\frac{1}{1}\right)=3x$

$x+1-(x-1)=3x$

$x+1-x+1=3x$

$2=3x$

$\frac{2}{3}=x$

51. **(a)** The problem on the left is an expression to be simplified while the problem on the right is an equation to be solved.

(b) Left: Convert each fraction to a fraction with the least common denominator; then perform the subtraction and addition; then simplify.

Right: Follow the steps in Problem 49(a).

(c) Left: $\frac{x}{3}-\frac{x}{4}+\frac{1}{x-1}=\frac{x\cdot 4(x-1)}{3\cdot 4(x-1)}-\frac{x\cdot 3(x-1)}{4\cdot 3(x-1)}+\frac{1\cdot 3\cdot 4}{(x-1)3\cdot 4}=\frac{4x(x-1)-3x(x-1)+12}{3\cdot 4(x-1)}$

$=\frac{4x^2-4x-3x^2+3x+12}{3\cdot 4(x-1)}=\frac{x^2-x+12}{12(x-1)}$

Right: $\frac{x}{3}-\frac{x}{4}=\frac{1}{x-1}$

Multiply both sides of the equation by the least common denominator, $3 \cdot 4(x-1)$.

$$3\cdot 4(x-1)\left[\frac{x}{3}-\frac{x}{4}\right]=\left[\frac{1}{x-1}\right]3\cdot 4(x-1)$$

$$3\cdot 4(x-1)\left[\frac{x}{3}\right]-3\cdot 4(x-1)\left[\frac{x}{4}\right]=\left[\frac{1}{1}\right]3\cdot 4$$

$$4(x-1)\left[\frac{x}{1}\right]-3(x-1)\left[\frac{x}{1}\right]=12$$

$$4x(x-1)-3x(x-1)=12$$

$$4x^2-4x-3x^2+3x=12$$

$$x^2-x-12=0$$

$$(x-4)(x+3)=0$$

$$x-4=0 \text{ or } x+3=0$$

$$x=4 \text{ or } x=-3$$

A check will show that 4 and –3 are both solutions to the equation.

Cumulative Review Exercises: 6.6

52. Let x = the number of bus rides Steve takes per month. Then $36 + 0.40x$ = total cost per month with the bus pass (in \$). $1.60x$ = total cost per month without the bus pass (in \$). Find x so that the total cost with and without the bus pass are equal.

$$36+0.40x=1.60x$$

$$36+\frac{40x}{100}=\frac{160x}{100}$$ Multiply both sides of the equation by 100.

$$100\left[36+\frac{40x}{100}\right]=\left[\frac{160x}{100}\right]100$$

$$100(36)+100\left[\frac{40x}{100}\right]=\left[\frac{160x}{100}\right]100$$

$$100(36)+\left(\frac{40x}{1}\right)=160x$$

$$3600+40x=160x$$

$$3600=120x$$

$$30=x$$

Steve would have to take 30 rides per month for the total cost with and without the bus pass to be the same.

53. Let x = the measure of the smaller angle in degrees. Then $2x + 30$ = the measure of the larger angle in degrees. Since the angles are supplementary, their sum is 180°.

$x + (2x + 30) = 180$

$3x + 30 = 180$

$3x = 150$

$x = 50$

$2x + 30 = 2(50) + 30 = 130$

The measures of the angles are 50° and 130°.

54. $\dfrac{600 \text{ gallons}}{8 \text{ gallons / minute}} = 75 \text{ minutes}$

55. A quadratic equation includes an x^2 term with nonzero coefficient. A linear equation does not.

Linear equation example: $3x + 5 = 12$

Quadratic equation example: $4x^2 + 3x - 6 = 0$

Group Activity/Challenge Problems

1. $\dfrac{1}{30} + \dfrac{1}{q} = \dfrac{1}{10}$

$\dfrac{1}{q} = \dfrac{1}{10} - \dfrac{1}{30}$

$\dfrac{1}{q} = \dfrac{2}{30}$

$2q = 30$

$q = 15$

The image will appear 15 cm from the mirror.

4. **(a)** $\frac{1}{x}+\frac{1}{3}=\frac{2}{x}$

$\frac{3+x}{3x}=\frac{2}{x}$

$(3+x)x=3x\cdot 2$

$3x+x^2=6x$

$x^2-3x=0$

$x(x-3)=0$

$x=3$

(x cannot equal zero in the original equation.)

Exercise Set 6.7

1. Let $h=$ height of triangle;

then base $=h+6$

Area $=\frac{1}{2}$base $\cdot$ height

$\frac{1}{2}(h+6)h=80$

$h^2+6h=160$

$h^2+6h-160=0$

$(h-10)(h+16)=0$

$h=10$ or $h=-16$

Since h must be positive,

height $=10$ cm

base $=10+6=16$ cm

3. Let one number be y; then the other number is $3y$.

$\frac{1}{3y}+\frac{1}{y}=\frac{4}{3}$

$\left(\frac{1}{3y}+\frac{1}{y}\right)3y=\left(\frac{4}{3}\right)3y$

$1+3=4y$

$1=y,\ 3y=3$

The numbers are 1 and 3.

5. $\frac{1}{3}+\frac{1}{5}=\frac{1}{x}$

$\frac{1}{3}+\frac{1}{5}=\frac{5}{15}+\frac{3}{15}=\frac{8}{15}$

which is the reciprocal of $\frac{15}{8}$.

7. Let c be the speed of the current.

Speed upstream is $4-c$

Speed downstream is $4+c$

$t=\frac{d}{r}$; time upstream = time downstream

$\frac{10}{4+c}=\frac{6}{4-c}$

$10(4-c)=6(4+c)$

$40-10c=24+6c$

$16=16c$

$1=c$

The current is 1 mph.

9. $t = \frac{d}{r}$

Time pulling younger son + time pulling older son = $\frac{1}{2}$ hr.

$\frac{d}{30} + \frac{d}{30} = \frac{1}{2}$

$\frac{2d}{30} = \frac{1}{2}$

$d = \frac{30}{4} = 7.5$ mi

11. Let s be the speed (rate) of the slower car. Then $s + 30$ is the rate of the faster car.

$t = \frac{d}{r}$

$\frac{250}{s} = \frac{400}{s+30}$

$250(s+30) = 400s$

$250s + 7500 = 400s$

$7500 = 150s$

$50 = s$

The slower car travels at 50 km/hr.

The faster car travels at $s + 30$ or 80 km/hr.

13. $t = \frac{d}{r}$

Time walking + time jogging = 1 hr.

$\frac{2}{r} + \frac{2}{2r} = 1$

$r\left(\frac{2}{r} + \frac{1}{r}\right) = 1 \cdot r$

$2 + 1 = r$

$3 = r$

Mario walks at 3 mph and jogs at $2r$ or 6 mph.

15. d = distance jogged,

$3 - d$ = distance walked,

$t = \frac{d}{r}$

$\frac{d}{5}$ = time jogged, $\frac{3-d}{2}$ = time walked.

Time walked + time jogged = 0.9 hr.

$\frac{d}{5} + \frac{3-d}{2} = 0.9$

$10\left(\frac{d}{5} + \frac{3-d}{2}\right) = 10(0.9)$

$2d + 5(3-d) = 9$

$2d + 15 - 5d = 9$

$15 - 3d = 9$

$-3d = -6$

$d = 2$, so he jogs 2 mi and walks $3 - d = 1$ mi.

Jogs $\frac{d}{5} = \frac{2}{5} = 0.4$ hr

Walks $\frac{3-2}{2} = \frac{1}{2} = 0.5$ hr

17. $t = \frac{d}{r}$

Sean's time = $\frac{d}{9}$

Scott's time = $\frac{d}{6}$

$\frac{d}{6} = \frac{d}{9} + 0.5$

$(6 \cdot 9)\frac{d}{6} = \left(\frac{d}{9} + 0.5\right)6 \cdot 9$

$9d = 6d + 27$

$3d = 27$

$d = 9$ mi

19. Rate of first truck = $\frac{1}{6}$

Rate of second truck = $\frac{1}{3}$

Rate together = $\frac{1}{6}+\frac{1}{3}$

Rate · time = part of task done

$\left(\frac{1}{6}+\frac{1}{3}\right)t=1$

$\frac{1}{2}t=1$

$t=2$

Working together, the trucks could meet the community's needs in 2 hr.

21. Rate of first hose = $\frac{1}{8}$

Rate of second hose = $\frac{1}{5}$

Rate together = $\frac{1}{8}+\frac{1}{5}$

$\left(\frac{1}{8}+\frac{1}{5}\right)t=1$

$\frac{5+8}{40}t=1$

$t=\frac{40}{13}=3\frac{1}{13}$ hr

23. Rate in = $\frac{1}{4}$, Rate out = $\frac{1}{5}$

Total rate = $\frac{1}{4}-\frac{1}{5}$

$\left(\frac{1}{4}-\frac{1}{5}\right)t=1$

$\frac{1}{20}t=1$

$t=20$ min

25. Rate for first = $\frac{1}{4}$

Rate for second = $\frac{1}{t}$

Rate together = $\frac{1}{4}+\frac{1}{t}=\frac{1}{3}$

$12t\left(\frac{1}{4}+\frac{1}{t}\right)=\left(\frac{1}{3}\right)12t$

$3t+12=4t$

$12 \text{ hr } = t$

27. Rate for first = $\frac{1}{12}$

Rate for second = $\frac{1}{15}$

Work done by first = $\frac{1}{12}\cdot 5=\frac{5}{12}$

Work done by second = $\frac{1}{15}\cdot t=\frac{t}{15}$

$\frac{5}{12}+\frac{t}{15}=1$

$60\left(\frac{5}{12}+\frac{t}{15}\right)=1\cdot 60$

$25+4t=60$

$4t=35$

$t=\frac{35}{4}=8\frac{3}{4}$ days

29. Rate of first $= \frac{1}{60}$

Rate of second $= \frac{1}{50}$

Rate of transfer $= \frac{1}{30}$

Rate together $= \frac{1}{60} + \frac{1}{50} - \frac{1}{30}$

[2 skimmers filling the tank, one valve emptying it (transferring)]

$$\left(\frac{1}{60} + \frac{1}{50} - \frac{1}{30}\right)t = 1$$

$$300\left(\frac{1}{60} + \frac{1}{50} - \frac{1}{30}\right)t = 300 \cdot 1$$

$$(5 + 6 - 10)t = 300$$

$$t = 300 \text{ hr}$$

31. Rate of $A = \frac{1}{300}$

Rate of $B = \frac{1}{100}$

Rate of $C = \frac{1}{200}$

Together the rate is $= \frac{1}{300} + \frac{1}{100} + \frac{1}{200}$

$$\left(\frac{1}{300} + \frac{1}{100} + \frac{1}{200}\right)t = 1$$

$$600\left(\frac{1}{300} + \frac{1}{100} + \frac{1}{200}\right)t = 600 \cdot 1$$

$$(2 + 6 + 3)t = 600$$

$$11t = 600$$

$$t = \frac{600}{11} = 54\frac{6}{11} \text{ yrs}$$

Cumulative Review Exercises: 6.7

33. $6 - [(3 - 5^2) \div 11]^2 + 18 \div 3$

$$= 6 - [(3 - 25) \div 11]^2 + 6$$

$$= 12 - [-22 \div 11]^2$$

$$= 12 - [-2]^2$$

$$= 12 - 4 = 8$$

34. $\frac{1}{2}(x + 3) - (2x + 6)$

$$= \frac{1}{2}x + \frac{3}{2} - 2x - 6$$

$$= \frac{x}{2} - \frac{4x}{2} + \frac{3}{2} - \frac{12}{2}$$

$$= -\frac{3x}{2} - \frac{9}{2}$$

35. $\frac{x^2 - 14x + 48}{x^2 - 5x - 24} \div \frac{2x^2 - 13x + 6}{2x^2 + 5x - 3}$

$$= \frac{x^2 - 14x + 48}{x^2 - 5x - 24} \cdot \frac{2x^2 + 5x - 3}{2x^2 - 13x + 6}$$

$$= \frac{(x - 6)(x - 8)}{(x + 3)(x - 8)} \cdot \frac{(2x - 1)(x + 3)}{(2x - 1)(x - 6)} = 1$$

36. $\frac{x}{6x^2-x-15}-\frac{5}{9x^2-12x-5}$

$=\frac{x}{(2x+3)(3x-5)}-\frac{5}{(3x+1)(3x-5)}$

$=\frac{x(3x+1)}{(2x+3)(3x-5)(3x+1)}$

$-\frac{5(2x+3)}{(2x+3)(3x-5)(3x+1)}$

$=\frac{3x^2+x-10x-15}{(2x+3)(3x-5)(3x+1)}$

$=\frac{3x^2-9x-15}{(2x+3)(3x-5)(3x+1)}$

Group Activity/Challenge Problems

1. $\frac{1}{x-3}=2\left(\frac{1}{2x-6}\right)$

$2x-6=2(x-3)$

$2x-6=2x-6$

All real numbers except 3.

3. $\frac{b}{3}=\frac{b}{6}+1.5$

$\frac{b}{3}=\frac{b+9}{6}$

$6b=3b+27$

$3b=27$

$b=9$

Each picked 9 buckets.

5. **(a)** $x=\frac{400}{10}$; $x=40$ minutes

(b) $x=\frac{400}{40}$; $x=10$ minutes

(c) $\frac{x}{40}+\frac{x}{10}=1$; $x=8$ minutes

Chapter 6 Review Exercises

1. $\frac{6}{2x-8}$

$2x-8=0,\ x\neq 4$

2. $\frac{5}{x^2-7x+12}$

$x^2-7x+12=(x-4)(x-3),\ x\neq 4, x\neq 3$

3. $2x^2-13x+15=0$

$(2x-3)(x-5)=0$

$x\neq\frac{3}{2},\ x\neq 5$

4. $\frac{x}{x-xy}=\frac{x}{x(1-y)}=\frac{1}{1-y}$

5. $\frac{x^3+4x^2+12x}{x}=\frac{x(x^2+4x+12)}{x}$

$=x^2+4x+12$

6. $\frac{9x^2+6xy}{3x}=\frac{3x(3x+2y)}{3x}=3x+2y$

7. $\frac{x^2+x-12}{x-3}=\frac{(x-3)(x+4)}{x-3}=x+4$

8. $\frac{x^2-4}{x-2}=\frac{(x-2)(x+2)}{x-2}=x+2$

9. $\frac{2x^2-7x+3}{3-x}=\frac{(2x-1)(x-3)}{-1(x-3)}=-(2x-1)$

10. $\frac{x^2-2x-24}{x^2+6x+8}=\frac{(x-6)(x+4)}{(x+2)(x+4)}=\frac{x-6}{x+2}$

11. $\frac{3x^2-8x-16}{x^2-8x+16}=\frac{(3x+4)(x-4)}{(x-4)^2}=\frac{3x+4}{x-4}$

12. $\frac{2x^2-21x+40}{4x^2-4x-15}=\frac{(x-8)(2x-5)}{(2x+3)(2x-5)}$

$=\frac{x-8}{2x+3}$

13. $\frac{4y}{3x}\cdot\frac{4x^2y}{2}=\frac{8xy^2}{3}$

14. $\frac{15x^2y^3}{3z}\cdot\frac{6z^3}{5xy^3}=\frac{15\cdot 6}{15}\left(\frac{x^2}{x}\right)\left(\frac{y^3}{y^3}\right)\left(\frac{z^3}{z}\right)$

$=6xz^2$

15. $\frac{40a^3b^4}{7c^3}\cdot\frac{14c^5}{5a^5b}=\frac{40}{5}\cdot\frac{14}{7}\cdot\frac{a^3}{a^5}\cdot\frac{b^4}{b}\cdot\frac{c^5}{c^3}$

$=16\frac{1}{a^2}b^3c^2=\frac{16b^3c^2}{a^2}$

16. $\frac{1}{x-2}\cdot\frac{2-x}{2}=\frac{1}{x-2}\cdot\frac{-(x-2)}{2}=-\frac{1}{2}$

17. $\frac{-x+2}{3}\cdot\frac{6x}{x-2}=\frac{-(x-2)\cdot 2x}{x-2}=-2x$

18. $\frac{a-2}{a+3}\cdot\frac{a^2+4a+3}{a^2-a-2}$

$=\frac{(a-2)(a+3)(a+1)}{(a+3)(a-2)(a+1)}=1$

19. $\frac{6y^3}{x}\div\frac{y^3}{6x}=\frac{6y^3}{x}\cdot\frac{6x}{y^3}=36$

20. $\frac{8xy^2}{z}\div\frac{x^4y^2}{4z^2}=\frac{8xy^2}{z}\cdot\frac{4z^2}{x^4y^2}=\frac{32z}{x^3}$

21. $\frac{3x+3y}{x^2}\div\frac{x^2-y^2}{x^2}$

$=\frac{3(x+3y)}{x^2}\cdot\frac{x^2}{(x+y)(x-y)}=\frac{3}{x-y}$

22. $\frac{1}{a^2+8a+15}\div\frac{3}{a+5}$

$=\frac{1}{(a+5)(a+3)}\cdot\frac{a+5}{3}=\frac{1}{3(a+3)}$

23. $(x+3)\div\frac{x^2-4x-21}{x-7}$

$=(x+3)\cdot\frac{x-7}{(x-7)(x+3)}=1$

24. $\frac{x^2-3xy-10y^2}{6x}\div\frac{x+2y}{12x^2}$

$=\frac{(x-5y)(x+2y)}{6x}\cdot\frac{12x^2}{x+2y}=2x(x-5y)$

25. $\frac{x}{x+2}-\frac{2}{x+2}=\frac{x-2}{x+2}$

26. $\frac{4x}{x+2}+\frac{8}{x+2}=\frac{4x+8}{x+2}$

$=\frac{4(x+2)}{x+2}=4$

27. $\frac{9x-4}{x+8}+\frac{76}{x+8}=\frac{9x+72}{x+8}$

$=\frac{9(x+8)}{x+8}=9$

28. $\frac{7x-3}{x^2+7x-30}-\frac{3x+9}{x^2+7x-30}$

$=\frac{4x-12}{x^2+7x-30}=\frac{4(x-3)}{(x+10)(x-3)}$

$=\frac{4}{x+10}$

29. $\frac{4x^2-11x+4}{x-3}-\frac{x^2-4x+10}{x-3}$

$=\frac{3x^2-7x-6}{x-3}=\frac{(3x+2)(x-3)}{x-3}$

$=3x+2$

30. $\frac{6x^2-4x}{2x-3}-\frac{(-3x+12)}{2x-3}$

$=\frac{6x^2-4x+3x-12}{2x-3}=\frac{6x^2-x-12}{2x-3}$

$=\frac{(3x+4)(2x-3)}{2x-3}=3x+4$

31. Denominators are 3, 8; $3\cdot 8=24$

32. $3x$, $5x^2$;
Least common denominator = $15x^2$

33. Denominators are $3x^3y^4$, $10x^6y^2$,
Least common denominator = $30x^6y^4$

34. Denominators are $x+1$, x;
Least common denominator = $x(x+1)$

35. $x+2$, $x-3$;
Least common denominator
$=(x+2)(x-3)$

36. $x^2+x=x(x+1)$;
= Least common denominator

37. $x^2-y^2=(x+y)(x-y)$;
= Least common denominator

38. Least common denominator = $x-7$

39. $x^2+2x-35=(x-5)(x+7)$
$x^2+9x+14=(x+2)(x+7)$;
Least common denominator
$=(x+7)(x-5)(x+2)$

40. $\frac{4}{2x}+\frac{x}{x^2}=\frac{2}{x}+\frac{1}{x}=\frac{3}{x}$

41. $\frac{1}{4x}+\frac{6x}{xy}=\frac{y}{4xy}+\frac{24x}{4xy}=\frac{24x+y}{4xy}$

42. $\frac{5x}{3xy}-\frac{4}{x^2}=\frac{5x^2}{3x^2y}-\frac{12y}{3x^2y}=\frac{5x^2-12y}{3x^2y}$

43. $5-\frac{3}{x+3}=\frac{5(x+3)}{x+3}-\frac{3}{x+3}$

$=\frac{5x+15-3}{x+3}=\frac{5x+12}{x+3}$

44. $\frac{a+c}{c}-\frac{a-c}{a}=\frac{a(a+c)}{ac}-\frac{c(a-c)}{ac}$

$=\frac{a^2+ac-ac+c^2}{ac}=\frac{a^2+c^2}{ac}$

45. $\dfrac{3}{x+3}+\dfrac{4}{x}=\dfrac{3x}{x(x+3)}+\dfrac{4(x+3)}{x(x+3)}$

$=\dfrac{3x+4x+12}{x(x+3)}=\dfrac{7x+12}{x(x+3)}$

46. $\dfrac{2}{3x}-\dfrac{3}{3x-6}=\dfrac{2}{3x}-\dfrac{3}{3(x-2)}$

$=\dfrac{2(x-2)}{3x(x-2)}-\dfrac{3x}{3x(x-2)}$

$=\dfrac{2x-4-3x}{3x(x-2)}=\dfrac{-x-4}{3x(x-2)}$

47. $\dfrac{4}{x+5}+\dfrac{6}{(x+5)^2}=\dfrac{4(x+5)}{(x+5)^2}+\dfrac{6}{(x+5)^2}$

$=\dfrac{4x+20+6}{(x+5)^2}=\dfrac{4x+26}{(x+5)^2}$

48. $\dfrac{x+2}{x^2-x-6}+\dfrac{x-3}{x^2-8x+15}$

$=\dfrac{x+2}{(x-3)(x+2)}+\dfrac{x-3}{(x-3)(x-5)}$

$=\dfrac{1}{x-3}+\dfrac{1}{x-5}$

$=\dfrac{x-5}{(x-3)(x-5)}+\dfrac{x-3}{(x-5)(x-3)}$

$=\dfrac{2x-8}{(x-5)(x-3)}$

49. $\dfrac{x+4}{x+3}-\dfrac{x-3}{x+4}$

$=\dfrac{(x+4)^2}{(x+3)(x+4)}-\dfrac{(x-3)(x+3)}{(x+3)(x+4)}$

$=\dfrac{x^2+8x+16-(x^2-9)}{(x+3)(x+4)}$

$=\dfrac{8x+25}{(x+3)(x+4)}$

50. $6+\dfrac{x}{x+2}=\dfrac{6(x+2)}{x+2}+\dfrac{x}{x+2}$

$=\dfrac{6x+12+x}{x+2}=\dfrac{7x+12}{x+2}$

51. $\dfrac{4x}{a+2}\div\dfrac{8x^2}{a-2}=\dfrac{4x}{a+2}\cdot\dfrac{a-2}{8x^2}=\dfrac{a-2}{2x(a+2)}$

52. $\dfrac{x+3}{x^2-9}+\dfrac{2}{x+3}=\dfrac{x+3}{(x-3)(x+3)}+\dfrac{2}{x+3}$

$=\dfrac{(x+3)}{(x-3)(x+3)}+\dfrac{2(x-3)}{(x-3)(x+3)}$

$=\dfrac{x+3+2x-6}{(x-3)(x+3)}=\dfrac{3x-3}{(x-3)(x+3)}$

53. $\dfrac{4x+4y}{x^2y}\cdot\dfrac{y^3}{8x}=\dfrac{4(x+y)y^2}{8x^3}=\dfrac{(x+y)y^2}{2x^3}$

54. $\dfrac{4}{(x+2)(x-3)}-\dfrac{4}{(x-2)(x+2)}$

$=\dfrac{4(x-2)-4(x-3)}{(x+2)(x-3)(x-2)}$

$=\dfrac{4x-8-4x+12}{(x+2)(x-3)(x-2)}$

$=\dfrac{4}{(x+2)(x-3)(x-2)}$

55. $\dfrac{x+5}{x^2-15x+50}-\dfrac{x-2}{x^2-25}$

$=\dfrac{x+5}{(x-5)(x-10)}-\dfrac{x-2}{(x-5)(x+5)}$

$=\dfrac{(x+5)(x+5)}{(x-5)(x-10)(x+5)}$

$-\dfrac{(x-2)(x-10)}{(x-5)(x-10)(x+5)}$

$=\dfrac{x^2+10x+25-(x^2-12x+20)}{(x-5)(x-10)(x+5)}$

$=\dfrac{22x+5}{(x-5)(x-10)(x+5)}$

56. $\dfrac{x^2-y^2}{x-y}\cdot\dfrac{x+y}{xy+x^2}=\dfrac{(x-y)(x+y)\cdot(x+y)}{(x-y)x(y+x)}$

$=\dfrac{x+y}{x}$

57. $\dfrac{4x^2-16y^2}{9}\div\dfrac{(x+2y)^2}{12}$

$=\dfrac{4x^2-16y^2}{9}\cdot\dfrac{12}{(x+2y)^2}$

$=\dfrac{4(x^2-4y^2)}{9}\cdot\dfrac{12}{(x+2y)^2}$

$=\dfrac{4(x-2y)(x+2y)4}{3(x+2y)^2}=\dfrac{16(x-2y)}{3(x+2y)}$

58. $\dfrac{a^2-9a+20}{a-4}\cdot\dfrac{a^2-8a+15}{a^2-10a+25}$

$=\dfrac{(a-4)(a-5)}{a-4}\cdot\dfrac{(a-5)(a-3)}{(a-5)^2}=a-3$

59. $\dfrac{4x^2-16}{x^2+7x+10}\div\dfrac{x^2-2x}{2x^2+6x-20}$

$=\dfrac{4(x^2-4)}{x^2+7x+10}\cdot\dfrac{2x^2+6x-20}{x^2-2x}$

$=\dfrac{4(x-2)(x+2)}{(x+2)(x+5)}\cdot\dfrac{2(x^2+3x-10)}{x(x-2)}$

$=\dfrac{8(x+5)(x-2)}{(x+5)x}=\dfrac{8(x-2)}{x}$

60. $\dfrac{x}{x^2-1}-\dfrac{2}{3x^2-2x-5}$

$=\dfrac{x}{(x-1)(x+1)}-\dfrac{2}{(3x-5)(x+1)}$

$=\dfrac{x(3x-5)-2(x-1)}{(x-1)(x+1)(3x-5)}$

$=\dfrac{3x^2-5x-2x+2}{(x-1)(x+1)(3x-5)}$

$=\dfrac{3x^2-7x+2}{(x-1)(x+1)(3x-5)}$

61. $\dfrac{1+\dfrac{5}{12}}{\dfrac{3}{8}}=\left(1+\dfrac{5}{12}\right)\dfrac{8}{3}=\dfrac{8}{3}+\dfrac{10}{9}$

$=\dfrac{24}{9}+\dfrac{10}{9}=\dfrac{34}{9}$

62. $\dfrac{\left(4-\dfrac{9}{16}\right)16}{\left(1+\dfrac{5}{8}\right)16}=\dfrac{64-9}{16+10}=\dfrac{55}{26}$

63. $\dfrac{\dfrac{15xy}{6z}}{\dfrac{3x}{z^2}}=\dfrac{15xy}{6z}\cdot\dfrac{z^2}{3x}=\dfrac{5yz}{6}$

64. $\dfrac{36x^4y^2}{\dfrac{9xy^2}{4z^2}}=\dfrac{36x^4y^2\cdot 4z^2}{\dfrac{9xy^5}{4z^2}\cdot 4z^2}=\dfrac{16x^3z^2}{y^3}$

65. $\dfrac{\left(x+\dfrac{1}{y}\right)y}{(y^2)y}=\dfrac{xy+1}{y^3}$

66. $\dfrac{\left(x-\dfrac{x}{y}\right)y}{\left(\dfrac{1+x}{y}\right)y}=\dfrac{xy-x}{1+x}$

67. $\dfrac{\left(\dfrac{4}{x}+\dfrac{2}{x^2}\right)x^2}{\left(6-\dfrac{1}{x}\right)x^2}=\dfrac{4x+2}{6x^2-x}=\dfrac{4x+2}{x(6x-1)}$

68. $\dfrac{\dfrac{x}{x+y}}{\dfrac{x^2}{2x+2y}}=\dfrac{x}{x+y}\cdot\dfrac{2x+2y}{x^2}$

$=\dfrac{1}{x+y}\cdot\dfrac{2(x+y)}{x}=\dfrac{2}{x}$

69. $\dfrac{\dfrac{1}{a}}{\dfrac{1}{a^2}}=\dfrac{1}{a}\cdot\dfrac{a^2}{1}=a$

70. $\dfrac{\dfrac{1}{a}+2}{\dfrac{1}{a}+\dfrac{1}{a}}=\dfrac{\dfrac{1}{a}+2}{\dfrac{2}{a}}=\dfrac{\left(\dfrac{1}{a}+2\right)a}{\left(\dfrac{2}{a}\right)a}=\dfrac{1+2a}{2}$

71. $\dfrac{\dfrac{1}{x^2}+\dfrac{1}{x}}{\dfrac{1}{x^2}-\dfrac{1}{x}}=\dfrac{x^2\left(\dfrac{1}{x^2}+\dfrac{1}{x}\right)}{x^2\left(\dfrac{1}{x^2}-\dfrac{1}{x}\right)}=\dfrac{1+x}{1-x}$

72. $\dfrac{\left(\dfrac{3x}{y}-x\right)xy}{\left(\dfrac{y}{x}-1\right)xy}=\dfrac{3x^2-x^2y}{y^2-xy}=\dfrac{x^2(3-y)}{y(y-x)}$

73. $\dfrac{3}{x}=\dfrac{8}{24}$

$3\cdot 24=8\cdot x$

$3\cdot 3=x$

$9=x$

74. $\frac{4}{a}=\frac{16}{4}$

$\frac{4}{a}=4$

$\frac{1}{a}=1$

$a=1$

75. $\frac{x+3}{5}=\frac{9}{5}$

$x+3=9$

$x=6$

76. $\frac{x}{6}=\frac{x-4}{2}$

$x=6\left(\frac{x-4}{2}\right)=3(x-4)$

$x=3x-12$

$12=2x$

$x=6$

77. $\frac{3x+4}{5}=\frac{2x-8}{3}$

$3(3x+4)=5(2x-8)$

$9x+12=10x-40$

$52=x$

78. $\frac{x}{5}+\frac{x}{2}=-14$

$\frac{2x+5x}{10}=-14$

$7x=-140$

$x=-20$

79. $\left(4-\frac{5}{x+5}\right)(x+5)=\frac{x}{x+5}(x+5)$

$4(x+5)-5=x$

$4x+20-5=x$

$3x=-15$

$x=-5$

Not a solution. (Causes division by 0.)

No solution.

80. $6x\left(\frac{4}{x}-\frac{1}{6}\right)=6x\left(\frac{1}{x}\right)$

$24-x=6$

$x=18$

81. $\frac{1}{x-2}+\frac{1}{x+2}=\frac{1}{x^2-4}$

$\frac{(x+2)}{(x-2)(x+2)}+\frac{x-2}{(x+2)(x-2)}=\frac{1}{x^2-4}$

$x+2+x-2=1$

$2x=1$

$x=\frac{1}{2}$

82. $\frac{x-3}{x-2}+\frac{x+1}{x+3}=\frac{2x^2+x+1}{x^2+x-6}$

$(x-2)(x+3)\left(\frac{x-3}{x-2}+\frac{x+1}{x+3}\right)$

$=(x-2)(x+3)\left(\frac{2x^2+x+1}{x^2+x-6}\right)$

So $(x+3)(x-3)+(x-2)(x+1)$

$=2x^2+x+1$

and $x^2-9+x^2-x-2=2x^2+x+1$

$-12=2x$

$-6=x$

83. $\frac{x}{x^2-9}+\frac{2}{x+3}=\frac{4}{x-3}$

$\left(\frac{x}{(x-3)(x+3)}+\frac{2}{x+3}\right)(x-3)(x+3)$

$=\frac{4}{x-3}(x-3)(x+3)$

$x+2(x-3)=4(x+3)$

$x+2x-6=4x+12$

$-x=18$

$x=-18$

84. $\frac{t}{5}+\frac{t}{4}=1$

$\frac{4t+5t}{20}=1$

$9t=20$

$t=\frac{20}{9}=2\frac{2}{9}$ hrs

85. $\frac{t}{7}-\frac{t}{12}=1$

$\frac{12t-7t}{84}=1$

$5t=84$

$t=\frac{84}{5}=16\frac{4}{5}$ hrs

86. $\frac{1}{x}+\frac{1}{4x}=\frac{1}{2}$

$\frac{4}{4x}+\frac{1}{4x}=\frac{1}{2}$

$\frac{5}{4x}=\frac{1}{2}$

$10=4x$

$x=\frac{10}{4}=\frac{5}{2}$

$4x=4\left(\frac{5}{2}\right)=10$

87. $t=\frac{d}{r}$

$\frac{400}{r}=\frac{600}{r+40}$

$\frac{2}{r}=\frac{3}{r+40}$

$2(r+40)=3r$

$2r+80=3r$

$80=r$

bus: 80 km / hr

$80+40=120$ km / hr train

Chapter 6 Practice Test

1. $\frac{3x^2y}{4z^2}\cdot\frac{8xz^3}{9y^4}=\frac{2x^3z}{3y^3}$

2. $\frac{a^2-9a+14}{a-2}\cdot\frac{a^2-4a-21}{(a-7)^2}$

$=\frac{(a-7)(a-2)(a-7)(a+3)}{(a-2)(a-7)^2}=a+3$

3. $\frac{x^2-9y^2}{3x+6y} \div \frac{x+3y}{x+2y}$

$= \frac{(x-3y)(x+3y)}{3(x+2y)} \cdot \frac{x+2y}{x+3y} = \frac{x-3y}{3}$

4. $\frac{16}{y^2+2y-15} \div \frac{4y}{y-3}$

$= \frac{16}{(y-3)(y+5)} \cdot \frac{y-3}{4y} = \frac{4}{y(y+5)}$

5. $\frac{6x+3}{2y} + \frac{x-5}{2y} = \frac{7x-2}{2y}$

6. $\frac{7x^2-4}{x+3} - \frac{6x+7}{x+3} = \frac{7x^2-6x-11}{x+3}$

7. $\frac{5}{x} + \frac{3}{2x^2} = \frac{10x}{2x^2} + \frac{3}{2x^2} = \frac{10x+3}{2x^2}$

8. $5 - \frac{6x}{x+2} = \frac{5(x+2)}{x+2} - \frac{6x}{x+2}$

$= \frac{5x+10-6x}{x+2} = \frac{-x+10}{x+2}$

9. $\frac{x-5}{x^2-16} - \frac{x-2}{x^2+2x-8}$

$= \frac{x-5}{(x-4)(x+4)} - \frac{x-2}{(x-2)(x+4)}$

$= \frac{x-5}{(x-4)(x+4)} - \frac{1}{x+4}$

$= \frac{x-5-(x-4)}{(x-4)(x+4)} = \frac{-1}{(x-4)(x+4)}$

10. $\frac{\left(3+\frac{5}{8}\right)8}{\left(2-\frac{3}{4}\right)8} = \frac{24+5}{16-6} = \frac{29}{10}$

11. $\frac{x+\frac{x}{y}}{\frac{1}{x}} = \left(x+\frac{x}{y}\right)\frac{x}{1} \cdot \frac{y}{y}$

$= \frac{yx^2+x^2}{y} = \frac{x^2(1+y)}{y}$

12. $\frac{x}{3} - \frac{x}{4} = 5$

$\frac{4x-3x}{12} = 5$

$x = 60$

13. $\frac{x}{x-8} + \frac{6}{x-2} = \frac{x^2}{x^2-10x+16}$

$(x-8)(x-2)\left(\frac{x}{x-8} + \frac{6}{x-2}\right)$

$= (x-8)(x-2)\frac{x^2}{(x-8)(x-2)}$

$x(x-2) + 6(x-8) = x^2$

$x^2 - 2x + 6x - 48 = x^2$

$4x - 48 = 0$

$x = \frac{48}{4} = 12$

14. $\frac{t}{8}+\frac{t}{5}=1$

$\frac{5t+8t}{40}=1$

$13t=40$

$t=\frac{40}{13}=3\frac{1}{13}$ hrs.

Chapter 6 Cumulative Review Test

1. $3x^2-2xy-7, x=-3, y=5$

$3(-3)^2-2(-3)5-7$

$=27+30-7=50$

2. $-4-[2(-6\div 3)^2]\div 2$

$=-4-[2(-2)^2]\div 2=-4-[8]\div 2$

$=-4-4=-8$

3. $4y+3=-2(y+6)$

$4y+3=-2y-12$

$6y=-15$

$y=\frac{-15}{6}=-\frac{5}{2}$

4. $\left(\frac{6x^2y^3}{2x^5y}\right)^3=\left(\frac{3y^2}{x^3}\right)^3=\frac{27y^6}{x^9}$

5. $P=2E+3R$

$P-2E=3R$

$R=\frac{P-2E}{3}$

6. $(6x^2-3x-5)-(3x^2+8x-9)$

$=6x^2-3x-5-3x^2-8x+9$

$=3x^2-11x+4$

7. $(4x^2-6x+3)(3x-5)$

$=4x^2(3x-5)-6x(3x-5)+3(3x-5)$

$=12x^3-20x^2-18x^2+30x+9x-15$

$=12x^3-38x^2+39x-15$

8. $6a^2-6a-5a+5=6a(a-1)-5(a-1)$

$=(6a-5)(a-1)$

9. $10x^2-5x+5=5(2x^2-x+1)$

10. $x^2-10x+24=(x-6)(x-4)$

11. $6x^2-11x-10=(3x+2)(2x-5)$

12. $2x^2=11x-12$

$2x^2-11x+12=0$

$(x-4)(2x-3)=0$

$x=4,$ or $x=\frac{3}{2}$

13. $\frac{x^2-9}{x^2-x-6}\cdot\frac{x^2-2x-8}{2x^2-7x-4}$

$=\frac{(x-3)(x+3)}{(x-3)(x+2)}\cdot\frac{(x-4)(x+2)}{(2x+1)(x-4)}=\frac{x+3}{2x+1}$

14. $\frac{x}{x+4}-\frac{3}{x-5}=\frac{x(x-5)-3(x+4)}{(x+4)(x-5)}$

$=\frac{x^2-5x-3x-12}{(x+4)(x-5)}=\frac{x^2-8x-12}{(x+4)(x-5)}$

15. $\dfrac{4}{x^2-3x-10}+\dfrac{2}{x^2+5x+6}$

$=\dfrac{4}{(x-5)(x+2)}+\dfrac{2}{(x+2)(x+3)}$

$=\dfrac{4(x+3)+2(x-5)}{(x-5)(x+2)(x+3)}$

$=\dfrac{4x+12+2x-10}{(x-5)(x+2)(x+3)}$

$=\dfrac{6x+2}{(x-5)(x+2)(x+3)}$

16. $\dfrac{x}{6}-\dfrac{x}{4}=\dfrac{1}{8}$

$\dfrac{2x}{12}-\dfrac{3x}{12}=\dfrac{1}{8}$

$\dfrac{-x}{12}=\dfrac{1}{8}$

$-x=\dfrac{12}{8}$

$x=-\dfrac{3}{2}$

17. $\dfrac{1}{x-4}+\dfrac{2}{x-3}=\dfrac{4}{x^2-7x+12}$

$(x-4)(x-3)\left(\dfrac{1}{x-4}+\dfrac{2}{x-3}\right)$

$=(x-4)(x-3)\dfrac{4}{(x-4)(x-3)}$

$x-3+2(x-4)=4$

$x-3+2x-8=4$

$3x=15$

$x=5$

18. $0.10x=100+0.05x$

$0.05x=100$

$x=\$2000$

19.

	Weight	Price	Cost
Chippy	6	3	18
Hippy	h	4	$4h$
	$h+6$	3.20	$3.2(h+6)$

$(h+6)(3.20)=18+4h$

$3.2h+19.2=18+4h$

$1.2=0.8h$

$\dfrac{12}{8}=h$

$h=\dfrac{3}{2}=1.5$ lb

20. d = distance at 4 mph

$6-d$ = distance at 12 mph

time at 4 mph is $\dfrac{d}{4}$

time at 12 mph is $\dfrac{6-d}{12}$

$\dfrac{d}{4}+\dfrac{6-d}{12}=1$

$\left(\dfrac{d}{4}+\dfrac{6-d}{12}\right)12=12$

$3d+6-d=12$

$2d=6$

$d=3$ mi at 4 mph

$6-d=3$ mi at 12 mph

CHAPTER 7

Exercise Set 7.1

1. **(a)** 53.5% of replacement tires were manufactured by Goodyear, Michelin, or Bridgestone.

(b) No, as only percents are indicated.

(c) No. Once again, only percents are indicated.

(d) No. The chart only gives information about tire sales, not profit.

3. **(a)** Money collected from ticket sales: $(0.75)(3.6) = 2.7$ million dollars.

(b) Money collected from fees from concessionaires and vendors: $(0.15)(3.6) = 0.54$ million dollars.

(c)

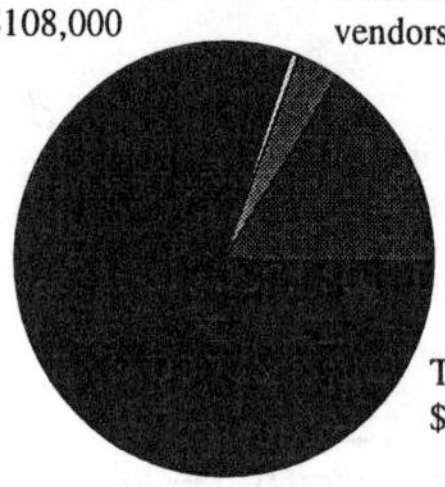

Fees for competition entries and stable fees: $108,000
Concessionaires and vendors: $540,000
Payment from sponsors: $252,000
Ticket sales: $2,700,000

State Fair 1993 Revenue $3.6 million

5. **(a)** 15 years

(b) It took a total of 23 years for the record to be reduced from 9.95 seconds to 9.90 seconds, a total of five one-hundredths of a second.

(c) From 9.95 seconds to 9.85 seconds, a one-tenth second decrease in the world record time took 15 years + 5 years + 3 years + $2\frac{1}{2}$ months + 3 years, which is more than 26 years.

(d) The new record time will be 9.82 seconds.

7. **(a)** The number of cinema screens in Western Europe has decreased by 14,464.

(b) The number of cinema screens in Canada/U.S. has increased by 5554.

(c) The total number of cinema screens in 1950 was 67,148. The total number in 1991 was 55,247.

9. **(a)** About 30,000

(b) About 60,000

(c) About 160,000

(d) About 960,000

11. **(a)** The savings for Japan is about 31%. The savings for the European Union is about 20%. The savings for the United States is about 18.5%.

(b) It has varied between 18.5% and 14%, but has remained relatively stable.

(c) It has varied between 31% and 35%. It increased up until 1990 - 1991, then began decreasing.

13. **(a)** About $320,000

(b) After 1989

(c) About $450,000

(d) About $650,000

(e) Percent increase $= \dfrac{650{,}000}{450{,}000} = 1.44$ or 144%

(f) $\dfrac{144}{5} = 28.8\%$

15. **(a)** 3 billion

(b) 7 billion

(c) 800,000,000

(d) 1 billion

(e) 2 billion

(f) The world population is expected to increase from about 5 billion to 11 billion from 1990 to 2100, with the poorest accounting for most of the increase.

17. A bar graph shows the number or amount of each item.

19. Answers will vary.

21. Answers will vary.

Cumulative Review Exercises: 7.1

22. $3x-\frac{2}{3}+\frac{3}{5}x-\frac{1}{2}$

$=\frac{15}{5}x+\frac{3}{5}x-\frac{4}{6}-\frac{3}{6}$

$=\frac{18}{5}x-\frac{7}{6}$

23.

$$\begin{array}{r} 3x+2 \\ x-2\overline{)3x^2-4x+6} \\ \underline{-3x^2+6x} \\ 2x+6 \\ \underline{-2x+4} \\ 10 \end{array}$$

Thus, $\frac{3x^2-4x+6}{x-2}=3x+2+\frac{10}{x-2}$

24. $4x^2-16=4(x^2-4)=4(x^2-2^2)$

$=4(x+2)(x-2)$

25. $x-\frac{2}{3}=\frac{3}{5}x-\frac{1}{2}$

Multiply through by 30

$30x-20=18x-15$

$12x=5$

$x=\frac{5}{12}$

Group Activity/Challenge Problems

1. Theme parks and resorts $=45\%$, $(0.45)(360°)=162°$; Films $=35\%$, $(0.35)(360°)=126°$; Consumer products $=20\%$, $(0.20)(360°)=72°$

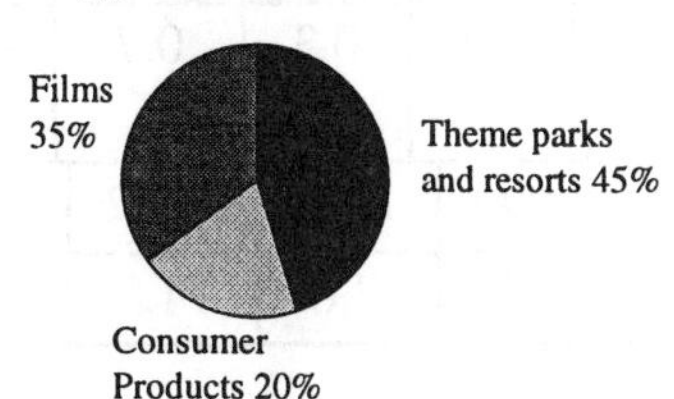

3. (a) Belgium: $\frac{3,400,000}{3,700,000}=0.92$ or 92%

Netherlands: $\frac{4,900,000}{6,100,000}=0.80$ or 80%

Canada: $\frac{7,537,000}{9,756,000}=0.77$ or 77%

United States: $\frac{54,890,000}{92,740,000}=0.59$ or 59%

Germany: $\frac{9,900,000}{33,281,000}=0.30$ or 30%

Japan: $\frac{6,170,000}{40,000,000}=0.15$ or 15%

United Kingdom: $\frac{490,000}{21,600,000}=0.02$ or 2%

(b)

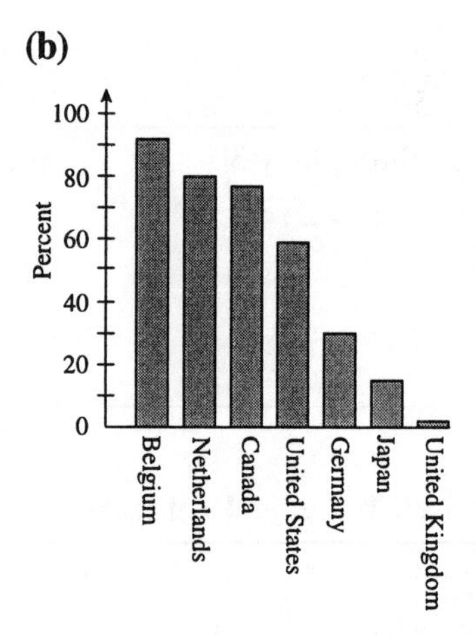

4.

Year	1984	1985	1986	1987	1988	1989	1990	1991	1992	1993
Subscribers (in millions)	0.1	0.3	0.7	1.2	2.1	3.5	5.3	7.6	11.0	16.0

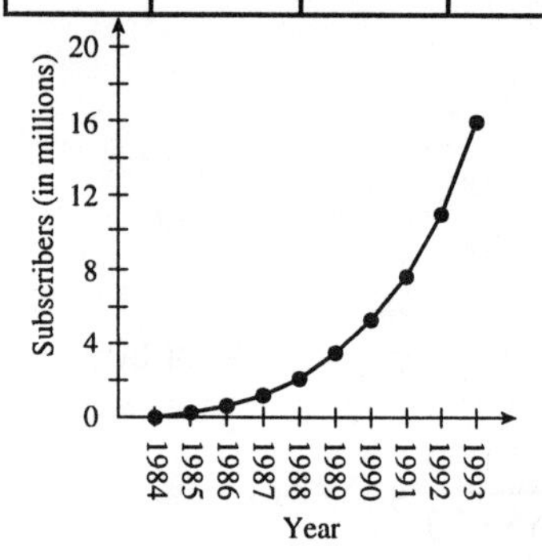

Exercise Set 7.2

1. I

3. III

5. II

7. I

9. III

11. II

13. $A(3, 1)$; $B(-3, 0)$; $C(1, -3)$; $D(-2, -3)$; $E(0, 3)$; $F\left(\frac{3}{2}, -1\right)$

15.

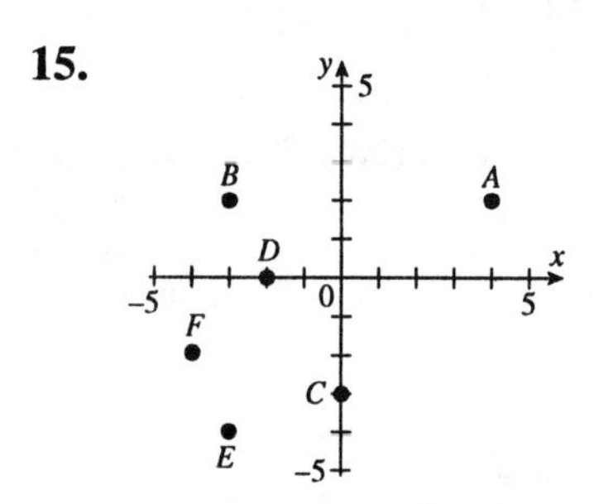

17.

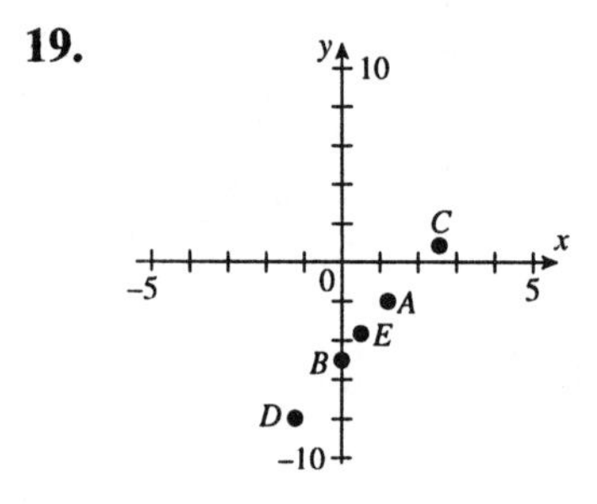

19.

C is not collinear.

21. **(a)** $y = x + 1$ $\quad$ $y = x + 1$
$1 = 0 + 1$ $\quad$ $0 = -1 + 1$
$1 = 1$ True $\quad$ $0 = 0$ True

$y = x + 1$ $\quad$ $y = x + 1$
$3 = 2 + 1$ $\quad$ $1 = 1 + 1$
$3 = 3$ True $\quad$ $1 = 2$ False

Points A, B, and C satisfy the equation.

(b)

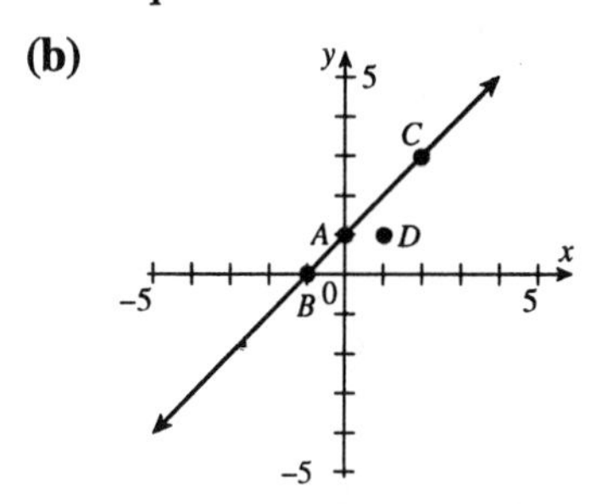

(c) Yes, the three points which are solutions are collinear, and the point (1, 1), which is not a solution, is not collinear with the other points.

(d) The straight line represents all solutions of $y = x + 1$.

23. **(a)** $3x - 2y = 6$ $\quad$ $3x - 2y = 6$
$3(4) - 2(0) = 6$ $\quad$ $3(2) - 2(0) = 6$
$12 = 6$ False $\quad$ $6 = 6$ True

$3x - 2y = 6$ $\quad$ $3x - 2y = 6$
$3\left(\frac{2}{3}\right) - 2(-2) = 6$ $\quad$ $3\left(\frac{4}{3}\right) - 2(-1) = 6$
$2 + 4 = 6$ $\quad$ $4 + 2 = 6$
$6 = 6$ True $\quad$ $6 = 6$ True

Points B, C, and D satisfy the equation.

(b)

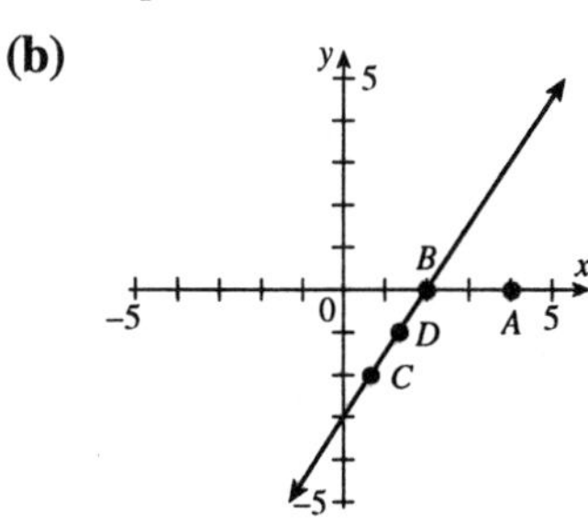

(c) Yes, the three points which are solutions are collinear, and the point (4, 0), which is not a solution, is not collinear with the other points.

(d) The straight line represents all solutions of $3x - 2y = 6$.

25. **(a)** $2x + 4y = 4$ $\quad$ $2x + 4y = 4$

$2(2) + 4(3) = 4$ $\quad$ $2(1) + 4\left(\frac{1}{2}\right) = 4$

$4 + 12 = 4$ $\quad$ $4 = 4$ True

$16 = 4$ False

$2x + 4y = 4$ $\quad$ $2x + 4y = 4$

$2(0) + 4(1) = 4$ $\quad$ $2(-2) + 4(1) = 4$

$4 = 4$ True $\quad$ $4 = 4$ True

Points B, C, and D satisfy the equation.

(b)

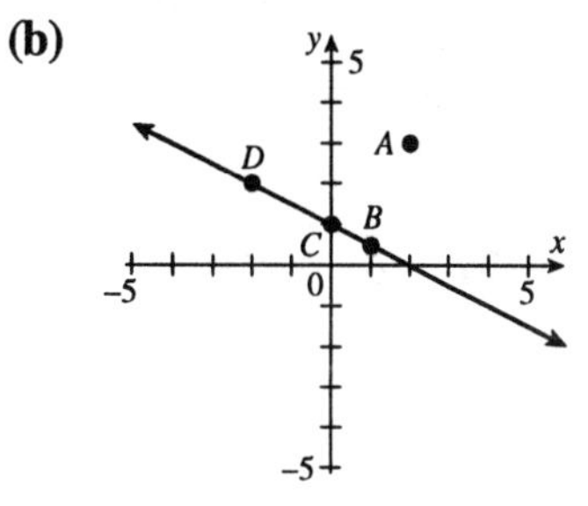

(c) Yes, the three points which are solutions are collinear, and the point (2, 3), which is not a solution, is not collinear with the other points.

(d) The straight line represents all solutions of $2x + 4y = 4$.

27. The x coordinate is always listed first.

29. **(a)** The horizontal axis is the x-axis.

(b) The vertical axis is the y-axis.

31. Axis is singular, while axes is plural.

33. The graph of a linear equation is an illustration of the set of points whose coordinates satisfy the equation.

35. The graph of a linear equation looks like a straight line.

37. $ax + by = c$

Cumulative Review Exercises: 7.2

39. $-2x^2 - 5x + 9 - (6x^2 - 4x + 5)$

$= -2x^2 - 5x + 9 - 6x^2 + 4x - 5$

$= -8x^2 - x + 4$

40. $(3x^2 - 4x + 5)(2x - 3)$

$= 3x^2(2x - 3) - 4x(2x - 3) + 5(2x - 3)$

$= 6x^3 - 9x^2 - 8x^2 + 12x + 10x - 15$

$= 6x^3 - 17x^2 + 22x - 15$

41. $x^2 - 2x + 3xy - 6y = x(x - 2) + 3y(x - 2)$

$= (x + 3y)(x - 2)$

42. $\dfrac{3}{x+2} - \dfrac{4}{x+1}$

$= \dfrac{3(x+1)}{(x+2)(x+1)} - \dfrac{4(x+2)}{(x+1)(x+2)}$

$= \dfrac{3x+3}{(x+2)(x+1)} - \dfrac{4x+8}{(x+1)(x+2)}$

$= \dfrac{3x+3-4x-8}{(x+1)(x+2)} = \dfrac{-x-5}{(x+1)(x+2)}$

Group Activity/Challenge Problems

1.

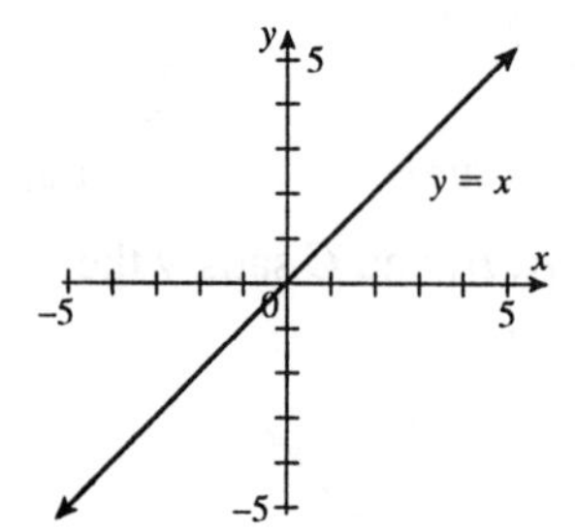

3.

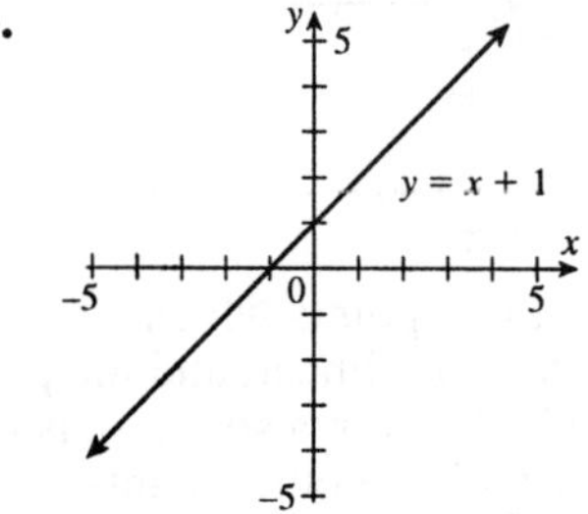

Exercise Set 7.3

1. $2x + y = 6$

$2 \cdot 2 + y = 6$

$4 + y = 6$

$y = 2$

3. $2x + y = 6$

$2x - 5 = 6$

$2x = 11$

$x = \dfrac{11}{2}$

5. $2x + y = 6$

$2x + 0 = 6$

$2x = 6$

$x = 3$

7. $3x - 2y = 8$

$3 \cdot 2 - 2y = 8$

$6 - 2y = 8$

$-2y = 2$

$y = -1$

9. $3x - 2y = 8$

$3x - 2 \cdot 0 = 8$

$3x = 8$

$x = \frac{8}{3}$

11. $3x - 2y = 8$

$3(-3) - 2y = 8$

$-9 - 2y = 8$

$-2y = 17$

$y = -\frac{17}{2}$

13.

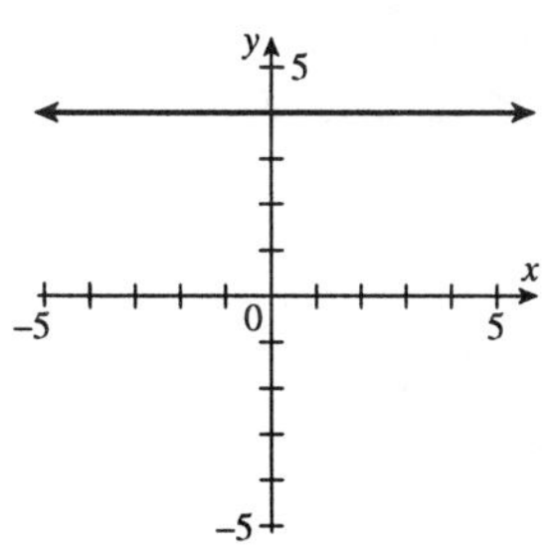

15.

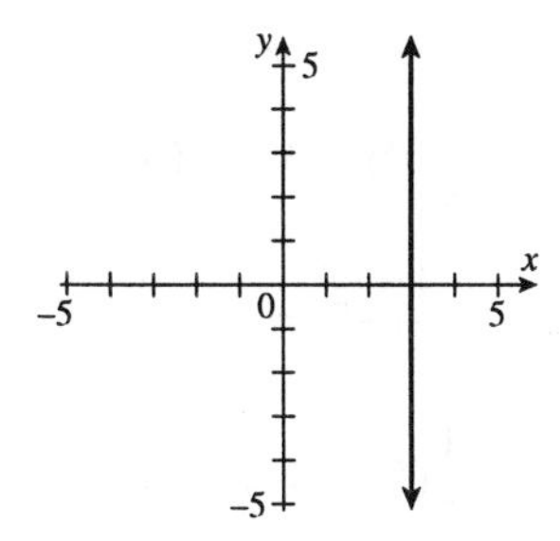

17.

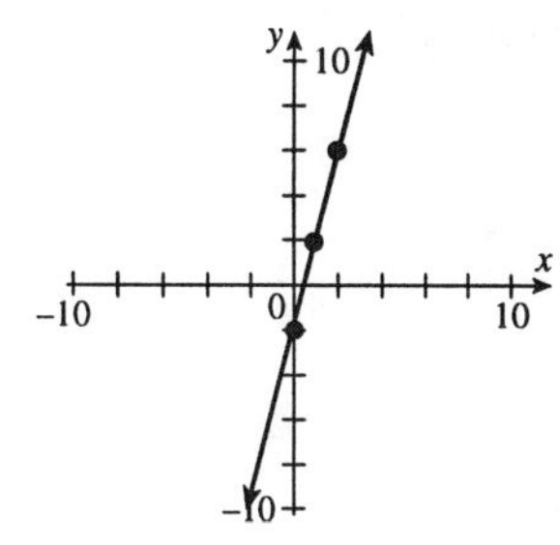

Let $x = 0$, $y = 4 \cdot 0 - 2 = -2$, $(0, -2)$

Let $x = 1$, $y = 4 \cdot 1 - 2 = 2$, $(1, 2)$

Let $x = 2$, $y = 4 \cdot 2 - 2 = 6$, $(2, 6)$

19.

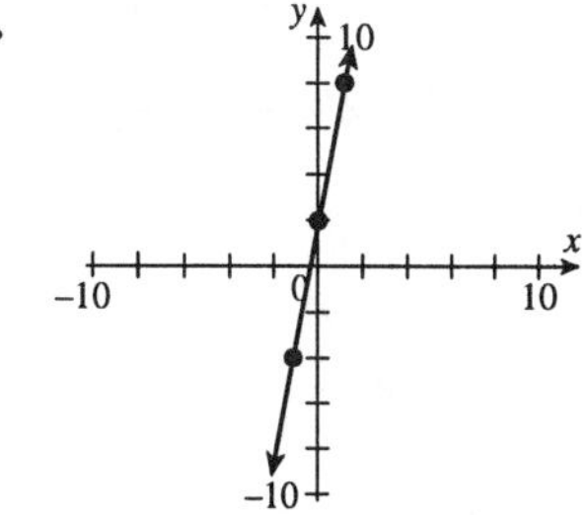

Let $x = -1$, $y = 6(-1) + 2 = -4$, $(-1, -4)$

Let $x = 0$, $y = 6(0) + 2 = 2$, $(0, 2)$

Let $x = 1$, $y = 6 \cdot 1 + 2 = 8$, $(1, 8)$

21.

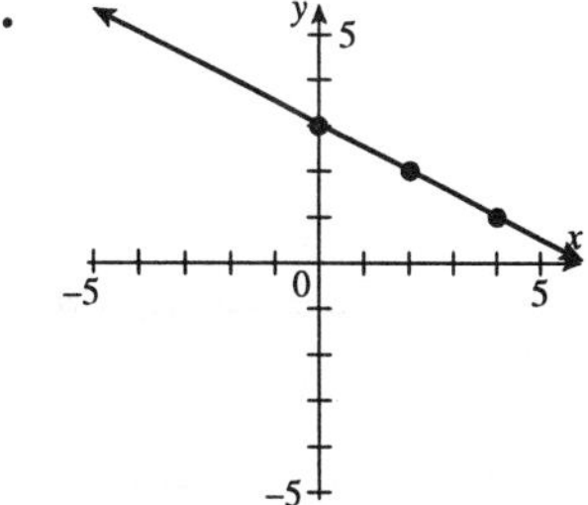

Let $x = 0$, $y = -\frac{1}{2}(0) + 3 = 3$, $(0, 3)$

Let $x = 2$, $y = -\frac{1}{2}(2) + 3 = 2$, $(2, 2)$

Let $x = 4$, $y = -\frac{1}{2}(4) + 3 = 1$, $(4, 1)$

23.

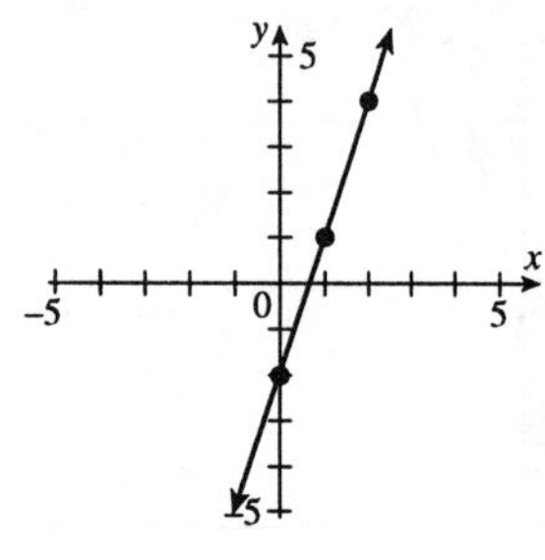

$6x - 2y = 4$

$-2y = -6x + 4$

$y = 3x - 2$

Let $x = 0$, $y = 3 \cdot 0 - 2 = -2$, $(0, -2)$

Let $x = 1$, $y = 3 \cdot 1 - 2 = 1$, $(1, 1)$

Let $x = 2$, $y = 3 \cdot 2 - 2 = 4$, $(2, 4)$

25.

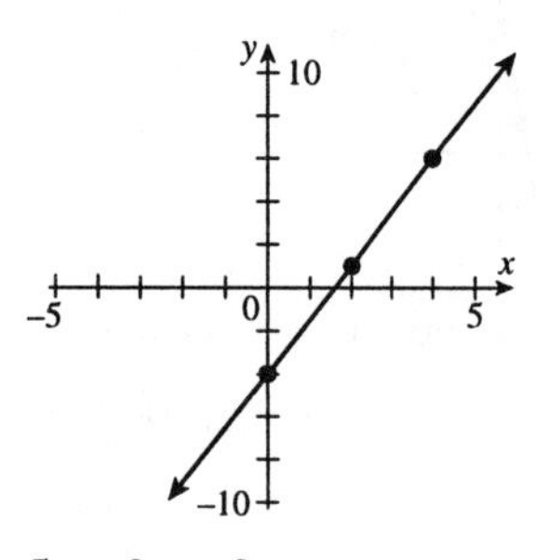

$5x - 2y = 8$

$-2y = -5x + 8$

$y = \frac{5}{2}x - 4$

Let $x = 0$, $y = \frac{5}{2}(0) - 4 = -4$, $(0, -4)$

Let $x = 2$, $y = \frac{5}{2}(2) - 4 = 1$, $(2, 1)$

Let $x = 4$, $y = \frac{5}{2}(4) - 4 = 6$, $(4, 6)$

27.

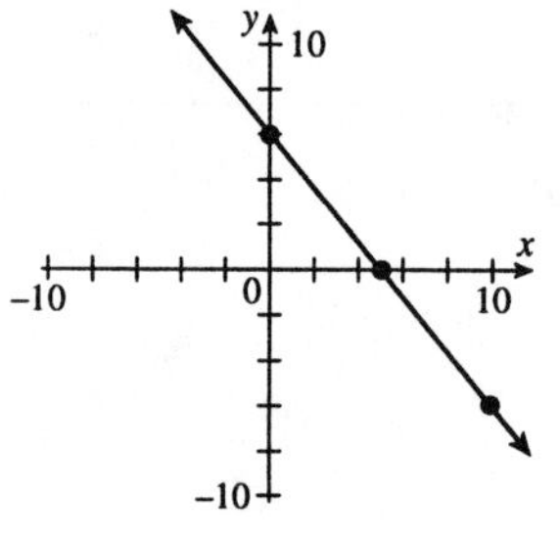

$6x + 5y = 30$

$5y = -6x + 30$

$y = -\frac{6}{5}x + 6$

Let $x = 0$, $y = -\frac{6}{5}(0) + 6 = 6$, $(0, 6)$

Let $x = 5$, $y = -\frac{6}{5}(5) + 6 = 0$, $(5, 0)$

Let $x = 10$, $y = -\frac{6}{5}(10) + 6 = -6$, $(10, -6)$

29.

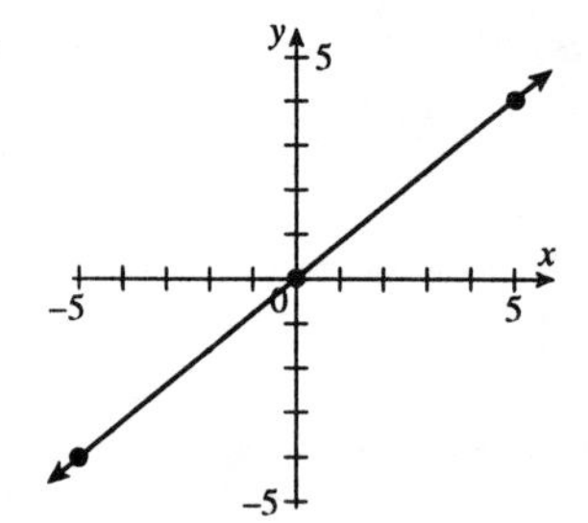

$-4x + 5y = 0$

$5y = 4x$

$y = \frac{4}{5}x$

Let $x = 0$, $y = \frac{4}{5}(0) = 0$, $(0, 0)$

Let $x = -5$, $y = \frac{4}{5}(-5) = -4$, $(-5, -4)$

Let $x = 5$, $y = \frac{4}{5}(5) = 4$, $(5, 4)$

31. $y = 20x + 40$

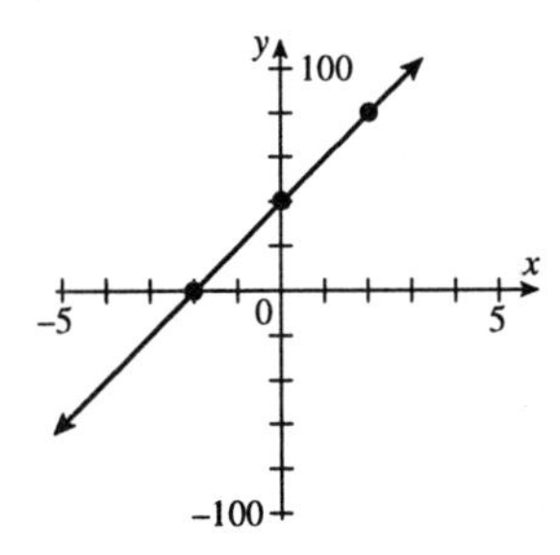

Let $x = -2$, $y = 20(-2) + 40 = 0$, $(-2, 0)$

Let $x = 0$. $y = 20 \cdot 0 + 40 = 40$, $(0, 40)$

Let $x = 2$, $y = 20 \cdot 2 + 40 = 80$, $(2, 80)$

33. $y = \frac{2}{3}x$

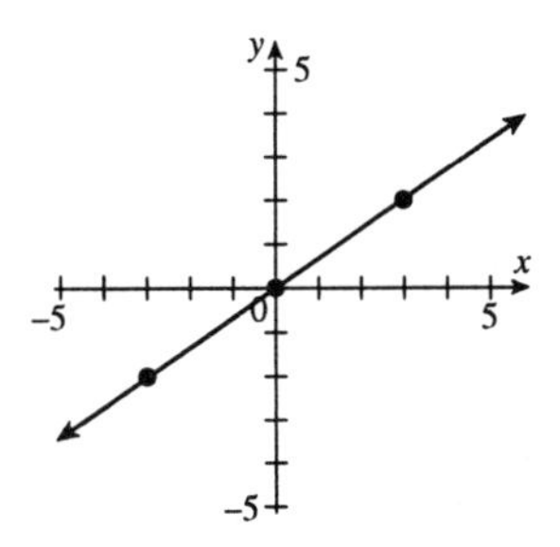

Let $x = -3$, $y = \frac{2}{3}(-3) = -2$, $(-3, -2)$

Let $x = 0$, $y = \frac{2}{3}(0) = 0$, $(0, 0)$

Let $x = 3$, $y = \frac{2}{3}(3) = 2$, $(3, 2)$

35.

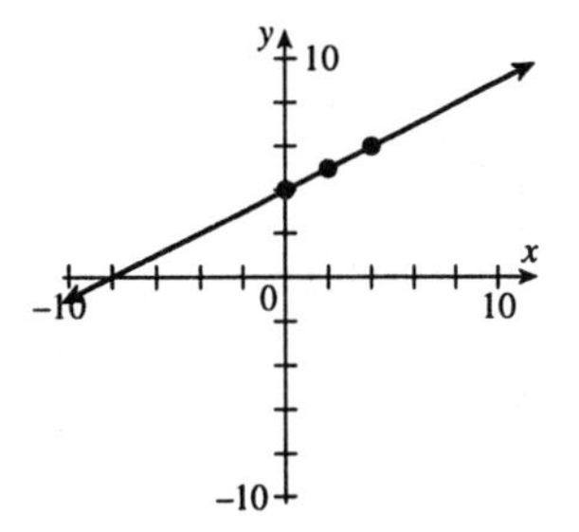

Let $x = 0$, $y = \frac{1}{2}(0) + 4 = 4$, $(0, 4)$

Let $x = 2$, $y = \frac{1}{2}(2) + 4 = 5$, $(2, 5)$

Let $x = 4$, $y = \frac{1}{2}(4) + 4 = 6$, $(4, 6)$

37.

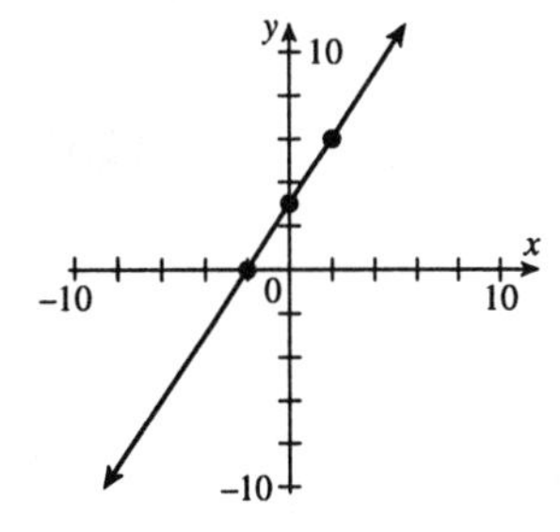

$2y = 3x + 6$

$y = \frac{3}{2}x + 3$

Let $x = -2$, $y = \frac{3}{2}(-2) + 3 = 0$, $(-2, 0)$

Let $x = 0$, $y = \frac{3}{2}(0) + 3 = 3$, $(0, 3)$

Let $x = 2$, $y = \frac{3}{2}(2) + 3 = 6$, $(2, 6)$

39.

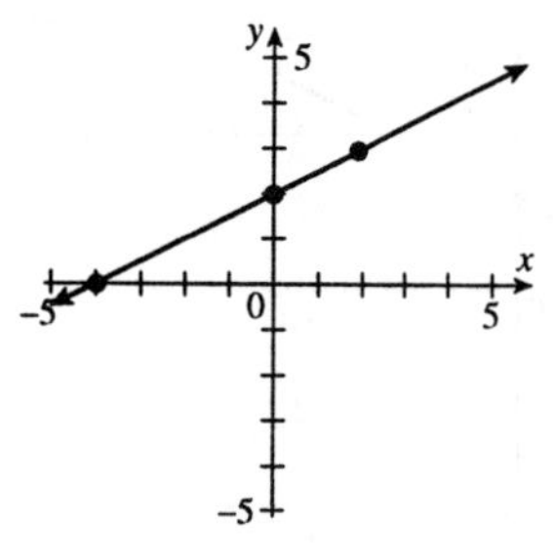

$-4x + 8y = 16$

$8y = 4x + 16$

$y = \frac{1}{2}x + 2$

Let $x = -4,\ y = \frac{1}{2}(-4) + 2 = 0,\ (-4, 0)$

Let $x = 0,\ y = \frac{1}{2}(0) + 2 = 2,\ (0, 2)$

Let $x = 2,\ y = \frac{1}{2}(2) + 2 = 3,\ (2, 3)$

41.

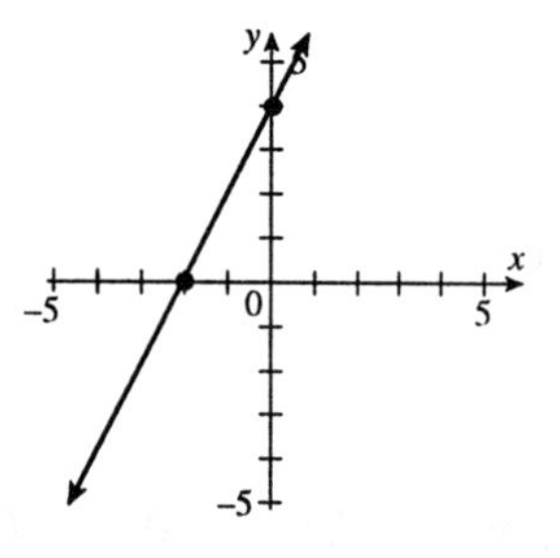

Let $x = 0$

$y = 2x + 4$

$y = 2(0) + 4$

$y = 4$

Let $y = 0$

$y = 2x + 4$

$0 = 2x + 4$

$-2x = 4$

$x = -2$

43.

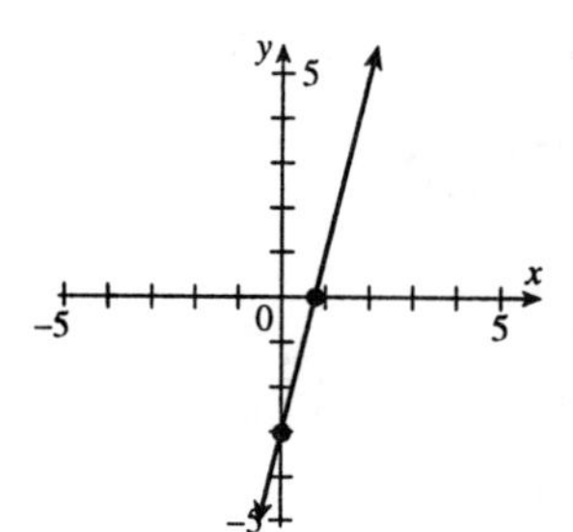

Let $x = 0$

$y = 4x - 3$

$y = 4 \cdot 0 - 3$

$y = -3$

Let $y = 0$

$y = 4x - 3$

$0 = 4x - 3$

$-4x = -3$

$x = \frac{3}{4}$

45.

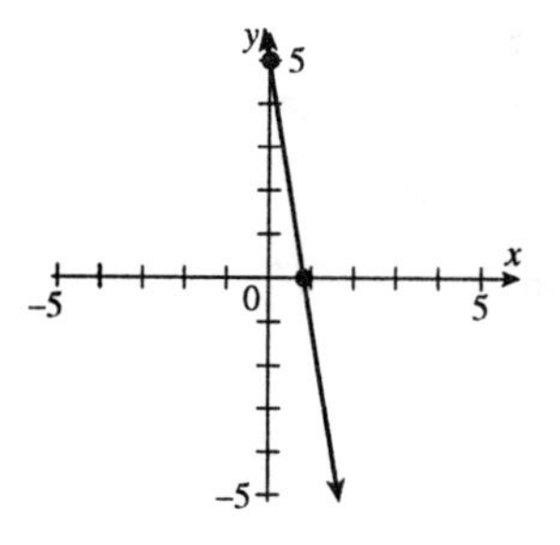

Let $x = 0$

$y = -6x + 5$

$y = -6 \cdot 0 + 5$

$y = 5$

Let $y = 0$

$y = -6x + 5$

$0 = -6x + 5$

$6x = 5$

$x = \frac{5}{6}$

47.

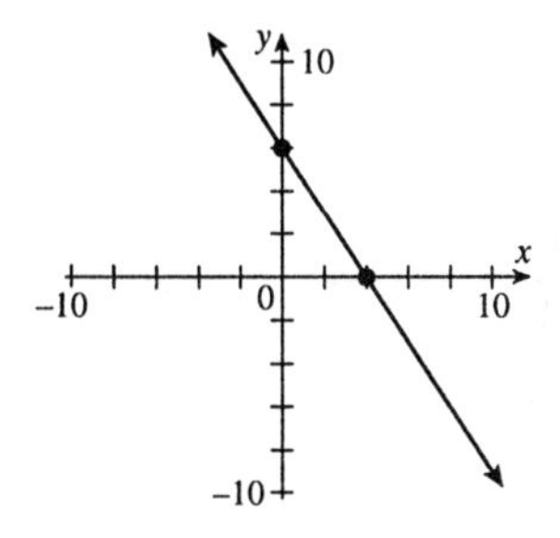

Let $x = 0$

$2y + 3x = 12$

$2y + 3 \cdot 0 = 12$

$2y = 12$

$y = 6$

Let $y = 0$

$2y + 3x = 12$

$2 \cdot 0 + 3x = 12$

$3x = 12$

$x = 4$

49.

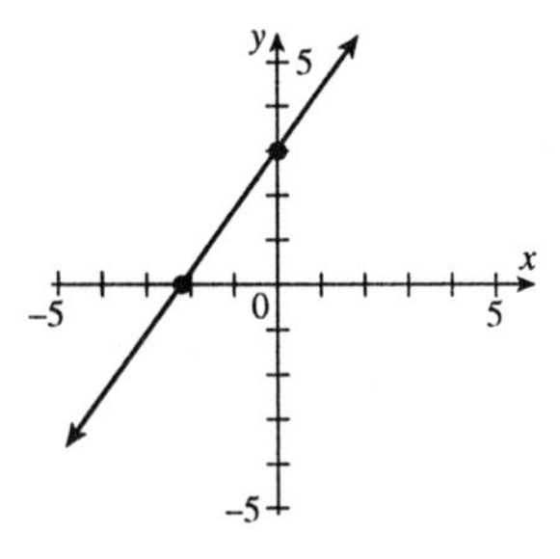

Let $x = 0$

$4x = 3y - 9$

$4 \cdot 0 = 3y - 9$

$0 = 3y - 9$

$-3y = -9$

$y = 3$

Let $y = 0$

$4x = 3y - 9$

$4x = 3 \cdot 0 - 9$

$4x = -9$

$x = -\dfrac{9}{4}$

51.

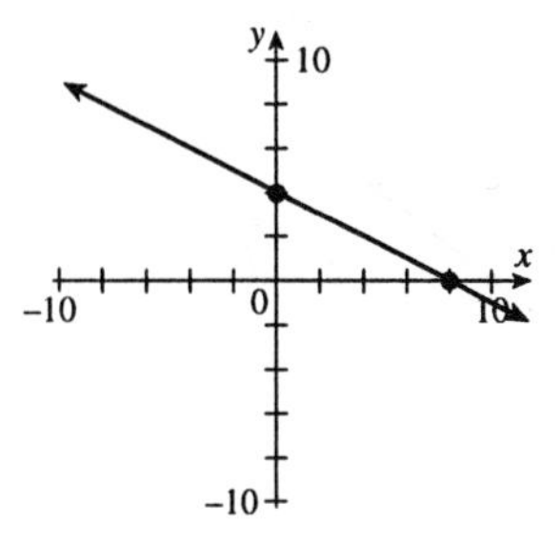

Let $x = 0$

$\frac{1}{2}x + y = 4$

$\frac{1}{2}(0) + y = 4$

$y = 4$

Let $y = 0$

$\frac{1}{2}x + y = 4$

$\frac{1}{2}x = 4$

$x = 8$

53.

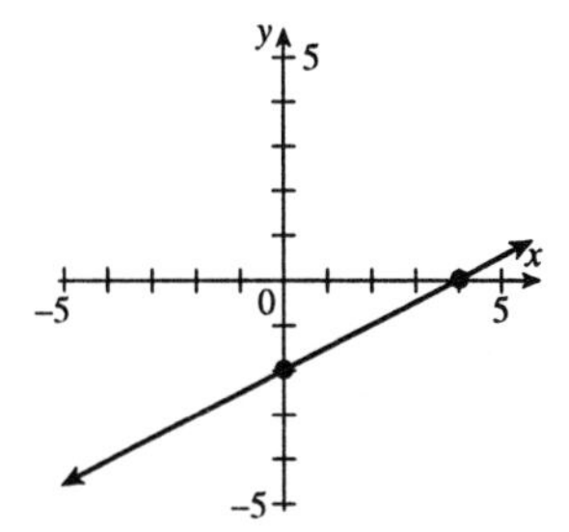

Let $x = 0$

$6x - 12y = 24$

$6 \cdot 0 - 12y = 24$

$-12y = 24$

$y = -2$

Let $y = 0$

$6x - 12y = 24$

$6x - 12 \cdot 0 = 24$

$6x = 24$

$x = 4$

55.

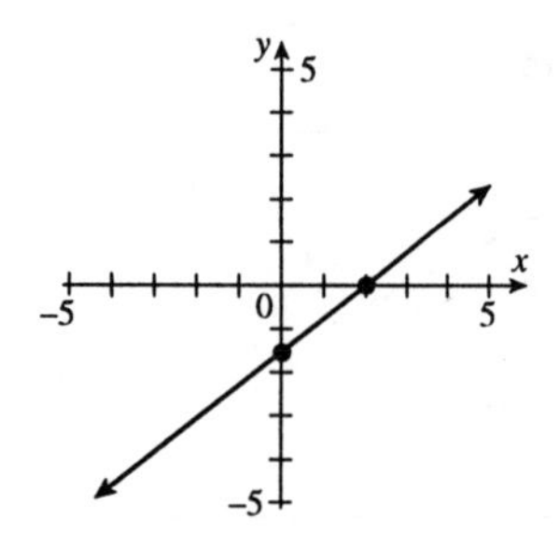

Let $x=0$

$8y=6x-12$

$8y=6\cdot 0-12$

$8y=-12$

$y=-\dfrac{3}{2}$

Let $y=0$

$8y=6x-12$

$8\cdot 0=6x-12$

$0=6x-12$

$-6x=-12$

$x=2$

57.

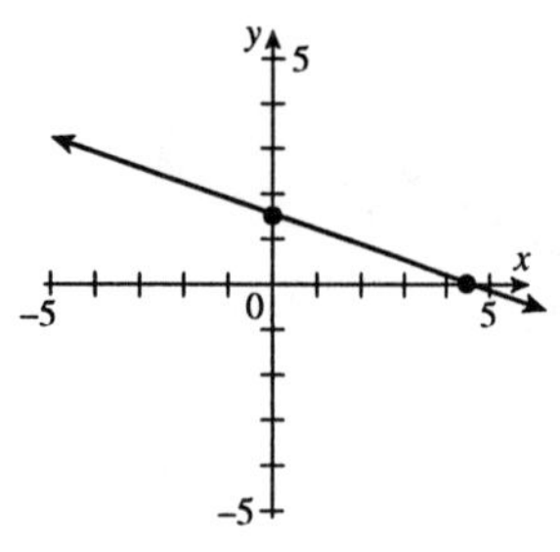

Let $x=0$

$30y+10x=45$

$30y+10\cdot 0=45$

$30y=45$

$y=\dfrac{3}{2}$

Let $y=0$

$30y+10x=45$

$30\cdot 0+10x=45$

$10x=45$

$x=\dfrac{9}{2}$

59.

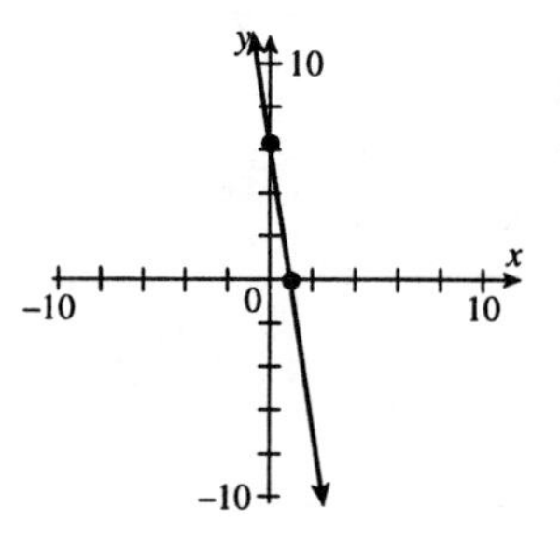

Let $x=0$

$40x+6y=40$

$40\cdot 0+6y=40$

$6y=40$

$y=\dfrac{20}{3}$

Let $y=0$

$40x+6y=40$

$40x+6\cdot 0=40$

$40x=40$

$x=1$

61.

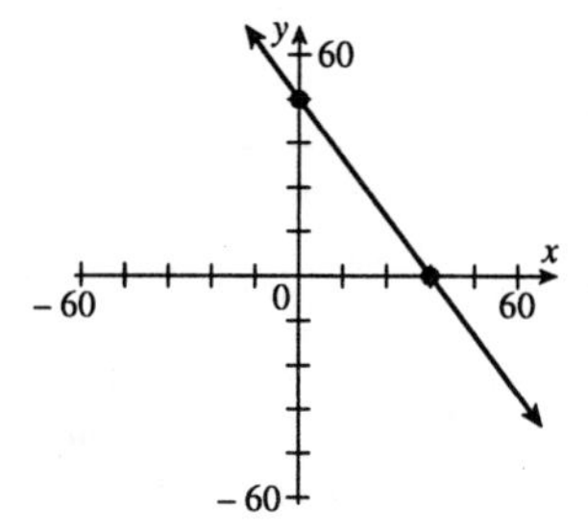

Let $x=0$

$\dfrac{1}{3}x+\dfrac{1}{4}y=12$

$\dfrac{1}{3}(0)+\dfrac{1}{4}y=12$

$\dfrac{1}{4}y=12$

$y=48$

Let $y=0$

$\dfrac{1}{3}x+\dfrac{1}{4}y=12$

$\dfrac{1}{3}x+\dfrac{1}{4}(0)=12$

$\dfrac{1}{3}x=12$

$x=36$

63.

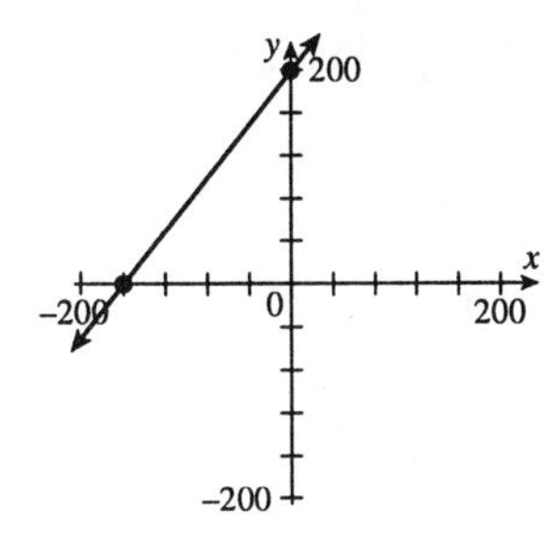

Let $x = 0$

$\frac{1}{2}x = \frac{2}{5}y - 80$

$\frac{1}{2}(0) = \frac{2}{5}y - 80$

$0 = \frac{2}{5}y - 80$

$-\frac{2}{5}y = -80$

$y = 200$

Let $y = 0$

$\frac{1}{2}x = \frac{2}{5}y - 80$

$\frac{1}{2}x = \frac{2}{5}(0) - 80$

$\frac{1}{2}x = -80$

$x = -160$

65. $x = -3$

67. $y = 3$

69. **(a)** $C = m + 25$

(b)

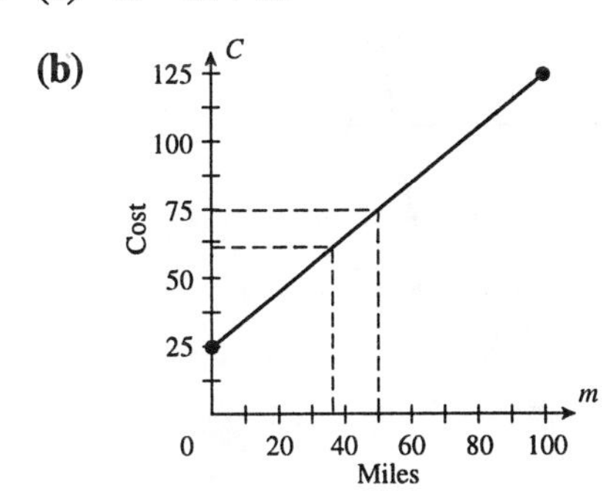

(c) $75

(d) 35 miles

71. **(a)**

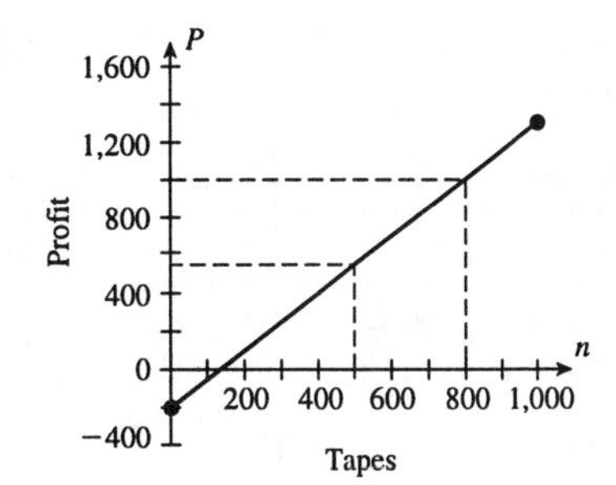

(b) $550

(c) 800 tapes

73. To find the x intercept, substitute 0 for y and find the corresponding value of x. To find the y intercept, substitute 0 for x and find the corresponding value of y.

75. The graph of $y = b$ is a horizontal line.

77. You may not be able to read exact answers from a graph.

79. Since each shaded area multiplied by the corresponding intercept must equal 20, the coefficients are 5 and 4 respectively.

81. Since each shaded area multiplied by the corresponding intercept must equal –12, the coefficients are 6 and 4 respectively.

Cumulative Review Exercises: 7.3

83. $4(x-2) - (3-x) = 2x + 4$

$4x - 8 - 3 + x = 2x + 4$

$5x - 11 = 2x + 4$

$3x - 11 = 4$

$3x = 15$

$x = 5$

84. Let r be the speed of the first cyclist. Then the speed of the other is $r + 3$.

Cyclist	Rate	Time	Distance
First	r	1.5	$1.5r$
Second	$r+3$	1.5	$1.5(r+3)$

$1.5r + 1.5(r + 3) = 18$

$1.5r + 1.5r + 4.5 = 18$

$3r + 4.5 = 18$

$3r = 13.5$

$r = 4.5$

The speed of the first cyclist is 4.5 mph, and the speed of the second is 7.5 mph.

85. $2x^2 = -23x + 12$

$2x^2 + 23x - 12 = 0$

$(2x - 1)(x + 12) = 0$

$2x - 1 = 0$ or $x + 12 = 0$

$2x = 1$ or $x = -12$

$x = \frac{1}{2}$

86. $x - 14 = \frac{-48}{x}$

$x^2 - 14x = -48$

$x^2 - 14x + 48 = 0$

$(x - 6)(x - 8) = 0$

$x - 6 = 0$ or $x - 8 = 0$

$x = 6$ or $x = 8$

Group Activity/Challenge Problems

1.

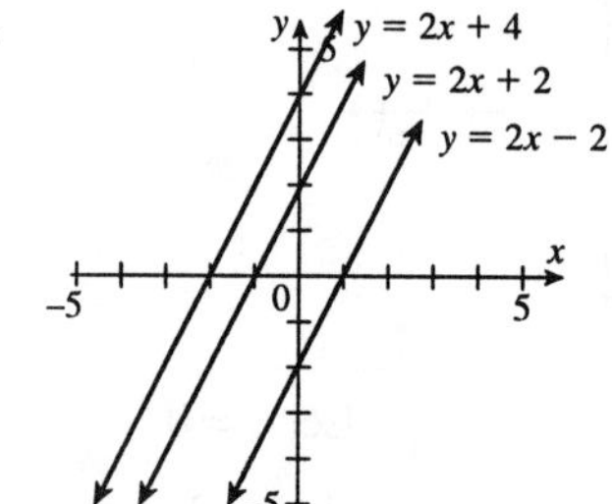

(b) They appear to be parallel lines.

(c) Individual Answer.

3.

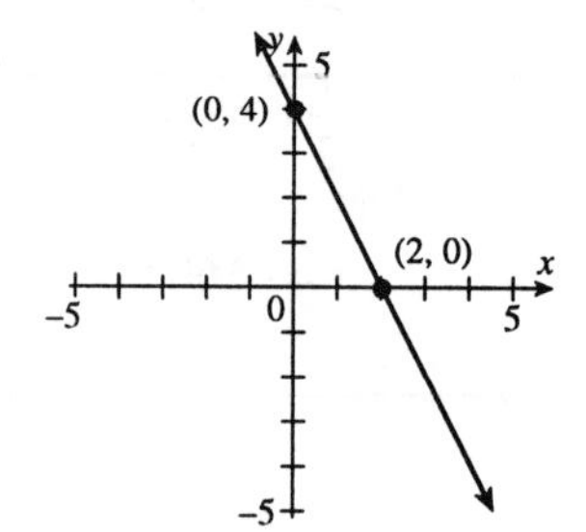

(b) When $x = 2$, $y = 0$ and when $x = 0$, $y = 4$.
$y - 4 = -2x$
$y = -2x + 4$

5.

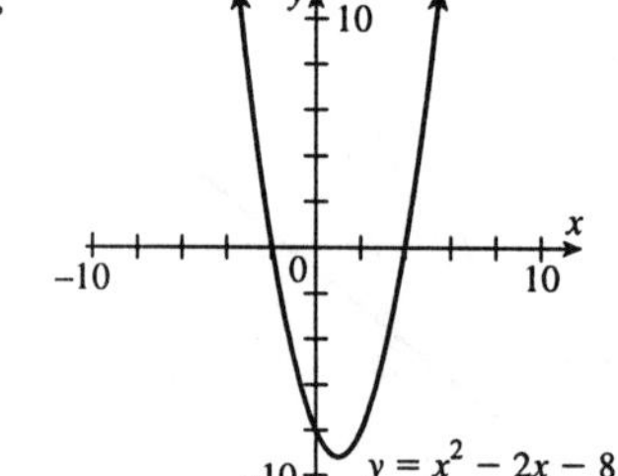

Exercise Set 7.4

1. $m = \frac{6-1}{5-4} = \frac{5}{1} = 5$

3. $m = \frac{-2-0}{5-9} = \frac{-2}{-4} = \frac{1}{2}$

5. $m = \dfrac{8-8}{-3-3} = \dfrac{0}{-6} = 0$

7. $m = \dfrac{6-6}{-2-(-4)} = \dfrac{0}{2} = 0$

9. $m = \dfrac{-2-4}{3-3} = \dfrac{-6}{0}$ undefined

11. $m = \dfrac{-3-2}{5-(-4)} = \dfrac{-5}{9} = -\dfrac{5}{9}$

13. $m = \dfrac{-3-7}{4-(-1)} = \dfrac{-10}{5} = -2$

15. $m = \dfrac{4}{2} = 2$

17. $m = \dfrac{3}{-2} = -\dfrac{3}{2}$

19. $m = \dfrac{6}{-4} = -\dfrac{3}{2}$

21. $m = \dfrac{7}{4}$

23. $m = \dfrac{0}{3} = 0$

25. Slope is undefined.

27. The slope of a line is the ratio of the vertical change to the horizontal change between any two points on a line.

29. The values of y increase as the values of x increase.

31. Lines that rise from the left to right have a positive slope. Lines that fall from left to right have a negative slope.

33. No, since we cannot divide by 0, the slope is undefined.

Cumulative Review Exercises: 7.4

34. **(a)** A linear equation in one variable is an equation that contains only one variable and that variable has an exponent of 1.

(b) $2x + 3 = 5x - 6$. (Answers will vary)

35. **(a)** A quadratic equation in one variable is an equation that contains only one variable and the greatest exponent on that variable is 2.

(b) $x^2 + 2x - 3 = 0$ (Answers will vary)

36. **(a)** A rational equation in one variable is an equation that contains only one variable and one or more fractions.

(b) $\dfrac{x}{3} + \dfrac{x}{4} = 12$. (Answers will vary)

37. **(a)** A linear equation in two variables is an equation that contains two variables where the exponent on both variables is 1.

(b) $y = 3x - 2$ (Answers will vary)

Group Activity/Challenge Problems

1. $m = \dfrac{-\frac{7}{2}-\left(-\frac{3}{8}\right)}{-\frac{4}{9}-\frac{1}{2}} = \dfrac{-\frac{28}{8}+\frac{3}{8}}{-\frac{8}{18}-\frac{9}{18}} = \dfrac{-\frac{25}{8}}{-\frac{17}{18}}$

$= \left(-\dfrac{25}{8}\right)\left(-\dfrac{18}{17}\right) = \dfrac{(-25)(-9)}{(4)(17)} = \dfrac{225}{68}$

3. $m = -\dfrac{3}{4} = \dfrac{2-(-7)}{-5-x}$

$-\dfrac{3}{4} = \dfrac{9}{-5-x}$

$-3(-5-x) = (4)(9)$

$15 + 3x = 36$

$3x = 21$

$x = 7$

5.

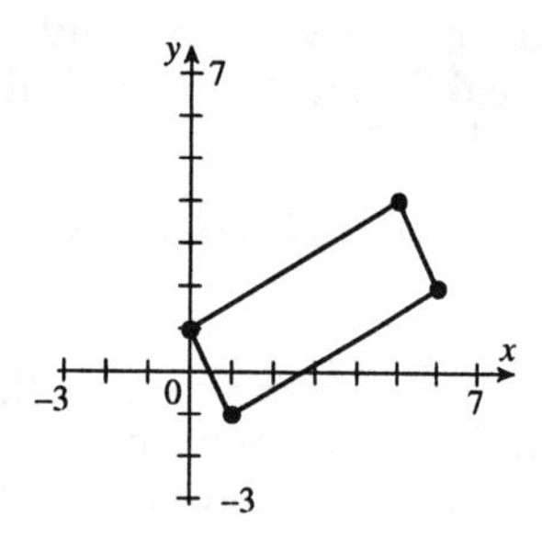

(b) $AC;\ m = \dfrac{4-1}{5-0} = \dfrac{3}{5}$

$CB;\ m = \dfrac{4-2}{5-6} = \dfrac{2}{-1} = -2$

$DB;\ m = \dfrac{2-(-1)}{6-1} = \dfrac{3}{5}$

$AD;\ m = \dfrac{-1-1}{1-0} = \dfrac{-2}{1} = -2$

(c) Yes; Opposite sides are parallel.

Exercise Set 7.5

1. $m = 2, b = -1$

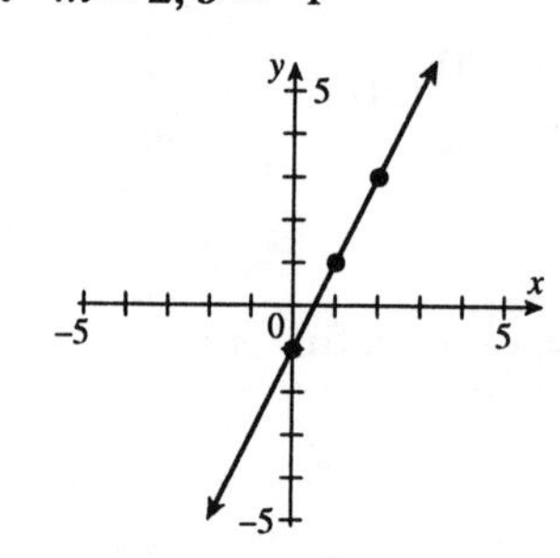

3. $m = -1, b = 5$

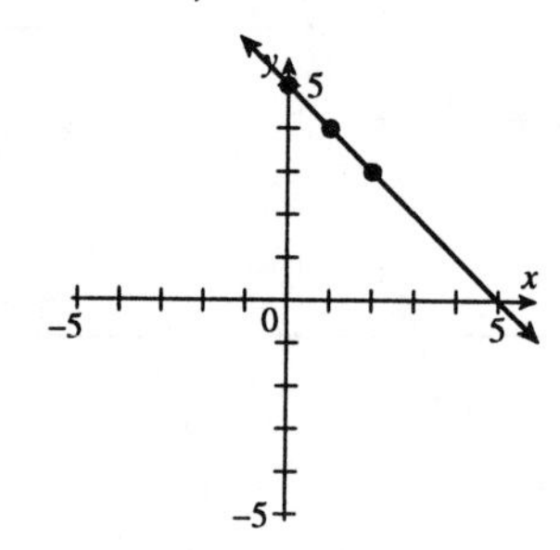

5. $m = -4, b = 0$

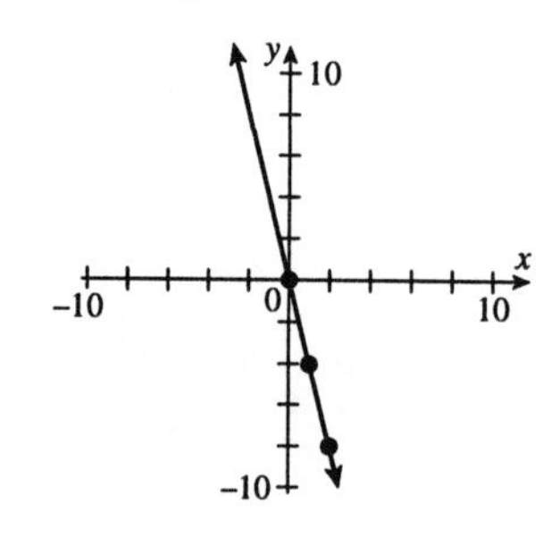

7. $-2x + y = -3$

$y = 2x - 3$

$m = 2, b = -3$

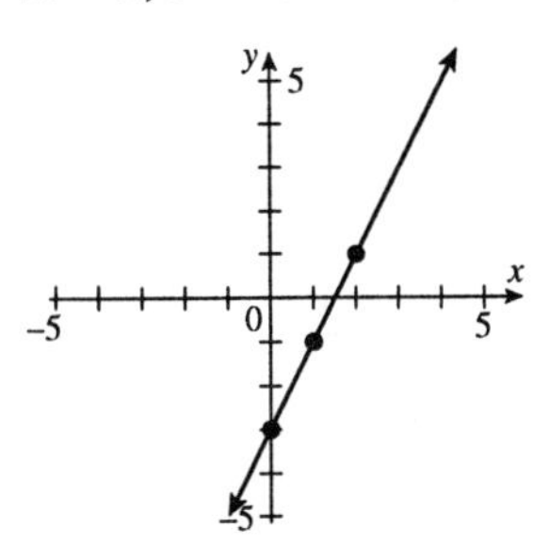

9. $3x + 3y = 9$

$3y = -3x + 9$

$y = -x + 3$

$m = -1, b = 3$

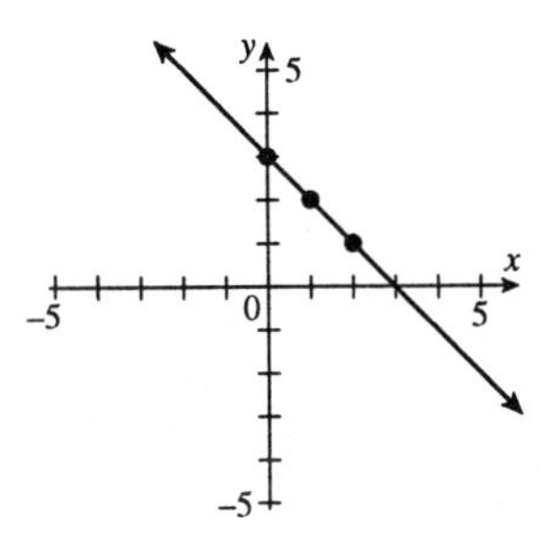

11. $-x + 2y = 8$

$2y = x + 8$

$y = \frac{1}{2}x + 4$

$m = \frac{1}{2}, b = 4$

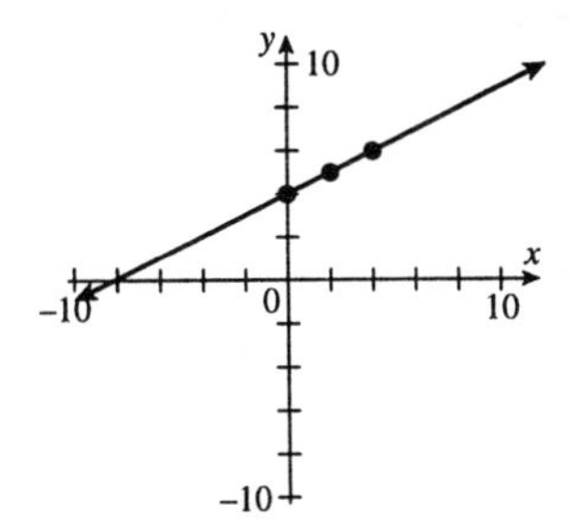

13. $4x = 6y + 9$

$-6y = -4x + 9$

$y = \frac{2}{3}x - \frac{3}{2}$

$m = \frac{2}{3},\ b = -\frac{3}{2}$

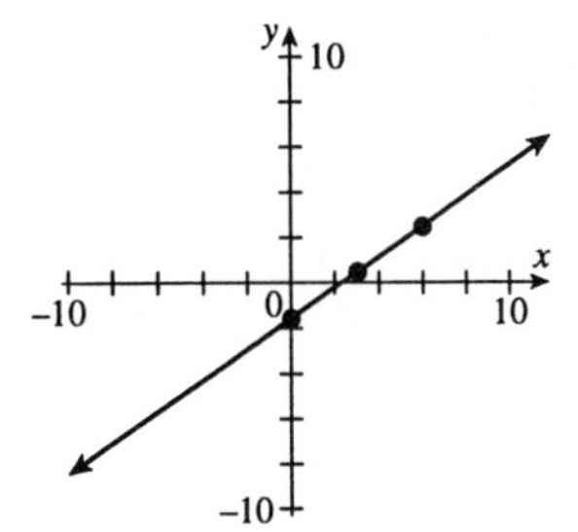

15. $-6x = -2y + 8$

$2y = 6x + 8$

$y = 3x + 4$

$m = 3, b = 4$

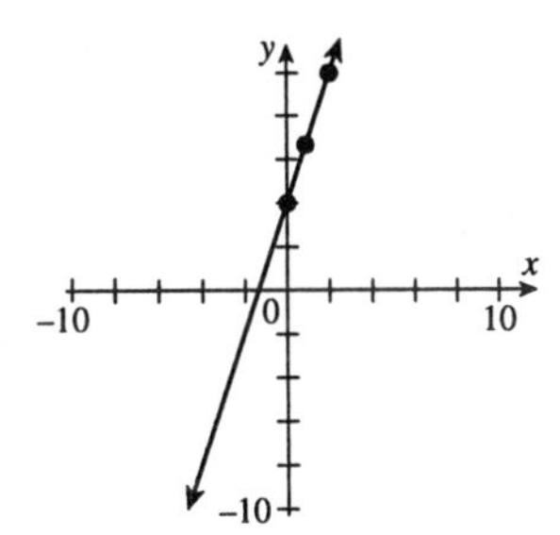

17. $-3x+8y=-8$

$8y=3x-8$

$y=\frac{3}{8}x-1$

$m=\frac{3}{8},\ b=-1$

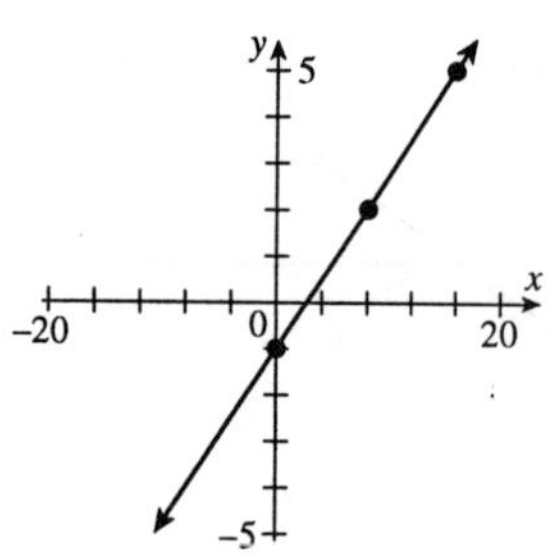

19. $3x=2y-4$

$-2y=-3x-4$

$y=\frac{3}{2}x+2$

$m=\frac{3}{2},\ b=2$

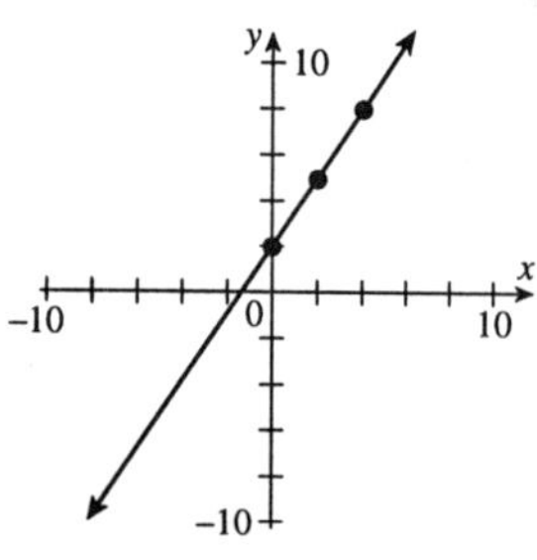

21. $m=\frac{2}{1}=2,\ b=0$

$y=2x$

23. $m=\frac{-3}{1}=-3,\ b=-5$

$y=-3x-5$

25. $m=\frac{5}{15}=\frac{1}{3},\ b=5$

$y=\frac{1}{3}x+5$

27. $m=\frac{2}{1}=2,\ b=-1$

$y=2x-1$

29. Yes.

31. $4x+2y=9$

$2y=-4x+9$

$y=-2+\frac{9}{2}$

$8x=4-4y$

$4y=-8x+4$

$y=-2x+1$

Yes.

33. $2x+5y=9$

$5y=-2x+9$

$y=-\frac{2}{5}x+\frac{9}{5}$

$-x+3y=9$

$3y=x+9$

$y=\frac{1}{3}x+3$

No.

35. $y=\frac{1}{2}x-6$

$3y=6x+9$

$y=2x+3$

No.

37. $y-4=5(x-0)$

$y-4=5x$

$y=5x+4$

39. $y-5=-2[x-(-4)]$

$y-5=-2x-8$

$y=-2x-3$

41. $y-(-5)=\frac{1}{2}[x-(-1)]$

$y+5=\frac{1}{2}x+\frac{1}{2}$

$y=\frac{1}{2}x-\frac{9}{2}$

43. $y=\frac{3}{5}x+7$

45. $m=\frac{4-(-2)}{-2-(-4)}=\frac{6}{2}=3$

$y-(-2)=3[x-(-4)]$

$y+2=3x+12$

$y=3x+10$

47. $m=\frac{-6-6}{4-(-4)}=\frac{-12}{8}=-\frac{3}{2}$

$y-6=-\frac{3}{2}[x-(-4)]$

$y-6=-\frac{3}{2}x-6$

$y=-\frac{3}{2}x$

49. $m=\frac{-2-3}{0-10}=\frac{-5}{-10}=\frac{1}{2}$

$y-3=\frac{1}{2}(x-10)$

$y-3=\frac{1}{2}x-5$

$y=\frac{1}{2}x-2$

51. $y=5.2x-1.6$

53. $3x-2y=4$

$-2y=-3x+4$

$y=\frac{3}{2}x-2$

(a) Let $x=0$, $y=\frac{3}{2}(0)-2=-2$, $(0,-2)$

Let $x=2$, $y=\frac{3}{2}(2)-2=1$, $(2, 1)$

Let $x=4$, $y=\frac{3}{2}(4)-2=4$, $(4, 4)$

(b)

Let $x=0$	Let $y=0$
$3x-2y=4$	$3x-2y=4$
$3\cdot 0-2y=4$	$3x-2\cdot 0=4$
$-2y=4$	$3x=4$
$y=-2$	$x=\frac{4}{3}$

(c) $m=\frac{3}{2}$, $b=-2$

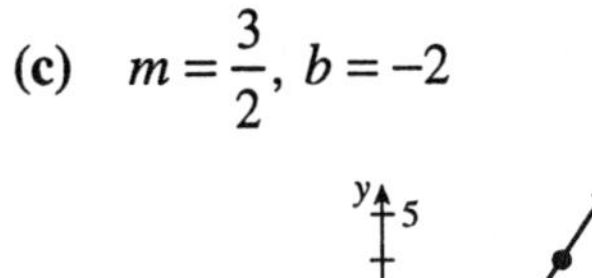

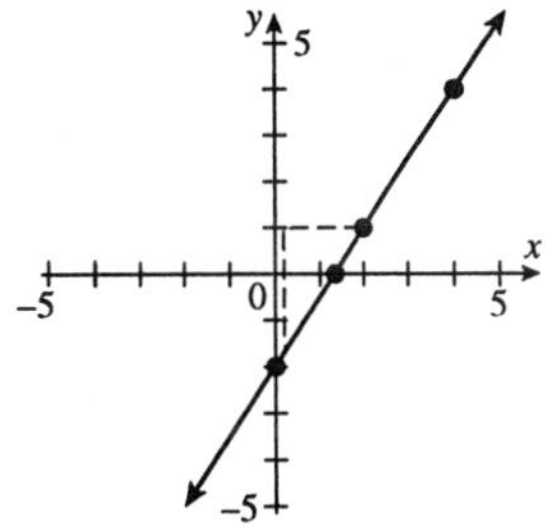

55. $2x - 3y = -6$

$-3y = -2x - 6$

$y = \frac{2}{3}x + 2$

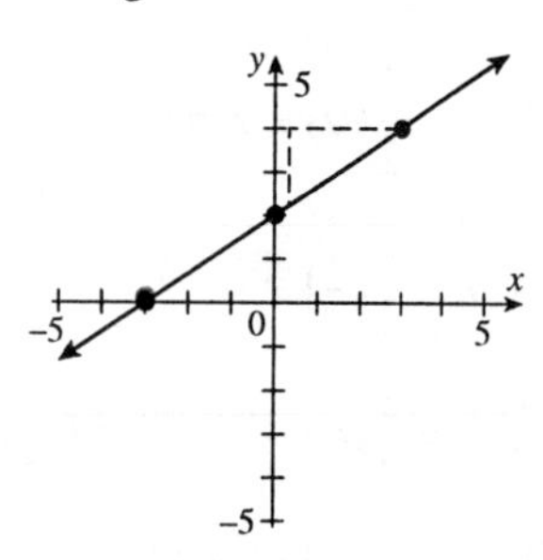

(a) Let $x = -3$, $y = \frac{2}{3}(-3) + 2 = 0$, $(-3, 0)$

Let $x = 0$, $y = \frac{2}{3}(0) + 2 = 2$, $(0, 2)$

Let $x = 3$, $y = \frac{2}{3}(3) + 2 = 4$, $(3, 4)$

(b) Let $x = 0$ | Let $y = 0$

$2x - 3y = -6$ | $2x - 3y = -6$

$2 \cdot 0 - 3y = -6$ | $2x - 3 \cdot 0 = -6$

$-3y = -6$ | $2x = -6$

$y = 2$ | $x = -3$

(c) $m = \frac{2}{3}$, $b = 2$

57. Compare their slopes: If slopes are the same and their y intercepts are different, the lines are parallel.

59. **(a)** $ax + by = c$

(b) $y = mx + b$

(c) $y - y_1 = m(x - x_1)$

61. **(a)** Answers will vary.

(b) $-3x + 2y = 4$

$2y = 3x + 4$

$y = \frac{3}{2}x + 2$

$m = \frac{3}{2}$, $b = 2$

Let $x = -2$, $y = \frac{3}{2}(-2) + 2 = -1$, $(-2, -1)$

Let $x = 0$, $y = \frac{3}{2}(0) + 2 = 2$, $(0, 2)$

Let $x = 2$, $y = \frac{3}{2}(2) + 2 = 5$, $(2, 5)$

Let $x = 0$ | Let $y = 0$

$y = \frac{3}{2}x + 2$ | $y = \frac{3}{2}x + 2$

$y = \frac{3}{2}(0) + 2$ | $0 = \frac{3}{2}x + 2$

$y = 2$ | $-\frac{3}{2}x = 2$

$x = -\frac{4}{3}$

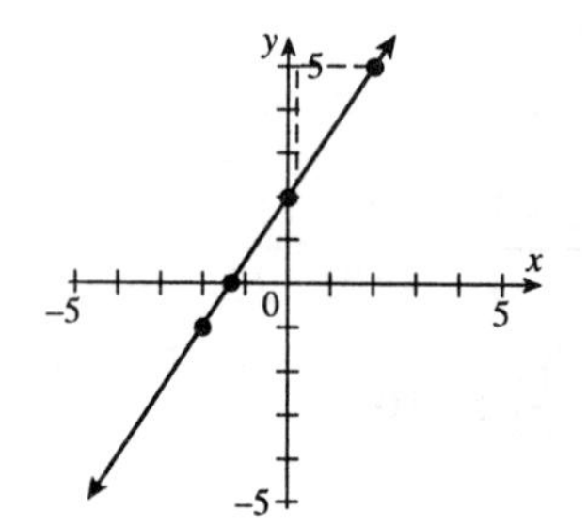

Cumulative Review Exercises: 7.5

62. $3x+2\overline{)9x^3-3x^2-9x+4}$ with quotient $3x^2-3x-1$

$$\begin{array}{r} -9x^3-6x^2 \\ \hline -9x^2-9x \\ 9x^2+6x \\ \hline -3x+4 \\ 3x+2 \\ \hline 6 \end{array}$$

Thus, $\dfrac{9x^3-3x^2-9x+4}{3x+2}$

$$=3x^2-3x-1+\frac{6}{3x+2}$$

63. Let x be the amount of pure water needed.

Solution	Strength	Liters	Amount
5%	0.05	1	0.05
Water	0	x	0
Mixture	0.02	$1+x$	$(1+x)(0.02)$

$0.05+0=(1+x)(0.02)$

$0.05=0.02+0.02x$

$0.03=0.02x$

$1.5=x$

Consuella needs 1.5 liters of pure water.

64. $\dfrac{x^2+2x-8}{x^2-16}\div\dfrac{2x^2-5x-3}{x^2-7x+12}$

$$=\frac{x^2+2x-8}{x^2-16}\cdot\frac{x^2-7x+12}{2x^2-5x-3}$$

$$=\frac{(x+4)(x-2)}{(x+4)(x-4)}\cdot\frac{(x-3)(x-4)}{(2x+1)(x-3)}$$

$$=\frac{(x-2)}{1}\cdot\frac{1}{(2x+1)}=\frac{x-2}{2x+1}$$

65. $\dfrac{3}{x-2}+\dfrac{5}{x+2}$

$$=\frac{3(x+2)}{(x-2)(x+2)}+\frac{5(x-2)}{(x+2)(x-2)}$$

$$=\frac{3x+6}{(x-2)(x+2)}+\frac{5x-10}{(x+2)(x-2)}$$

$$=\frac{3x+6+5x-10}{(x+2)(x-2)}=\frac{8x-4}{(x+2)(x-2)}$$

$$=\frac{4(2x-1)}{(x+2)(x-2)}$$

66. $x+\dfrac{30}{x}=11$

$x^2+30=11x$

$x^2-11x+30=0$

$(x-5)(x-6)=0$

$x-5=0$ or $x-6=0$

$x=5$ or $x=6$

67. Let b be the base of the triangle. Then the height is $2b-7$.

$A=\frac{1}{2}b\cdot h$

$36=\frac{1}{2}b(2b-7)$

$72=2b^2-7b$

$0=2b^2-7b-72$

$0=(2b+9)(b-8)$

$2b+9=0$ or $b-8=0$

$2b=-9$ or $b=8$

$b=-\frac{9}{2}$

Since all lengths are positive, the base is 8 feet and the height is 9 feet.

Group Activity/Challenge Problems

3. $m = \frac{90-30}{20-60} = \frac{60}{-40} = -\frac{3}{2}$

   Line with points (2, 0) and (0, 3);

   $m = \frac{3-0}{0-2} = \frac{3}{-2} = -\frac{3}{2}$

   Since slopes are the same, they are parallel.

4. $3x - 4y = 6$

   $4y = 3x - 6$

   $y = \frac{3}{4}x - \frac{3}{2}$

   $m = \frac{3}{4}$

   $y - (-1) = \frac{3}{4}(x - (-4))$

   $y + 1 = \frac{3}{4}(x + 4)$

   $y = \frac{3}{4}x + 3 - 1$

   Answer: $y = \frac{3}{4}x + 2$

5. $-5x + 2y = -4$

   $2y = 5x - 4$

   $y = \frac{5}{2}x - 2$

   $m = \frac{5}{2}$

   Therefore, slope perpendicular is $-\frac{2}{5}$

   $y - \frac{1}{2} = -\frac{2}{5}(x - 2)$

   $y - \frac{1}{2} = -\frac{2}{5}x + \frac{4}{5}$

   $y = -\frac{2}{5}x + \frac{4}{5} + \frac{1}{2}$

   Answer: $y = -\frac{2}{5}x + \frac{13}{10}$

Exercise Set 7.6

1. Function, Domain {1, 2, 3, 4, 5}, Range {1, 2, 3, 4, 5}

3. Relation, Domain {1, 2, 3, 5, 7}, Range {–2, 0, 2, 4, 5}

5. Relation, Domain {0, 1, 3, 5}, Range {–4, –1, 0, 1, 2}

7. Relation, Domain {0, 1, 4} Range {–3, 0, 2, 5}

9. Function, Domain {0, 1, 2, 3, 4}, Range {3}

11. **(a)** {(1, 4), (2, 5), (3, 5), (4, 7)}
    (b) Function

13. **(a)** {(–4, 5), (0, 7), (6, 9), (6, 3)}
    (b) Not a function

15. Function

17. Function

19. Function

21. Function

23. Not a function

25. Function

27. **(a)** $f(2) = 2 \cdot 2 + 3 = 7$

(b) $f(-2) = 2(-2) + 3 = -1$

29. **(a)** $f(2) = 2^2 - 4 = 0$

(b) $f(3) = 3^2 - 4 = 5$

31. **(a)** $f(0) = 3 \cdot 0^2 - 0 + 5 = 5$

(b) $f(2) = 3 \cdot 2^2 - 2 + 5 = 15$

33. **(a)** $f(2) = \dfrac{2+4}{2} = \dfrac{6}{2} = 3$

(b) $f(6) = \dfrac{6+4}{2} = \dfrac{10}{2} = 5$

35.

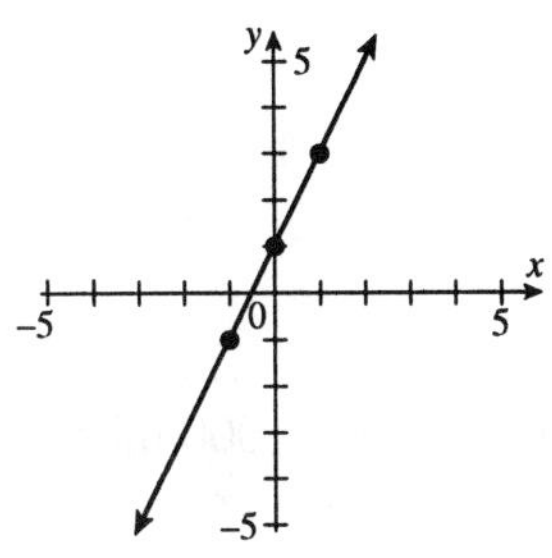

Let $x = -1, f(-1) = 2(-1) + 1 = -1, (-1, -1)$

Let $x = 0, f(0) = 2 \cdot 0 + 1 = 1, (0, 1)$

Let $x = 1, f(1) = 2 \cdot 1 + 1 = 3, (1, 3)$

37.

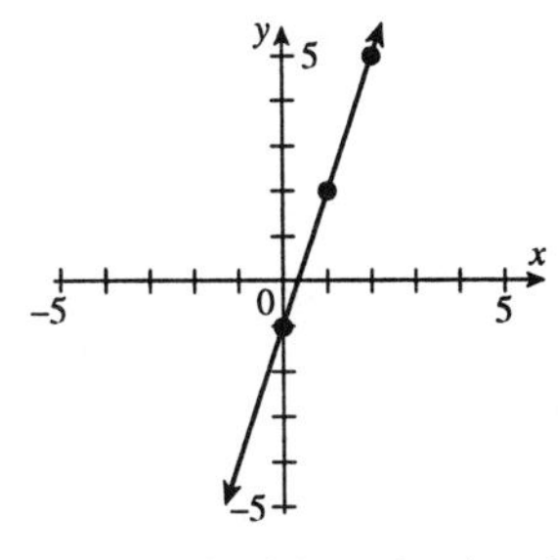

Let $x = 0, f(0) = 3 \cdot 0 - 1 = -1, (0, -1)$

Let $x = 1, f(1) = 3 \cdot 1 - 1 = 2, (1, 2)$

Let $x = 2, f(2) = 3 \cdot 2 - 1 = 5, (2, 5)$

39.

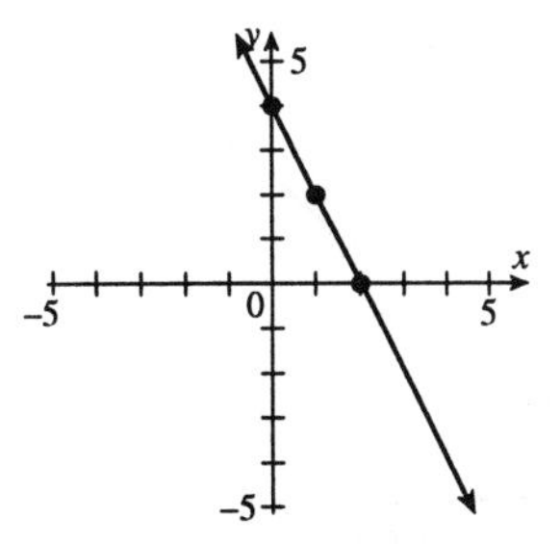

Let $x = 0, f(0) = -2 \cdot 0 + 4 = 4, (0, 4)$

Let $x = 1, f(1) = -2 \cdot 1 + 4 = 2, (1, 2)$

Let $x = 2, f(2) = -2 \cdot 2 + 4 = 0, (2, 0)$

41.

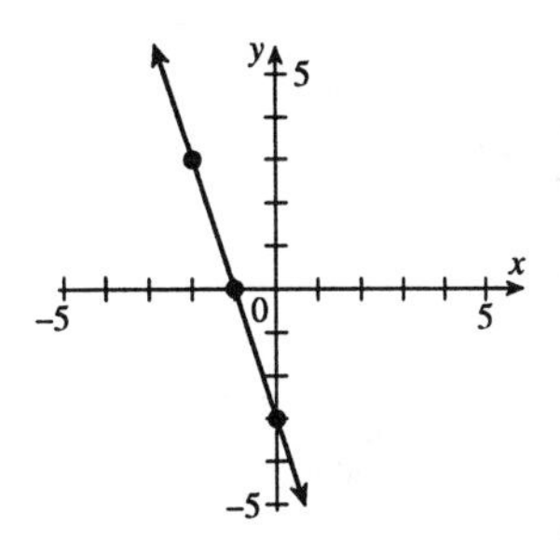

Let $x = -2, f(-2) = -3(-2) - 3 = 3, (-2, 3)$

Let $x = -1, f(-1) = -3(-1) - 3 = 0, (-1, 0)$

Let $x = 0, f(0) = -3 \cdot 0 - 3 = -3, (0, -3)$

43. **(a)**

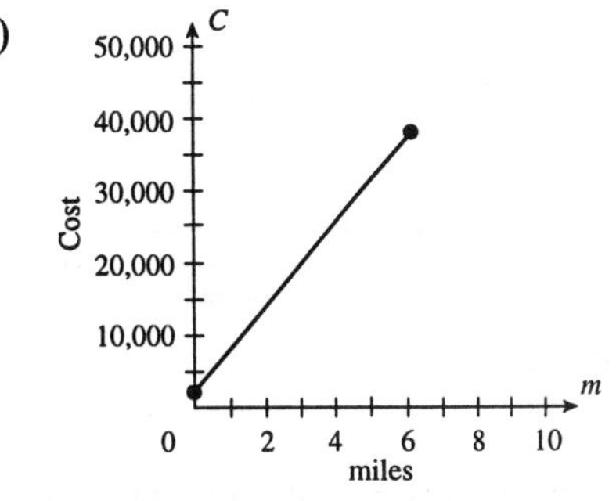

(b) \$14,000

45. (a)

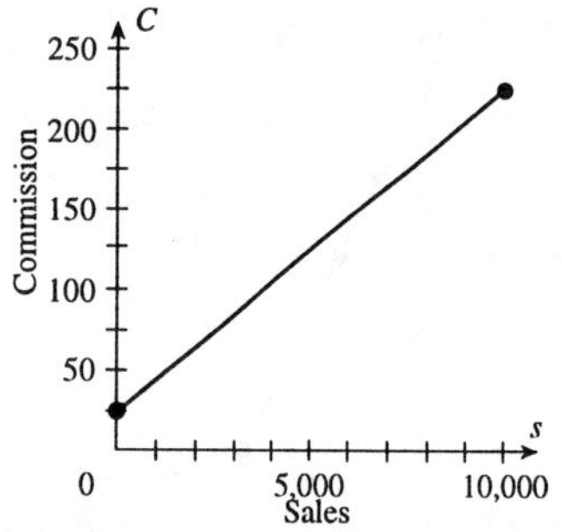

(b) \$185

47. (a)

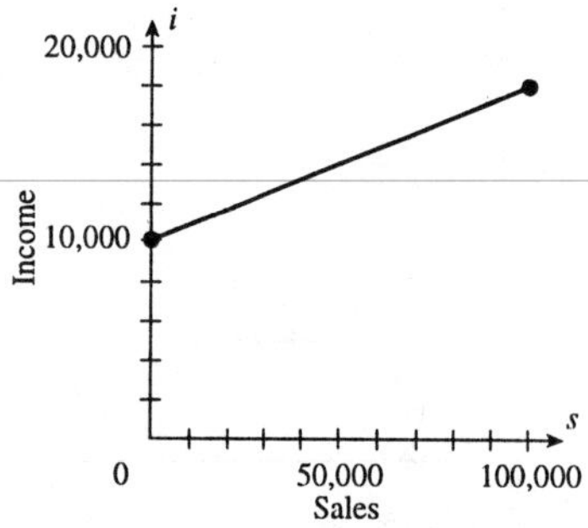

(b) \$11,600

49. (a)

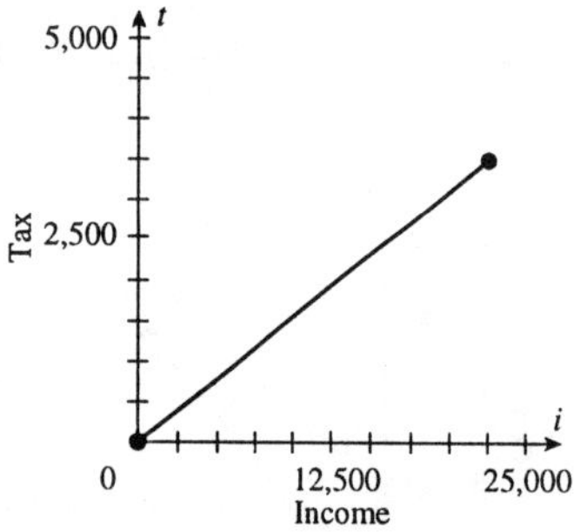

(b) \$2250

51. A function is a relation in which no two ordered pairs have the same first coordinate and different second coordinates.

53. The range of a relation or function is the set of values that represent the dependent variable.

55. No, each x must have a unique y for it to be a function.

57. Yes, each x may have a unique value of y.

59. Function

61. Not a function

Cumulative Review Exercises: 7.6

63. $3x-4=5(x-2)-1$

$3x-4=5x-10-1$

$3x-4=5x-11$

$-2x-4=-11$

$-2x=-7$

$x=\dfrac{-7}{-2}=\dfrac{7}{2}$

64. Let x be the cost to travel 1000 miles.

$\dfrac{x}{1000}=\dfrac{15}{280}$

$x=\dfrac{15{,}000}{280}=53.57$

It will cost \$53.57 to travel 1000 miles.

65. $3x+9<-x+5$

$4x+9<5$

$4x<-4$

$x<-1$

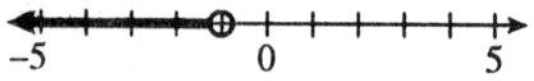

66. $A = 2l + 2w$

$2w = A - 2l$

$w = \dfrac{A - 2l}{2}$

Group Activity/Challenge Problems

3. $f(x) = x^2 + 2x - 3$

(a) $f(1) = (1)^2 + 2(1) - 3$

$= 1 + 2 - 3$

$= 3 - 3 = 0$

(b) $f(2) = (2)^2 + 2(2) - 3$

$= 4 + 4 - 3$

$= 8 - 3 = 5$

(c) $f(a) = a^2 + 2a - 3$

4. **(a)** $2x + 1 > 5$ contains one variable, x.

(b) $2x + y > 5$

(c) $2x + 1 > 5$

$2x > 4$

$x > 2$

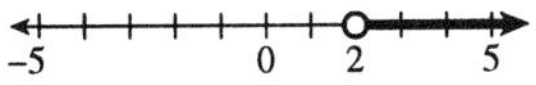

(d)

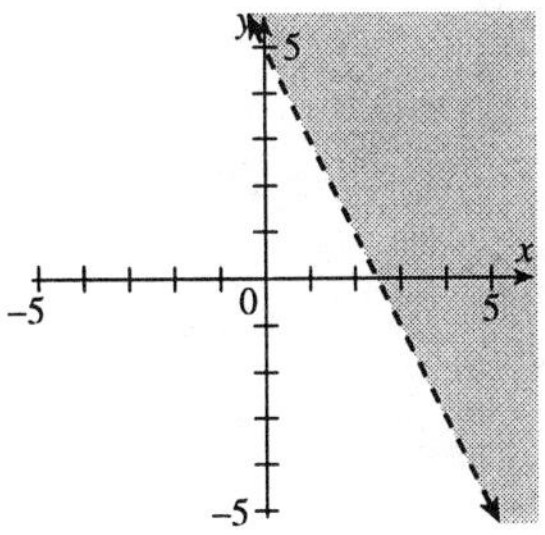

5.

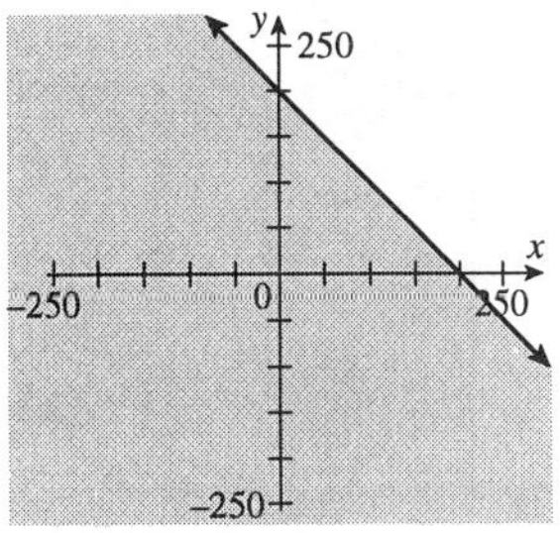

7. **(a)** $2x + y \leq 10$

(b)

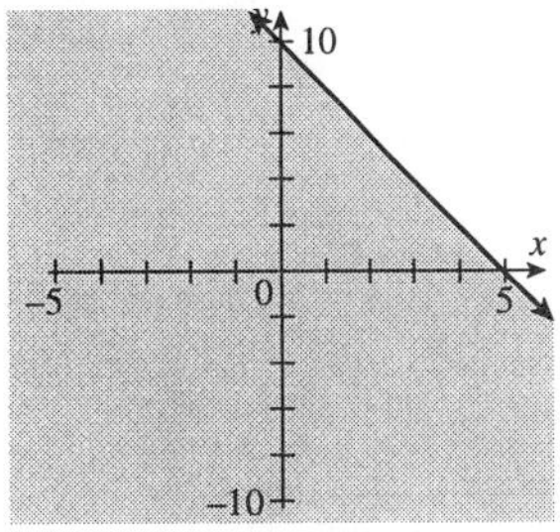

Exercise Set 7.7

1.

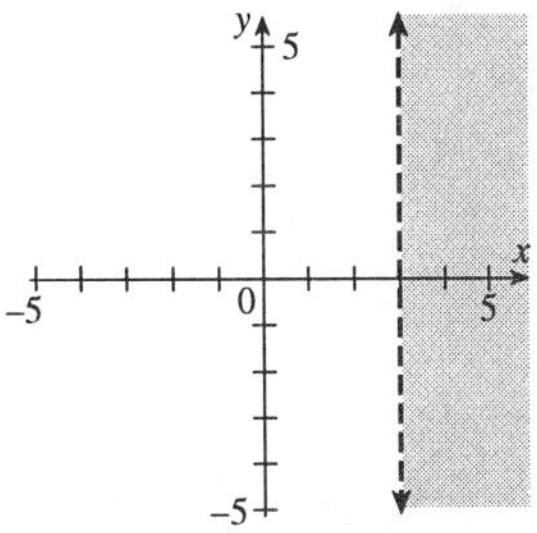

Check (0, 0): $x > 3$

$0 > 3$ False

3.

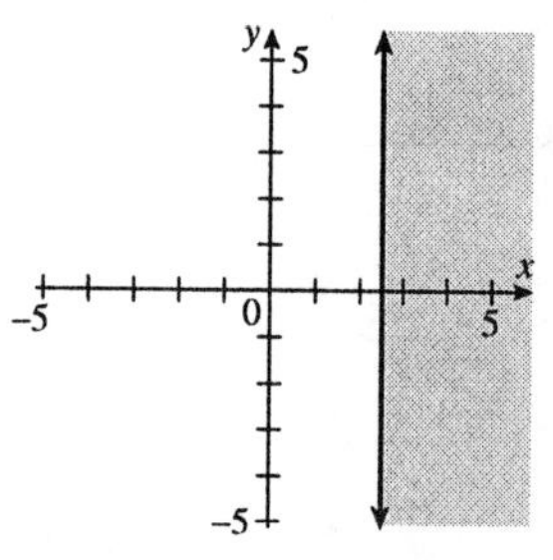

Check (0, 0): $x \geq \frac{5}{2}$

$0 \geq \frac{5}{2}$ False

5.

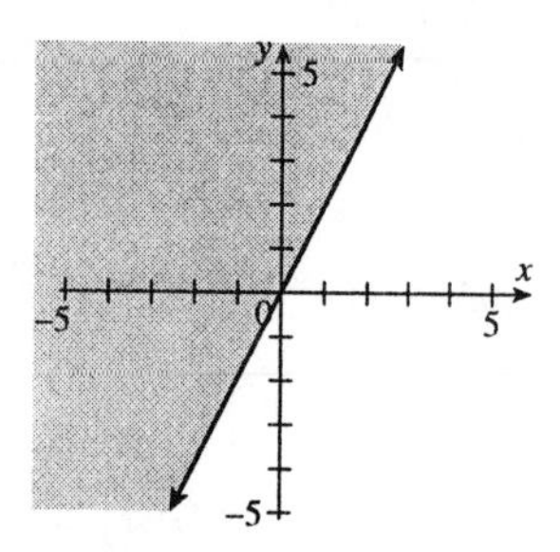

Check (1, 0): $y \geq 2x$

$0 \geq 2 \cdot 1$

$0 \geq 2$ False

7.

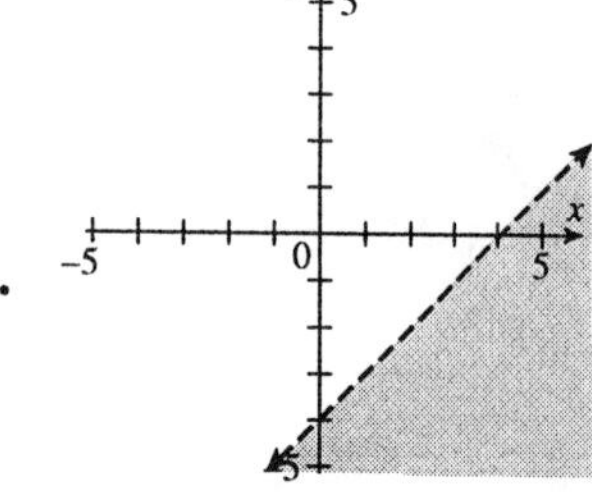

Check (0, 0): $y < x - 4$

$0 < 0 - 4$

$0 < -4$ False

9.

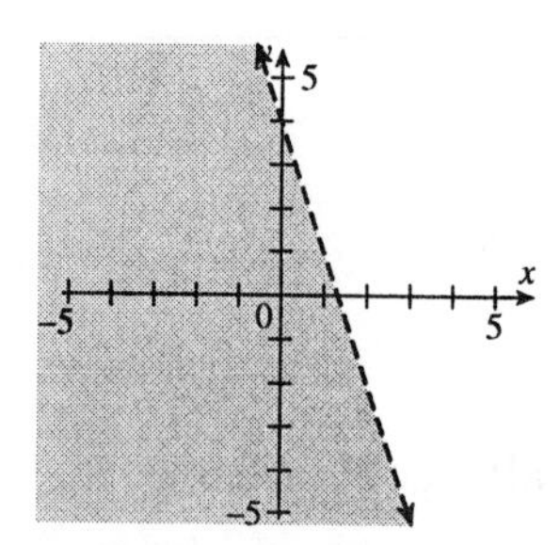

Check (0, 0): $y < -3x + 4$

$0 < -3 \cdot 0 + 4$

$0 < 4$ True

11.

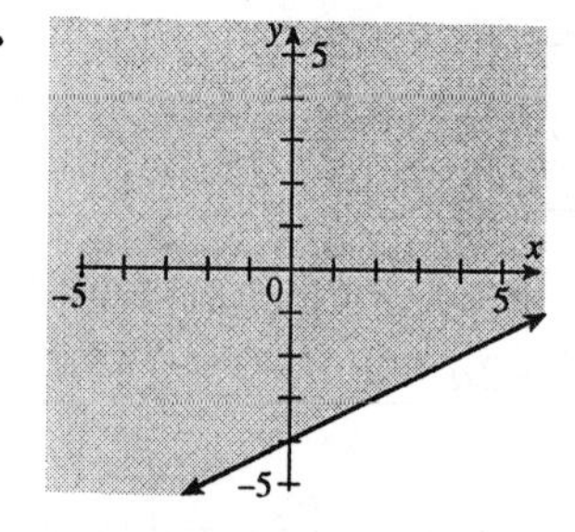

Check (0, 0): $y \geq \frac{1}{2}x - 4$

$0 \geq \frac{1}{2}(0) - 4$

$0 \geq -4$ True

13.

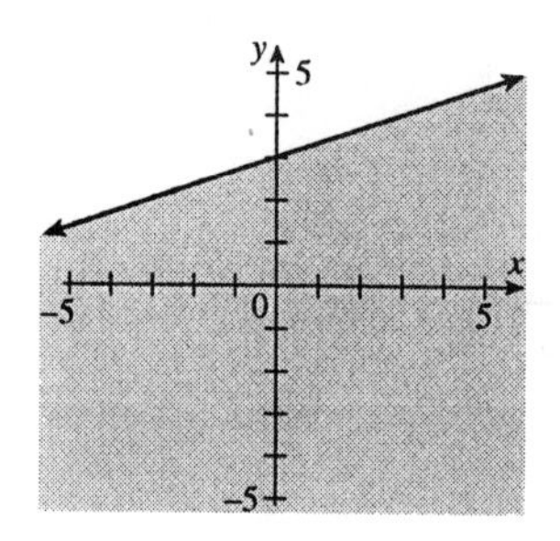

Check (0, 0): $y \leq \frac{1}{3}x + 3$

$0 \leq \frac{1}{3}(0) + 3$

$0 \leq 3$ True

15.

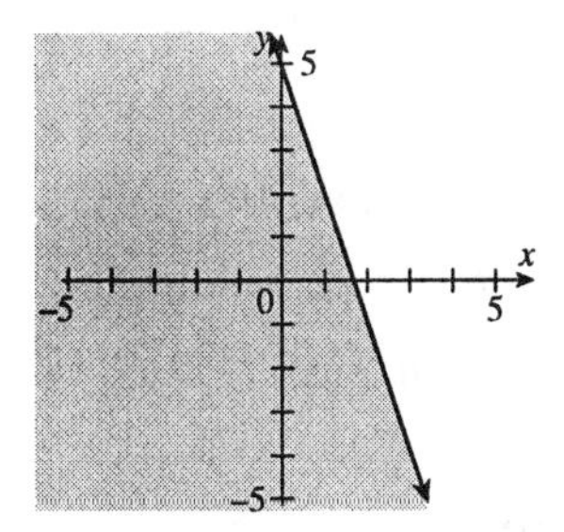

Check (0, 0): $3x + y \le 5$

$3 \cdot 0 + 0 \le 5$

$0 \le 5$ True

17.

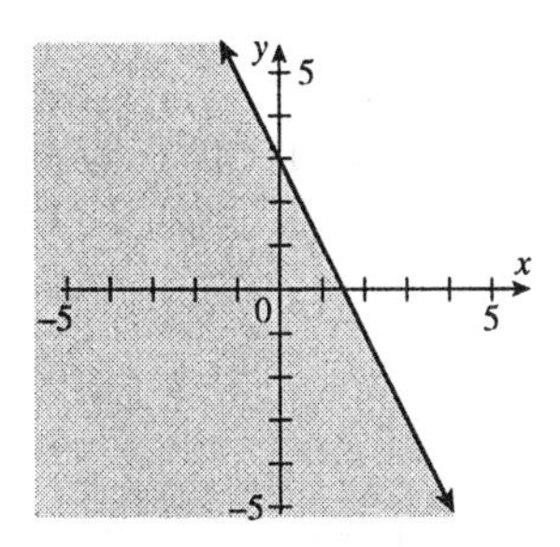

Check (0, 0): $2x + y \le 3$

$2 \cdot 0 + 0 \le 3$

$0 \le 3$ True

19.

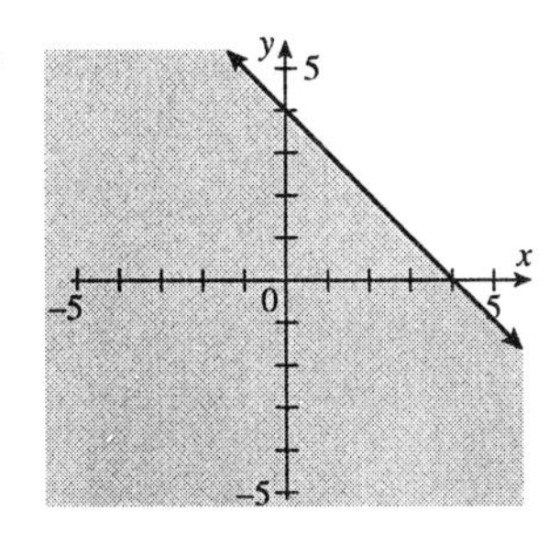

Check (0, 0): $y - 4 \le -x$

$0 - 4 \le -0$

$-4 \le 0$ True

21.

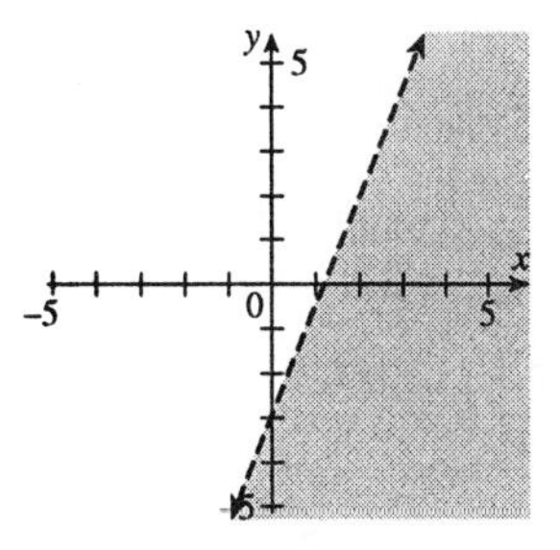

Check (0, 0): $2y - 5x < -6$

$2 \cdot 0 - 5 \cdot 0 < -6$

$0 < -6$ False

23.

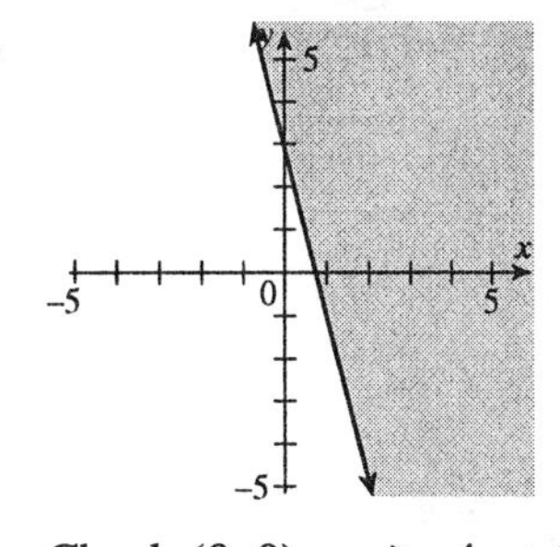

Check (0, 0): $y \ge -4x + 3$

$0 \ge -4 \cdot 0 + 3$

$0 \ge 3$ False

25. Points on the line satisfy the = part of the inequality.

Cumulative Review Exercises: 7.7

27. $2(x - 3) < 4(x - 2) - 4$

$2x - 6 < 4x - 8 - 4$

$2x - 6 < 4x - 12$

$-2x - 6 < -12$

$-2x < -6$

$x > 3$

-5 0 5

28. $i = prt$

$300 = p(0.08)(3)$

$300 = 0.24p$

$1250 = p$

The principal is \$1250.

29. $C = \frac{5}{9}(F - 32)$

$\frac{9}{5}C = F - 32$

$\frac{9}{5}C + 32 = F$ or $F = \frac{9}{5}C + 32$

30. $6x - 5y = 9$

$-5y = -6x + 9$

$y = \frac{6}{5}x - \frac{9}{5}$

$m = \frac{6}{5},\ b = -\frac{9}{5}$

Chapter 7 Review Exercises

1. **(a)** 38%

(b) 62%

(c) $(635)(0.16) = \$101.6$ billion

(d)

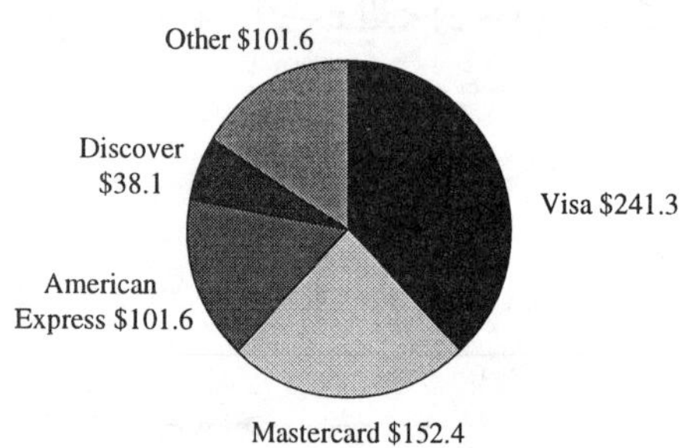

2. **(a)** 48%

(b) 18%

(c) Greater, by about 30%

(d) Superstores

(e) Answers will vary.

3. **(a)** 68%, 10%

(b) About the third quarter of 1991

(c) About 94%

(d) About 87%

4. **(a)** About 135 per 100,000 in 1979 and about 350 per 100,000 in 1993, for an increase of about 215 per 100,000.

(b) 25%, 60%

(c) 100%

(d) 100%

(e) Answers will vary.

5.

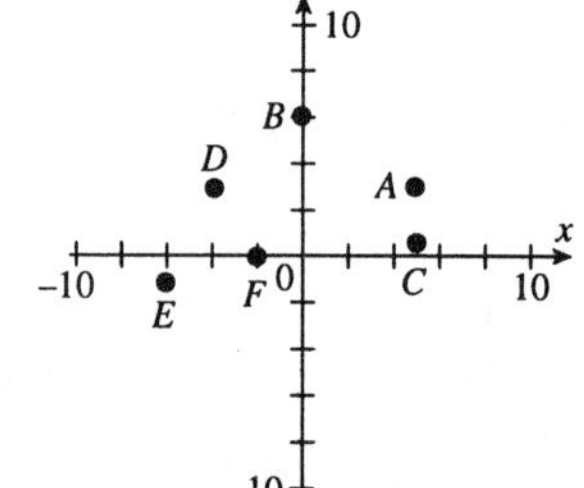

6.

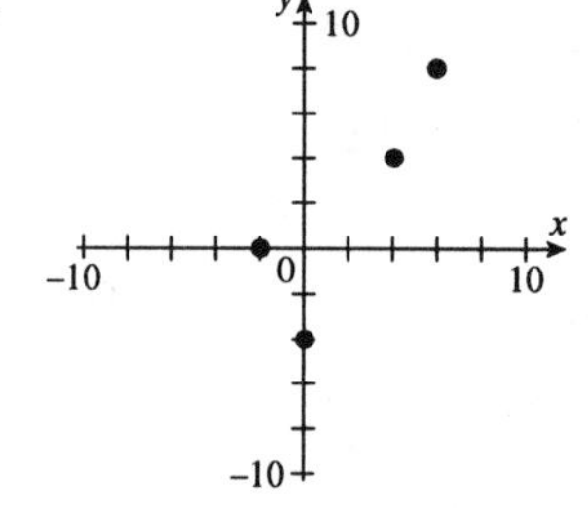

The points are not collinear.

7. **(a)** $2x + 3y = 9$

$2 \cdot 4 + 3 \cdot 3 = 9$

$8 + 9 = 9$

$17 = 9$ False

(b) $2x + 3y = 9$

$2 \cdot 0 + 3 \cdot 3 = 9$

$9 = 9$ True

(c) $2x + 3y = 9$

$2(-1) + 3 \cdot 4 = 9$

$-2 + 12 = 9$

$10 = 9$ False

(d) $2x + 3y = 9$

$2 \cdot 2 + 3\left(\frac{5}{3}\right) = 9$

$4 + 5 = 9$

$9 = 9$ True

8. **(a)** $3x - 2y = 8$

$3 \cdot 2 - 2y = 8$

$6 - 2y = 8$

$-2y = 2$

$y = -1$

(b) $3x - 2y = 8$

$3 \cdot 0 - 2y = 8$

$-2y = 8$

$y = -4$

(c) $3x - 2y = 8$

$3x - 2 \cdot 4 = 8$

$3x - 8 = 8$

$3x = 16$

$x = \frac{16}{3}$

(d) $3x - 2y = 8$

$3x - 2 \cdot 0 = 8$

$3x = 8$

$x = \frac{8}{3}$

9.

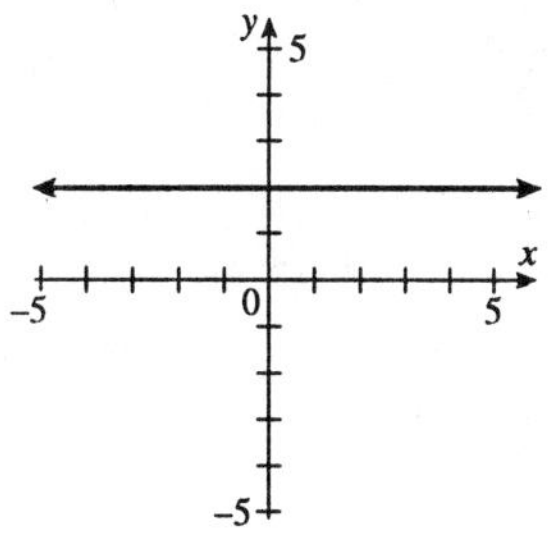

10.

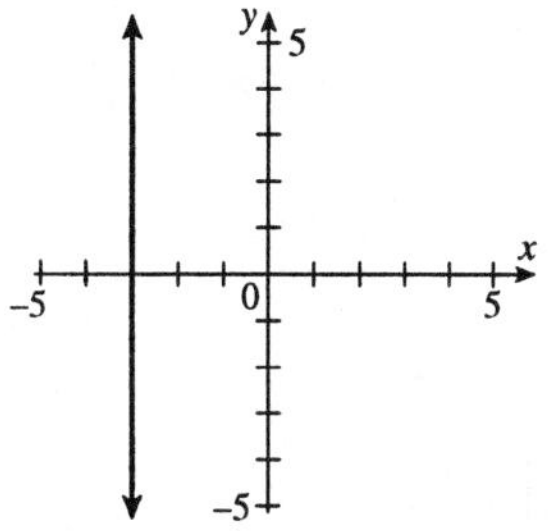

11.

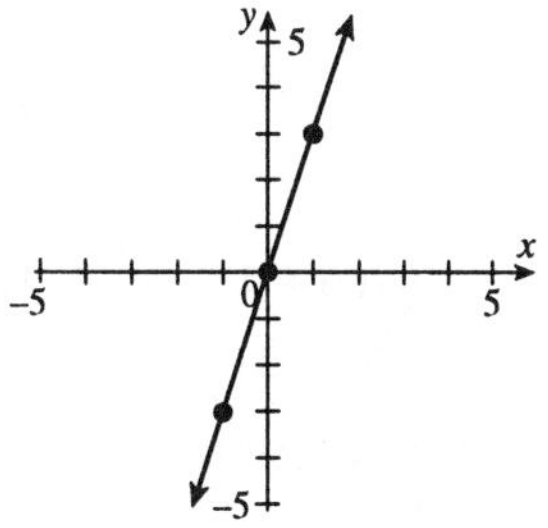

Let $x = -1$, $y = 3(-1) = -3$, $(-1, -3)$

Let $x = 0$, $y = 3 \cdot 0 = 0$, $(0, 0)$

Let $x = 1$, $y = 3 \cdot 1 = 3$, $(1, 3)$

12.

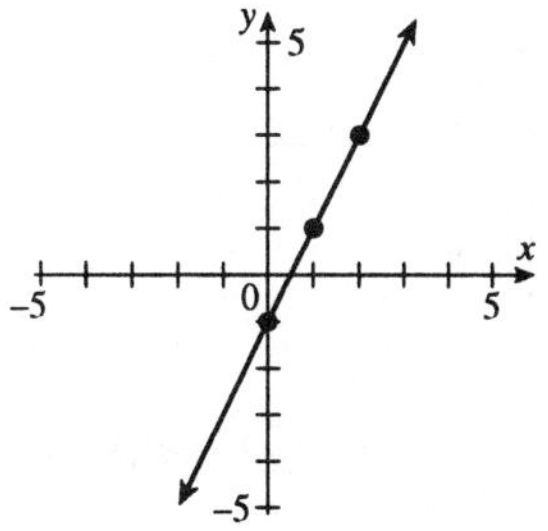

Let $x = 0$, $y = 2 \cdot 0 - 1 = -1$, $(0, -1)$

Let $x = 1$, $y = 2 \cdot 1 - 1 = 1$, $(1, 1)$

Let $x = 2$, $y = 2 \cdot 2 - 1 = 3$, $(2, 3)$

13.

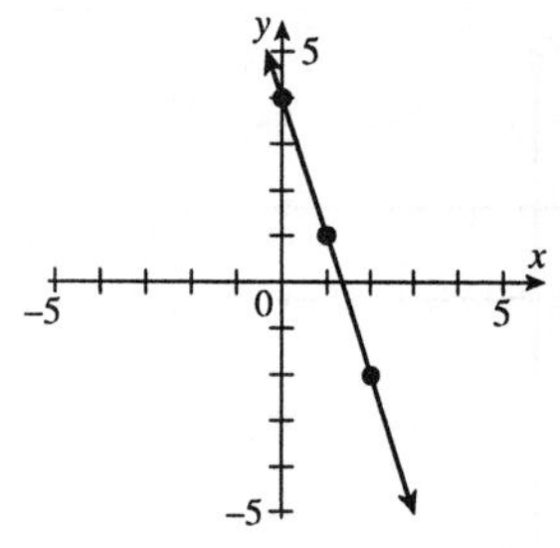

Let $x = 0$, $y = -3 \cdot 0 + 4 = 4$, (0, 4)

Let $x = 1$, $y = -3 \cdot 1 + 4 = 1$, (1, 1)

Let $x = 2$, $y = -3 \cdot 2 + 4 = -2$, (2, −2)

14.

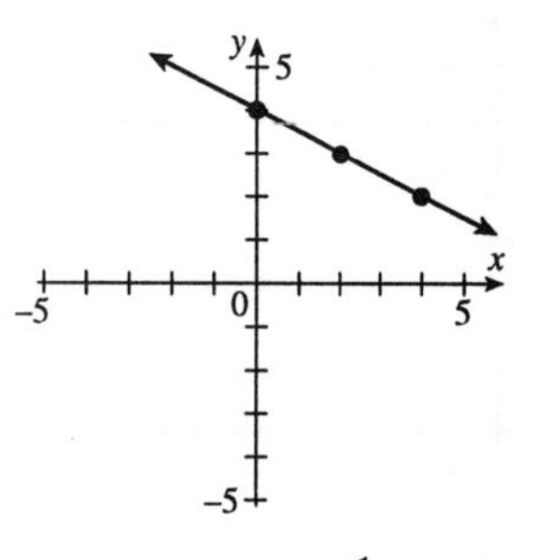

Let $x = 0$, $y = -\frac{1}{2}(0) + 4 = 4$, (0, 4)

Let $x = 2$, $y = -\frac{1}{2}(2) + 4 = 3$, (2, 3)

Let $x = 4$, $y = -\frac{1}{2}(4) + 4 = 2$, (4, 2)

15.

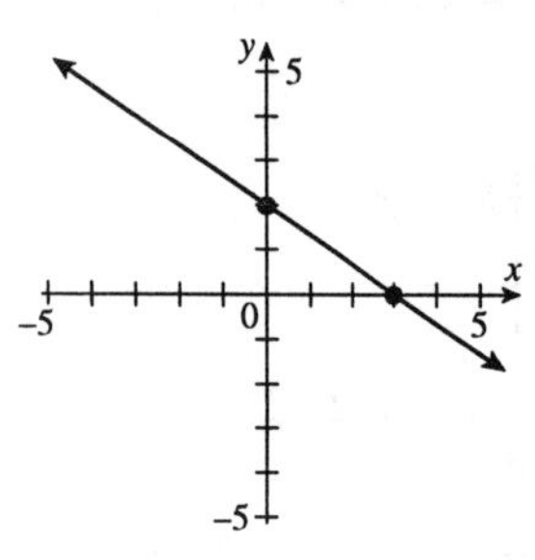

Let $x = 0$	Let $y = 0$
$2x + 3y = 6$	$2x + 3y = 6$
$2 \cdot 0 + 3y = 6$	$2x + 3 \cdot 0 = 6$
$3y = 6$	$2x = 6$
$y = 2$	$x = 3$

16.

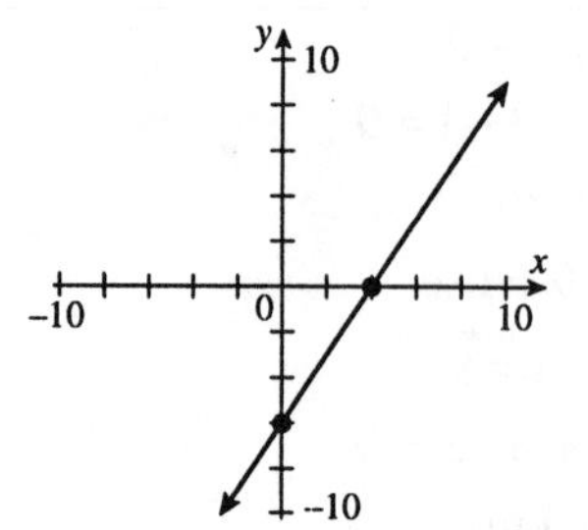

Let $x = 0$	Let $y = 0$
$3x - 2y = 12$	$3x - 2y = 12$
$3 \cdot 0 - 2y = 12$	$3x - 2 \cdot 0 = 12$
$-2y = 12$	$3x = 12$
$y = -6$	$x = 4$

17.

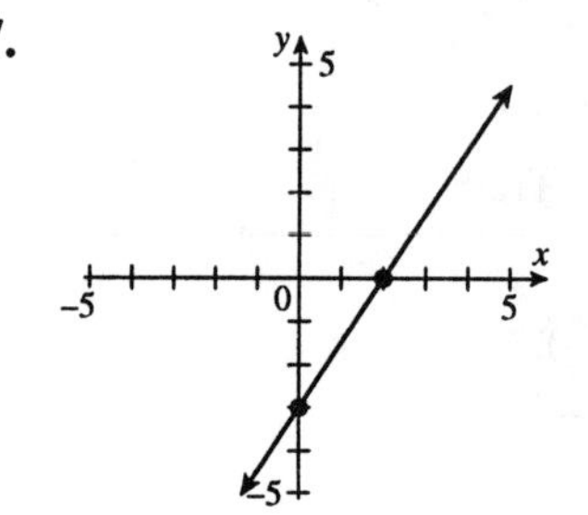

Let $x = 0$	Let $y = 0$
$2y = 3x - 6$	$2y = 3x - 6$
$2y = 3 \cdot 0 - 6$	$2 \cdot 0 = 3x - 6$
$2y = -6$	$0 = 3x - 6$
$y = -3$	$-3x = -6$
	$x = 2$

18.

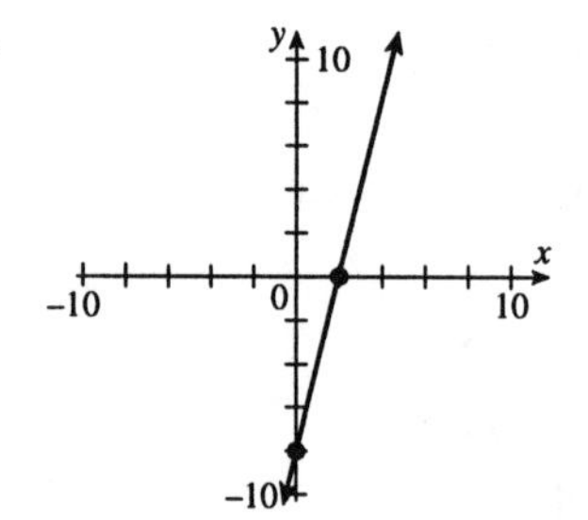

Let $x=0$	Let $y=0$
$4x-y=8$	$4x-y=8$
$4\cdot 0-y=8$	$4x-0=8$
$-y=8$	$4x=8$
$y=-8$	$x=2$

19.

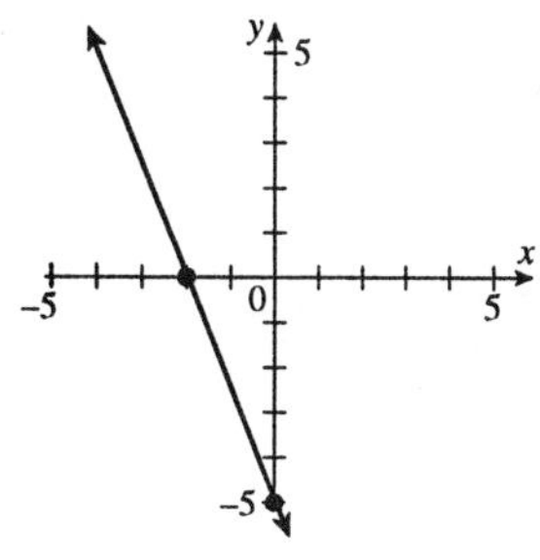

Let $x=0$	Let $y=0$
$-5x-2y=10$	$-5x-2y=10$
$-5\cdot 0-2y=10$	$-5x-2\cdot 0=10$
$-2y=10$	$-5x=10$
$y=-5$	$x=-2$

20.

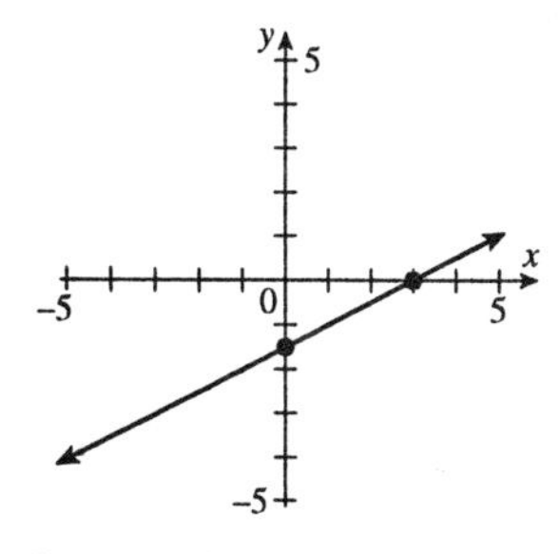

Let $x=0$	Let $y=0$
$3x=6y+9$	$3x=6y+9$
$3\cdot 0=6y+9$	$3x=6\cdot 0+9$
$0=6y+9$	$3x=9$
$-6y=9$	$x=3$
$y=-\dfrac{3}{2}$	

21.

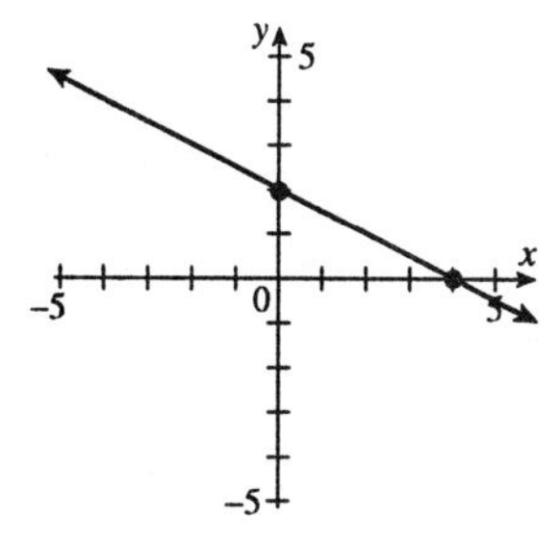

Let $x=0$	Let $y=0$
$25x+50y=100$	$25x+50y=100$
$25\cdot 0+50y=100$	$25x+50\cdot 0=100$
$50y=100$	$25x=100$
$y=2$	$x=4$

22.

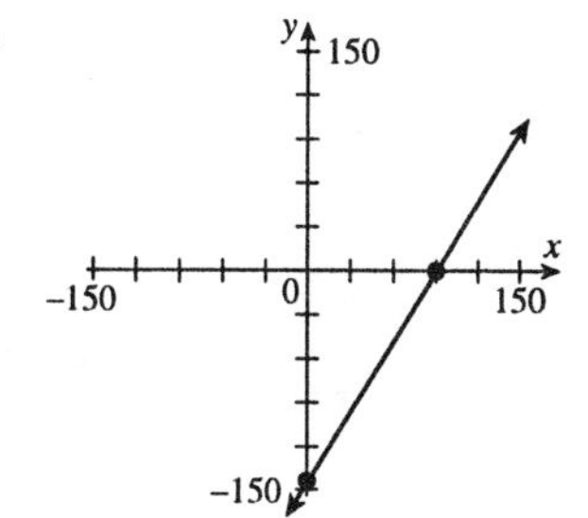

Let $x=0$	Let $y=0$
$3x-2y=270$	$3x-2y=270$
$3\cdot 0-2y=270$	$3x-2\cdot 0=270$
$-2y=270$	$3x=270$
$y=-135$	$x=90$

23.

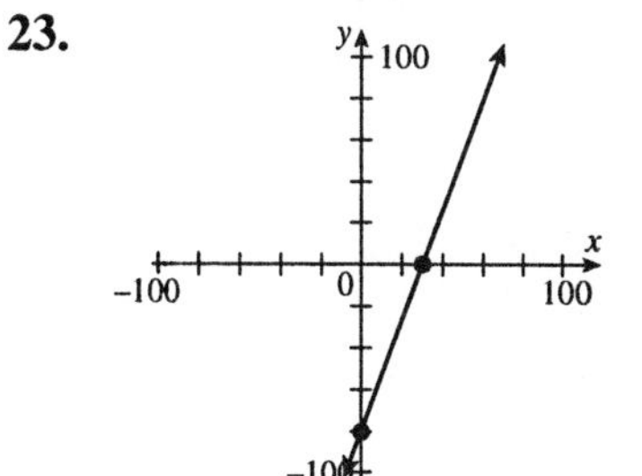

Let $x=0$	Let $y=0$
$\frac{2}{3}x=\frac{1}{4}y+20$	$\frac{2}{3}x=\frac{1}{4}y+20$
$\frac{2}{3}\cdot 0=\frac{1}{4}y+20$	$\frac{2}{3}x=\frac{1}{4}\cdot 0+20$
$0=\frac{1}{4}y+20$	$\frac{2}{3}x=20$
$-\frac{1}{4}y=20$	$x=30$
$y=-80$	

24. $m=\frac{5-(-7)}{-2-3}=\frac{12}{-5}=-\frac{12}{5}$

25. $m=\frac{-3-(-2)}{8-(-4)}=\frac{-1}{12}=-\frac{1}{12}$

26. $m=\frac{3-(-1)}{-4-(-2)}=\frac{4}{-2}=-2$

27. The slope of a horizontal line is 0.

28. The slope of a vertical line is undefined.

29. The slope of a straight line is the ratio of the vertical change to the horizontal change between any two points on it.

30. $m=\frac{-5}{7}=-\frac{5}{7}$

31. $m=\frac{9}{4}$

32. $m=\frac{2}{8}=\frac{1}{4}$

33. $m=-1, b=4$

34. $m=-4,\ b=\frac{1}{2}$

35. $2x+3y=8$

$3y=-2x+8$

$y=-\frac{2}{3}x+\frac{8}{3}$

$m=-\frac{2}{3},\ b=\frac{8}{3}$

36. $3x+6y=9$

$6y=-3x+9$

$y=-\frac{1}{2}x+\frac{3}{2}$

$m=-\frac{1}{2},\ b=\frac{3}{2}$

37. $4y=6x+12$

$y=\frac{3}{2}x+3$

$m=\frac{3}{2},\ b=3$

38. $3x+5y=12$

$5y=-3x+12$

$y=-\frac{3}{5}x+\frac{12}{5}$

$m=-\frac{3}{5},\ b=\frac{12}{5}$

39. $9x+7y=15$

$7y=-9x+15$

$y=-\frac{9}{7}x+\frac{15}{7}$

$m=-\frac{9}{7},\ b=\frac{15}{7}$

40. $4x-8=0$

$4x=8$

$x=2$

Slope is undefined, no y intercept.

41. $3y+9=0$

$3y=-9$

$y=-3$

$m=0, b=-3$

42. $m=\frac{2}{1}=2,\ b=2$

$y=2x+2$

43. $m=\frac{2.5}{2.5}=1,\ b=-2.5=-\frac{5}{2}$

$y=x-\frac{5}{2}$

44. $m=\frac{-2}{4}=-\frac{1}{2},\ b=2$

$y=-\frac{1}{2}x+2$

45. $y=3x-6$ $\quad$ $6y=18x+6$

$y=3x+1$

The lines are parallel.

46. $2x-3y=9$ $\quad$ $3x-2y=6$

$-3y=-2x+9$ $\quad$ $-2y=-3x+6$

$y=\frac{2}{3}x-3$ $\quad$ $y=\frac{3}{2}x-3$

The lines are not parallel.

47. $y=\frac{4}{9}x+5$ $\quad$ $4x=9y+2$

$9y=4x-2$

$y=\frac{4}{9}x-\frac{2}{9}$

The lines are parallel.

48. $4x=6y+3$ $\quad$ $-2x=-3y+10$

$6y=4x-3$ $\quad$ $3y-2x=10$

$y=\frac{2}{3}x-\frac{1}{2}$ $\quad$ $3y=2x+10$

$y=\frac{2}{3}x+\frac{10}{3}$

The lines are parallel.

49. $y-4=2(x-3)$

$y-4=2x-6$

$y=2x-2$

50. $y-5=-3[x-(-1)]$

$y-5=-3x-3$

$y=-3x+2$

51. $y-2=-\frac{2}{3}(x-3)$

$y-2=-\frac{2}{3}x+2$

$y=-\frac{2}{3}x+4$

52. $y - 2 = 0(x - 4)$

$y - 2 = 0$

$y = 2$

53. $x = 3$

54. $y = -2x - 4$

55. $m = \dfrac{-4 - 3}{0 - (-2)} = \dfrac{-7}{2} = -\dfrac{7}{2}$

$y - 3 = -\dfrac{7}{2}[x - (-2)]$

$y - 3 = -\dfrac{7}{2}x - 7$

$y = -\dfrac{7}{2}x - 4$

56. $m = \dfrac{3 - (-2)}{-4 - (-4)} = \dfrac{5}{0} =$ undefined

$x = -4$

57. Function, Domain $\{0, 1, 2, 4, 6\}$,
Range $\{-3, -1, 2, 4, 5\}$

58. Not a function, Domain $\{3, 4, 6, 7\}$,
Range $\{0, 1, 2, 5\}$

59. Not a function, Domain $\{3, 4, 5, 6\}$,
Range $\{-3, 1, 2\}$

60. Function, Domain $\{-2, 3, 4, 5, 9\}$,
Range $\{-2\}$

61. **(a)** $\{(1, 3), (4, 5), (7, 2), (9, 2)\}$
(b) Function

62. **(a)** $\{(4, 1), (6, 3), (6, 5), (8, 7)\}$
(b) Not a function

63. Function

64. Not a function

65. Function

66. Function

67. **(a)** $f(2) = 3 \cdot 2 - 4 = 2$
(b) $f(-5) = 3(-5) - 4 = -19$

68. **(a)** $f(-4) = -4(-4) - 5 = 11$
(b) $f(8) = -4 \cdot 8 - 5 = -37$

69. **(a)** $f(3) = \dfrac{1}{3}(3) - 5 = -4$

(b) $f(-9) = \dfrac{1}{3}(-9) - 5 = -8$

70. **(a)** $f(3) = 2 \cdot 3^2 - 4 \cdot 3 + 6 = 12$
(b) $f(-5) = 2 \cdot (-5)^2 - 4(-5) + 6 = 76$

71.

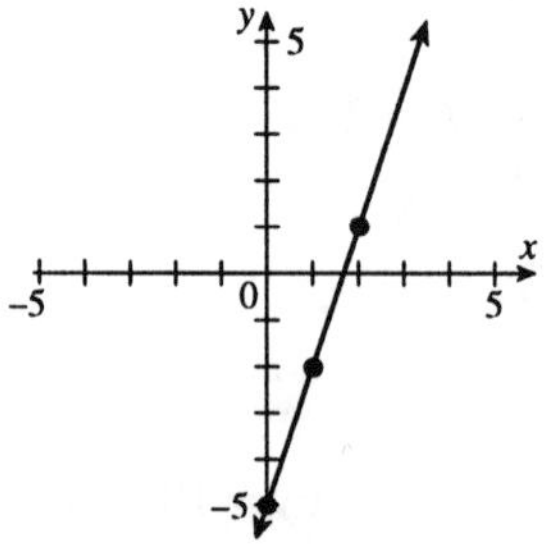

Let $x = 0$, $f(0) = 3 \cdot 0 - 5 = -5$, $(0, -5)$
Let $x = 1$, $f(1) = 3 \cdot 1 - 5 = -2$, $(1, -2)$
Let $x = 2$, $f(2) = 3 \cdot 2 - 5 = 1$, $(2, 1)$

72.

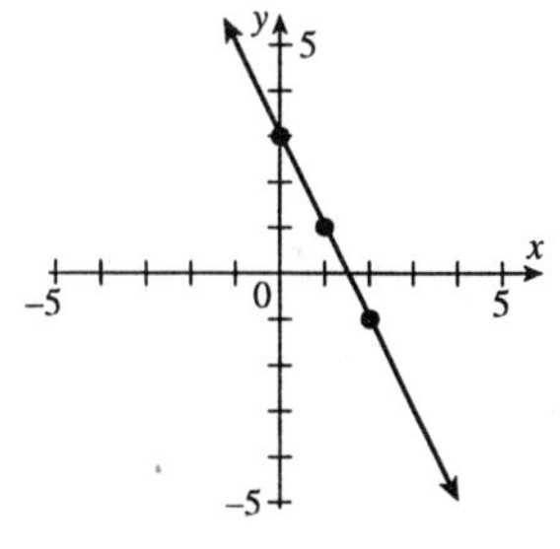

Let $x = 0$, $f(0) = -2 \cdot 0 + 3 = 3$, $(0, 3)$

Let $x = 1$, $f(1) = -2 \cdot 1 + 3 = 1$, $(1, 1)$

Let $x = 2$, $f(2) = -2 \cdot 2 + 3 = -1$, $(2, -1)$

73.

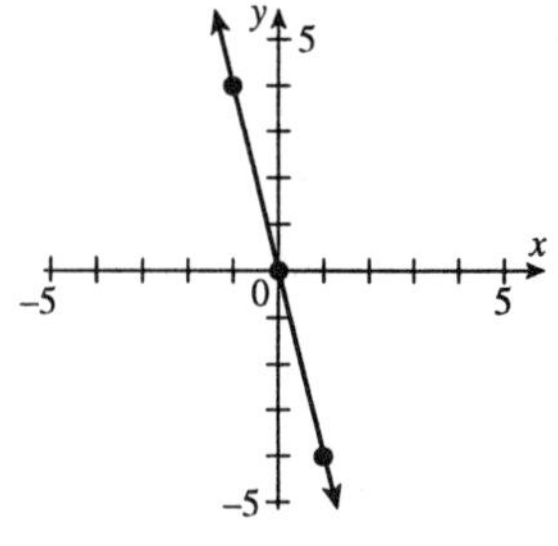

Let $x = -1$, $f(-1) = -4(-1) = 4$, $(-1, 4)$

Let $x = 0$, $f(0) = -4 \cdot 0 = 0$, $(0, 0)$

Let $x = 1$, $f(1) = -4 \cdot 1 = -4$, $(1, -4)$

74.

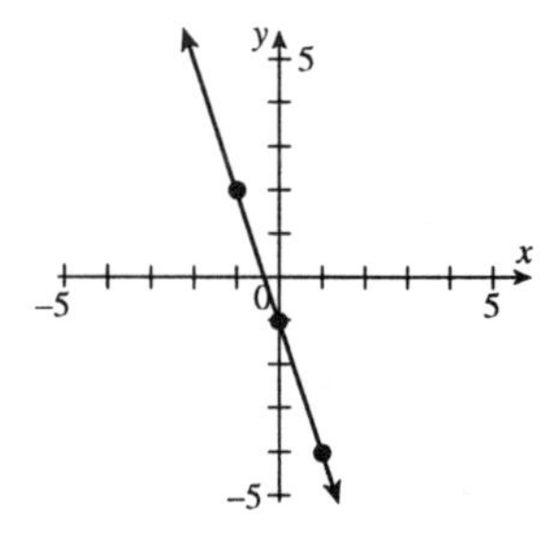

Let $x = -1$, $f(-1) = -3(-1) - 1 = 2$, $(-1, 2)$

Let $x = 0$, $f(0) = -3 \cdot 0 - 1 = -1$, $(0, -1)$

Let $x = 1$, $f(1) = -3 \cdot 1 - 1 = -4$, $(1, -4)$

75. **(a)**

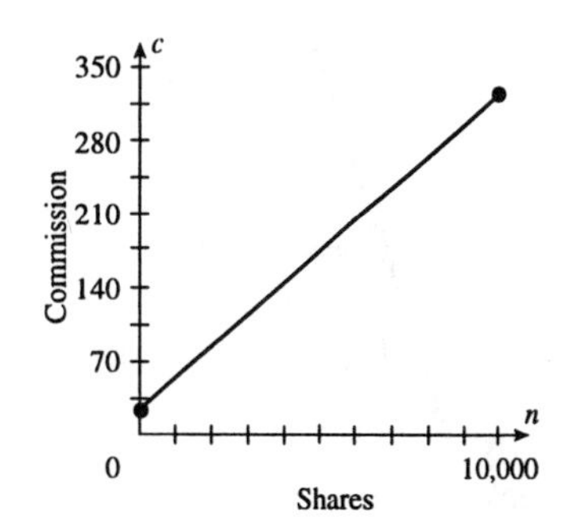

(b) \$55

76. **(a)**

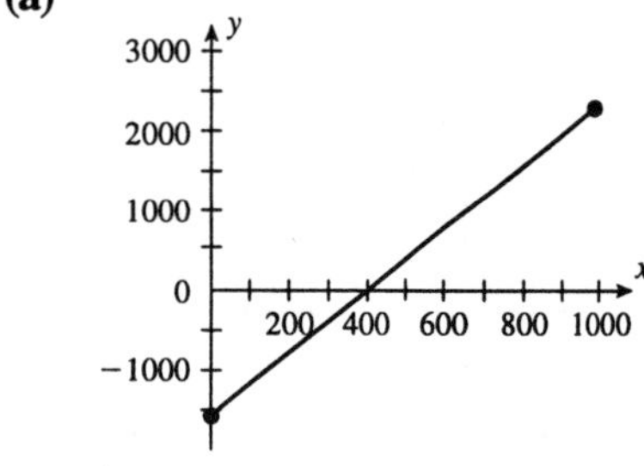

(b) \$0

77.

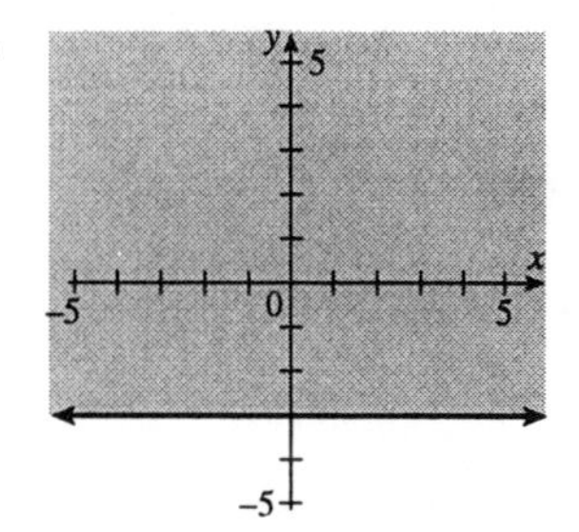

Check (0, 0): $y \geq -3$

$0 \geq -3$ True

78.

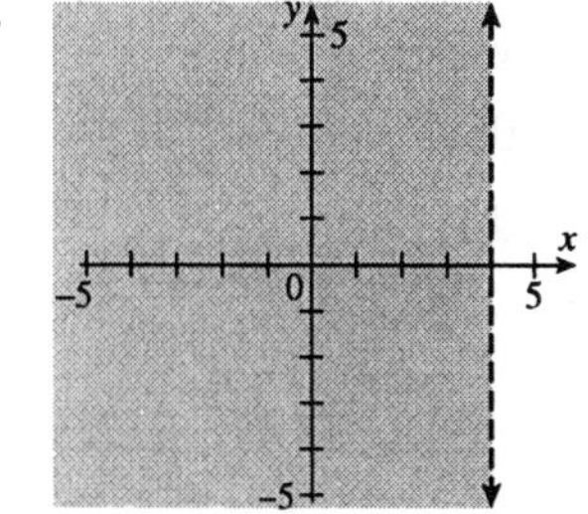

Check (0, 0): $x < 4$

$0 < 4$ True

79.

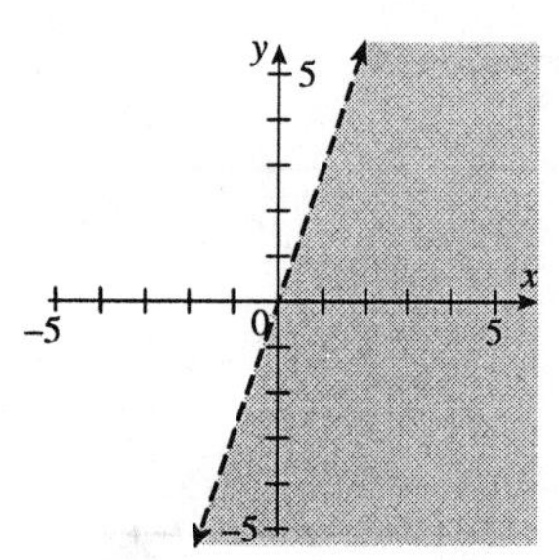

Check (1, –1): $y < 3x$

$-1 < 3 \cdot 1$

$-1 < 3$ True

80.

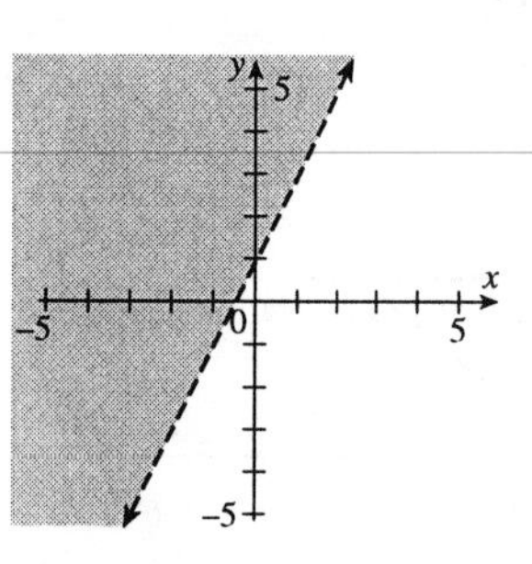

Check (0, 0): $y > 2x + 1$

$0 > 2 \cdot 0 + 1$

$0 > 1$ False

81.

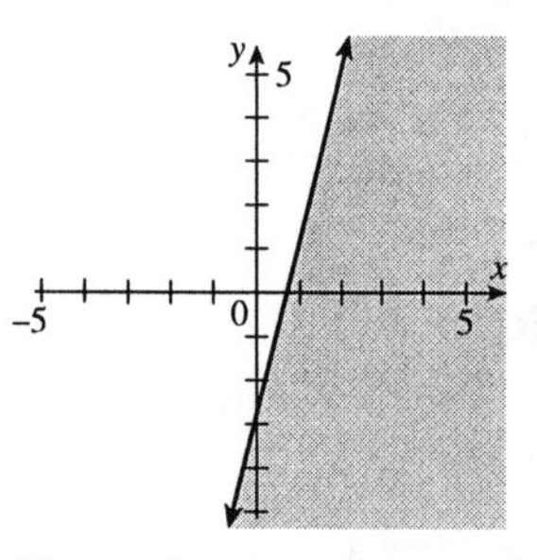

Check (0, 0): $y \leq 4x - 3$

$0 \leq 4 \cdot 0 - 3$

$0 \leq -3$ False

82.

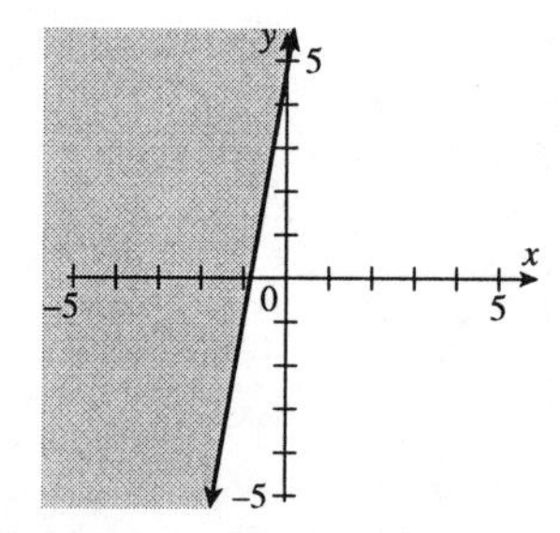

Check (0, 0): $-6x + y \geq 5$

$-6 \cdot 0 + 0 \geq 5$

$0 \geq 5$ False

83.

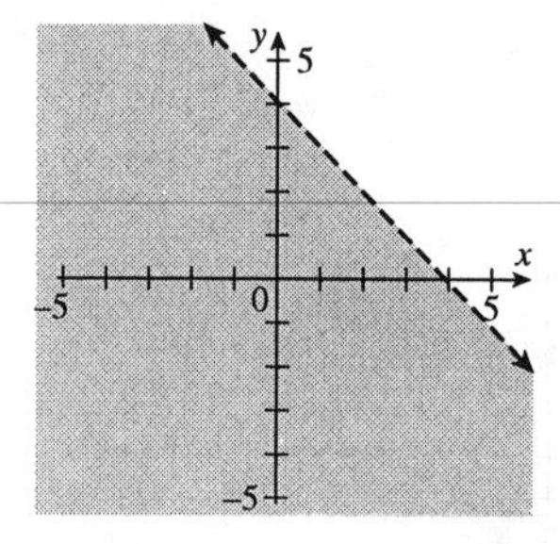

Check (0, 0): $y < -x + 4$

$0 < -0 + 4$

$0 < 4$ True

84.

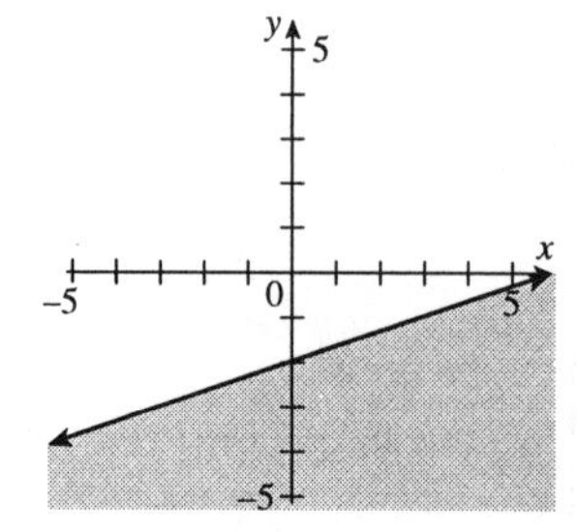

Check (0, 0): $3y + 6 \leq x$

$3 \cdot 0 + 6 \leq 0$

$6 \leq 0$ False

Chapter 7 Practice Test

1. (a) The bar for "other" would correspond to 36.7% since the total of all bars must equal 100%.

(b) 5.7%

(c) $(2.6)(0.312) = 0.8112$. The annual dollar sales of Tylenol is \$0.8112 billion or \$811.2 million.

2. (a) $3y = 5x - 9$

$3 \cdot 2 = 5 \cdot 3 - 9$

$6 = 15 - 9$

$6 = 6$ Yes

(b) $3y = 5x - 9$

$3 \cdot 0 = 5 \cdot \frac{9}{5} - 9$

$0 = 9 - 9$

$0 = 0$ Yes

(c) $3y = 5x - 9$

$3(-6) = 5(-2) - 9$

$-18 = -10 - 9$

$-18 = -19$ No

(d) $3y = 5x - 9$

$3 \cdot 3 = 5 \cdot 0 - 9$

$9 = -9$ No

3. $m = \frac{3-(-5)}{-4-2} = \frac{8}{-6} = -\frac{4}{3}$

4. $4x - 9y = 15$

$-9y = -4x + 15$

$y = \frac{4}{9}x - \frac{5}{3}$

$m = \frac{4}{9},\ b = -\frac{5}{3}$

5. $m = \frac{-1}{1} = -1,\ b = -1$

$y = -x - 1$

6. $y - 3 = 3(x - 1)$

$y - 3 = 3x - 3$

$y = 3x$

7. $m = \frac{2-(-1)}{-4-3} = \frac{3}{-7} = -\frac{3}{7}$

$y - (-1) = \frac{-3}{7}(x - 3)$

$y + 1 = -\frac{3}{7}x + \frac{9}{7}$

$y = -\frac{3}{7}x + \frac{2}{7}$

8. $2y = 3x - 6$ $\qquad$ $y - \frac{3}{2}x = -5$

$y = \frac{3}{2}x - 3$ $\qquad$ $y = \frac{3}{2}x - 5$

The lines are parallel since they have the same slope.

9.

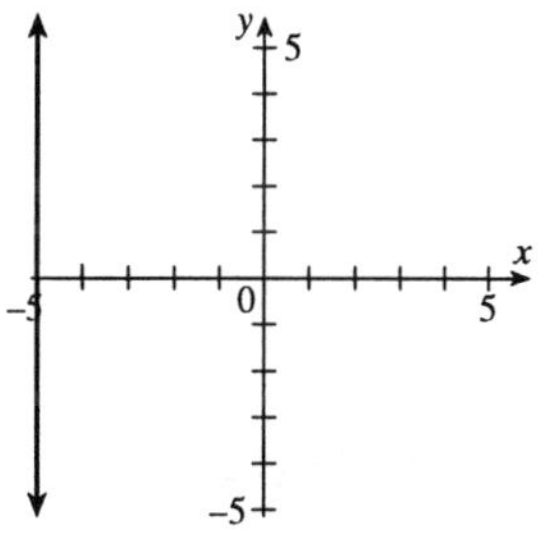

10.

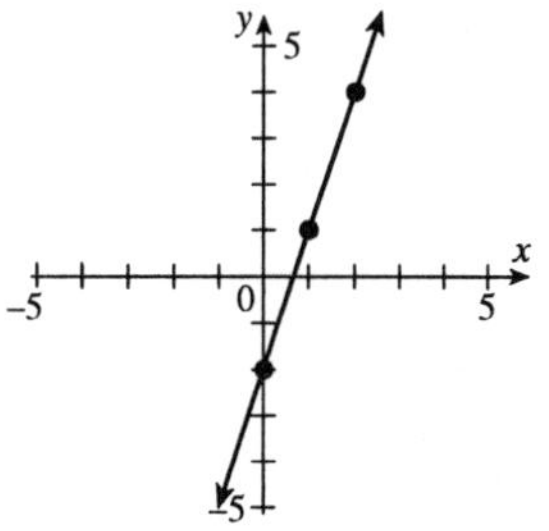

Let $x = 0$, $y = 3 \cdot 0 - 2 = -2$, $(0, -2)$

Let $x = 1$, $y = 3 \cdot 1 - 2 = 1$, $(1, 1)$

Let $x = 2$, $y = 3 \cdot 2 - 2 = 4$, $(2, 4)$

11.

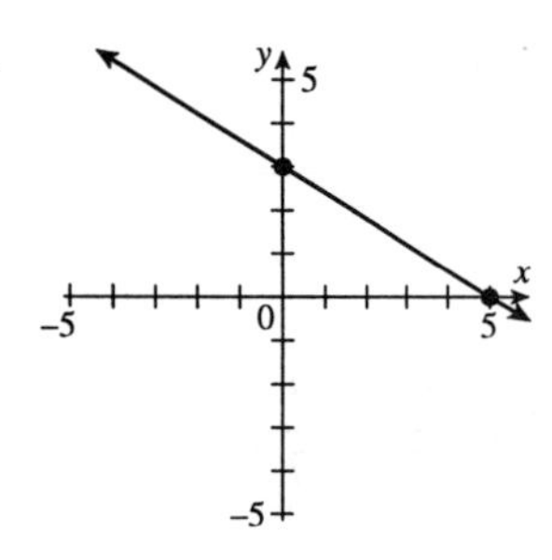

Let $x = 0$	Let $y = 0$
$3x + 5y = 15$	$3x + 5y = 15$
$3 \cdot 0 + 5y = 15$	$3x + 5 \cdot 0 = 15$
$5y = 15$	$3x = 15$
$y = 3$	$x = 5$

12.

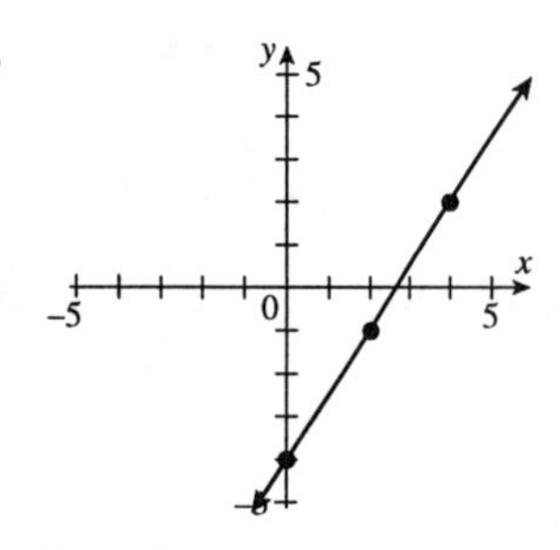

$3x - 2y = 8$

$-2y = -3x + 8$

$y = \frac{3}{2}x - 4$

Let $x = 0$, $y = \frac{3}{2}(0) - 4 = -4$, $(0, -4)$

Let $x = 2$, $y = \frac{3}{2}(2) - 4 = -1$, $(2, -1)$

Let $x = 4$, $y = \frac{3}{2}(4) - 4 = 2$, $(4, 2)$

13. A function is a relation in which no two ordered pairs have the same first element and a different second element.

14. **(a)** Not a function

(b) Domain $\{1, 3, 5, 6\}$
Range $\{-4, 0, 2, 3, 5\}$

15. **(a)** Function

(b) Not a function

16.

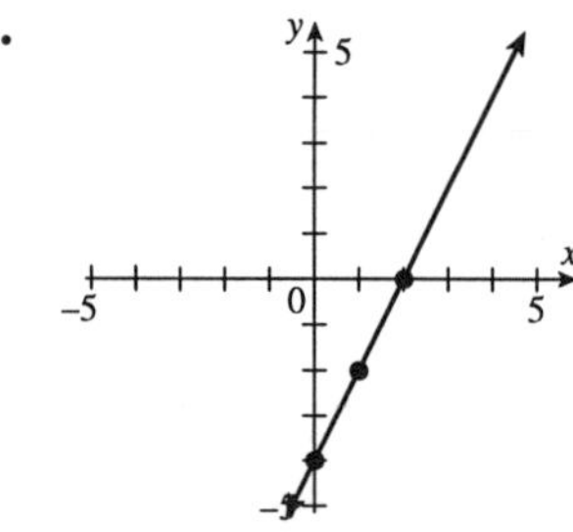

Let $x = 0$, $f(0) = 2 \cdot 0 - 4 = -4$, $(0, -4)$

Let $x = 1$, $f(1) = 2 \cdot 1 - 4 = -2$, $(1, -2)$

Let $x = 2$, $f(2) = 2 \cdot 2 - 4 = 0$, $(2, 0)$

17.

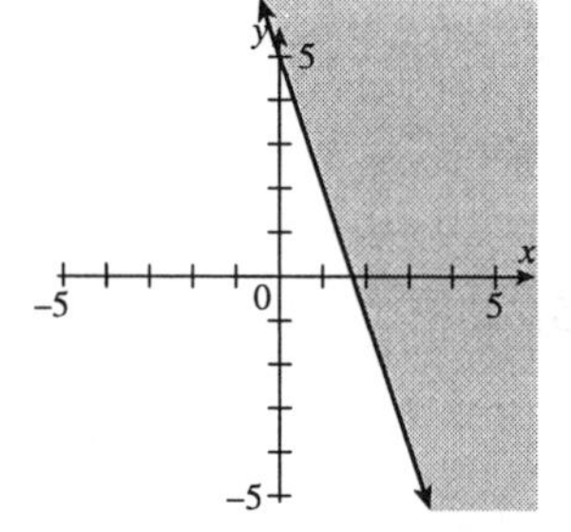

Check (0, 0): $y \geq -3x + 5$

$0 \geq -3 \cdot 0 + 5$

$0 \geq 5$ False

18.

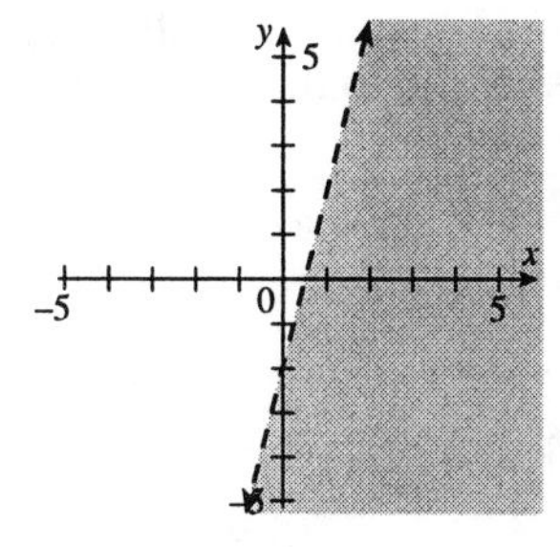

Check (0, 0): $y < 4x - 2$

$0 < 4 \cdot 0 - 2$

$0 < -2$ False

CHAPTER 8

Exercise Set 8.1

1. **(a)** Substitute –2 for x and 2 for y in each equation.

$y = 3x - 4$ | $y = -x + 4$
$2 = 3(-2) - 4$ | $2 = -2 + 4$
$2 = -6 - 4$ | $2 = 2$ True
$2 = -10$ False

Since (–2, 2) does not satisfy both equations, it is not a solution to the system of equations.

(b) Substitute –4 for x and –8 for y in each equation.

$y = 3x - 4$ | $y = -x + 4$
$-8 = 3(-4) - 4$ | $-8 = -(-4) + 4$
$-8 = -12 - 4$ | $-8 = 4 + 4$
$-8 = -16$ False | $-8 = 8$ False

Since (–4, –8) does not satisfy both equations, it is not a solution to the system of equations.

(c) Substitute 2 for x and 2 for y in each equation.

$y = 3x - 4$ | $y = -x + 4$
$2 = 3(2) - 4$ | $2 = -2 + 4$
$2 = 6 - 4$ | $2 = 2$ True
$2 = 2$ True

Since (2, 2) satisfies both equations, it is a solution to the system of equations.

3. **(a)** Substitute 8 for x and 13 for y in each equation.

$y = 2x - 3$ | $y = x + 5$
$13 = 2(8) - 3$ | $13 = 8 + 5$
$13 = 16 - 3$ | $13 = 13$ True
$13 = 13$ True

Since (8, 13) satisfies both equations, it is a solution to the system of equations.

(b) Substitute 4 for x and 5 for y in each equation.

$y = 2x - 3$ | $y = x + 5$
$5 = 2(4) - 3$ | $5 = 4 + 5$
$5 = 8 - 3$ | $5 = 9$ False
$5 = 5$ True

Since (4, 5) does not satisfy both equations, it is not a solution to the system of equations.

(c) Substitute 4 for x and 9 for y in each equation.

$y = 2x - 3$ | $y = x + 5$
$9 = 2(4) - 3$ | $9 = 4 + 5$
$9 = 8 - 3$ | $9 = 9$ True
$9 = 5$ False

Since (4, 9) does not satisfy both equations, it is not a solution to the system of equations.

5. **(a)** Substitute 3 for x and 3 for y in each equation.

$3x - y = 6$ | $2x + y = 9$
$3(3) - 3 = 6$ | $2(3) + 3 = 9$
$9 - 3 = 6$ | $6 + 3 = 9$
$6 = 6$ True | $9 = 9$ True

Since (3, 3) satisfies both equations, it is a solution to the system of equations.

(b) Substitute 4 for x and –2 for y in each equation.

$3x - y = 6$ | $2x + y = 9$
$3(4) - (-2) = 6$ | $2(4) + (-2) = 9$
$12 + 2 = 6$ | $8 + (-2) = 9$
$14 = 6$ False | $6 = 9$ False

Since (4, –2) does not satisfy both equations, it is not a solution to the system of equations.

(c) Substitute –6 for x and 3 for y in each equation.

$3x - y = 6 \qquad 2x + y = 9$

$3(-6) - 3 = 6 \qquad 2(-6) + 3 = 9$

$-18 - 3 = 6 \qquad -12 + 3 = 9$

$-21 = 6$ False $\qquad -9 = 9$ False

Since (–6, 3) does not satisfy both equations, it is not a solution to the system of equations.

7. **(a)** Substitute 3 for x and 0 for y in each equation.

$2x - 3y = 6 \qquad y = \frac{2}{3}x - 2$

$2(3) - 3(0) = 6 \qquad 0 = \frac{2}{3}(3) - 2$

$6 - 0 = 6 \qquad 0 = 2 - 2$

$6 = 6$ True $\qquad 0 = 0$ True

Since (3, 0) satisfies both equations, it is a solution to the system of equations.

(b) Substitute 3 for x and –2 for y in each equation.

$2x - 3y = 6 \qquad y = \frac{2}{3}(x) - 2$

$2(3) - 3(-2) = 6 \qquad -2 = \frac{2}{3}(3) - 2$

$6 + 6 = 6 \qquad -2 = 2 - 2$

$12 = 6$ False $\qquad -2 = 0$ False

Since (3, –2) does not satisfy both equations, it is not a solution to the system of equations.

(c) Substitute 6 for x and 2 for y in each equation.

$2x - 3y = 6 \qquad y = \frac{2}{3}(x) - 2$

$2(6) - 3(2) = 6 \qquad 2 = \frac{2}{3}(6) - 2$

$12 - 6 = 6 \qquad 2 = 4 - 2$

$6 = 6$ True $\qquad 2 = 2$ True

Since (6, 2) satisfies both equations, it is a solution to the system of equations.

9. **(a)** Substitute 0 for x and –2 for y in each equation.

$3x - 4y = 8 \qquad 2y = \frac{3}{2}x - 4$

$3(0) - 4(-2) = 8 \qquad 2(-2) = \frac{3}{2}(0) - 4$

$0 + 8 = 8 \qquad -4 = 0 - 4$

$8 = 8$ True $\qquad -4 = -4$ True

Since (0, –2) satisfies both equations, it is a solution to the system of equations.

(b) Substitute 1 for x and –6 for y in each equation.

$3x - 4y = 8 \qquad 2y = \frac{3}{2}x - 4$

$3(1) - 4(-6) = 8 \qquad 2(-6) = \frac{3}{2}(1) - 4$

$3 + 24 = 8 \qquad -12 = \frac{3}{2} - 4$

$27 = 8$ False $\qquad -12 = -\frac{5}{2}$ False

Since (1, –6) does not satisfy both equations, it is not a solution to the system of equations.

(c) Substitute $-\frac{1}{3}$ for x and $-\frac{9}{4}$ for y in each equation.

$3x - 4y = 8$

$3\left(-\frac{1}{3}\right) - 4\left(-\frac{9}{4}\right) = 8$

$-1 + 9 = 8$

$8 = 8$ True

$2y = \frac{3}{2}x - 4$

$$2\left(-\frac{9}{4}\right)=\frac{3}{2}\left(-\frac{1}{3}\right)-4$$

$$-\frac{9}{2}=-\frac{1}{2}-4$$

$$-\frac{9}{2}=-\frac{9}{2} \text{ True}$$

Since $\left(-\frac{1}{3},-\frac{9}{4}\right)$ satisfies both equations, it is a solution to the system of equations.

11. The lines are not parallel and intersect at exactly one point. The system is consistent and has exactly one solution.

13. Lines 1 and 2 are the same line. The system is dependent and has an infinite number of solutions.

15. The lines are not parallel and intersect at exactly one point. The system is consistent and has exactly one solution.

17. Lines 1 and 2 are the same line. The system is dependent and has an infinite number of solutions.

19. $2y=4x-6$ $\qquad$ $y=3x-2$

$$y=\frac{4x-6}{2}$$

$$y=2x-3$$

The equations have different slopes and the lines are therefore not parallel. The system of equations has exactly one solution.

21. $3y=2x+3$ $\qquad$ $y=\frac{2}{3}x-2$

$$y=\frac{2x+3}{3}$$

$$y=\frac{2}{3}x+1$$

The equations have the same slope and different y-intercepts and the lines are therefore parallel. The system of equations has no solution.

23. $4x=y-6$ $\qquad$ $3x=4y+5$

$$4x+6=y \qquad 3x-5=4y$$

$$y=4x+6 \qquad \frac{3x-5}{4}=y$$

$$y=\frac{3}{4}x-\frac{5}{4}$$

Since the equations have different slopes, the lines are not parallel. The system of equations has exactly one solution.

25. $2x=3y+4$ $\qquad$ $6x-9y=12$

$$2x-4=3y \qquad -9y=-6x+12$$

$$\frac{2x-4}{3}=y \qquad y=\frac{-6x+12}{-9}$$

$$y=\frac{2}{3}x-\frac{4}{3} \qquad y=\frac{2}{3}x-\frac{4}{3}$$

Since the equations have the same slope and the same y-intercept, they represent the same line. The system of equations has an infinite number of solutions.

27. $3x - 2y = -\frac{1}{2}$ $\qquad y = \frac{3}{2}x + \frac{1}{2}$

$-2y = -3x - \frac{1}{2}$

$y = \frac{-3x - \frac{1}{2}}{-2}$

$y = \frac{3}{2}x + \frac{1}{4}$

The equations have the same slope and different y-intercepts. The lines are therefore parallel and the system of equations has no solution.

29. $3x + 5y = -7$ $\qquad -3x - 5y = -7$

$5y = -3x - 7$ $\qquad -5y = 3x - 7$

$y = \frac{-3x - 7}{5}$ $\qquad y = \frac{3x - 7}{-5}$

$y = -\frac{3}{5}x - \frac{7}{5}$ $\qquad y = -\frac{3}{5}x + \frac{7}{5}$

Since the equations have the same slope and different y-intercepts, the lines are parallel. The system of equations has no solution.

31. $y = x + 2$

	Ordered Pair
Let $x = 0$, then $y = 2$	(0, 2)
Let $y = 0$, then $x = -2$	(–2, 0)

$y = -x + 2$

	Ordered Pair
Let $x = 0$, then $y = 2$	(0, 2)
Let $y = 0$, then $x = 2$	(2, 0)

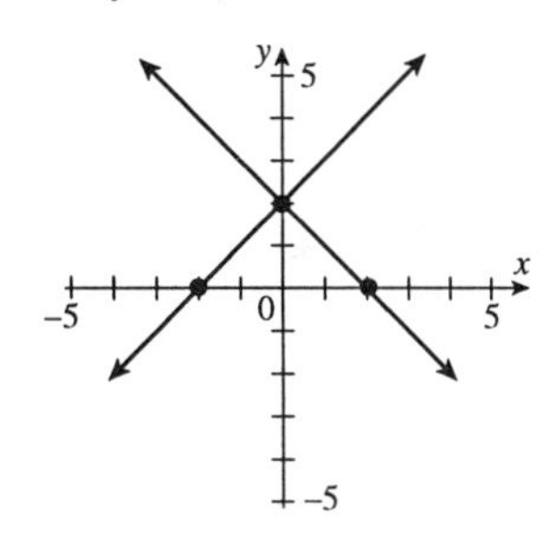

The graphs intersect at the point (0, 2).

Check: $y = x + 2$ $\qquad y = -x + 2$

$2 = 0 + 2$ $\qquad 2 = -0 + 2$

$2 = 2$ True $\qquad 2 = 2$ True

Since (0, 2) satisfies both equations, it is a solution to the system of equations.

33. $y = 3x - 6$

	Ordered Pair
Let $x = 0$, then $y = -6$	(0, –6)
Let $y = 0$, then $x = 2$	(2, 0)

$y = -x + 6$

	Ordered Pair
Let $x = 0$, then $y = 6$	(0, 6)
Let $y = 0$, then $x = 6$	(6, 0)

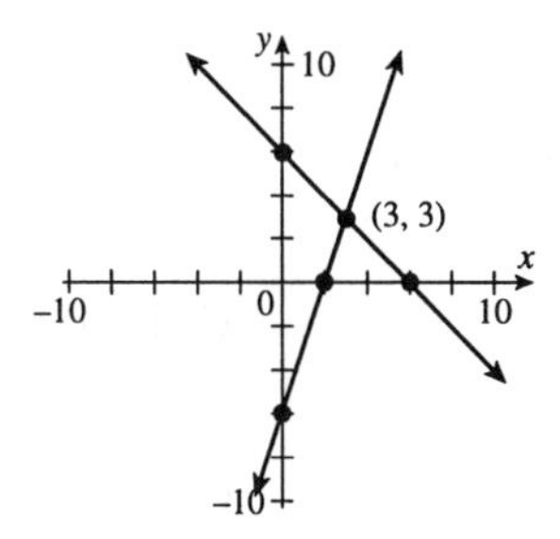

The graphs intersect at the point (3, 3).

Check: $y = 3x - 6$ $\qquad y = -x + 6$

$3 = 3(3) - 6$ $\qquad 3 = -3 + 6$

$3 = 9 - 6$ $\qquad 3 = 3$ True

$3 = 3$ True

Since (3, 3) satisfies both equations, it is a solution to the system of equations.

35. $2x = 4$ represents the same line as $x = 2$, $x = 2$ is a vertical line and $y = -3$ is a horizontal line.

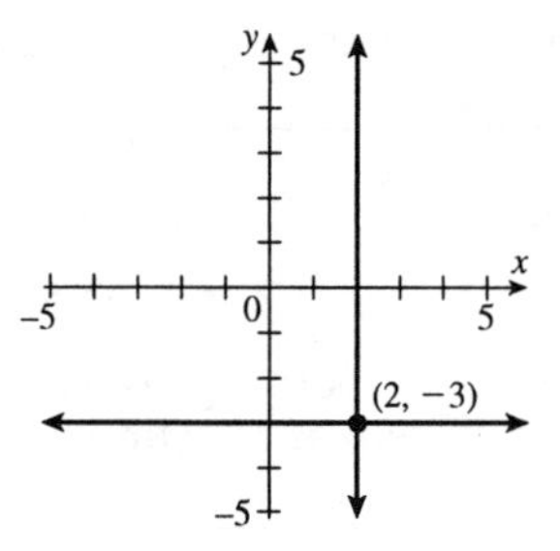

The graphs intersect at the point (2, –3).

Check: $2x = 4$ $\quad$ $y = -3$

$2(2) = 4$ $\quad$ $-3 = -3$ True

$4 = 4$ True

Since (2, –3) satisfies both equations, it is a solution to the system of equations.

37. $y = x + 2$ $\quad$ Ordered Pair

Let $x = 0$, then $y = 2$ $\quad$ (0, 2)

Let $y = 0$, then $x = -2$ $\quad$ (–2, 0)

$x + y = 4$ $\quad$ Ordered Pair

Let $x = 0$, then $y = 4$ $\quad$ (0, 4)

Let $y = 0$, then $x = 4$ $\quad$ (4, 0)

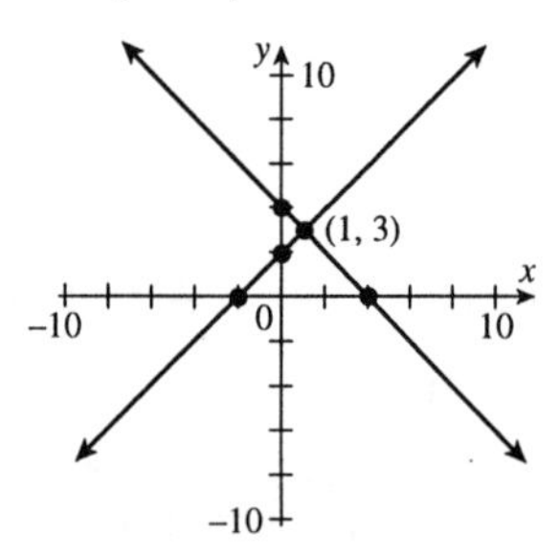

The graphs intersect at the point (1, 3).

Check: $y = x + 2$ $\quad$ $x + y = 4$

$3 = 1 + 2$ $\quad$ $1 + 3 = 4$

$3 = 3$ True $\quad$ $4 = 4$ True

Since (1, 3) satisfies both equations, it is a solution to the system of equations.

39. $y = -\frac{1}{2}x + 4$ $\quad$ Ordered Pair

Let $x = 0$, then $y = 4$ $\quad$ (0, 4)

Let $y = 0$, then $x = 8$ $\quad$ (8, 0)

$x + 2y = 6$ $\quad$ Ordered Pair

Let $x = 0$, then $y = 3$ $\quad$ (0, 3)

Let $y = 0$, then $x = 6$ $\quad$ (6, 0)

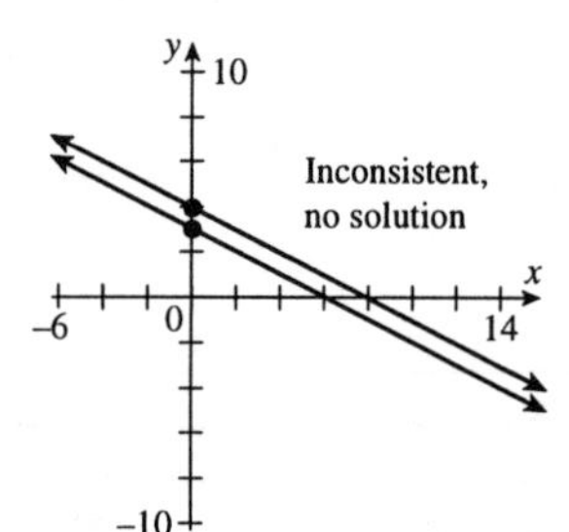

The lines appear parallel.

Check: $y = -\frac{1}{2}x + 4$ $\quad$ $x + 2y = 6$

$2y = -x + 6$

$y = -\frac{1}{2}x + 3$

Since the equations have the same slope and different y-intercepts, the lines are parallel. The system of equations is inconsistent and has no solution.

41. $x + 2y = 8$ $\quad$ Ordered Pair

Let $x = 0$, then $y = 4$ $\quad$ (0, 4)

Let $y = 0$, then $x = 8$ $\quad$ (8, 0)

$2x - 3y = 2$ $\quad$ Ordered Pair

Let $x = 0$, then $y = -\frac{2}{3}$ $\quad$ $\left(0, -\frac{2}{3}\right)$

Let $y = 0$, then $x = 1$ $\quad$ (1, 0)

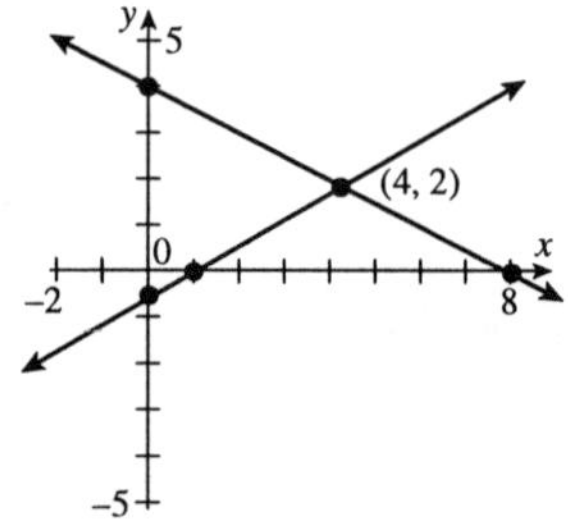

The graphs intersect at (4, 2).

Check: $x+2y=8$ $\quad$ $2x-3y=2$

$4+2(2)=8$ $\quad$ $2(4)-3(2)=2$

$4+4=8$ $\quad$ $8-6=2$

$8=8$ True $\quad$ $2=2$ True

Since (4, 2) satisfies both equations, it is a solution to the system of equations.

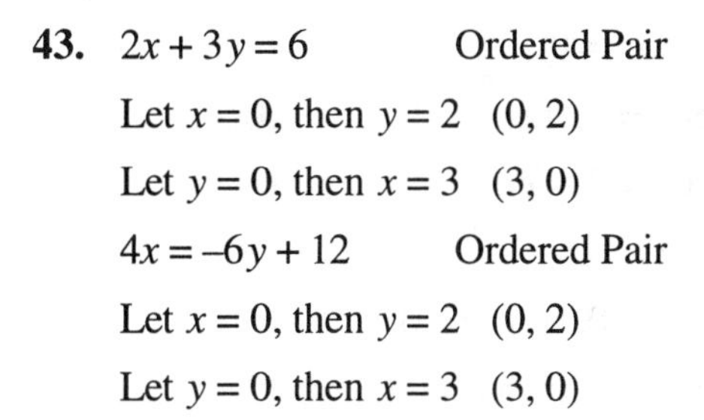

43. $2x+3y=6$ $\quad$ Ordered Pair

Let $x=0$, then $y=2$ $\quad$ (0, 2)

Let $y=0$, then $x=3$ $\quad$ (3, 0)

$4x=-6y+12$ $\quad$ Ordered Pair

Let $x=0$, then $y=2$ $\quad$ (0, 2)

Let $y=0$, then $x=3$ $\quad$ (3, 0)

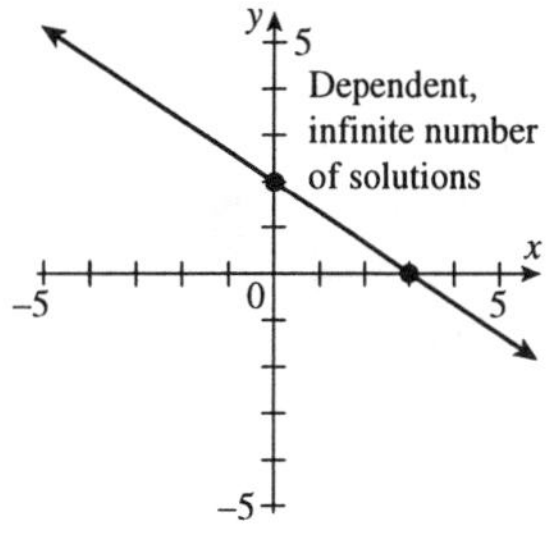

Both equations represent the same line.

Check: $2x+3y=6$ $\quad$ $4x=-6y+12$

$3y=-2x+6$ $\quad$ $4x-12=-6y$

$y=-\frac{2}{3}x+2$ $\quad$ $-\frac{2}{3}x+2=y$

The system of equations is dependent and has an infinite number of solutions.

45. $y=2x-3$ $\quad$ Ordered Pair

Let $x=0$, then $y=-3$ $\quad$ (0, −3)

Let $y=0$, then $x=\frac{3}{2}$ $\quad$ $\left(\frac{3}{2},0\right)$

$y=3$

horizontal line

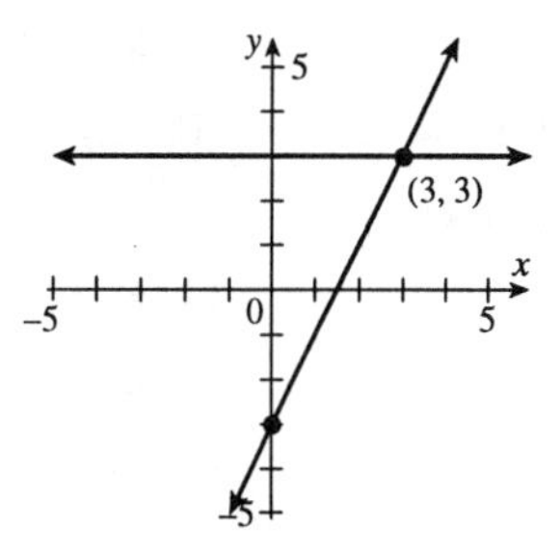

The graphs intersect at the point (3, 3).

Check: $y=2x-3$ $\quad$ $y=3$

$3=2(3)-3$ $\quad$ $3=3$ True

$3=6-3$

$3=3$ True

Since (3, 3) satisfies both equations, it is a solution to the system of equations.

47. $x-2y=4$ $\quad$ Ordered Pair

Let $x=0$, then $y=-2$ $\quad$ (0, −2)

Let $y=0$, then $x=4$ $\quad$ (4, 0)

$2x-4y=8$ $\quad$ Ordered Pair

Let $x=0$, then $y=-2$ $\quad$ (0, −2)

Let $y=0$, then $x=4$ $\quad$ (4, 0)

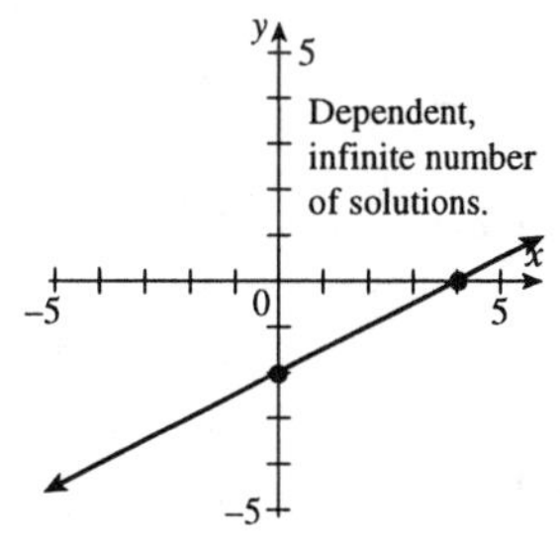

Both equations represent the same line.

Check: $x-2y=4$ $\quad$ $2x-4y=8$

$-2y=-x+4$ $\quad$ $-4y=-2x+8$

$y=\frac{1}{2}x-2$ $\quad$ $y=\frac{1}{2}x-2$

The system of equations is dependent and has an infinite number of solutions.

49. $2x + y = -2$ — Ordered Pair

Let $x = 0$, then $y = -2$ — $(0, -2)$

Let $y = 0$, then $x = -1$ — $(-1, 0)$

$6x + 3y = 6$ — Ordered Pair

Let $x = 0$, then $y = 2$ — $(0, 2)$

Let $y = 0$, then $x = 1$ — $(1, 0)$

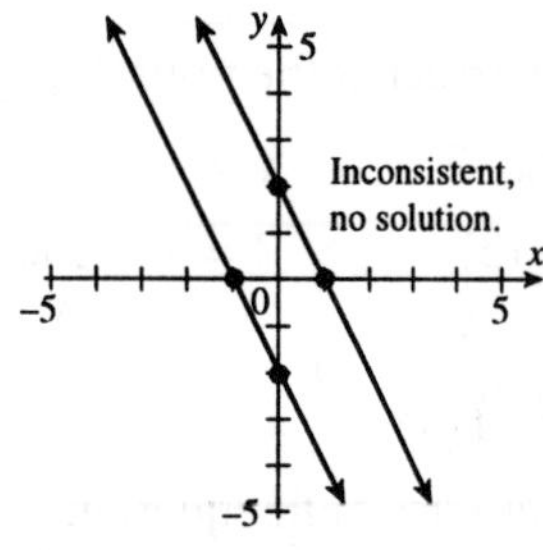

The lines appear to be parallel.

Check: $2x + y = -2$ $\quad$ $6x + 3y = 6$

$y = -2x - 2$ $\quad$ $3y = -6x + 6$

$y = -2x + 2$

Since the equations have the same slope and different y-intercepts, the lines are parallel. The system of equations is inconsistent and has no solution.

51. $4x - 3y = 6$ — Ordered Pair

Let $x = 0$, then $y = -2$ — $(0, -2)$

Let $y = 0$, then $x = \frac{3}{2}$ — $\left(\frac{3}{2}, 0\right)$

$2x + 4y = 14$ — Ordered Pair

Let $x = 0$, then $y = \frac{7}{2}$ — $\left(0, \frac{7}{2}\right)$

Let $y = 0$, then $x = 7$ — $(7, 0)$

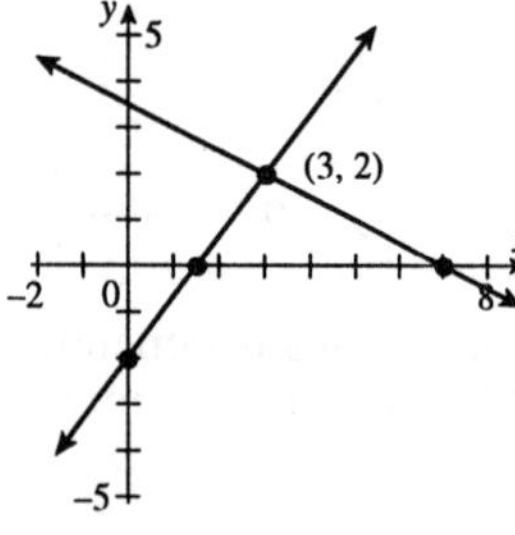

The graphs intersect at the point (3, 2).

Check: $4x - 3y = 6$ $\quad$ $2x + 4y = 14$

$4(3) - 3(2) = 6$ $\quad$ $2(3) + 4(2) = 14$

$12 - 6 = 6$ $\quad$ $6 + 8 = 14$

$6 = 6$ True $\quad$ $14 = 14$ True

Since (3, 2) satisfies both equations, it is a solution to the system of equations.

53. $2x - 3y = 0$ — Ordered Pair

Let $x = 0$, then $y = 0$ — $(0, 0)$

Let $x = 3$, then $y = 2$ — $(3, 2)$

$x + 2y = 0$ — Ordered Pair

Let $x = 0$, then $y = 0$ — $(0, 0)$

Let $x = 2$, then $y = -1$ — $(2, -1)$

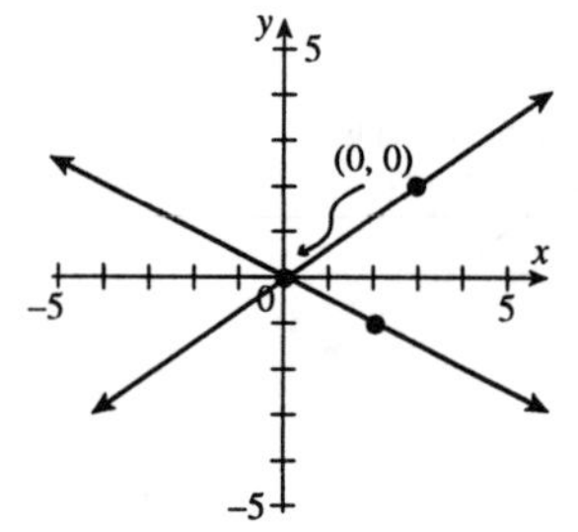

The graphs intersect at the point (0, 0).

Check: $2x - 3y = 0$ $\quad$ $x + 2y = 0$

$2(0) - 3(0) = 0$ $\quad$ $0 + 2(0) = 0$

$0 - 0 = 0$ $\quad$ $0 + 0 = 0$

$0 = 0$ True $\quad$ $0 = 0$ True

Since (0, 0) satisfies both equations, it is a solution to the system of equations.

55. The cost of each system can be represented as follows:

Moneywell: $c = 4400 + 15n$

Doile: $c = 3400 + 25n$

Graph each equation.

$c = 4400 + 15n$

Let $n = 0$ $\quad$ $c = 4400 + 15(0) = 4400$

Let $n = 160$ $\quad$ $c = 4400 + 15(160) = 6800$

n	c
0	4400
160	6800

$c = 3400 + 25n$

Let $n = 0$ $\quad c = 3400 + 25(0) = 3400$

Let $n = 160$ $\quad c = 3400 + 25(160) = 7400$

n	c
0	3400
160	7400

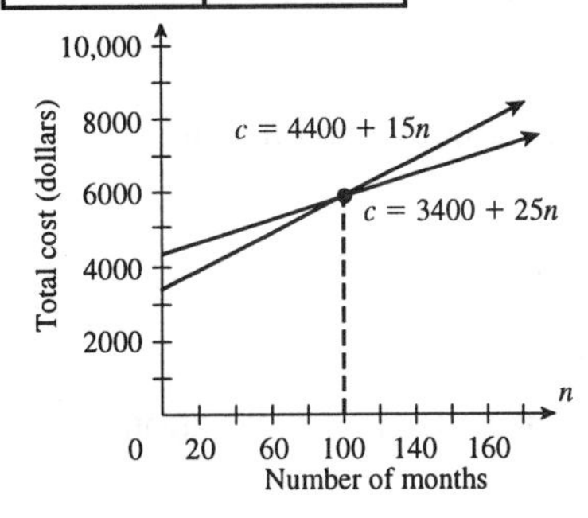

The graphs intersect at the point (100, 5900).

The total cost of the two systems is the same after 100 months or 8 years 4 months.

57. Graph each equation.

$c = 200 + 60h$

Let $h = 0$, then $c = 200 + 60(0) = 200$

Let $h = 10$, then $c = 200 + 60(10) = 800$

h	c
0	200
10	800

$c = 300 + 40h$

Let $h = 0$, then $c = 300 + 40(0) = 300$

Let $h = 10$, then $c = 300 + 40(10) = 700$

h	c
0	300
10	700

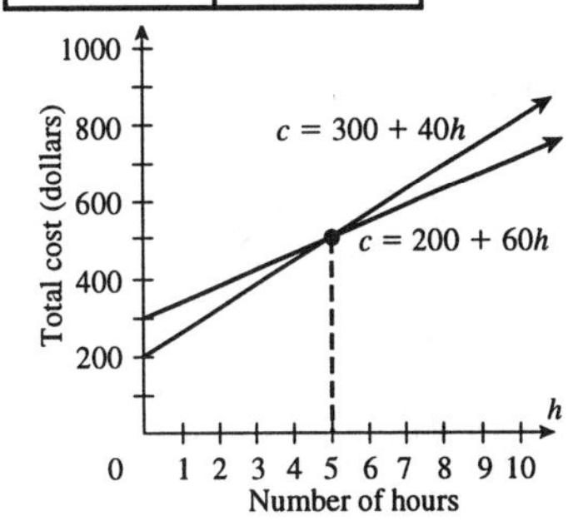

The graphs intersect at the point (5, 500).

The two services have the same total cost if the number of hours of labor is 5.

59. Write the equations in slope-intercept form.

$5x - 4y = 10$ $\qquad$ $12y = 15x - 20$

$-4y = -5x + 10$ $\qquad$ $y = \frac{5}{4}x - \frac{5}{3}$

$y = \frac{5}{4}x - \frac{5}{2}$

Since the slopes are the same, but the y-intercepts are different, the lines will be parallel.

61. Write both equations in slope-intercept form and then compare their slopes and y-intercepts. If the slopes are different there is one solution. If the slopes are the same but the y-intercepts are different there is no solution If both the slopes and y-intercepts are the same there are an infinite number of solutions.

Cumulative Review Exercises: 8.1

63. (a) 6

(b) 6, 0

(c) 6, –4, 0

(d) $6,\ -4,\ 0,\ 2\frac{1}{2},\ -\frac{9}{5},\ 4.22$

(e) $\sqrt{3},\ -\sqrt{7}$

(f) $6, -4, 0, \sqrt{3}, 2\frac{1}{2}, -\frac{9}{5}, 4.22, -\sqrt{7}$

64. Factor by grouping

$xy - 4x + 3y - 12 = x(y-4) + 3(y-4)$

$= (x+3)(y-4)$

65. Factor by grouping

Find two numbers whose product is $8(-15) = -120$ and whose sum is 2. The numbers are 12 and –10.

$8x^2 + 2x - 15 = 8x^2 + 12x - 10x - 15$

$= 4x(2x+3) - 5(2x+3)$

$= (4x-5)(2x+3)$

66. The least common denominator is $x + 3$.

$$\frac{x}{x+3} - 2 = \frac{x}{x+3} - \frac{x+3}{x+3}\cdot 2$$

$$= \frac{x}{x+3} - \frac{2x+6}{x+3} = \frac{-x-6}{x+3}$$

67. $\frac{x}{x+3} - 2 = 0$

Multiply both sides of the equation by the least common denominator, $(x+3)$.

$$(x+3)\left[\frac{x}{x+3} - 2\right] = (x+3)\cdot 0$$

$x - 2(x+3) = 0$

$x - 2x - 6 = 0$

$-x - 6 = 0$

$-6 = x$

Group Activity/Challenge Problems

1. 0; the lines are parallel.

2. Infinite number; the lines are the same.

Exercise Set 8.2

1. Solve for x in the first equation.

$x + 2y = 4$

$x = -2y + 4$

Substitute $-2y + 4$ for x in the other equation.

$2x - 3y = 1$

$2(-2y+4) - 3y = 1$

$-4y + 8 - 3y = 1$

$-7y + 8 = 1$

$-7y = -7$

$y = 1$

Find the value of x.

$x = -2y + 4$

$x = -2(1) + 4$

$x = -2 + 4 = 2$

The solution is the ordered pair (2, 1).

3. Solve for y in the first equation.

$x + y = -2$

$y = -x - 2$

Substitute $-x - 2$ for y in the other equation.

$x - y = 0$

$x - (-x - 2) = 0$

$x + x + 2 = 0$

$2x + 2 = 0$

$2x = -2$

$x = -1$

Find the value of y.

$y = -x - 2$

$y = -(-1) - 2 = 1 - 2 = -1$

The solution is the ordered pair $(-1, -1)$.

5. Solve for y in the first equation.

$2x + y = 3$

$y = -2x + 3$

Substitute $-2x + 3$ for y in the other equation.

$2x + y + 5 = 0$

$2x + (-2x + 3) + 5 = 0$

$2x - 2x + 3 + 5 = 0$

$8 = 0$

Since the statement $8 = 0$ is false, the system is inconsistent and has no solution.

7. Substitute 4 for x in the second equation.

$x + y + 5 = 0$

$4 + y + 5 = 0$

$y + 9 = 0$

$y = -9$

Find the value of x.

From the first equation $x = 4$.

The solution is the ordered pair $(4, -9)$.

9. Substitute $2x - 4$ for y in the first equation.

$$x - \frac{1}{2}y = 2$$

$$x - \frac{1}{2}(2x - 4) = 2$$

$$x - x + 2 = 2$$

$$2 = 2$$

Since the statement $2 = 2$ is true, the system is dependent and has an infinite number of solutions.

11. Substitute $3x + 5$ for y in the first equation.

$3x + y = -1$

$3x + (3x + 5) = -1$

$3x + 3x + 5 = -1$

$6x + 5 = -1$

$6x = -6$

$x = -1$

Find the value of y.

$y = 3x + 5$

$y = 3(-1) + 5$

$y = -3 + 5 = 2$

The solution is the ordered pair $(-1, 2)$.

13. Substitute $\frac{1}{3}x - 2$ for y in the second equation.

$x - 3y = 6$

$$x - 3\left(\frac{1}{3}x - 2\right) = 6$$

$$x - x + 6 = 6$$

$$6 = 6$$

Since the statement $6 = 6$ is true, the system is dependent and has an infinite number of solutions.

15. Solve for y in the second equation.

$6x - 2y = 10$

$-2y = -6x + 10$

$y = 3x - 5$

Substitute $3x - 5$ for y in the first equation.

$2x + 3y = 7$

$2x + 3(3x - 5) = 7$

$2x + 9x - 15 = 7$

$11x - 15 = 7$

$11x = 22$

$x = 2$

Find the value of y.

$y = 3x - 5$

$y = 3(2) - 5$

$y = 6 - 5 = 1$

The solution is the ordered pair (2, 1).

17. Solve for y in the first equation.

$3x - y = 14$

$-y = -3x + 14$

$y = 3x - 14$

Substitute $3x - 14$ for y in the second equation.

$6x - 2y = 10$

$6x - 2(3x - 14) = 10$

$6x - 6x + 28 = 10$

$28 = 10$ False

Since the statement $28 = 10$ is false, the system is inconsistent and has no solution.

19. Solve for x in the second equation.

$3x = 2y - 3$

$$x = \frac{2y - 3}{3}$$

$$x = \frac{2}{3}y - 1$$

Substitute $\frac{2}{3}y - 1$ for x in the first equation.

$4x - 5y = -4$

$$4\left(\frac{2}{3}y - 1\right) - 5y = -4$$

$$\frac{8}{3}y - 4 - 5y = -4$$

$$-\frac{7}{3}y - 4 = -4$$

$$-\frac{7}{3}y = 0$$

$y = 0$

Find the value of x.

$$x = \frac{2}{3}y - 1$$

$$x = \frac{2}{3}(0) - 1$$

$x = 0 - 1 = -1$

The solution is the ordered pair (–1, 0).

21. Solve for x in the second equation.

$$x-\frac{5}{3}y=-2$$

$$x=\frac{5}{3}y-2$$

Substitute $\frac{5}{3}y-2$ for x in the first equation.

$$5x+4y=-7$$

$$5\left(\frac{5}{3}y-2\right)+4y=-7$$

$$\frac{25}{3}y-10+4y=-7$$

$$\frac{37}{3}y-10=-7$$

$$\frac{37}{3}y=3$$

$$y=\frac{9}{37}$$

Find the value of x.

$$x=\frac{5}{3}y-2$$

$$x=\frac{5}{3}\left(\frac{9}{37}\right)-2$$

$$x=\frac{15}{37}-\frac{74}{37}$$

$$x=-\frac{59}{37}$$

The solution is the ordered pair $\left(-\frac{59}{37}, \frac{9}{37}\right)$.

23. The situation can be represented by the system of equations

$$T=82-2t$$

$$T=64+2.5t$$

(a) Substitute $82-2t$ for T in the second equation.

$$T=64+2.5t$$

$$82-2t=64+2.5t$$

$$18-2t=2.5t$$

$$18=4.5t$$

$$4=t$$

Both cities will have the same temperature after 4 hours.

(b) Substitute 4 for t in the first equation.

$$T=82-2t$$

$$T=82-2(4)$$

$$T=82-8=74$$

When both cities have the same temperature (after 4 hours) that temperature is $74°$F.

25. The situation can be represented by the system of equations

$$m=100+65t$$

$$m=115+60t$$

(a) Substitute $100+65t$ for m in the second equation.

$$m=115+60t$$

$$100+65t=115+60t$$

$$65t=15+60t$$

$$5t=15$$

$$t=3$$

It will take 3 hours for Jim to catch Kathy.

(b) Substitute 3 for t in the first equation.

$m = 100 + 65t$

$m = 100 + 65(3)$

$m = 100 + 195 = 295$

They will meet at the 295 mile marker.

27. You will obtain a false statement, such as $3 = 0$.

Cumulative Review Exercises: 8.2

29. $-3^3 = -(3)(3)(3) = -27$

30. $(-3)^3 = (-3)(-3)(-3) = -27$

31. $-3^4 = -(3)(3)(3)(3) = -81$

32. $(-3)^4 = (-3)(-3)(-3)(-3) = 81$

33. $\dfrac{3 \text{ cups brownie mix}}{2 \text{ eggs}}$

$= \dfrac{9 \text{ cups brownie mix}}{x \text{ eggs}}$

$\dfrac{3}{2} = \dfrac{9}{x}$

$3x = 2 \cdot 9$

$3x = 18$

$x = 6$

6 eggs should be added to 9 cups of brownie mix.

34. $\dfrac{6}{8} = \dfrac{4}{x}$

$6x = 8 \cdot 4$

$6x = 32$

$x = \dfrac{32}{6} = 5\dfrac{1}{3}$

The length of side x is $5\frac{1}{3}$ inches.

35. $\left(\dfrac{3x^2y^4}{x^3y^2}\right)^2 = \left(3 \cdot \dfrac{x^2}{x^3} \cdot \dfrac{y^4}{y^2}\right)^2$

$= \left(\dfrac{3y^2}{x}\right)^2 = \dfrac{9y^4}{x^2}$

36. $(4x^2y^3)^3(3x^4y^5)^2$

$= (4)^3(x^2)^3(y^3)^3(3)^2(x^4)^2(y^5)^2$

$= 64x^6y^9 9x^8y^{10} = 576x^{14}y^{19}$

37. $\dfrac{3}{x-12} + \dfrac{5}{x-5} = \dfrac{5}{x^2 - 17x + 60}$

$\dfrac{3}{x-12} + \dfrac{5}{x-5} = \dfrac{5}{(x-12)(x-5)}$

Multiply both sides by the least common denominator, $(x-12)(x-5)$.

$(x-12)(x-5)\left[\dfrac{3}{x-12} + \dfrac{5}{x-5}\right]$

$= \dfrac{5}{(x-12)(x-5)} \cdot (x-12)(x-5)$

$3(x-5) + 5(x-12) = 5$

$3x - 15 + 5x - 60 = 5$

$8x - 75 = 5$

$8x = 80$

$x = 10$

Group Activity/Challenge Problems

1. **(a)** $y = ax + b, y = cx + d$

$ax + b = cx + d$

$ax - cx = d - b$

$x(a - c) = d - b$

$x = \dfrac{d-b}{a-c}$

(b) $y = 3x + 2, y = x + 6$

$x = \dfrac{d-b}{a-c} = \dfrac{6-2}{3-1} = \dfrac{4}{2} = 2$

(c) $y = 3x + 2$

$y = 3(2) + 2$

$y = 6 + 2 = 8$

$(2, 8)$

3. **(a)** $y = 4x + 600, y = 2x + 1200$

$4x + 600 = 2x + 1200$

$2x = 600$

$x = 300$

$y = 4(300) + 600 = 1800$

$(300, 1800)$

(b) x value remains the same, while the y value is halved.

(c) $y = 2x + 300, y = x + 600$

$2x + 300 = x + 600$

$x = 300$

$y = 2(300) + 300 = 600 + 300 = 900$

$(300, 900)$

5. $x = 4, 2x - y = 6, -x + y + z = -3$

$2(4) - y = 6, 8 - y = 6, y = 8 - 6 = 2$

$-4 + 2 + z - 3, z = -3 + 4 - 2 = -1$

$(4, 2, -1)$

Exercise Set 8.3

1. $x + y = 8$

$\underline{x - y = 4}$

$2x \quad = 12$

$x = 6$

Solve for y: $x + y = 8$

$6 + y = 8$

$y = 2$

The solution is $(6, 2)$.

3. $-x + y = 5$

$\underline{x + y = 1}$

$2y = 6$

$y = 3$

Solve for x: $x + y = 1$

$x + 3 = 1$

$x = -2$

The solution is $(-2, 3)$.

5. $x + 2y = 15$

$\underline{x - 2y = -7}$

$2x \quad = 8$

$x \quad = 4$

Solve for y: $x + 2y = 15$

$4 + 2y = 15$

$2y = 11$

$y = \dfrac{11}{2}$

The solution is $\left(4, \dfrac{11}{2}\right)$.

7. $2[4x+y=6]$ gives $8x+2y=12$
$-8x-2y=20$ $\quad -8x-2y=20$

$$\begin{aligned} 8x+2y&=12\\ \underline{-8x-2y}&\underline{=20}\\ 0&=32 \quad \text{False} \end{aligned}$$

Since a false statement is obtained, the system is inconsistent and has no solution.

9. $-1[-5x+y=14]$ gives $5x-y=-14$
$-3x+y=-2$ $\quad -3x+y=-2$

$$\begin{aligned} 5x-y&=-14\\ \underline{-3x+y}&\underline{=-2}\\ 2x\quad&=-16\\ x&=-8 \end{aligned}$$

Solve for y: $-5x+y=14$
$$\begin{aligned} -5(-8)+y&=14\\ 40+y&=14\\ y&=-26 \end{aligned}$$

The solution is $(-8, -26)$.

11. $2[3x+y=10]$ gives $6x+2y=20$
$3x-2y=16$ $\quad 3x-2y=16$

$$\begin{aligned} 6x+2y&=20\\ \underline{3x-2y}&\underline{=16}\\ 9x\quad&=36\\ x&=4 \end{aligned}$$

Solve for y: $3x+y=10$
$$\begin{aligned} 3(4)+y&=10\\ 12+y&=10\\ y&=-2 \end{aligned}$$

The solution is $(4, -2)$.

13. $4x-3y=8$ gives $4x-3y=8$
$3[2x+y=14]$ $\quad 6x+3y=42$

$$\begin{aligned} 4x-3y&=8\\ \underline{6x+3y}&\underline{=42}\\ 10x\quad&=50\\ x&=5 \end{aligned}$$

Solve for y: $2x+y=14$
$$\begin{aligned} 2(5)+y&=14\\ 10+y&=14\\ y&=4 \end{aligned}$$

The solution is $(5, 4)$.

15. $4[5x+3y=6]$ gives $20x+12y=24$
$3[2x-4y=5]$ $\quad 6x-12y=15$

$$\begin{aligned} 20x+12y&=24\\ \underline{6x-12y}&\underline{=15}\\ 26x\quad&=39\\ x&=\frac{3}{2} \end{aligned}$$

Solve for y: $5x+3y=6$
$$\begin{aligned} 5\left(\frac{3}{2}\right)+3y&=6\\ \frac{15}{2}+3y&=6\\ 3y&=-\frac{3}{2}\\ y&=-\frac{1}{2} \end{aligned}$$

The solution is $\left(\frac{3}{2}, -\frac{1}{2}\right)$.

17. Align the x and y terms on the left side of the equation.

$$y = 2x - 3$$
$$-2x + y = -3$$

$4x - 2y = 6$ gives $4x - 2y = 6$

$2[-2x + y = -3]$ $\quad -4x + 2y = -6$

$$\begin{array}{l} 4x - 2y = 6 \\ \underline{-4x + 2y = -6} \\ \quad 0 = 0 \quad \text{True} \end{array}$$

Since a true solution is obtained, the system is dependent and has an infinite number of solutions.

19. Align the x and y terms on the left side of the equation.

$$3y = 2x + 4$$
$$-2x + 3y = 4$$

$2[3x - 2y = -2]$ gives $6x - 4y = -4$

$3[-2x + 3y = 4]$ $\quad -6x + 9y = 12$

$$\begin{array}{l} 6x - 4y = -4 \\ \underline{-6x + 9y = 12} \\ \quad 5y = 8 \\ \quad y = \frac{8}{5} \end{array}$$

Solve for x: $3x - 2y = -2$

$$3x - 2\left(\frac{8}{5}\right) = -2$$
$$3x - \frac{16}{5} = -2$$
$$3x = \frac{6}{5}$$
$$x = \frac{2}{5}$$

The solution is $\left(\frac{2}{5}, \frac{8}{5}\right)$.

21. $2[3x - 4y = 6]$ gives $6x - 8y = 12$

$-6x + 8y = -4$ $\quad -6x + 8y = -4$

$$\begin{array}{l} 6x - 8y = 12 \\ \underline{-6x + 8y = -4} \\ \quad 0 = 8 \quad \text{False} \end{array}$$

Since a false statement is obtained, the system is inconsistent and has no solution.

23. $3[3x - 5y = 0]$ gives $9x - 15y = 0$

$5[2x + 3y = 0]$ $\quad 10x + 15y = 0$

$$\begin{array}{l} 9x - 15y = 0 \\ \underline{10x + 15y = 0} \\ 19x \qquad = 0 \\ \quad x = 0 \end{array}$$

Solve for y: $2x + 3y = 0$

$$2(0) + 3y = 0$$
$$0 + 3y = 0$$
$$3y = 0$$
$$y = 0$$

The solution is (0, 0).

25. $5x - 4y = 20$ gives $5x - 4y = 20$

$2[-3x + 2y = -15]$ $\quad -6x + 4y = -30$

$$\begin{array}{l} 5x - 4y = 20 \\ \underline{-6x + 4y = -30} \\ -x \qquad = -10 \\ \quad x = 10 \end{array}$$

Solve for y: $5x - 4y = 20$

$$5(10) - 4y = 20$$
$$50 - 4y = 20$$
$$-4y = -30$$
$$y = \frac{15}{2}$$

The solution is $\left(10, \frac{15}{2}\right)$.

27. Align the x and y terms on the left side of the equation.

$$3y = 5x - 8$$
$$-5x + 3y = -8$$

$3[6x + 2y = 5]$ gives $18x + 6y = 15$

$-2[-5x + 3y = -8]$ $\quad 10x - 6y = 16$

$$18x + 6y = 15$$
$$\underline{10x - 6y = 16}$$
$$28x \quad = 31$$
$$x = \frac{31}{28}$$

Solve for y: $3y = 5x - 8$

$$3y = 5\left(\frac{31}{28}\right) - 8$$
$$3y = \frac{155}{28} - 8$$
$$3y = -\frac{69}{28}$$
$$y = -\frac{23}{28}$$

The solution is $\left(\frac{31}{28}, -\frac{23}{28}\right)$.

29. Align the x and y terms on the left side of the equation.

$$3x = 6y + 4$$
$$3x - 6y = 4$$

$3[4x + 5y = 0]$ gives $12x + 15y = 0$

$-4[3x - 6y = 4]$ $\quad -12x + 24y = -16$

$$12x + 15y = 0$$
$$\underline{-12x + 24y = -16}$$
$$39y = -16$$
$$y = -\frac{16}{39}$$

Solve for x: $3x = 6y + 4$

$$3x = 6\left(-\frac{16}{39}\right) + 4$$
$$3x = \frac{60}{39}$$
$$x = \frac{20}{39}$$

The solution is $\left(\frac{20}{39}, -\frac{16}{39}\right)$.

31. $2\left[x - \frac{1}{2}y = 4\right]$ gives $2x - y = 8$

$3x + y = 6$ $\quad 3x + y = 6$

$$2x - y = 8$$
$$\underline{3x + y = 6}$$
$$5x \quad = 14$$
$$x = \frac{14}{5}$$

Solve for y: $3x + y = 6$

$$3\left(\frac{14}{5}\right) + y = 6$$
$$\frac{42}{5} + y = 6$$
$$y = -\frac{12}{5}$$

The solution is $\left(\frac{14}{5}, -\frac{12}{5}\right)$.

33. You will obtain a false solution like $0 = 6$.

35. **(a)** Answers will vary.

(b) $5[3x - 2y = 10]$ gives $15x - 10y = 50$

$2[2x + 5y = 13]$ $\quad 4x + 10y = 26$

$$\begin{array}{r} 15x - 10y = 50 \\ \underline{4x + 10y = 26} \\ 19x \qquad = 76 \\ x = 4 \end{array}$$

Solve for y: $3x - 2y = 10$

$3(4) - 2y = 10$

$12 - 2y = 10$

$-2y = -2$

$y = 1$

The solution is (4, 1).

Cumulative Review Exercises: 8.3

36. $\left(\dfrac{16x^4y^6z^3}{4x^6y^5z^7}\right)^3 = \left(\dfrac{16}{4} \cdot \dfrac{x^4}{x^6} \cdot \dfrac{y^6}{y^5} \cdot \dfrac{z^3}{z^7}\right)^3$

$= \left(\dfrac{4y}{x^2z^4}\right)^3 = \dfrac{64y^3}{x^6z^{12}}$

37. $\dfrac{110 \text{ feet}}{1 \text{ day}} = \dfrac{2420 \text{ feet}}{x \text{ days}}$

$\dfrac{110}{1} = \dfrac{2420}{x}$

$110x = 2420$

$x = \dfrac{2420}{110} = 22$

It will take the crew 22 days to pave 2420 feet.

38. $\dfrac{2x^2 - x - 6}{x^2 - 7x + 10} \div \dfrac{2x^2 - 7x - 4}{x^2 - 9x + 20}$

$= \dfrac{2x^2 - x - 6}{x^2 - 7x + 10} \cdot \dfrac{x^2 - 9x + 20}{2x^2 - 7x - 4}$

$= \dfrac{(2x+3)(x-2)}{(x-2)(x-5)} \cdot \dfrac{(x-5)(x-4)}{(2x+1)(x-4)}$

$= \dfrac{2x+3}{2x+1}$

39. $\dfrac{x}{x^2 - 1} - \dfrac{3}{x^2 - 16x + 15}$

$= \dfrac{x}{(x+1)(x-1)} - \dfrac{3}{(x-1)(x-15)}$

The least common denominator $(x+1)(x-1)(x-15)$.

$\dfrac{(x-15)}{(x-15)} \cdot \dfrac{x}{(x+1)(x-1)}$

$- \dfrac{(x+1)}{(x+1)} \cdot \dfrac{3}{(x-1)(x-15)}$

$= \dfrac{x^2 - 15x}{(x+1)(x-1)(x-15)}$

$- \dfrac{3x+3}{(x+1)(x-1)(x-15)}$

$= \dfrac{x^2 - 18x - 3}{(x+1)(x-1)(x-15)}$

Group Activity/Challenge Problems

1. There are many possible answers. Write the x and y with any rational coefficients, then substitute $x = 2$ and $y = 3$ to obtain the constant. Repeat the process to get the second equation.

3. $\frac{x+2}{2} - \frac{y+4}{3} = 4$

$6\left(\frac{x+2}{2} - \frac{y+4}{3}\right) = (4)6$

$3(x+2) - 2(y+4) = 24$

$3x + 6 - 2y - 8 = 24$

$3x - 2y = 24 - 6 + 8$

$3x - 2y = 26$

$\frac{x+y}{2} = \frac{1}{2} + \frac{x-y}{3}$

$6\left(\frac{x+y}{2}\right) = \left(\frac{1}{2} + \frac{x-y}{3}\right)6$

$3(x+y) = 3 + 2(x-y)$

$3x + 3y = 3 + 2x - 2y$

$3x - 2x + 3y + 2y = 3$

$x + 5y = 3$

$$\begin{aligned} 3x - 2y &= 26 \\ \underline{-3x - 15y} &\underline{= -9} \\ -17y &= 17 \\ y &= -1 \end{aligned}$$

$3x - 2(-1) = 26$

$3x + 2 = 26$

$3x = 24$

$x = 8$

The solution is $(8, -1)$

4. $\frac{5x}{2} + 3y = \frac{9}{2} + y$

$2\left(\frac{5x}{2} + 3y\right) = \left(\frac{9}{2} + y\right)2$

$5x + 6y = 9 + 2y$

$5x + 6y - 2y = 9$

$5x + 4y = 9$

$\frac{1}{4}x - \frac{1}{2}y = 6x + 12$

$4\left(\frac{1}{4}x - \frac{1}{2}y\right) = (6x + 12)4$

$x - 2y = 24x + 48$

$-x + 24x + 2y = -48$

$23x + 2y = -48$

$$\begin{aligned} 5x + 4y &= 9 \\ \underline{-46x - 4y} &\underline{= 96} \\ -41x \quad &= 105 \\ x \quad &= -\frac{105}{41} \end{aligned}$$

$5\left(-\frac{105}{41}\right) + 4y = 9$

$-\frac{525}{41} + 4y = 9$

$4y = \frac{369}{41} + \frac{525}{41}$

$y = \frac{894}{41} \cdot \frac{1}{4}$

$y = \frac{447}{82}$

The solution is $\left(-\frac{105}{41}, \frac{447}{82}\right)$

6. $x+2y-z=2, 2x-y+z=3, 2x+y+z=7$

$x+2y-z=2$

$\underline{2x-y+z=3}$

$3x+y=5$

$2x-y+z=3$

$\underline{-2x-y-z=-7}$

$-2y=-4$

$y=2$

$3x+2=5$

$3x=3$

$x=1$

$1+2(2)-z=2$

$1+4-z=2$

$5-z=2$

$z=3$

The solution is (1, 2, 3).

Exercise Set 8.4

1. Let the two integers be x and y. Then the system of equations is

$x+y=26$

Sum of integers is 26.

$x=2y+2$

One integer is two more than twice the other.

Substitute $2y+2$ for x in the first equation.

$x+y=26$

$(2y+2)+y=26$

$3y+2=26$

$3y=24$

$y=8$

Solve for x: $x=2y+2$

$x=2(8)+2$

$x=16+2=18$

Thus the two integers are 8 and 18.

3. The system of equations is

$A+B=90$

Angles are complementary.

$B=A+18$

B is 18° greater than A.

Subtract A from both sides of the second equation and add.

$A+B=90$

$\underline{-A+B=18}$

$2B=108$

$B=54$

Solve for A: $A+B=90$

$A+54=90$

$A=36$

Thus $A=36°$ and $B=54°$.

5. The system of equations is

$A+B=180$

A and B are supplementary.

$A=B+52$

A is 52° greater than B.

Subtract B from both sides of the second equation and add.

$A+B=180$

$\underline{A-B=52}$

$2A=232$

$A=116$

Solve for B $A+B=180$

$116+B=180$

$B=64$

Thus A = 116° and B = 64°.

7. Let l = length, w = width.

The system of equations is

$800=2l+2w$

Perimeter is 800.

$l = w + 100$

Length is 100 greater than width.

Substitute $w + 100$ for l in the first equation.

$800 = 2(w + 100) + 2w$

$800 = 2w + 200 + 2w$

$600 = 4w$

$150 = w$

Therefore the width is 150 feet and the length is $150 + 100 = 250$ feet.

9. Let x = number of Eagles,

y = number of Maple Leafs

The system of equations is

$x + y = 14$

Robin has 14 coins.

$480x + 460y = 6560$

The total value of the collection is $6560.

Solve for x in the first equation.

$x = 14 - y$

Substitute $14 - y$ for x in the second equation.

$480(14 - y) + 460y = 6560$

$6720 - 480y + 460y = 6560$

$-20y = -160$

$y = 8$

Solve for x: $x + y = 14$

$x + 8 = 14$

$x = 6$

Robin has 6 Eagles, 8 Maple Leafs.

11. Let B = number of BancOne shares,

M = number of Microsoft shares.

The system of equations is

$B = 5M$ Karla bought five times as many BancOne shares as Microsoft shares

$37B + 75M = 7800$

Total cost of stock = 7800.

Substitute $5M$ for B in the second equation.

$37(5M) + 75M = 7800$

$185M + 75M = 7800$

$260M = 7800$

$M = 30$

Karla bought 30 Microsoft shares and $5(30) = 150$ BancOne shares.

13. Let k = speed of Kayak in still water,

c = speed of current.

The system of equations is

$k + c = 4.5$

Speed of kayak with current.

$k - c = 3.2$

Speed of kayak against current.

$k + c = 4.5$

$\underline{k - c = 3.2}$

$2k = 7.7$

$k = 3.85$

Solve for c: $k + c = 4.5$

$3.85 + c = 4.5$

$c = 0.65$

The speed of the kayak in still water is 3.85 mph, the speed of the current is 0.65 mph.

15. Let p = amount of merchandise in dollars,

c = total cost.

The system of equations is

$c = 50 + 0.85p$

Cost under plan A.

$c = 100 + 0.80p$

Cost under plan B.

Cost under plan A = cost under plan B

$50 + 0.85p = 100 + 0.80p$

$0.85p = 50 + 0.80p$

$0.05p = 50$

$p = 1000$

To pay the same amount under both plans one would have to purchase \$1000 of merchandise.

17. Note: The solution in the text uses s for sales and c for salary. The equations and the answers are the same, except that c in the text is s here, and vice versa.

Let s = total salary,

c = amount of sales (in dollars).

The system of equations is

$s = 0.40c$

Salary as an employee.

$s = 1.00c - 1500$

Salary of Jerry's own company.

(a) Salary as an employee = salary of Jerry's own company

$0.40c = 1.00c - 1500$

$-0.60c = -1500$

$c = 2500$

Jerry's company would have to make \$2500 in sales

(b) Let n = number of months to recover the \$6000

Then $n(3000 - 1500) = 6000$

$1500n = 6000$

$n = 4$

It would take Jerry 4 months to recover the \$6000.

19. Let x = principal invested at 10%,

y = principal invested at $5\frac{1}{4}\%$.

	Principal	Rate	Time	Interest
10% account	x	0.10	1	$0.10x$
$5\frac{1}{4}\%$ account	y	0.0525	1	$0.0525y$

The system of equations is

$x + y = 12{,}500$ Total principal invested is \$12,500.

$0.10x + 0.0525y = 1200$ Total interest is \$1200.

$100[0.10x + 0.0525y = 1200]$ gives $10x + 5.25y = 120{,}000$

$-10[x + y = 12{,}500]$ $\underline{-10x - 10y = -125{,}000}$

$-4.75y = -5000$

$y = 1052.63$

Solve for x: $x + y = 12{,}500$

$x + 1052.63 = 12{,}500$

$x = 11{,}447.37$

The Websters should invest at least \$11,447.37 in the 10% account.

21. Let T = speed of Teresa's boat, J = speed of Jill's boat.

Boat	Rate	Time	Distance
Teresa	T	3	$3T$
Jill	J	3.2	$3.2J$

The system of equations is

$T = J + 4$ Teresa's boat travels 4 mph faster than Jill's.

$3T = 3.2J$ Both boats travel the same distance.

Substitute $J + 4$ for T in the second equation.

$3T = 3.2J$

$3(J + 4) = 3.2J$

$3J + 12 = 3.2J$

$12 = 0.2J$

$60 = J$

The speed of Jill's boat is 60 mph. The speed of Teresa's boat is $60 + 4 = 64$ mph.

23. Let U = speed of United jet, D = speed of Delta jet

Plane	Rate	Time	Distance
United	U	3	$3U$
Delta	D	3	$3D$

The system of equations is

$U = D + 100$ United jet travels 100 mph faster than Delta jet.

$3U + 3D = 2700$ Distance traveled by United jet + distance traveled by Delta jet = 2700.

Substitute $D + 100$ for U in the second equation.

$$3U + 3D = 2700$$
$$3(D+100) + 3D = 2700$$
$$3D + 300 + 3D = 2700$$
$$6D = 2400$$
$$D = 400$$

The speed of the Delta jet is 400 mph, the speed of the United jet is 400 + 100 = 500 mph.

25. Let m = time Micki jogs, p = time Petra jogs.

Jogger	Rate	Time	Distance
Micki	5	m	$5m$
Petra	8	p	$8p$

The system of equations is

$m = p + 0.3$ Micki starts 0.3 hours before Petra.

$5m = 8p$ When Petra catches up, they have both traveled the same distance.

Substitute $p + 0.3$ for m in the second equation.

$$5m = 8p$$
$$5(p + 0.3) = 8p$$
$$1.5 = 3p$$
$$0.5 = p$$

It will take Petra 0.5 hour to catch up to Micki.

27. Let x = number of liters of 25% solution, y = number of liters of 50% solution.

	Number of liters	Concentration	Acid Content
25% solution	x	0.25	$0.25x$
50% solution	y	0.50	$0.50y$
Mixture	10	0.40	0.40(10)

The system of equations is

$x + y = 10$ Total volume is 10 liters.

$0.25x + 0.5y = 0.4(10)$ Acid content of 25% solution + acid content of 50% solution = acid content of mixture.

Solve for y in the first equation.

$x + y = 10$

$y = 10 - x$

Substitute $10 - x$ for y in the second equation.

$0.25x + 0.5y = 0.4(10)$

$0.25x + 0.5(10 - x) = 4$

$0.25x + 5 - 0.5x = 4$

$1 = 0.25x$

$4 = x$

Solve for y: $x + y = 10$

$4 + y = 10$

$y = 6$

She should mix 4 liters of 25% solution and 6 liters of 50% solution.

29. Let x = number of pounds of \$5 coffee, y = number of pounds of \$7 coffee.

Value of the \$5 coffee = $5x$

Value of the \$7 coffee = $7y$

The system of equations is

$x + y = 30$ Total weight = 30.

$5x + 7y = 160$ Value of \$5 coffee + value of \$7 coffee = 160.

Solve for y in the first equation.

$x + y = 30$

$y = 30 - x$

Substitute $30 - x$ for y in the second equation.

$5x + 7y = 160$

$5x + 7(30 - x) = 160$

$5x + 210 - 7x = 160$

$50 = 2x$

$25 = x$

Solve for y: $x + y = 30$

$25 + y = 30$

$y = 5$

She should use 25 pounds of the \$5 coffee and 5 pounds of the \$7 coffee.

31. Let x = number of ounces of apple juice, y = number of ounces of apple drink.

	Price	Number of ounces	Value
Juice	12	x	$12x$
Drink	6	y	$6y$
Mixture	10	8	10(8)

The system of equations is

$x + y = 8$ Total volume = 8 ounces.

$12x + 6y = 10(8)$ Value of juice + value of drink = value of mixture.

Solve for y in the first equation.

$x + y = 8$

$y = 8 - x$

Substitute $8 - x$ for y in the second equation.

$12x + 6y = 10(8)$

$12x + 6(8 - x) = 80$

$12x + 48 - 6x = 80$

$6x = 32$

$x = 5\frac{1}{3}$

Solve for y: $x + y = 8$

$$5\frac{1}{3} + y = 8$$

$$y = 2\frac{2}{3}$$

$2\frac{2}{3}$ ounces of apple drink and $5\frac{1}{3}$ ounces of apple juice will be used.

Cumulative Review Exercises: 8.4

33. Substitute 3 for x and -2 for y.

$3(4x-3)^2 - 2y^2 - 1$
$= 3[4(3)-3]^2 - 2(-2)^2 - 1$
$= 3[12-3]^2 - 2(4) - 1 = 3(9)^2 - 8 - 1$
$= 3(81) - 8 - 1 = 243 - 8 - 1$
$= 235 - 1 = 234$

34. $(3x^4)^{-2} = 3^{-2}x^{(4)(-2)} = \frac{1}{3^2}x^{-8} = \frac{1}{9x^8}$

35. **(a)** Yes, third.

(b) Yes, fourth.

(c) No, a polynomial cannot have a negative exponent on the variable.

36.

$$\begin{array}{r} 2x - 3 \\ x+4\overline{)\,2x^2 + 5x - 10} \\ \underline{-2x^2 - 8x} \\ -3x - 10 \\ \underline{+3x + 12} \\ 2 \end{array}$$

$(2x^2 + 5x - 10) \div (x+4) = 2x - 3 + \frac{2}{x+4}$

Group Activity/Challenge Problems

1. $P = 200 - 2m$

$P = 20 + 2m$

$200 - 2m = 20 + 2m$

$4m = 180$

$m = 45$

It will take 45 minutes.

3. $9t = d$ or $t = \frac{d}{9}$

$5t = d - \frac{1}{2}$ or $t = \frac{d}{5} - \frac{1}{10}$

$\frac{d}{9} = \frac{d}{5} - \frac{1}{10}$

$90\left(\frac{d}{9}\right) = \left(\frac{d}{5} - \frac{1}{10}\right)90$

$10d = 18d - 9$

$9 = 8d$

$d = \frac{9}{8} = 1.125$

The school is 1.125 miles from the boys' house.

4. Use any two of the following three equations to solve:

$x + y = 300$, $0.7x + 0.4y = 0.6(300)$,

$0.3x + 0.6y = 0.4(300)$

$x = 300 - y$

$0.7(300 - y) + 0.4y = 0.6(300)$

$210 - 0.7y + 0.4y = 180$

$0.3y = 30$

$y = 100$

$x = 300 - 100 = 200$

200 g of 1st alloy

100 g of 2nd alloy

Exercises Set 8.5

1. Graph both inequalities on the same set of axes.

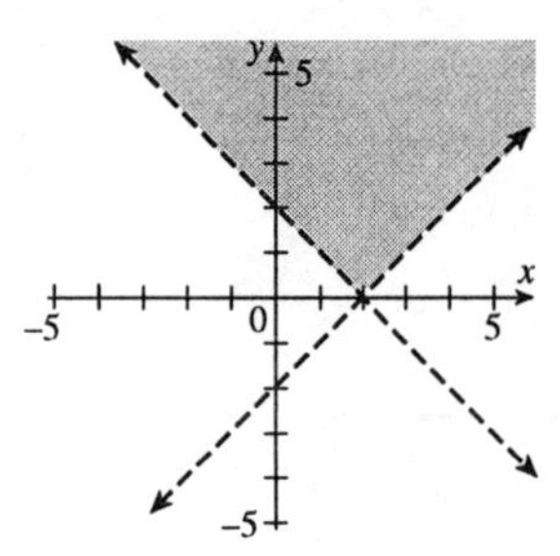

The solution is the part of the graph that contains shading.

3. Graph both inequalities on the same set of axes.

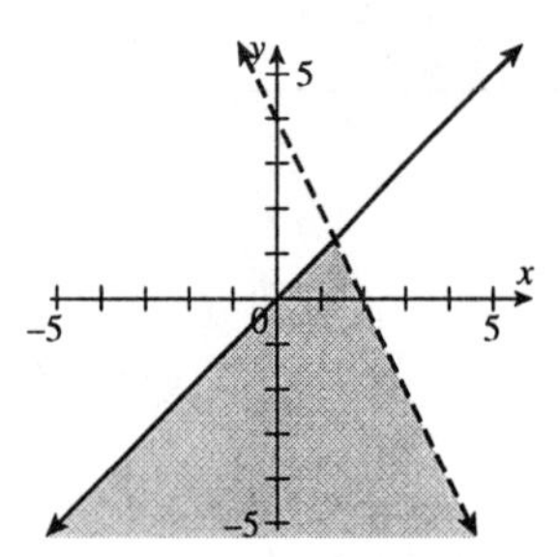

The solution is the part of the graph that contains the shading and the part of the solid line than satisfies both inequalities.

5. Graph both inequalities on the same set of axes.

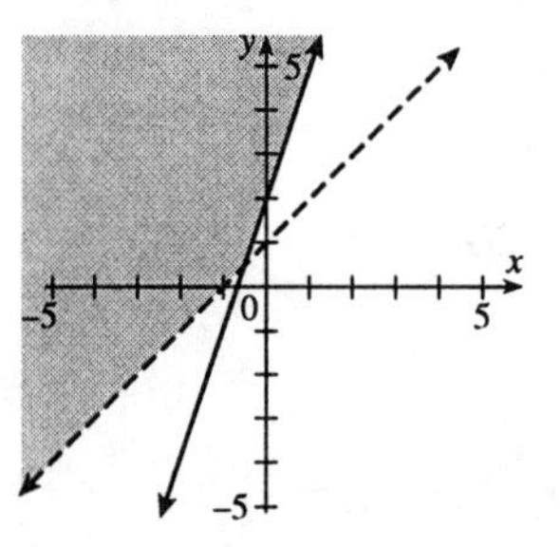

The solution is the part of the graph that contains the shading and the part of the solid line than satisfies both inequalities.

7. Graph both inequalities on the same set of axes.

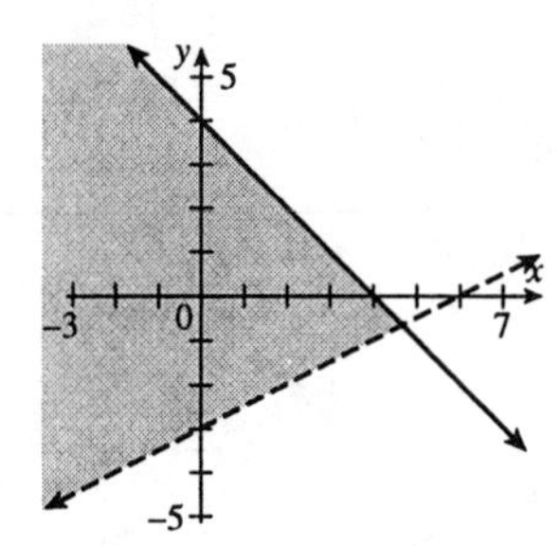

The solution is the part of the graph that contains the shading and the part of the solid line than satisfies both inequalities.

9. Graph both inequalities on the same set of axes.

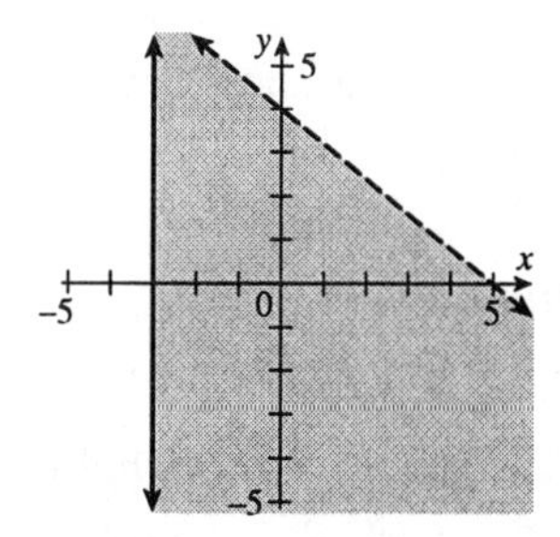

The solution is the part of the graph that contains the shading and the part of the solid line than satisfies both inequalities.

11. Graph both inequalities on the same set of axes.

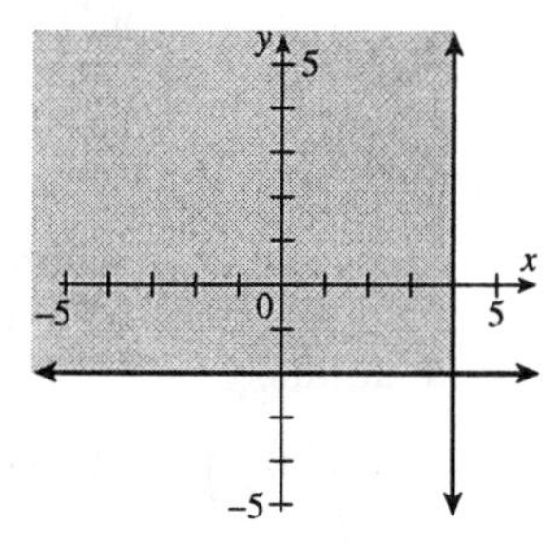

The solution is the part of the graph that contains the shading and the parts of the solid lines that satisfy both inequalities.

13. Graph both inequalities on the same set of axes.

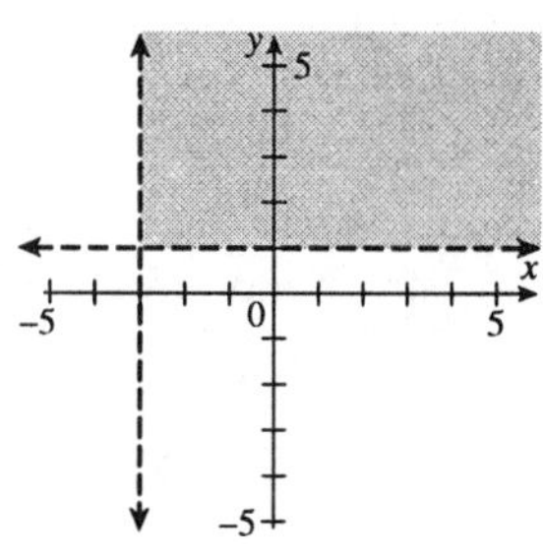

The solution is the part of the graph that contains the shading.

15. Graph both inequalities on the same set of axes.

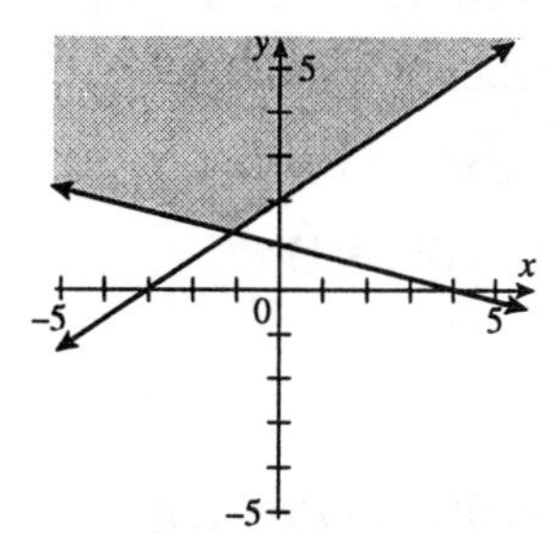

The solution is the part of the graph that contains the shading and the parts of the solid lines that satisfy both inequalities.

17. No, if two lines are not parallel then they intersect. When one side of each line is shaded, the shading must overlap; this is the solution.

Cumulative Review Exercises: 8.5

18. $(x^2 - 2x + 7)(3x - 4)$

$= x^2(3x - 4) - 2x(3x - 4) + 7(3x - 4)$

$= 3x^3 - 4x^2 - 6x^2 + 8x + 21x - 28$

$= 3x^3 - 10x^2 + 29x - 28$

19. Factor by grouping

$xy + x - 3y - 3 = x(y + 1) - 3(y + 1)$

$= (x - 3)(y + 1)$

20. Find two numbers whose product is 42 and whose sum is -13.

The two numbers are -6 and -7.

Thus, $x^2 - 13x + 42 = (x - 6)(x - 7)$.

21. Find two numbers whose product is $6(-2) = -12$ and whose sum is -1.

The two numbers are -4 and 3.

$6x^2 - x - 2 = 6x^2 + 3x - 4x - 2$

$= 3x(2x + 1) - 2(2x + 1)$

$= (3x - 2)(2x + 1)$

Group Activity/Challenge Problems

1. $x + 2y \le 6$

$2x - y < 2$

$y > 2$

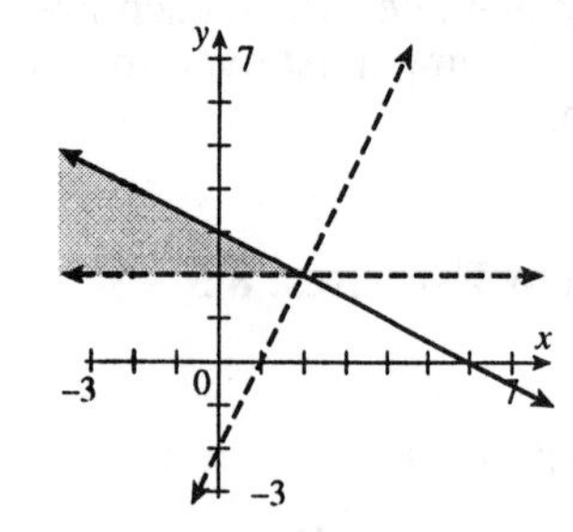

3. $x \ge 0,\ y \ge 0$

$5x + y \le 40$

$10x + y \le 60$

$x \le 4$

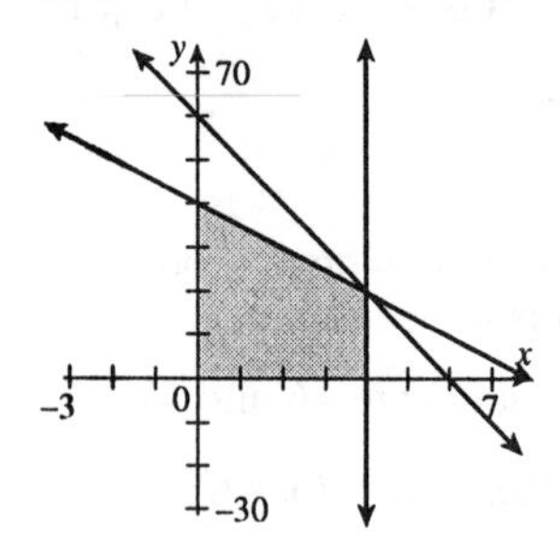

Chapter 8 Review Exercises

1. **(a)** Substitute 0 for x and -2 for y in each equation.

$y = 3x - 2$	$2x + 3y = 5$
$-2 = 3(0) - 2$	$2(0) + 3(-2) = 5$
$-2 = 0 - 2$	$0 + (-6) = 5$
$-2 = -2$ True	$-6 = 5$ False

Since $(0, -2)$ does not satisfy both equations, it is not a solution to the system of equations.

(b) Substitute 2 for x and 4 for y in each equation.

$y = 3x - 2$	$2x + 3y = 5$
$4 = 3(2) - 2$	$2(2) + 3(4) = 5$
$4 = 6 - 2$	$4 + 12 = 5$
$4 = 4$ True	$16 = 5$ False

Since $(2, 4)$ does not satisfy both equations, it is not a solution to the system of equations.

(c) Substitute 1 for x and 1 for y in each equation.

$y = 3x - 2$	$2x + 3y = 5$
$1 = 3(1) - 2$	$2(1) + 3(1) = 5$
$1 = 3 - 2$	$2 + 3 = 5$
$1 = 1$ True	$5 = 5$ True

Since $(1, 1)$ satisfies both equations, it is a solution to the system of equations.

2. **(a)** Substitute $\frac{5}{2}$ for x and $\frac{3}{2}$ for y in both equations.

$y = -x + 4$	$3x + 5y = 15$
$\frac{3}{2} = -\frac{5}{2} + 4$	$3\left(\frac{5}{2}\right) + 5\left(\frac{3}{2}\right) = 15$
$\frac{3}{2} = \frac{3}{2}$ True	$\frac{15}{2} + \frac{15}{2} = 15$
	$15 = 15$ True

Since $\left(\frac{5}{2}, \frac{3}{2}\right)$ satisfies both equations, it is a solution to the system of equations.

(b) Substitute 0 for x and 4 for y in each equation.

$y = -x + 4$

$4 = -0 + 4$

$4 = 4$ True

$3x + 5y = 15$

$3(0) + 5(4) = 15$

$0 + 20 = 15$

$20 = 15$ False

Since (0, 4) does not satisfy both equations, it is not a solution to the system of equations.

(c) Substitute $\frac{1}{2}$ for x and $\frac{3}{5}$ for y in each equation.

$y = -x + 4$

$\frac{3}{5} = -\frac{1}{2} + 4$

$\frac{3}{5} = \frac{7}{2}$ False

$3x + 5y = 15$

$3\left(\frac{1}{2}\right) + 5\left(\frac{3}{5}\right) = 15$

$\frac{3}{2} + 3 = 15$

$\frac{9}{2} = 15$ False

Since $\left(\frac{1}{2}, \frac{3}{5}\right)$ does not satisfy both equations, it is not a solution to the system of equations.

3. The lines are not parallel and intersect at one point. The system is consistent and has exactly one solution.

4. Lines 1 and 2 are two different parallel lines and do not intersect. The system is inconsistent and has no solution.

5. Lines 1 and 2 are the same line. The system is dependent and has an infinite number of solutions.

6. The lines are not parallel and intersect at one point. The system is consistent and has exactly one solution.

7. $x + 2y = 8$

$2y = -x + 8$

$y = -\frac{1}{2}x + 4$

$3x + 6y = 12$

$6y = -3x + 12$

$y = -\frac{1}{2}x + 2$

The equations have the same slope and different y-intercepts. The lines are thus parallel and the system of equations has no solution.

8. $y = -3x - 6$

$2x + 3y = 8$

$3y = -2x + 8$

$y = -\frac{2}{3}x + \frac{8}{3}$

The equations have different slopes and the lines are therefore not parallel. The system of equations has exactly one solution.

9. $y = \frac{1}{2}x - 4$

$x - 2y = 8$

$-2y = -x + 8$

$y = \frac{1}{2}x - 4$

The equations have the same slope and the same y-intercept and thus represent the same line. The system of equations has an infinite number of solutions.

10. $6x = 4y - 8$

$6x + 8 = 4y$

$\frac{3}{2}x + 2 = y$

$4x = 6y + 8$

$4x - 8 = 6y$

$\frac{2}{3}x - \frac{4}{3} = y$

The equations have different slopes and the lines are therefore not parallel. The system of equations has exactly one solution.

11. $y = x + 3$ — Ordered Pair

Let $x = 0$, then $y = 3$ — $(0, 3)$

Let $y = 0$, then $x = -3$ — $(-3, 0)$

$y = 2x + 5$ — Ordered Pair

Let $x = 0$, then $y = 5$ — $(0, 5)$

Let $y = 0$, then $x = -\frac{5}{2}$ — $\left(-\frac{5}{2}, 0\right)$

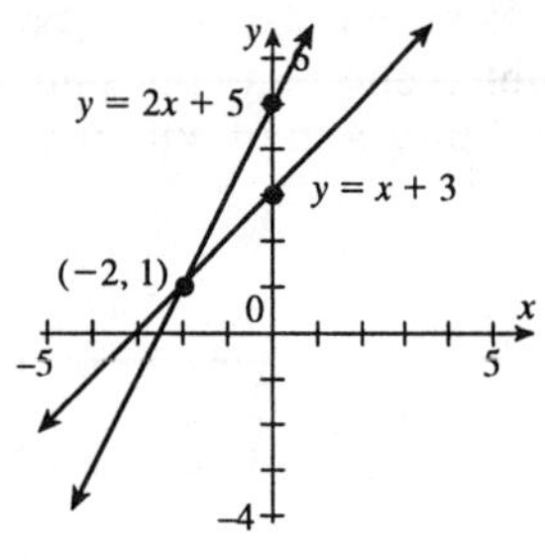

The graphs intersect at the point $(-2, 1)$.

Check: $y = x + 3$ $y = 2x + 5$

$1 = -2 + 3$ $1 = 2(-2) + 5$

$1 = 1$ True $1 = -4 + 5$

$1 = 1$ True

Since $(-2, 1)$ satisfies both equations, it is a solution to the system of equations.

12. $x = -2$ is a vertical line.

$y = 3$ is a horizontal line.

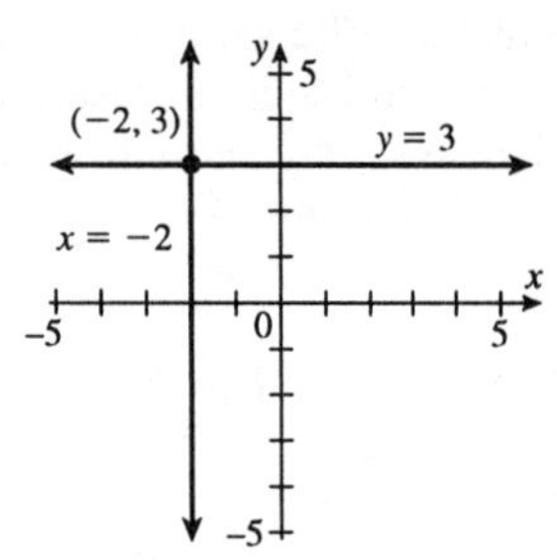

The graphs intersect at the point $(-2, 3)$.

Check: $x = -2$ $y = 3$

$-2 = -2$ True $3 = 3$ True

Since $(-2, 3)$ satisfies both equations, it is a solution to the system of equations.

13. $y = 3$, horizontal line.

$y = -2x + 5$ — Ordered Pair

Let $x = 0$, then $y = 5$ — $(0, 5)$

Let $y = 0$, then $x = \frac{5}{2}$ — $\left(\frac{5}{2}, 0\right)$

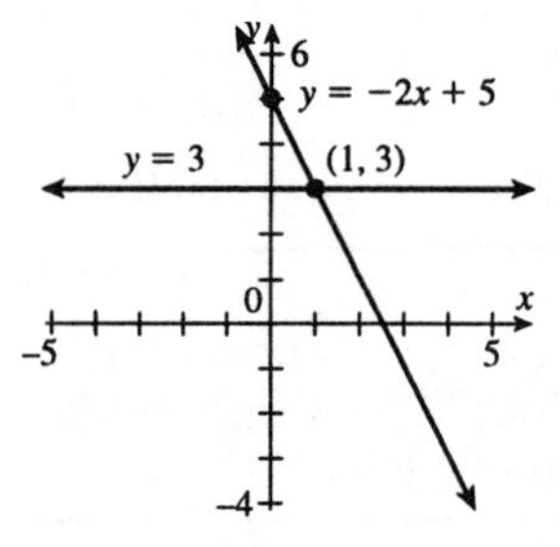

The graphs intersect at the point $(1, 3)$.

Check: $y = 3$ $y = -2x + 5$

$3 = 3$ True $3 = -2(1) + 5$

$3 = -2 + 5$

$3 = 3$ True

Since $(1, 3)$ satisfies both equations, it is a solution to the system of equations.

14. $x + 3y = 6$ — Ordered Pair

Let $x = 0$, then $y = 2$ — $(0, 2)$

Let $y = 0$, then $x = 6$ — $(6, 0)$

$y = 2$, horizontal line.

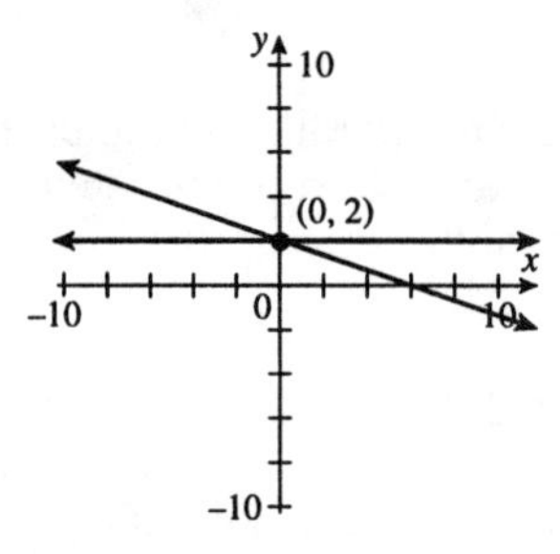

The graphs intersect at $(0, 2)$.

Check: $x + 3y = 6$ $y = 2$

$0 + 3(2) = 6$ $2 = 2$ True

$6 = 6$ True

Since $(0, 2)$ satisfies both equations, it is a solution to the system of equations.

15. $x+2y=8$ | Ordered Pair

Let $x=0$, then $y=4$ | $(0, 4)$

Let $y=0$, then $x=8$ | $(8, 0)$

$2x-y=-4$ | Ordered Pair

Let $x=0$, then $y=4$ | $(0, 4)$

Let $y=0$, then $x=-2$ | $(-2, 0)$

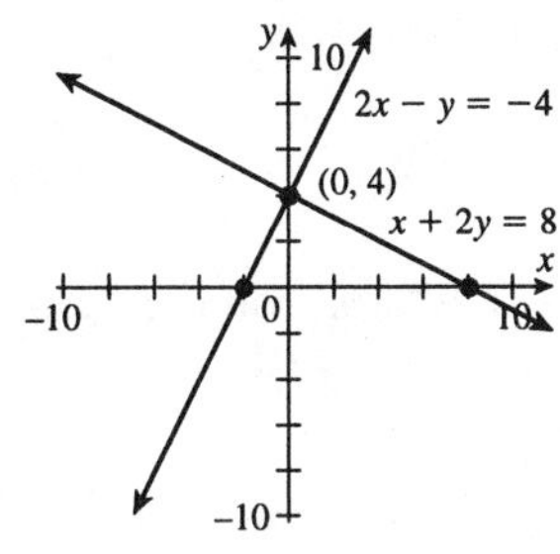

The graphs intersect at the point (0, 4).

Check: $x+2y=8$ | $2x-y=-4$

$0+2(4)=8$ | $2(0)-4=-4$

$8=8$ True | $-4=-4$ True

Since (0, 4) satisfies both equations, it is a solution to the system of equations.

16. $y=x-3$ | Ordered Pair

Let $x=0$, then $y=-3$ | $(0, -3)$

Let $y=0$, then $x=3$ | $(3, 0)$

$2x-2y=6$ | Ordered Pair

Let $x=0$, then $y=-3$ | $(0, -3)$

Let $y=0$, then $x=3$ | $(3, 0)$

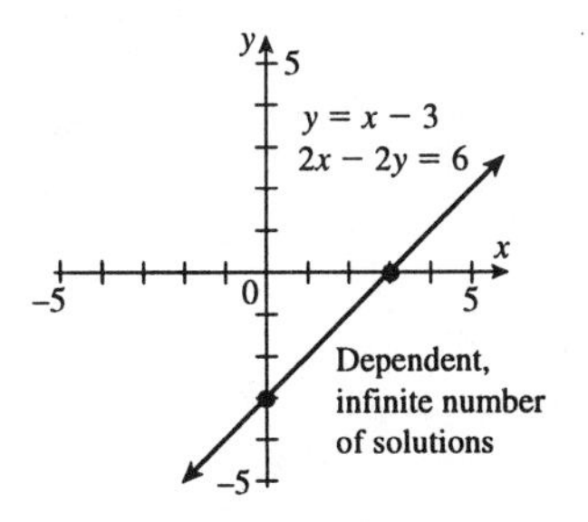

The lines are the same.

Check: $y=x-3$ | $2x-2y=6$

$-2y=-2x+6$

$y=x-3$

The system of equations is dependent and has an infinite number of solutions.

17. $2x+y=0$ | Ordered Pair

Let $x=0$, then $y=0$ | $(0, 0)$

Let $x=-1$, then $y=2$ | $(-1, 2)$

$4x-3y=10$ | Ordered Pair

Let $x=0$, then $y=-\frac{10}{3}$ | $\left(0,\ -\frac{10}{3}\right)$

Let $y=0$, then $x=\frac{5}{2}$ | $\left(\frac{5}{2},\ 0\right)$

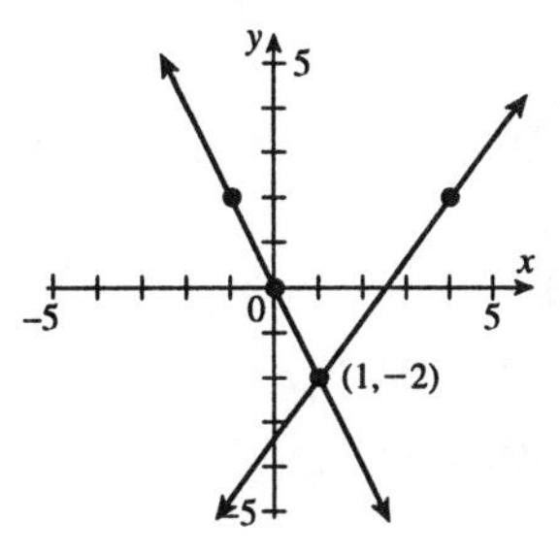

The graphs intersect at the point (1, −2).

Check: $2x+y=0$ | $4x-3y=10$

$2(1)+(-2)=0$ | $4(1)-3(-2)=10$

$2-2=0$ | $4+6=10$

$0=0$ True | $10=10$ True

Since (1, −2) satisfies both equations, it is a solution to the system of equations.

18. $x+2y=4$ | Ordered Pair

Let $x=0$, then $y=2$ | $(0, 2)$

Let $y=0$, then $x=4$ | $(4, 0)$

$\frac{1}{2}x+y=-2$ | Ordered Pair

Let $x=0$, then $y=-2$ | $(0, -2)$

Let $y=0$, then $x=-4$ | $(-4, 0)$

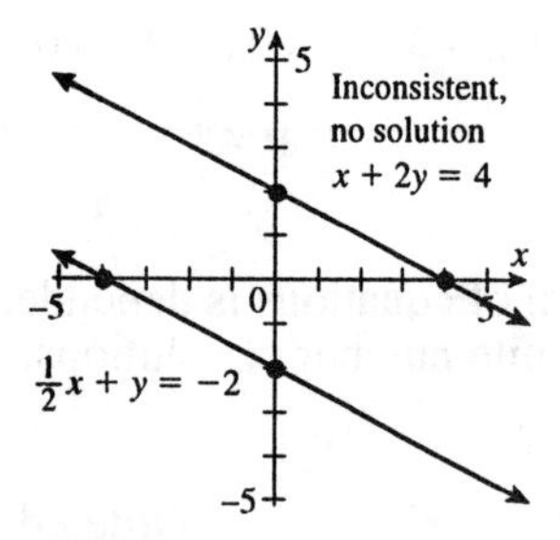

The lines do not intersect.

Check: $x+2y=4$ $\quad \frac{1}{2}x+y=-2$

$2y=-x+4 \quad y=-\frac{1}{2}x-2$

$y=-\frac{1}{2}x+2$

The system of equations is inconsistent and has no solution.

19. Substitute $2x-8$ for y in the second equation.

$2x-5y=0$

$2x-5(2x-8)=0$

$2x-10x+40=0$

$40=8x$

$5=x$

Solve for y.

$y=2x-8$

$y=2(5)-8$

$y=10-8=2$

The solution is (5, 2).

20. Substitute $3y-9$ for x in the second equation.

$x+2y=1$

$(3y-9)+2y=1$

$3y-9+2y=1$

$5y=10$

$y=2$

Solve for x.

$x=3y-9$

$x=3(2)-9$

$x=6-9=-3$

The solution is (–3, 2).

21. Solve for y in the first equation.

$2x+y=5$

$y=-2x+5$

Substitute $-2x+5$ for y in the second equation.

$3x+2y=8$

$3x+2(-2x+5)=8$

$3x-4x+10=8$

$-x+10=8$

$-x=-2$

$x=2$

Solve for y.

$y=-2x+5$

$y=-2(2)+5$

$y=-4+5=1$

The solution is (2, 1).

22. Solve for x in the second equation.

$x+2y=13$

$x=13-2y$

Substitute $13-2y$ for x in the first equation.

$2x-y=6$

$2(13-2y)-y=6$

$26-4y-y=6$

$26-5y=6$

$-5y=-20$

$y=4$

Solve for x.

$x = 13 - 2y$

$x = 13 - 2(4)$

$x = 13 - 8 = 5$

The solution is (5, 4).

23. Solve for y in the first equation.

$3x + y = 17$

$y = -3x + 17$

Substitute $-3x + 17$ for y in the second equation.

$2x - 3y = 4$

$2x - 3(-3x + 17) = 4$

$2x + 9x - 51 = 4$

$11x - 51 = 4$

$11x = 55$

$x = 5$

Solve for y.

$y = -3x + 17$

$y = -3(5) + 17$

$y = -15 + 17 = 2$

The solution is (5, 2).

24. Substitute $-3y$ for x in the second equation.

$x + 4y = 6$

$(-3y) + 4y = 6$

$-3y + 4y = 6$

$y = 6$

Solve for x.

$x = -3y$

$x = -3(6) = -18$

The solution is (−18, 6).

25. Substitute $2x + 3$ for y in the first equation.

$4x - 2y = 10$

$4x - 2(2x + 3) = 10$

$4x - 4x - 6 = 10$

$-6 = 10$ False

Since the statement $-6 = 10$ is false, the system is inconsistent and has no solution.

26. Solve for x in the first equation.

$2x + 4y = 8$

$2x = 8 - 4y$

$x = 4 - 2y$

Substitute $4 - 2y$ for x in the second equation.

$4x + 8y = 16$

$4(4 - 2y) + 8y = 16$

$16 - 8y + 8y = 16$

$16 = 16$

Since the statement $16 = 16$ is true, the system is dependent and has an infinite number of solutions.

27. Solve for x in the first equation.

$2x - 3y = 8$

$2x = 3y + 8$

$x = \frac{3}{2}y + 4$

Substitute $\frac{3}{2}y + 4$ for x in the second equation.

$6x + 5y = 10$

$6\left(\frac{3}{2}y + 4\right) + 5y = 10$

$9y + 24 + 5y = 10$

$14y + 24 = 10$

$14y = -14$

$y = -1$

Solve for x.

$x = \frac{3}{2}y + 4$

$x = \frac{3}{2}(-1) + 4$

$x = -\frac{3}{2} + 4 = \frac{5}{2}$

The solution is $\left(\frac{5}{2}, -1\right)$.

28. Solve for x in the second equation.

$x + 2y = 8$

$x = -2y + 8$

Substitute $-2y + 8$ for x in the first equation.

$4x - y = 6$

$4(-2y + 8) - y = 6$

$-8y + 32 - y = 6$

$-9y + 32 = 6$

$-9y = -26$

$y = \frac{26}{9}$

Solve for x.

$x = -2y + 8$

$x = -2\left(\frac{26}{9}\right) + 8$

$x = -\frac{52}{9} + 8 = \frac{20}{9}$

The solution is $\left(\frac{20}{9}, \frac{26}{9}\right)$.

29.

$$\begin{array}{l} x + y = 6 \\ \underline{x - y = 10} \\ 2x \quad = 16 \end{array}$$

$x = 8$

Solve for y.

$x + y = 6$

$8 + y = 6$

$y = -2$

The solution is $(8, -2)$.

30.

$$\begin{array}{l} x + 2y = -3 \\ \underline{2x - 2y = \ 6} \\ 3x \quad = 3 \end{array}$$

$x = 1$

Solve for y.

$x + 2y = -3$

$1 + 2y = -3$

$2y = -4$

$y = -2$

The solution is $(1, -2)$.

31.

$$\begin{array}{lcr} 2x + 3y = 4 & \text{gives} & 2x + 3y = \ 4 \\ -2[x + 2y = -6] & & \underline{-2x - 4y = \ 12} \\ & & -y = \ 16 \\ & & y = -16 \end{array}$$

Solve for x.

$2x + 3y = 4$

$2x + 3(-16) = 4$

$2x - 48 = 4$

$2x = 52$

$x = 26$

The solution is $(26, -16)$.

32. $-1[x + y = 12]$ gives $-x - y = -12$

$$\begin{array}{ll} 2x + y = 5 & \underline{2x + y = \quad 5} \\ & x \quad = -7 \end{array}$$

Solve for y.

$x + y = 12$

$-7 + y = 12$

$y = 19$

The solution is $(-7, 19)$.

33. $4x - 3y = 8$ gives $4x - \quad 3y = \quad 8$

$$\begin{array}{ll} -2[2x + 5y = 8] & \underline{-4x - 10y = -16} \\ & -13y = \ -8 \\ & y = \dfrac{8}{13} \end{array}$$

Solve for x.

$4x - 3y = 8$

$4x - 3\left(\dfrac{8}{13}\right) = 8$

$4x - \dfrac{24}{13} = 8$

$4x = \dfrac{128}{13}$

$x = \dfrac{32}{13}$

The solution is $\left(\dfrac{32}{13}, \dfrac{8}{13}\right)$.

34. $-1[-2x + 3y = 15]$ gives $2x - 3y = -15$

$$\begin{array}{ll} 3x + 3y = 10 & \underline{3x + 3y = \ 10} \\ & 5x \quad = -5 \\ & x = -1 \end{array}$$

Solve for y.

$3x + 3y = 10$

$3(-1) + 3y = 10$

$-3 + 3y = 10$

$3y = 13$

$y = \dfrac{13}{3}$

The solution is $\left(-1, \dfrac{13}{3}\right)$.

35. $2[2x + y = 9]$ gives $4x + 2y = 18$

$$\begin{array}{ll} -4x - 2y = 4 & \underline{-4x - 2y = \ 4} \\ & 0 = 22 \end{array}$$

Since a false statement is obtained, the system is inconsistent and has no solution.

36. Align the x and y terms on the left side of the equation.

$y = 4x - 3$

$-4x + y = -3$

$2[2x + 2y = 8]$ gives $4x + 4y = 16$

$$\begin{array}{ll} -4x + y = -3 & \underline{-4x + y = -3} \\ & 5y = 13 \\ & y = \dfrac{13}{5} \end{array}$$

Solve for x.

$2x + 2y = 8$

$2x + 2\left(\dfrac{13}{5}\right) = 8$

$2x + \dfrac{26}{5} = 8$

$2x = \dfrac{14}{5}$

$x = \dfrac{7}{5}$

The solution is $\left(\dfrac{7}{5}, \dfrac{13}{5}\right)$.

37. $2[3x+4y=10]$ gives $6x+8y=20$

$-6x-8y=-20 \qquad \underline{-6x-8y=-20}$

$0=0$

Since a true statement is obtained, the system is dependent and has an infinite number of solutions.

38. $3[2x-5y=12]$ gives $6x-15y=36$

$-2[3x-4y=-6] \qquad \underline{-6x+8y=12}$

$-7y=48$

$y=-\frac{48}{7}$

Solve for x.

$2x-5y=12$

$2x-5\left(-\frac{48}{7}\right)=12$

$2x+\frac{240}{7}=12$

$2x=-\frac{156}{7}$

$x=-\frac{78}{7}$

The solution is $\left(-\frac{78}{7}, -\frac{48}{7}\right)$.

39. Let x and y be the two integers.

The system of equations is

$x+y=48$

The sum of the integers is 48.

$y=2x-3$

The larger is 3 less than twice the smaller.

Substitute $2x-3$ for y in the first equation.

$x+y=48$

$x+(2x-3)=48$

$3x-3=48$

$3x=51$

$x=17$

Solve for y.

$y=2x-3$

$y=2(17)-3$

$y=34-3=31$

Thus the two integers are 17 and 31.

40. Let p = speed of the plane in still air, w = speed of the wind.

The system of equations is

$p+w=600$

Speed of plane with wind.

$p-w=530$

Speed of plane against wind.

$p+w=600$

$\underline{p-w=530}$

$2p \quad =1130$

$p \quad = 565$

Solve for w.

$p+w=600$

$565+w=600$

$w=35$

Speed of plane is still air = 565 mph, speed of wind = 35 mph.

41. Let m = number of miles,

c = total cost.

The system of equations is

$c=20+0.50m$

Cost with ABC.

$c=35+0.40m$

Cost with Murtz.

Cost with ABC = cost with Murtz

$20+0.50m=35+0.40m$

$0.50m=15+0.40m$

$0.10m=15$

$m=150$

You would have to travel 150 miles for the total cost of both companies to be the same.

42. Let x = principal invested at 4%, y = principal invested at 6%.

Interest earned in 1 year from 4% account $= 0.04x$

Interest earned in 1 year from 6% account $= 0.06y$

The system of equations is $x + y = 16{,}000$

Total principal invested is \$16,000.

$0.04x + 0.06y = 760$

Total interest is \$760.

$100[0.04x + 0.06y = 760]$ gives

$-4[x + y = 16{,}000]$

$$\begin{aligned} 4x + 6y &= 76{,}000 \\ \underline{-4x - 4y} &= \underline{-64{,}000} \\ 2y &= 12{,}000 \\ y &= 6000 \end{aligned}$$

Solve for x.

$x + y = 16{,}000$

$x + 6000 = 16{,}000$

$x = 10{,}000$

They invested \$10,000 at 4% and \$6000 at 6%.

43. Let R = speed of Ron's car, A = speed of Audra's car.

Distance traveled by Ron in 5 hours $= 5R$, distance traveled by Audra in 5 hours $= 5A$. The system of equations is

$A = R + 6$

Audra's speed was 6 mph greater than Ron's speed.

$5R + 5A = 600$

Distance traveled by Ron + distance traveled by Audra = 600.

Substitute $R + 6$ for A in the second equation.

$5R + 5A = 600$

$5R + 5(R + 6) = 600$

$5R + 5R + 30 = 600$

$10R = 570$

$R = 57$

The average speed of Ron's car is 57 mph and the average speed of Audra's car is $57 + 6 = 63$ mph.

44. Let G = number of pounds of Green Turf, A = number of pounds of Agway.

Price of Green Turf seed is $0.60G$.

Price of Agway seed is $0.45A$

The system of equations is

$G + A = 40$

Total weight = 40.

$0.60G + 0.45A = 20.25$

Price of Green Turf seed + price of Agway seed = 20.25

Solve for G in the first equation.

$G + A = 40$

$G = 40 - A$

Substitute $40 - A$ for G in the second equation.

$0.60G + 0.45A = 20.25$

$0.60(40 - A) + 0.45A = 20.25$

$24 - 0.6A + 0.45A = 20.25$

$3.75 = 0.15A$

$25 = A$

Solve for G.

$G + A = 40$

$G + 25 = 40$

$G = 15$

15 lb of Green Turf and 25 lb of Agway were used.

45. Let x = number of liters of 30% solution, y = number of liters of 50% solution.

Acid content of 30% solution = $0.3x$

Acid content of 50% solution = $0.5y$

Acid content of 40% mixture = $0.40(6)$

The system of equations is

$x + y = 6$

Total volume is 6 liters

$0.3x + 0.5y = 0.4(6)$

Acid content of 30% solution + acid content of 50% solution = acid content of mixture.

Solve for y in the first equation.

$x + y = 6$

$y = 6 - x$

Substitute $6 - x$ for y in the second equation.

$0.3x + 0.5y = 0.4(6)$

$0.3x + 0.5(6 - x) = 2.4$

$0.3x + 3 - 0.5x = 2.4$

$3 - 0.2x = 2.4$

$0.6 = 0.2x$

$3 = x$

Solve for y.

$x + y = 6$

$3 + y = 6$

$y = 3$

3 liters of each solution should be mixed.

46. Graph both inequalities on the same set of axes.

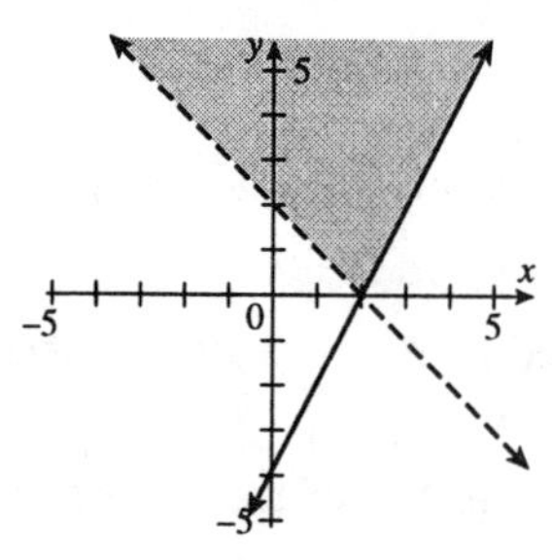

The solution is the part of the graph that contains the shading and the part of the solid line that satisfies both inequalities.

47. Graph both inequalities on the same set of axes.

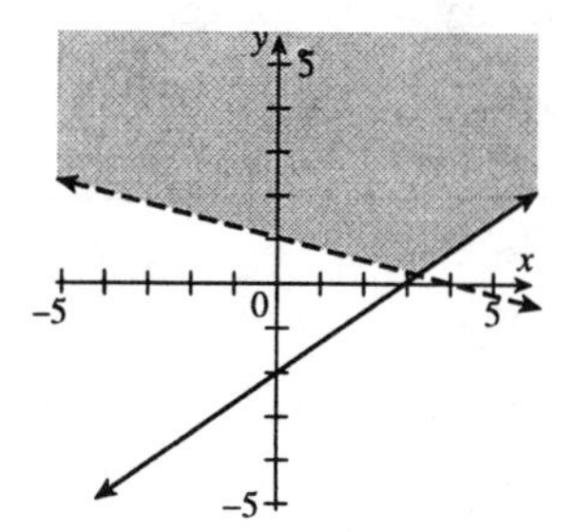

The solution is the part of the graph that contains the shading and the part of the solid line that satisfies both inequalities.

48. Graph both inequalities on the same set of axes.

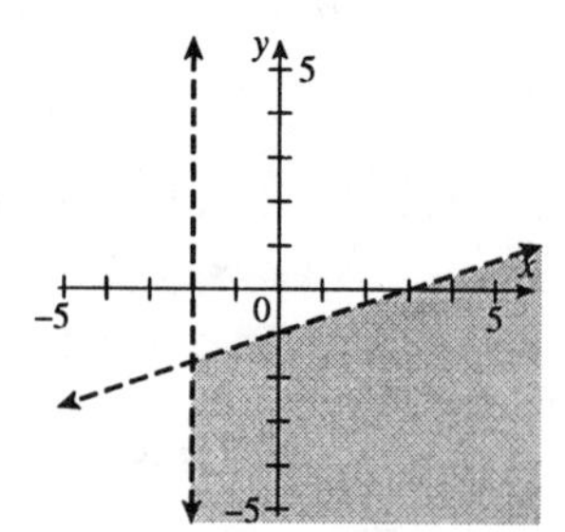

The solution is the part of the graph that contains the shading.

49. Graph both inequalities on the same set of axes.

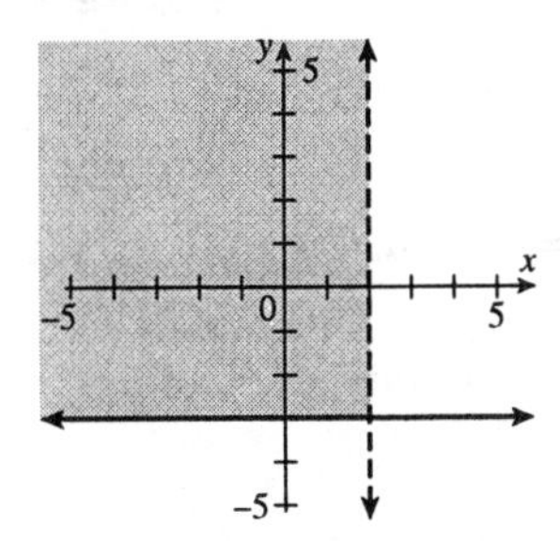

The solution is the part of the graph that contains the shading and the part of the solid line that satisfies both inequalities.

Chapter 8 Practice Test

1. **(a)** Substitute 0 for x and –6 for y in each equation.

$x+2y=-6$	$3x+2y=-12$
$0+2(-6)=-6$	$3(0)+2(-6)=-12$
$0-12=-6$	$0-12=-12$
$-12=-6$ False	$-12=-12$ True

Since (0, –6) does not satisfy both equations, it is not a solution to the system of equations.

(b) Substitute –3 for x and $-\frac{3}{2}$ for y in each equation.

$x+2y=-6$

$-3+2\left(-\frac{3}{2}\right)=-6$

$-3+(-3)=-6$

$-6=-6$ True

$3x+2y=-12$

$3(-3)+2\left(-\frac{3}{2}\right)=-12$

$-9+(-3)=-12$

$-12=-12$ True

Since $\left(-3,\ -\frac{3}{2}\right)$ satisfies both equations, it is a solution to the system of equations.

(c) Substitute 2 for x and –4 for y in each equation.

$x+2y=-6$	$3x+2y=-12$
$2+2(-4)=-6$	$3(2)+2(-4)=-12$
$2+(-8)=-6$	$6+(-8)=-12$
$-6=-6$ True	$-2=-12$ False

Since (2, –4) does not satisfy both equations, it is not a solution to the system of equations.

2. Lines 1 and 2 are two different parallel lines and do not intersect. The system is inconsistent and has no solution.

3. The lines are not parallel and intersect at one point. The system is consistent and has exactly one solution.

4. Lines 1 and 2 are the same line. The system is dependent and has an infinite number of solutions.

5.

$3y=6x-9$	$2x-y=6$
$y=2x-3$	$-y=-2x+6$
	$y=2x-6$

The equations have the same slope and different y-intercepts. The lines are therefore parallel and the system of equations has no solution.

6.

$3x+2y=10$	$3x-2y=10$
$2y=-3x+10$	$-2y=-3x+10$
$y=-\frac{3}{2}x+5$	$y=\frac{3}{2}x-5$

The equations have different slopes so the lines are not parallel. The system of equations has exactly one solution.

7. $y = 3x - 2$ — Ordered Pair

Let $x = 0$, then $y = -2$ — $(0, -2)$

Let $y = 0$, then $x = \frac{2}{3}$ — $\left(\frac{2}{3}, 0\right)$

$y = -2x + 8$ — Ordered Pair

Let $x = 0$, then $y = 8$ — $(0, 8)$

Let $y = 0$, then $x = 4$ — $(4, 0)$

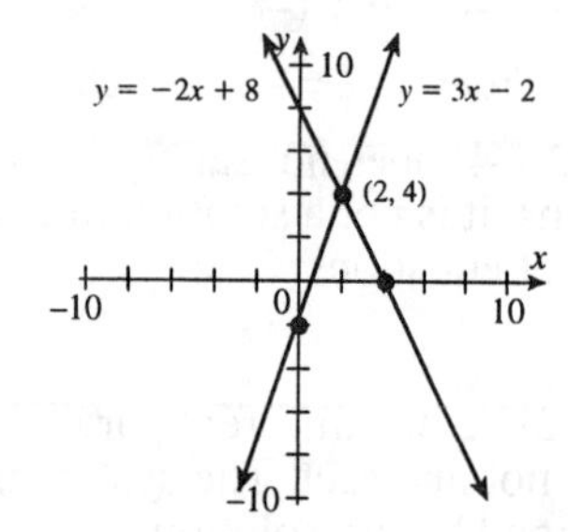

The graphs intersect at the point (2, 4).

Check: $y = 3x - 2$ — $y = -2x + 8$

$4 = 3(2) - 2$ — $4 = -2(2) + 8$

$4 = 6 - 2$ — $4 = -4 + 8$

$4 = 4$ True — $4 = 4$ True

Since (2, 4) satisfies both equations, it is a solution to the system of equations.

8. $3x - 2y = -3$ — Ordered Pair

Let $x = 0$, then $y = \frac{3}{2}$ — $\left(0, \frac{3}{2}\right)$

Let $y = 0$, then $x = -1$ — $(-1, 0)$

$3x + y = 6$ — Ordered Pair

Let $x = 0$, then $y = 6$ — $(0, 6)$

Let $y = 0$, then $x = 2$ — $(2, 0)$

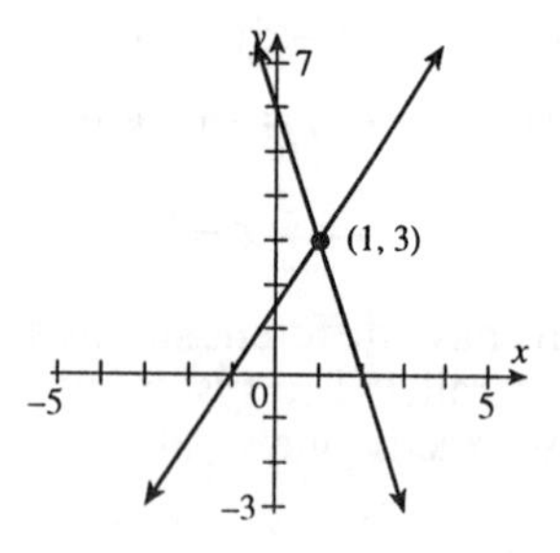

The graphs intersect at the point (1, 3).

Check: $3x - 2y = -3$ — $3x + y = 6$

$3(1) - 2(3) = -3$ — $3(1) + 3 = 6$

$3 - 6 = -3$ — $3 + 3 = 6$

$-3 = -3$ True — $6 = 6$ True

Since (1, 3) satisfies both equations, it is a solution to the system of equations.

9. Solve for y in the first equation.

$3x + y = 8$

$y = -3x + 8$

Substitute $-3x + 8$ for y in the second equation.

$x - y = 6$

$x - (-3x + 8) = 6$

$x + 3x - 8 = 6$

$4x = 14$

$x = \frac{7}{2}$

Solve for y.

$x - y = 6$

$\frac{7}{2} - y = 6$

$-y = \frac{5}{2}$

$y = -\frac{5}{2}$

The solution is $\left(\frac{7}{2}, -\frac{5}{2}\right)$.

10. Solve for x in the first equation.

$4x - 3y = 9$

$4x = 3y + 9$

$x = \frac{3}{4}y + \frac{9}{4}$

Substitute $\frac{3}{4}y + \frac{9}{4}$ for x in the second equation.

$2x + 4y = 10$

$2\left(\frac{3}{4}y + \frac{9}{4}\right) + 4y = 10$

$\frac{3}{2}y + \frac{9}{2} + 4y = 10$

$\frac{11}{2}y = \frac{11}{2}$

$y = 1$

Solve for x.

$4x - 3y = 9$

$4x - 3(1) = 9$

$4x - 3 = 9$

$4x = 12$

$x = 3$

The solution is (3, 1).

11. $2x + y = 5$ gives $2x + y = 5$

$-2[x + 3y = -10]$ $\underline{-2x - 6y = 20}$

$-5y = 25$

$y = -5$

Solve for x.

$2x + y = 5$

$2x + (-5) = 5$

$2x = 10$

$x = 5$

The solution is (5, –5).

12. $2[3x + 2y = 12]$ gives $6x + 4y = 24$

$3[-2x + 5y = 8]$ $\underline{-6x + 15y = 24}$

$19y = 48$

$y = \frac{48}{19}$

Solve for x.

$3x + 2y = 12$

$3x + 2\left(\frac{48}{19}\right) = 12$

$3x + \frac{96}{19} = 12$

$3x = \frac{132}{19}$

$x = \frac{44}{19}$

The solution is $\left(\frac{44}{19}, \frac{48}{19}\right)$.

13. Let x = number of miles, c = total cost.

The system of equations is

$c = 40 + 0.08x$

Total cost with Budget.

$c = 45 + 0.03x$

Total cost with Hertz.

cost with Budget = cost with Hertz

$40 + 0.08x = 45 + 0.03x$

$0.08x = 5 + 0.03x$

$0.05x = 5$

$x = 100$

100 miles have to be driven for the cost with the two companies to be the same.

14. Let x = number of pounds of cashews, y = number of pounds of peanuts.

The price of the cashews = $6x$.

The price of the peanuts = $4.5y$.

The price of the mixture = $5(20)$.

The system of equations is

$x + y = 20$

Total weight is 20 pounds.

$6x + 4.5y = 5(20)$

Price of cashews + price of peanuts = price of mixture

Solve for y in the first equation.

$x + y = 20$

$y = 20 - x$

Substitute $20 - x$ for y in the second equation.

$6x + 4.5(20 - x) = 5(20)$

$6x + 90 - 4.5x = 100$

$1.5x = 10$

$x = 6\frac{2}{3}$

Solve for y.

$x + y = 20$

$6\frac{2}{3} + y = 20$

$y = 13\frac{1}{3}$

Albert must mix $13\frac{1}{3}$ pounds of peanuts and $6\frac{2}{3}$ pounds of cashews.

15. Graph both inequalities on the same set of axes.

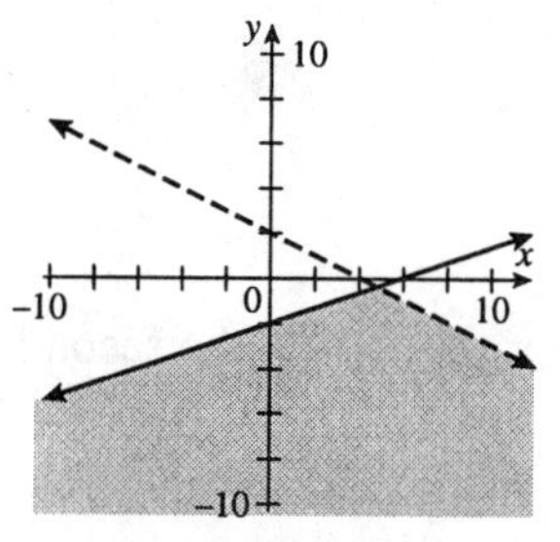

The solution is the part of the graph that contains the shading and the part of the solid line that satisfies both inequalities.

Chapter 8 Cumulative Review Test

1. $\frac{|-4|+|-16| \div 2^2}{3-[2-(4 \div 2)]} = \frac{4+16 \div 4}{3-[2-2]}$

$= \frac{4+4}{3-0} = \frac{8}{3}$

2. $4(x - 2) + 6(x - 3) = 2 - 4x$

$4x - 8 + 6x - 18 = 2 - 4x$

$10x - 26 = 2 - 4x$

$14x - 26 = 2$

$14x = 28$

$x = 2$

3. $3x^2 - 13x + 12 = 0$

$(3x - 4)(x - 3) = 0$

$3x - 4 = 0$ or $x - 3 = 0$

$3x = 4$ or $\quad x = 3$

$x = \frac{4}{3}$

The solutions are $\frac{4}{3}$ and 3.

4. $\frac{1}{3}(x+2)+\frac{1}{4}=8$

Multiply both sides of the equation by 12.

$4(x+2)+3=96$

Use the distributive property.

$4x+8+3=96$

$4x+11=96$

$4x=85$

$x=\frac{85}{4}$

5. $\frac{1}{x-3}+\frac{1}{x+3}=\frac{1}{x^2-9}$

$\frac{1}{x-3}+\frac{1}{x+3}=\frac{1}{(x+3)(x-3)}$

Multiply both sides of the equation by the least common denominator, $(x+3)(x-3)$.

$(x+3)(x-3)\left[\frac{1}{x-3}+\frac{1}{x+3}\right]=\frac{1}{(x+3)(x-3)}\cdot(x+3)(x-3)$

$(x+3)(x-3)\left[\frac{1}{x-3}\right]+(x+3)(x-3)\left[\frac{1}{x+3}\right]=\frac{1}{(x+3)(x-3)}\cdot(x+3)(x-3)$

$(x+3)\left[\frac{1}{1}\right]+(x-3)\left[\frac{1}{1}\right]=\frac{1}{1}$

$(x+3)+(x-3)=1$

$x+3+x-3=1$

$2x=1$

$x=\frac{1}{2}$

6. $\frac{6}{4}=\frac{10}{x}$

$6\cdot x=4\cdot 10$

$6x=40$

$x=\frac{40}{6}=\frac{20}{3}$

The length of side x is $\frac{20}{3}$ which is approximately 6.67 inches.

7. $(x^5y^3)^4(2x^3y^5) = (x^5)^4(y^3)^4(2x^3y^5) = x^{20}y^{12}(2x^3y^5) = 2x^{23}y^{17}$

8. Factor $6x^2 - 11x + 4$ by grouping.

Find two numbers whose product is $6 \cdot 4 = 24$ and whose sum is -11.

The two numbers are -3 and -8.

$6x^2 - 11x + 4 = 6x^2 - 3x - 8x + 4 = 3x(2x - 1) - 4(2x - 1) = (3x - 4)(2x - 1)$

9. $\dfrac{4}{x^2 - 9} - \dfrac{3}{x^2 - 9x + 18} = \dfrac{4}{(x+3)(x-3)} - \dfrac{3}{(x-6)(x-3)}$

The least common denominator is $(x + 3)(x - 3)(x - 6)$

$\dfrac{(x-6)}{(x-6)} \cdot \dfrac{4}{(x+3)(x-3)} - \dfrac{(x+3)}{(x+3)} \cdot \dfrac{3}{(x-6)(x-3)} = \dfrac{4x-24}{(x+3)(x-3)(x-6)} - \dfrac{3x+9}{(x+3)(x-3)(x+6)}$

$= \dfrac{x-33}{(x+3)(x-3)(x-6)}$

10. $\dfrac{x^2 - 7x + 12}{2x^2 - 11x + 12} \div \dfrac{x^2 - 9}{x^2 - 16} = \dfrac{x^2 - 7x + 12}{2x^2 - 11x + 12} \cdot \dfrac{x^2 - 16}{x^2 - 9} = \dfrac{(x-3)(x-4)}{(2x-3)(x-4)} \cdot \dfrac{(x-4)(x+4)}{(x+3)(x-3)}$

$= \dfrac{1}{(2x-3)} \cdot \dfrac{(x-4)(x+4)}{(x+3)} = \dfrac{(x-4)(x+4)}{(2x-3)(x+3)}$

11. $2x - 3y = 6$

Solve for y.

$-3y = -2x + 6$

$y = \dfrac{2}{3}x - 2$

Let $x = 0$, $y = \dfrac{2}{3}(0) - 2 = -2$

Let $x = 3$, $y = \dfrac{2}{3}(3) - 2 = 0$

Let $x = -3$, $y = \dfrac{2}{3}(-3) - 2 = -4$

x	y
0	−2
3	0
−3	−4

Plot the points and draw the line.

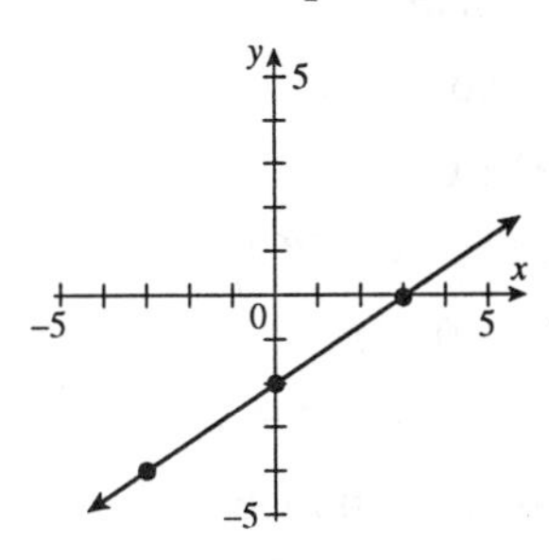

12. $3x+2y=9$

Find y-intercept	Find x-intercept
Let $x=0$	Let $y=0$
$3x+2y=9$	$3x+2y=9$
$3(0)+2y=9$	$3x+2(0)=9$
$2y=9$	$3x=9$
$y=\frac{9}{2}$	$x=3$

Plot the points and draw the graph.

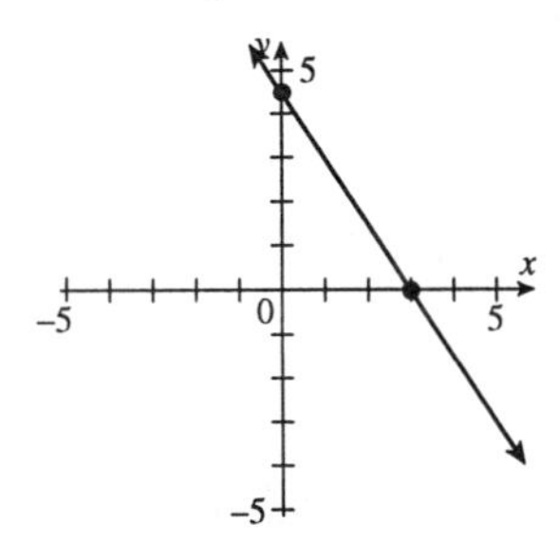

13. Graph the equation $2x-y=6$. Since the inequality is <, use a dashed line. Check whether (0, 0) satisfies the inequality.

$2x-y<6$

$2(0)-0<6$

$0<6$ True

Since the point (0, 0) satisfies the inequality, every point on the same side of the line as (0, 0) also satisfies the inequality.

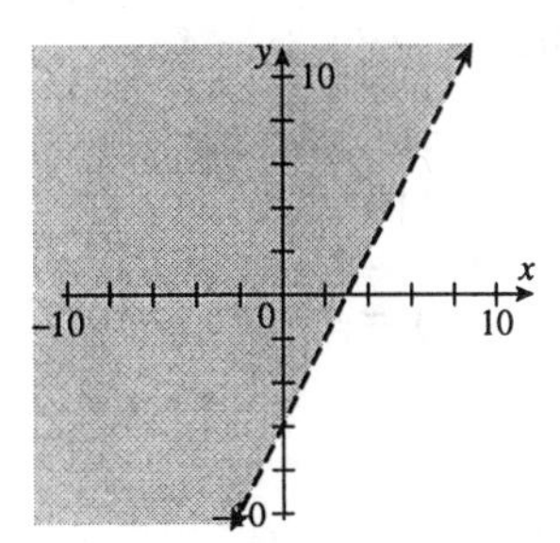

14. $3x=2y+8$

$3x-8=2y$

$\frac{3}{2}x-4=y$

$-4y=-6x+12$

$y=\frac{3}{2}x-3$

The equations have the same slope and different y-intercepts. The lines are therefore parallel and there is no solution.

15. $x+2y=2$ | Ordered Pair

$x+2y=2$	Ordered Pair
Let $x=0$, then $y=1$	(0, 1)
Let $y=0$, then $x=2$	(2, 0)
$2x-3y=-3$	Ordered Pair
Let $x=0$, then $y=1$	(0, 1)
Let $y=0$, then $x=-\frac{3}{2}$	$\left(-\frac{3}{2}, 0\right)$

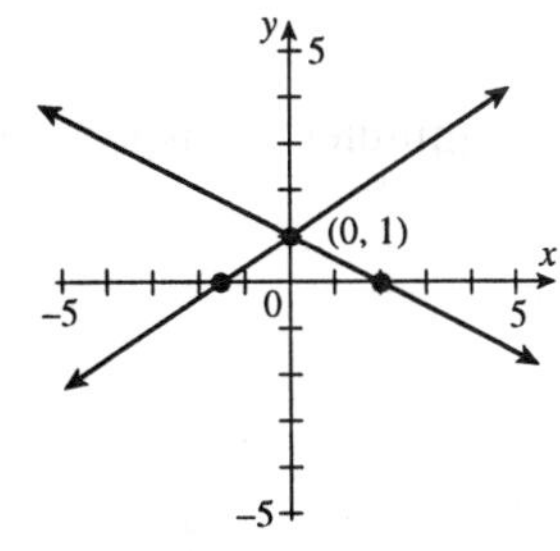

The lines intersect at (0, 1).

Check: $x+2y=2$ $2x-3y=-3$

$0+2(1)=2$ $2(0)-3(1)=-3$

$2=2$ True $-3=-3$ True

Since (0, 1) satisfies both equations, it is the solution.

16. $2x+3y=4$ gives $2x+3y=4$

$-2[x-4y]=6$ $\quad \underline{-2x+8y=-12}$

$$11y=-8$$

$$y=-\frac{8}{11}$$

Solve for x.

$$2x+3y=4$$

$$2x+3\left(-\frac{8}{11}\right)=4$$

$$2x-\frac{24}{11}=4$$

$$2x=\frac{68}{11}$$

$$x=\frac{34}{11}$$

The solution is $\left(\frac{34}{11}, -\frac{8}{11}\right)$.

17. $\dfrac{40 \text{ units}}{15 \text{ minutes}}=\dfrac{160 \text{ units}}{x \text{ minutes}}$

$$\frac{40}{15}=\frac{160}{x}$$

$$40x=15(160)$$

$$40x=2400$$

$$x=60$$

It will take her 60 minutes to inspect 160 units.

18. **(a)** Let i = total income, s = amount of sales (in dollars)

The system of equations is

$i=20{,}000+0.10s$

Total income from PCR.

$i=10{,}000+0.12s$

Total income from ARA.

Total income from PCR

= total income from ARA

$$20{,}000+0.10s=10{,}000+0.12s$$

$$10{,}000+0.10s=0.12s$$

$$10{,}000=0.02s$$

$$500{,}000=s$$

\$500,000 of sales are needed for the income to be the same from both companies.

(b) Income from PCR $=20{,}000+0.10s$

$=20{,}000+0.10(200{,}000)$

$=20{,}000+20{,}000$

$=\$40{,}000$

Income from ARA $=10{,}000+0.12s$

$=10{,}000+0.12(200{,}000)$

$=10{,}000+24{,}000$

$=\$34{,}000$

PCR would give a higher income.

19. Let x = number of liters of 20% solution, y = number of liters of 35% solution

Acid content of 20% solution = $0.2x$

Acid content of 35% solution = $0.35y$

Acid content of mixture = $0.25(10)$

The system of equations is $x + y = 10$. Total volume is 10 liters.

$0.2x + 0.35y = 0.25(10)$

Acid content of 20% solution + acid content of 35% solution = acid content of mixture.

$-20[x + y = 10]$ gives $-20x - 20y = -200$

$100[0.2x + 0.35y = 0.25(10)]$ gives $20x + 35y = 250$

$$15y = 50$$

$$y = 3\tfrac{1}{3}$$

Solve for x.

$x + y = 10$

$x + 3\tfrac{1}{3} = 10$

$x = 6\tfrac{2}{3}$. $6\tfrac{2}{3}$ liters of 20% solution should be mixed with $3\tfrac{1}{3}$ liters of 35% solution.

20. Let x = number of minutes worked by Mr. Pontilo, y = number of minutes worked by Mrs. Pontilo.

Fraction of shop cleaned by Mr. Pontilo in x minutes = $\frac{x}{50}$.

Fraction of shop cleaned by Mrs. Pontilo in y minutes = $\frac{y}{60}$.

The system of equations is $x = y$. Mr. and Mrs. Pontilo work together the same length of time, so $x = y$.

$\frac{x}{50} + \frac{y}{60} = 1$. Together they clean the whole shop.

Substitute y for x in the second equation.

$\frac{y}{50} + \frac{y}{60} = 1$

$6y + 5y = 300$

$11y = 300$

$y = \frac{300}{11}$

Working together, it will take them $\frac{300}{11}$ minutes or approximately 27.3 minutes.

CHAPTER 9

Exercise Set 9.1

1. $\sqrt{1}=1$ since $1^2=1$

3. $\sqrt{0}=0$ since $0^2=0$

5. $\sqrt{81}=9$ Take the opposite of both sides to get $-\sqrt{81}=-9$

7. $\sqrt{121}=11$ since $(11)^2=121$

9. $\sqrt{16}=4$ Take the opposite of both sides to get $-\sqrt{16}=-4$

11. $\sqrt{144}=12$ since $(12)^2=144$

13. $\sqrt{169}=13$ since $(13)^2=169$

15. $\sqrt{1}=1$ Take the opposite of both sides to get $-\sqrt{1}=-1$

17. $\sqrt{81}=9$ since $9^2=81$

19. $\sqrt{121}=11$ Take the opposite of both sides to get $-\sqrt{121}=-11$

21. $\sqrt{\frac{1}{4}}=\frac{1}{2}$ since $\left(\frac{1}{2}\right)^2=\frac{1}{4}$

23. $\sqrt{\frac{9}{16}}=\frac{3}{4}$ since $\left(\frac{3}{4}\right)^2=\frac{9}{16}$

25. $\sqrt{\frac{4}{25}}=\frac{2}{5}$ since $\left(\frac{2}{5}\right)^2=\frac{4}{25}$ Take the opposite of both sides to get $-\sqrt{\frac{4}{25}}=-\frac{2}{5}$

27. $\sqrt{\frac{36}{49}}=\frac{6}{7}$ since $\left(\frac{6}{7}\right)^2=\frac{36}{49}$

29. $\sqrt{8}\approx 2.8284271$

31. $\sqrt{15}\approx 3.8729833$

33. $\sqrt{80}\approx 8.9442719$

35. $\sqrt{81}=9$

37. $\sqrt{97}\approx 9.8488578$

39. $\sqrt{3}\approx 1.7320508$

41. True; since 25 is a perfect square, $\sqrt{25}$ is a rational number.

43. True; the square root of a negative number is not a real number.

45. False; since 9 is a perfect square, $\sqrt{9}$ is a rational number.

47. True; $\sqrt{\frac{4}{9}}=\frac{2}{3}$ which is a rational number.

49. False; since 125 is not a perfect square, $\sqrt{125}$ is an irrational number

51. True; $\sqrt{(18)^2} = \sqrt{324} = 18$ which is an integer.

53. $\sqrt{7} = 7^{1/2}$

55. $\sqrt{17} = (17)^{1/2}$

57. $\sqrt{5x} = (5x)^{1/2}$

59. $\sqrt{12x^2} = \left(12x^2\right)^{1/2}$

61. $\sqrt{19xy^2} = \left(19xy^2\right)^{1/2}$

63. $\sqrt{40x^3} = \left(40x^3\right)^{1/2}$

65. It is nonnegative. The square root of a negative number is not a real number.

67. No real number when squared will be a negative number.

Cumulative Review Exercises: 9.1

68. $\dfrac{2x}{x^2-4} + \dfrac{1}{x-2} = \dfrac{2}{x+2}$

$$\frac{2x}{(x-2)(x+2)} + \frac{1}{x-2} = \frac{2}{x+2}$$

$$(x-2)(x+2)\left[\frac{2x}{(x-2)(x+2)} + \frac{1}{(x-2)}\right]$$

$$= \frac{2}{x+2} \cdot (x-2)(x+2)$$

$$2x + (x+2) = 2(x-2)$$

$$2x + x + 2 = 2x - 4$$

$$3x + 2 = 2x - 4$$

$$3x = 2x - 6$$

$$x = -6$$

69. $\dfrac{4x}{x^2+6x+9} - \dfrac{2x}{x+3} = \dfrac{x+1}{x+3}$

$$\frac{4x}{(x+3)(x+3)} - \frac{2x}{x+3} = \frac{x+1}{x+3}$$

$$(x+3)(x+3)\left[\frac{4x}{(x+3)(x+3)} - \frac{2x}{x+3}\right]$$

$$= \frac{x+1}{x+3} \cdot (x+3)(x+3)$$

$$4x - (x+3)2x = (x+1)(x+3)$$

$$4x - 2x^2 - 6x = x^2 + 4x + 3$$

$$0 = 3x^2 + 6x + 3$$

$$0 = 3(x^2 + 2x + 1)$$

$$0 = 3(x+1)^2$$

$$0 = x + 1$$

$$x = -1$$

70. Let t = number of hours for the two belts together to fill the silo.

In t hours, the first belt can fill a fraction $\frac{t}{6}$ of the silo.

In t hours, the second belt can fill a fraction $\frac{t}{5}$ of the silo.

Together the two belts fill the whole silo in t hours.

Thus $\frac{t}{6} + \frac{t}{5} = 1$

$$30\left(\frac{t}{6}+\frac{t}{5}\right)=30$$
$$5t+6t=30$$
$$11t=30$$
$$t=\frac{30}{11}$$

It will take both belts together $\frac{30}{11}$ or $2\frac{8}{11}$ hours to fill the silo.

71. $m=\frac{\Delta y}{\Delta x}=\frac{7-3}{6-(-5)}=\frac{4}{11}$

72. $f(x)=x^2-4x-5$

$f(-3)=(-3)^2-4(-3)-5$

$=9+12-5=16$

Group Activity/Challenge Problems

1. **(a)** Yes

(b) No

(c) No

(d) Yes

(e) No

3. **(a)** Yes; $\sqrt{2^2}=2$

(b) Yes; $\sqrt{5^2}=5$

(c) $\sqrt{a^2}=a,\ a\geq 0$

(d) Individual answer

5. $\sqrt{x^3}=x^{3/2}$

7. $\sqrt[4]{(2x)^8}=(2x)^{8/4}=(2x)^2=4x^2$

Exercise Set 9.2

1. $\sqrt{16}=4$ since $4^2=16$

3. $\sqrt{8}=\sqrt{4\cdot 2}$

$=\sqrt{4}\cdot\sqrt{2}$

$=2\sqrt{2}$

5. $\sqrt{96}=\sqrt{16\cdot 6}$

$=\sqrt{16}\cdot\sqrt{6}$

$=4\sqrt{6}$

7. $\sqrt{32}=\sqrt{16\cdot 2}$

$=\sqrt{16}\cdot\sqrt{2}$

$=4\sqrt{2}$

9. $\sqrt{160}=\sqrt{16\cdot 10}$

$=\sqrt{16}\cdot\sqrt{10}$

$=4\sqrt{10}$

11. $\sqrt{48} = \sqrt{16 \cdot 3}$
$= \sqrt{16} \cdot \sqrt{3}$
$= 4\sqrt{3}$

13. $\sqrt{108} = \sqrt{36 \cdot 3}$
$= \sqrt{36} \cdot \sqrt{3}$
$= 6\sqrt{3}$

15. $\sqrt{156} = \sqrt{4 \cdot 39}$
$= \sqrt{4} \cdot \sqrt{39}$
$= 2\sqrt{39}$

17. $\sqrt{256} = 16$ since $(16)^2 = 256$

19. $\sqrt{900} = 30$ since $(30)^2 = 900$

21. $\sqrt{y^6} = y^3$

23. $\sqrt{x^2y^4} = \sqrt{x^2}\sqrt{y^4} = xy^2$

25. $\sqrt{x^9y^{12}} = \sqrt{x^8y^{12}} \cdot \sqrt{x}$
$= x^4y^6\sqrt{x}$

27. $\sqrt{a^2b^4c} = \sqrt{a^2b^4} \cdot \sqrt{c} = ab^2\sqrt{c}$

29. $\sqrt{3x^3} = \sqrt{x^2} \cdot \sqrt{3x} = x\sqrt{3x}$

31. $\sqrt{50x^2y^3} = \sqrt{25x^2y^2} \cdot \sqrt{2y} = 5xy\sqrt{2y}$

33. $\sqrt{200y^5z^{12}} = \sqrt{100y^4z^{12}} \cdot \sqrt{2y}$
$= 10y^2z^6\sqrt{2y}$

35. $\sqrt{243q^2b^3c} = \sqrt{81q^2b^2} \cdot \sqrt{3bc}$
$= 9qb\sqrt{3bc}$

37. $\sqrt{128x^3yz^5} = \sqrt{64x^2z^4} \cdot \sqrt{2xyz}$
$= 8xz^2\sqrt{2xyz}$

39. $\sqrt{250x^4yz} = \sqrt{25x^4} \cdot \sqrt{10yz}$
$= 5x^2\sqrt{10yz}$

41. $\sqrt{8} \cdot \sqrt{3} = \sqrt{24}$
$= \sqrt{4} \cdot \sqrt{6}$
$= 2\sqrt{6}$

43. $\sqrt{18} \cdot \sqrt{3} = \sqrt{54}$
$= \sqrt{9} \cdot \sqrt{6}$
$= 3\sqrt{6}$

45. $\sqrt{75} \cdot \sqrt{6} = \sqrt{450}$
$= \sqrt{225} \cdot \sqrt{2}$
$= 15\sqrt{2}$

47. $\sqrt{3x} \cdot \sqrt{5x} = \sqrt{15x^2}$
$= \sqrt{x^2} \cdot \sqrt{15}$
$= x\sqrt{15}$

49. $\sqrt{5x^2}\sqrt{8x^3} = \sqrt{40x^5}$
$= \sqrt{4x^4} \cdot \sqrt{10x}$
$= 2x^2\sqrt{10x}$

51. $\sqrt{12x^2y}\sqrt{6xy^3} = \sqrt{72x^3y^4}$
$= \sqrt{36x^2y^4}\sqrt{2x}$
$= 6xy^2\sqrt{2x}$

53. $\sqrt{18xy^4}\sqrt{3x^2y} = \sqrt{54x^3y^5}$
$= \sqrt{9x^2y^4} \cdot \sqrt{6xy}$
$= 3xy^2\sqrt{6xy}$

55. $\sqrt{15xy^6} \cdot \sqrt{6xyz} = \sqrt{90x^2y^7z}$
$= \sqrt{9x^2y^6} = \sqrt{10yz}$
$= 3xy^3\sqrt{10yz}$

57. $\sqrt{9x^4y^6}\sqrt{4x^2y^4} = \sqrt{36x^6y^{10}} = 6x^3y^5$

59. $\left(\sqrt{4x}\right)^2 = (4x)^{(1/2)\cdot 2} = (4x)^1 = 4x$

61. $\left(\sqrt{13x^4y^6}\right)^2 = \left(13x^4y^6\right)^{(1/2)\cdot 2}$
$= \left(13x^4y^6\right)^1 = 13x^4y^6$

63. $\left(\sqrt{4x}\right)^2\left(\sqrt{5x}\right)^2 = (4x)^{(1/2)\cdot 2}(5x)^{(1/2)\cdot 2}$
$= (4x)^1 \cdot (5x)^1$
$= (4x) \cdot (5x)$
$= 20x^2$

65. Exponent on $x = 2 \cdot 2 = 4$

67. Exponent on $x = 2 \cdot 3 = 6$

Exponent on $y = 2\left(\frac{5}{2}\right) = 5$

69. Coefficient = 8 because
$\sqrt{2} \cdot \sqrt{8} \cdot \sqrt{16} = 4$
Exponent on $x = 12$ because
$\sqrt{x^{12}} \cdot \sqrt{x^3} = \sqrt{x^{15}} = x^7\sqrt{x}$
Exponent on $y = 7$ because
$\sqrt{y^5} \cdot \sqrt{y^7} = \sqrt{y^{12}} = y^6$

71. $\sqrt{a} \cdot \sqrt{b} = \sqrt{a \cdot b}$ (when $a, b \geq 0$) means that the product of square roots is the square root of the product.

73. Radicands must be greater than or equal to 0.

75. **(b)** $\sqrt{x^9} = \sqrt{x^8} \cdot \sqrt{x} = x^4\sqrt{x}$

77. **(a)** There can be no perfect square factors nor any exponents greater than 1 in the radicand.

(b) $\sqrt{75x^5} = \sqrt{25x^4} \cdot \sqrt{3x} = 5x^2\sqrt{3x}$

Cumulative Review Exercises: 9.2

78. $\frac{3x^2 - 16x - 12}{3x^2 - 10x - 8} \div \frac{x^2 - 7x + 6}{3x^2 - 11x - 4}$
$= \frac{3x^2 - 16x - 12}{3x^2 - 10x - 8} \cdot \frac{3x^2 - 11x - 4}{x^2 - 7x + 6}$
$= \frac{(3x+2)(x-6)}{(3x+2)(x-4)} \cdot \frac{(3x+1)(x-4)}{(x-6)(x-1)}$
$= \frac{3x+1}{x-1}$

79. $3x + 6y = 9$

$6y = -3x + 9$

$\frac{6y}{6} = \frac{-3x+9}{6}$

$y = -\frac{1}{2}x + \frac{3}{2}$

$m = -\frac{1}{2}, b = \frac{3}{2}$

80. Graph the line $6x - 5y = 30$.

Since the inequality symbol is $\geq$, draw a solid line.

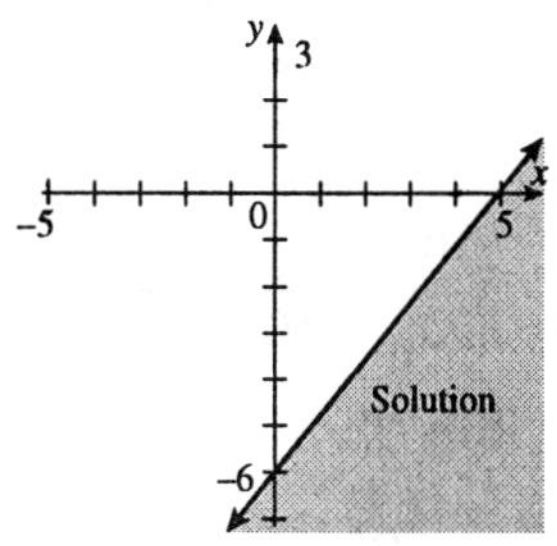

Check (0, 0): $6x - 5y \geq 30$

$6(0) - 5(0) \geq 30$

$0 - 0 \geq 30$

$0 \geq 30$ False

Since (0, 0) does not satisfy the inequality, all the points on the other side of the line from (0, 0) satisfy the inequality. The solution is the shaded region.

81. $5[3x - 4y = 6]$ gives $15x - 20y = 30$

$-3[5x - 3y = 5]$ $\underline{-15x + 9y = -15}$

$-11y = 15$

$y = -\frac{15}{11}$

Solve for x.

$3x - 4y = 6$

$3x - 4\left(-\frac{15}{11}\right) = 6$

$3x + \frac{60}{11} = 6$

$3x = \frac{6}{11}$

$x = \frac{2}{11}$

The solution is $\left(\frac{2}{11}, -\frac{15}{11}\right)$

Group Activity/Challenge Problems

4. **(a)** $(\sqrt{a} + \sqrt{b})(\sqrt{a} - \sqrt{b})$

$= \sqrt{a}\sqrt{a} - \sqrt{a}\sqrt{b} + \sqrt{a}\sqrt{b} - \sqrt{b}\sqrt{b}$

$= (\sqrt{a})^2 - (\sqrt{b})^2 = a - b$

(b) $(\sqrt{6} + \sqrt{3})(\sqrt{6} - \sqrt{3}) = 6 - 3 = 3$

Exercise Set 9.3

1. $\sqrt{\frac{12}{3}} = \sqrt{4} = 2$

3. $\sqrt{\frac{27}{3}} = \sqrt{9} = 3$

5. $\frac{\sqrt{18}}{\sqrt{2}} = \sqrt{\frac{18}{2}} = \sqrt{9} = 3$

7. $\sqrt{\frac{1}{25}} = \frac{\sqrt{1}}{\sqrt{25}} = \frac{1}{5}$

9. $\sqrt{\frac{9}{49}} = \frac{\sqrt{9}}{\sqrt{49}} = \frac{3}{7}$

11. $\frac{\sqrt{10}}{\sqrt{490}} = \sqrt{\frac{10}{490}} = \sqrt{\frac{1}{49}} = \frac{\sqrt{1}}{\sqrt{49}} = \frac{1}{7}$

13. $\sqrt{\frac{40x^3}{2x}} = \sqrt{20x^2} = \sqrt{4x^2} \cdot \sqrt{5} = 2x\sqrt{5}$

15. $\sqrt{\frac{9xy^4}{3y^3}} = \sqrt{3xy}$

17. $\sqrt{\frac{25x^6y}{45x^6y^3}} = \sqrt{\frac{5}{9y^2}} = \frac{\sqrt{5}}{\sqrt{9y^2}} = \frac{\sqrt{5}}{3y}$

19. $\sqrt{\frac{72xy}{72x^3y^5}} = \sqrt{\frac{1}{x^2y^4}} = \frac{\sqrt{1}}{\sqrt{x^2y^4}} = \frac{1}{xy^2}$

21. $\frac{\sqrt{32x^5}}{\sqrt{8x}} = \sqrt{\frac{32x^5}{8x}} = \sqrt{4x^4} = 2x^2$

23. $\frac{\sqrt{16x^4y}}{\sqrt{25x^6y^3}} = \sqrt{\frac{16x^4y}{25x^6y^3}} = \sqrt{\frac{16}{25x^2y^2}}$

$= \frac{\sqrt{16}}{\sqrt{25x^2y^2}} = \frac{4}{5xy}$

25. $\frac{\sqrt{45xy^6}}{\sqrt{9xy^4z^2}} = \sqrt{\frac{45xy^6}{9xy^4z^2}} = \sqrt{\frac{5y^2}{z^2}}$

$= \frac{\sqrt{5y^2}}{\sqrt{z^2}} = \frac{\sqrt{5}\sqrt{y^2}}{\sqrt{z^2}} = \frac{y\sqrt{5}}{z}$

27. $\frac{\sqrt{72x^{12}y^{20}}}{\sqrt{2x^2y^4}} = \sqrt{\frac{72x^{12}y^{20}}{2x^2y^4}}$

$= \sqrt{36x^{10}y^{16}} = 6x^5y^8$

29. $\frac{3}{\sqrt{2}} = \frac{3}{\sqrt{2}} \cdot \frac{\sqrt{2}}{\sqrt{2}} = \frac{3\sqrt{2}}{\sqrt{4}} = \frac{3\sqrt{2}}{2}$

31. $\frac{4}{\sqrt{8}} = \frac{4}{\sqrt{4}\sqrt{2}} = \frac{4}{2\sqrt{2}} = \frac{2}{\sqrt{2}}$

$= \frac{2}{\sqrt{2}} \cdot \frac{\sqrt{2}}{\sqrt{2}} = \frac{2\sqrt{2}}{2} = \sqrt{2}$

33. $\frac{5}{\sqrt{10}} = \frac{5}{\sqrt{10}} \cdot \frac{\sqrt{10}}{\sqrt{10}} = \frac{5\sqrt{10}}{10} = \frac{\sqrt{10}}{2}$

35. $\sqrt{\frac{2}{5}} = \frac{\sqrt{2}}{\sqrt{5}} = \frac{\sqrt{2}}{\sqrt{5}} \cdot \frac{\sqrt{5}}{\sqrt{5}} = \frac{\sqrt{10}}{5}$

37. $\sqrt{\frac{3}{15}} = \sqrt{\frac{1}{5}} = \frac{\sqrt{1}}{\sqrt{5}} = \frac{\sqrt{1} \cdot \sqrt{5}}{\sqrt{5} \cdot \sqrt{5}} = \frac{\sqrt{5}}{5}$

39. $\sqrt{\frac{x^2}{2}} = \frac{\sqrt{x^2}}{\sqrt{2}} = \frac{x}{\sqrt{2}} = \frac{x}{\sqrt{2}} \cdot \frac{\sqrt{2}}{\sqrt{2}} = \frac{x\sqrt{2}}{2}$

41. $\sqrt{\frac{x^2}{8}} = \frac{\sqrt{x^2}}{\sqrt{8}} = \frac{x}{\sqrt{4}\sqrt{2}} = \frac{x}{2\sqrt{2}}$

$= \frac{x}{2\sqrt{2}} \cdot \frac{\sqrt{2}}{\sqrt{2}} = \frac{x\sqrt{2}}{2 \cdot 2} = \frac{x\sqrt{2}}{4}$

43. $\sqrt{\frac{x^4}{5}} = \frac{\sqrt{x^4}}{\sqrt{5}} = \frac{x^2}{\sqrt{5}} = \frac{x^2}{\sqrt{5}} \cdot \frac{\sqrt{5}}{\sqrt{5}} = \frac{x^2\sqrt{5}}{5}$

45. $\sqrt{\frac{x^6}{15y}} = \frac{\sqrt{x^6}}{\sqrt{15y}} = \frac{x^3}{\sqrt{15y}}$

$= \frac{x^3}{\sqrt{15y}} \cdot \frac{\sqrt{15y}}{\sqrt{15y}} = \frac{x^3\sqrt{15y}}{15y}$

47. $\sqrt{\frac{8x^4y^2}{32x^2y^3}} = \sqrt{\frac{x^2}{4y}} = \frac{\sqrt{x^2}}{\sqrt{4y}} = \frac{x}{\sqrt{4}\sqrt{y}}$

$= \frac{x}{2\sqrt{y}} = \frac{x}{2\sqrt{y}} \cdot \frac{\sqrt{y}}{\sqrt{y}} = \frac{x\sqrt{y}}{2y}$

49. $\sqrt{\frac{18yz}{75x^4y^5z^3}} = \sqrt{\frac{6}{25x^4y^4z^2}}$

$= \frac{\sqrt{6}}{\sqrt{25x^4y^4z^2}} = \frac{\sqrt{6}}{5x^2y^2z}$

51. $\frac{\sqrt{90x^4y}}{\sqrt{2x^5y^5}} = \sqrt{\frac{90x^4y}{2x^5y^5}} = \sqrt{\frac{45}{xy^4}} = \frac{\sqrt{45}}{\sqrt{xy^4}}$

$= \frac{\sqrt{9}\sqrt{5}}{\sqrt{y^4}\sqrt{x}} = \frac{3\sqrt{5}}{y^2\sqrt{x}}$

$= \frac{3\sqrt{5}}{y^2\sqrt{x}} \cdot \frac{\sqrt{x}}{\sqrt{x}} = \frac{3\sqrt{5x}}{xy^2}$

53. **1.** No perfect square factors in any radicand.

2. No radicand contains a fraction.

3. No square roots in any denominator.

55. The radicand contains a fraction.

$\sqrt{\frac{1}{3}} = \frac{\sqrt{1}}{\sqrt{3}} = \frac{1}{\sqrt{3}} = \frac{1}{\sqrt{3}} \cdot \frac{\sqrt{3}}{\sqrt{3}} = \frac{\sqrt{3}}{3}$

57. Cannot be simplified: the radicand does not have a factor that is a perfect square, the radicand does not contain a fraction, and the denominator does not contain a square root.

59. The numerator and denominator have a common factor: $\frac{x^2\sqrt{2}}{x} = x\sqrt{2}$

61. Cannot be simplified: the radicand does not have a factor that is a perfect square, the radicand does not contain a fraction, and the denominator does not contain a square root.

63. $\frac{\sqrt{a}}{\sqrt{b}} = \sqrt{\frac{a}{b}}$ when $a \ge 0$, $b > 0$ the quotient of two square roots is equal to the square root of the quotient.

65. **(b)** $\frac{a}{\sqrt{b}} = \frac{a}{\sqrt{b}} \cdot \frac{\sqrt{b}}{\sqrt{b}} = \frac{a\sqrt{b}}{b}$

Cumulative Review Exercises: 9.3

66. $3x^2 - 12x - 96 = 3(x^2 - 4x - 32)$

To factor $x^2 - 4x - 32$, find two numbers whose product is -32 and whose sum is -4. The numbers are -8 and 4. Thus

$3x^2 - 12x - 96 = 3(x + 4)(x - 8)$.

67. $\frac{x-1}{x^2-1} = \frac{(x-1)}{(x+1)(x-1)} = \frac{1}{x+1}$

68. $x + \frac{24}{x} = 10$

Multiply both sides of the equation by x.

$$x\left(x + \frac{24}{x}\right) = 10 \cdot x$$
$$x^2 + 24 = 10x$$
$$x^2 - 10x + 24 = 0$$
$$(x-6)(x-4) = 0$$
$$x - 6 = 0 \text{ or } x - 4 = 0$$
$$x = 6 \text{ or } x = 4$$

69.

$y = 2x - 2$	Ordered Pair
Let $x = 0$, then $y = -2$	$(0, -2)$
Let $y = 0$, then $x = 1$	$(1, 0)$

$2x + 3y = 10$	Ordered Pair
Let $x = 0$, then $y = 3\frac{1}{3}$	$\left(0, 3\frac{1}{3}\right)$
Let $y = 0$, then $x = 5$	$(5, 0)$

Graph the equations.

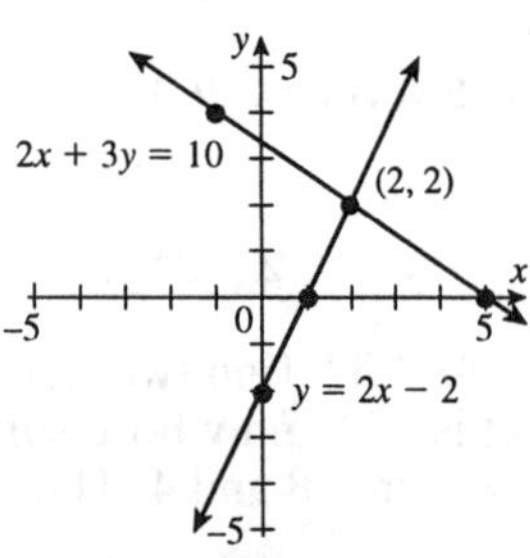

The graphs intersect at the point (2, 2).

Check:	$y = 2x - 2$	$2x + 3y = 10$
	$2 = 2(2) - 2$	$2(2) + 3(2) = 10$
	$2 = 4 - 2$	$4 + 6 = 10$
	$2 = 2$ True	$10 = 10$ True

The solution is (2, 2).

Group Activity/Challenge Problems

2. $$\frac{2}{\sqrt{6}+\sqrt{3}} = \frac{2(\sqrt{6}-\sqrt{3})}{(\sqrt{6}+\sqrt{3})(\sqrt{6}-\sqrt{3})}$$
$$= \frac{2(\sqrt{6}-\sqrt{3})}{(\sqrt{6})^2 - \sqrt{6}\sqrt{3} + \sqrt{6}\sqrt{3} - (\sqrt{3})^2}$$
$$= \frac{2(\sqrt{6}-\sqrt{3})}{6-2} = \frac{2(\sqrt{6}-\sqrt{3})}{3}$$

4. $$\sqrt{\frac{n}{4x^2}} = 4x^4$$
$$\left(\sqrt{\frac{n}{4x^2}}\right)^2 = (4x^4)^2$$
$$\frac{n}{4x^2} = 16x^8$$
$$n = (4x^2)(16x^8)$$
$$n = 64x^{10}$$

5. $$\frac{\sqrt{32x^5}}{\sqrt{n}} = 2x^2$$
$$\sqrt{\frac{32x^5}{n}} = 2x^2$$
$$\left(\sqrt{\frac{32x^5}{n}}\right)^2 = (2x^2)^2$$
$$\frac{32x^5}{n} = 4x^4$$
$$n = \frac{32x^5}{4x^4}$$
$$n = 8x$$

7. $\frac{3x}{\sqrt{n}} = \frac{3\sqrt{2x}}{2}$

$2(3x) = (\sqrt{n})3\sqrt{2x}$

$(6x)^2 = (3\sqrt{2xn})^2$

$36x^2 = 9(2xn)$

$36x^2 = 18xn$

$n = \frac{36x^2}{18x}$

$n = 2x$

Exercise Set 9.4

1. $5\sqrt{6} - 2\sqrt{6} = (5-2)\sqrt{6} = 3\sqrt{6}$

3. $6\sqrt{7} - 8\sqrt{7} = (6-8)\sqrt{7} = -2\sqrt{7}$

5. $2\sqrt{3} - 2\sqrt{3} - 4\sqrt{3} + 5$

$= (2-2-4)\sqrt{3} + 5$

$= -4\sqrt{3} + 5$

7. $4\sqrt{x} + \sqrt{x} = (4+1)\sqrt{x} = 5\sqrt{x}$

9. $-\sqrt{x} + 6\sqrt{x} - 2\sqrt{x} = (-1+6-2)\sqrt{x}$

$= 3\sqrt{x}$

11. $3\sqrt{y} - \sqrt{y} + 3 = (3-1)\sqrt{y} + 3 = 2\sqrt{y} + 3$

13. $\sqrt{x} + \sqrt{y} + x + 3\sqrt{y} = x + \sqrt{x} + 4\sqrt{y}$

Only $\sqrt{y}$ and $3\sqrt{y}$ can be combined

15. $3 + 4\sqrt{x} - 6\sqrt{x} = 3 - 2\sqrt{x}$

Only $4\sqrt{x}$ and $-6\sqrt{x}$ can be combined

17. $5 - 2\sqrt{z} - \sqrt{z} - 5\sqrt{z} = 5 + (-2-1-5)\sqrt{z}$

$= 5 - 8\sqrt{z}$

19. $\sqrt{8} - \sqrt{12} = \sqrt{4 \cdot 2} - \sqrt{4 \cdot 3}$

$= \sqrt{4}\sqrt{2} - \sqrt{4}\sqrt{3}$

$= 2\sqrt{2} - 2\sqrt{3}$

21. $\sqrt{200} - \sqrt{72} = \sqrt{100 \cdot 2} - \sqrt{36 \cdot 2}$

$= \sqrt{100}\sqrt{2} - \sqrt{36}\sqrt{2}$

$= 10\sqrt{2} - 6\sqrt{2}$

$= 4\sqrt{2}$

23. $\sqrt{125} + \sqrt{20} = \sqrt{25 \cdot 5} + \sqrt{4 \cdot 5}$

$= \sqrt{25}\sqrt{5} + \sqrt{4}\sqrt{5}$

$= 5\sqrt{5} + 2\sqrt{5} = 7\sqrt{5}$

25. $4\sqrt{50} - \sqrt{72} + \sqrt{8}$

$= 4\sqrt{25 \cdot 2} - \sqrt{36 \cdot 2} + \sqrt{4 \cdot 2}$

$= 4\sqrt{25}\sqrt{2} - \sqrt{36}\sqrt{2} + \sqrt{4}\sqrt{2}$

$= 4 \cdot 5\sqrt{2} - 6\sqrt{2} + 2\sqrt{2}$

$= 20\sqrt{2} - 6\sqrt{2} + 2\sqrt{2}$

$= 16\sqrt{2}$

27. $-6\sqrt{75} + 4\sqrt{125}$

$= -6\sqrt{25 \cdot 3} + 4\sqrt{25 \cdot 5}$

$= -6\sqrt{25}\sqrt{3} + 4\sqrt{25}\sqrt{5}$

$= -6 \cdot 5\sqrt{3} + 4 \cdot 5\sqrt{5}$

$= -30\sqrt{3} + 20\sqrt{5}$

29. $5\sqrt{250}-9\sqrt{80}$
$=5\sqrt{25\cdot 10}-9\sqrt{16\cdot 5}$
$=5\sqrt{25}\sqrt{10}-9\sqrt{16}\sqrt{5}$
$=5\cdot 5\sqrt{10}-9\cdot 4\sqrt{5}$
$=25\sqrt{10}-36\sqrt{5}$

31. $8\sqrt{64}-\sqrt{96}$
$=8(8)-\sqrt{16\cdot 6}$
$=8(8)-\sqrt{16}\sqrt{6}$
$=64-4\sqrt{6}$

33. $(3+\sqrt{2})(3-\sqrt{2})$
F O I L
$=3(3)+3(-\sqrt{2})+3\sqrt{2}+\sqrt{2}(-\sqrt{2})$
$=9-3\sqrt{2}+3\sqrt{2}-\sqrt{4}$
$=9-2$
$=7$

35. $(6-\sqrt{5})(6+\sqrt{5})$
F O I L
$=6(6)+6(\sqrt{5})+6(-\sqrt{5})+\sqrt{5}(-\sqrt{5})$
$=36+6\sqrt{5}-6\sqrt{5}-\sqrt{25}$
$=36-5$
$=31$

37. $(\sqrt{x}+3)(\sqrt{x}-3)$
F O I L
$=\sqrt{x}(\sqrt{x})+\sqrt{x}(-3)+3\sqrt{x}+3(-3)$
$=x-3\sqrt{x}+3\sqrt{x}-9$
$=x-9$

39. $(\sqrt{6}+x)(\sqrt{6}-x)$
F O I L
$=\sqrt{6}(\sqrt{6})+\sqrt{6}(-x)+\sqrt{6}(x)+x(-x)$
$=6-\sqrt{6}x+\sqrt{6}x-x^2$
$=6-x^2$

41. $(\sqrt{x}+y)(\sqrt{x}-y)$
F O I L
$=\sqrt{x}(\sqrt{x})+\sqrt{x}(-y)+y(\sqrt{x})+y(-y)$
$=x-y\sqrt{x}+y\sqrt{x}-y^2$
$=x-y^2$

43. $(2\sqrt{x}+3\sqrt{y})(2\sqrt{x}-3\sqrt{y})$
$=(2\sqrt{x})(2\sqrt{x})+(2\sqrt{x})(-3\sqrt{y})$
$+(3\sqrt{y})(2\sqrt{x})+(3\sqrt{y})(-3\sqrt{y})$
$=4x-6\sqrt{xy}+6\sqrt{xy}-9y$
$=4x-9y$

45. $\frac{4}{2+\sqrt{3}}=\frac{4}{2+\sqrt{3}}\cdot\frac{2-\sqrt{3}}{2-\sqrt{3}}$
$=\frac{4(2-\sqrt{3})}{4-3}=\frac{8-4\sqrt{3}}{1}$
$=8-4\sqrt{3}$

47. $\frac{3}{\sqrt{5}+2}=\frac{3}{\sqrt{5}+2}\cdot\frac{\sqrt{5}-2}{\sqrt{5}-2}$
$=\frac{3(\sqrt{5}-2)}{5-4}=\frac{3\sqrt{5}-6}{1}$
$=3\sqrt{5}-6$

49. $\frac{2}{\sqrt{2}+\sqrt{3}}=\frac{2}{\sqrt{2}+\sqrt{3}}\cdot\frac{\sqrt{2}-\sqrt{3}}{\sqrt{2}-\sqrt{3}}$
$=\frac{2(\sqrt{2}-\sqrt{3})}{2-3}$
$=\frac{2\sqrt{2}-2\sqrt{3}}{-1}=-2\sqrt{2}+2\sqrt{3}$

51. $\frac{8}{\sqrt{5}-\sqrt{8}}=\frac{8}{\sqrt{5}-\sqrt{8}}\cdot\frac{\sqrt{5}+\sqrt{8}}{\sqrt{5}+\sqrt{8}}$
$=\frac{8(\sqrt{5}+\sqrt{8})}{5-8}=\frac{8\sqrt{5}+8\sqrt{8}}{-3}$
$=\frac{8\sqrt{5}+8\sqrt{4}\sqrt{2}}{-3}=\frac{8\sqrt{5}+8\cdot2\sqrt{2}}{-3}$
$=\frac{8\sqrt{5}+16\sqrt{2}}{-3}=\frac{-8\sqrt{5}-16\sqrt{2}}{3}$

53. $\frac{2}{6+\sqrt{x}}=\frac{2}{6+\sqrt{x}}\cdot\frac{6-\sqrt{x}}{6-\sqrt{x}}=\frac{2(6-\sqrt{x})}{36-x}$
$=\frac{12-2\sqrt{x}}{36-x}$

55. $\frac{6}{4-\sqrt{y}}=\frac{6}{4-\sqrt{y}}\cdot\frac{4+\sqrt{y}}{4+\sqrt{y}}=\frac{6(4+\sqrt{y})}{16-y}$
$=\frac{24+6\sqrt{y}}{16-y}$

57. $\frac{4}{\sqrt{x}-y}=\frac{4}{\sqrt{x}-y}\cdot\frac{\sqrt{x}+y}{\sqrt{x}+y}=\frac{4(\sqrt{x}+y)}{x-y^2}$
$=\frac{4\sqrt{x}+4y}{x-y^2}$

59. $\frac{x}{\sqrt{x}+\sqrt{y}}=\frac{x}{\sqrt{x}+\sqrt{y}}\cdot\frac{\sqrt{x}-\sqrt{y}}{\sqrt{x}-\sqrt{y}}$
$=\frac{x(\sqrt{x}-\sqrt{y})}{x-y}=\frac{x\sqrt{x}-x\sqrt{y}}{x-y}$

61. $\frac{\sqrt{x}}{\sqrt{5}+\sqrt{x}}=\frac{\sqrt{x}}{\sqrt{5}+\sqrt{x}}\cdot\frac{\sqrt{5}-\sqrt{x}}{\sqrt{5}-\sqrt{x}}$
$=\frac{\sqrt{x}(\sqrt{5}-\sqrt{x})}{5-x}=\frac{\sqrt{5x}-x}{5-x}$

63. $\frac{8-4\sqrt{3}}{2}=\frac{4(2-\sqrt{3})}{2}=2(2-\sqrt{3})$

65. $\frac{6+24\sqrt{5}}{3}=\frac{6(1+4\sqrt{5})}{3}=2(1+4\sqrt{5})$

67. $\frac{4+3\sqrt{6}}{2}$ Cannot be simplified since the numerator and denominator have no common factors

69. $\frac{6+2\sqrt{75}}{3}=\frac{6+2\sqrt{25}\sqrt{3}}{3}$
$=\frac{6+2\cdot5\sqrt{3}}{3}$
$=\frac{2(3+5\sqrt{3})}{3}$

71. $\dfrac{-2+4\sqrt{80}}{10}=\dfrac{-2+4\sqrt{16}\sqrt{5}}{10}$

$=\dfrac{-2+4\cdot4\sqrt{5}}{10}=\dfrac{-2+16\sqrt{5}}{10}$

$=\dfrac{2\left(-1+8\sqrt{5}\right)}{10}$

$=\dfrac{-1+8\sqrt{5}}{5}$

73. $\dfrac{60-40\sqrt{18}}{100}=\dfrac{60-40\sqrt{9}\sqrt{2}}{100}$

$=\dfrac{60-40\cdot3\sqrt{2}}{100}$

$=\dfrac{60-120\sqrt{2}}{100}$

$=\dfrac{60\left(1-2\sqrt{2}\right)}{100}$

$=\dfrac{3\left(1-2\sqrt{2}\right)}{5}$

75. $\left(a-\sqrt{b}\right)\left(a+\sqrt{b}\right)$

F O I L

$=a\cdot a+a\sqrt{b}+\left(-\sqrt{b}\right)a+\left(-\sqrt{b}\right)\left(\sqrt{b}\right)$

$=a^2+a\sqrt{b}-a\sqrt{b}-b$

$=a^2-b$

77. $\left(\sqrt{a}-\sqrt{b}\right)\left(\sqrt{a}+\sqrt{b}\right)$

F O I L

$=\sqrt{a}\sqrt{a}+\sqrt{a}\sqrt{b}+\left(-\sqrt{b}\right)\left(\sqrt{a}\right)-\sqrt{b}\sqrt{b}$

$=a+\sqrt{ab}-\sqrt{ab}-b$

$=a-b$

79. When they have the same radicands.

Cumulative Review Exercises: 9.4

81. $3(2x-6)=4(x-9)+3x$

$6x-18=4x-36+3x$

$6x-18=7x-36$

$-18=x-36$

$18=x$

82. $2x^2-x-36=0$

$(2x-9)(x+4)=0$

$2x-9=0$ or $x+4=0$

$x=\dfrac{9}{2}$ or $x=-4$

83. $\dfrac{1}{x^2-4}-\dfrac{2}{x-2}=\dfrac{1}{(x-2)(x+2)}-\dfrac{2}{(x-2)}$

$=\dfrac{1}{(x-2)(x+2)}-\dfrac{2}{(x-2)}\cdot\dfrac{(x+2)}{(x+2)}$

$=\dfrac{1}{(x-2)(x+2)}-\dfrac{(2x+4)}{(x-2)(x+2)}$

$=\dfrac{-2x-3}{(x-2)(x+2)}$

84. Let t = number of minutes for Mrs. Moreno to stack wood. In 12 minutes, Mr. Moreno can stack a fraction, $\frac{12}{20}$, of the wood. In 12 minutes, Mrs. Moreno can stack a fraction, $\frac{12}{t}$, of the wood. Together they can stack all the wood in 12 minutes.

Thus $\dfrac{12}{20}+\dfrac{12}{t}=1$

$$20t\left(\frac{12}{20}+\frac{12}{t}\right)=20t$$
$$12t+240=20t$$
$$240=8t$$
$$30=t$$

It would take Mrs. Moreno 30 minutes to stack the wood by herself.

Group Activity/Challenge Problems

2. Perimeter:

$(\sqrt{5}+\sqrt{3})+(\sqrt{5}+\sqrt{3})+(\sqrt{5}+\sqrt{3})$
$+(\sqrt{5}+\sqrt{3})=4(\sqrt{5}+\sqrt{3})$ or about 15.87

Area:

$(\sqrt{5}+\sqrt{3})(\sqrt{5}+\sqrt{3})$
$=(\sqrt{5})^2+2\sqrt{3}\sqrt{5}+(\sqrt{3})^2$
$=5+2\sqrt{15}+3=8+2\sqrt{15}$ or about 15.75

3. Perimeter:

$(\sqrt{6}+2)+(\sqrt{6}-2)+(\sqrt{6}+2)+(\sqrt{6}-2)$
$=2(\sqrt{6}+2)+2(\sqrt{6}-2)$
$=2\sqrt{6}+4+2\sqrt{6}-4$
$=2(2\sqrt{6})=4\sqrt{6}$ or about 9.80

Area:

$(\sqrt{6}+2)(\sqrt{6}-2)=(\sqrt{6})^2-2^2=6-4=2$

5. $-5\sqrt{x}+2\sqrt{3}+3\sqrt{27}=-9\sqrt{3}$

$-5\sqrt{x}+2\sqrt{3}+9\sqrt{3}=-9\sqrt{3}$

$-5\sqrt{x}+11\sqrt{3}=-9\sqrt{3}$

$-5\sqrt{x}=-20\sqrt{3}$

$(-5\sqrt{x})^2=(-20\sqrt{3})^2$

$25x=1200$

$x=48$

Exercise Set 9.5

1. $\sqrt{x}=8$

$(\sqrt{x})^2=8^2$

$x=64$

Check: $\sqrt{x}=8$

$\sqrt{64}=8$

$8=8$ True

3. $\sqrt{x}=-3$

$\left(\sqrt{x}\right)^2=(-3)^2$

$x=9$

Check: $\sqrt{x}=-3$

$\sqrt{9}=-3$

$3=-3$ False

9 is an extraneous root and $\sqrt{x}=-3$ has no real solution

5. $\sqrt{x+5}=3$

$(\sqrt{x+5})^2=3^2$

$x+5=9$

$x+5-5=9-5$

$x=4$

Check: $\sqrt{x+5}=3$

$\sqrt{4+5}=3$

$\sqrt{9}=3$

$3=3$ True

7. $\sqrt{2x+4}=-6$

$(\sqrt{2x+4})^2=(-6)^2$

$2x+4=36$

$2x=32$

$x=16$

Check: $\sqrt{2x+4}=-6$

$\sqrt{2(16)+4}=-6$

$\sqrt{32+4}=-6$

$\sqrt{36}=-6$

$6=-6$ False

16 is an extraneous root and

$\sqrt{2x+4}=-6$ has no real solution.

9. $\sqrt{x}+3=5$

$\sqrt{x}+3-3=5-3$

$\sqrt{x}=2$

$(\sqrt{x})^2=2^2$

$x=4$

Check: $\sqrt{x}+3=5$

$\sqrt{4}+3=5$

$2+3=5$

$5=5$ True

11. $6=4+\sqrt{x}$

$6-4=4-4+\sqrt{x}$

$2=\sqrt{x}$

$2^2=(\sqrt{x})^2$

$4=x$

Check: $6=4+\sqrt{x}$

$6=4+\sqrt{4}$

$6=4+2$

$6=6$ True

13. $4+\sqrt{x}=2$

$4-4+\sqrt{x}=2-4$

$\sqrt{x}=-2$

$(\sqrt{x})^2=(-2)^2$

$x=4$

Check: $4+\sqrt{x}=2$

$4+\sqrt{4}=2$

$4+2=2$

$6=2$ False

4 is an extraneous root and

$4+\sqrt{x}=2$ has no real solution.

15. $\sqrt{2x-5}=x-4$

$(\sqrt{2x-5})^2=(x-4)^2$

$2x-5=x^2-8x+16$

$0=x^2-10x+21$

$0=(x-3)(x-7)$

$x-3=0$ or $x-7=0$

$x=3$ $\quad$ $x=7$

Check: $x=3$ $\quad$ $x=7$

$\sqrt{2x-5}=x-4$ $\quad$ $\sqrt{2x-5}=x-4$

$\sqrt{2(3)-5}=3-4$ $\quad$ $\sqrt{2(7)-5}=7-4$

$\sqrt{1}=-1$ $\quad$ $\sqrt{9}=3$

$1=-1$ False $\quad$ $3=3$ True

The solution is 7; 3 is not a solution.

17. $\sqrt{2x-6}=\sqrt{5x-27}$

$$\left(\sqrt{2x-6}\right)^2=\left(\sqrt{5x-27}\right)^2$$
$$2x-6=5x-27$$
$$-6=3x-27$$
$$21=3x$$
$$7=x$$

Check: $\sqrt{2x-6}=\sqrt{5x-27}$

$$\sqrt{2(7)-6}=\sqrt{5(7)-27}$$
$$\sqrt{8}=\sqrt{8}\quad\text{True}$$

19. $\sqrt{3x+3}=\sqrt{5x-1}$

$$\left(\sqrt{3x+3}\right)^2=\left(\sqrt{5x-1}\right)^2$$
$$3x+3=5x-1$$
$$3=2x-1$$
$$4=2x$$
$$2=x$$

Check: $\sqrt{3x+3}=\sqrt{5x-1}$

$$\sqrt{3(2)+3}=\sqrt{5(2)-1}$$
$$\sqrt{9}=\sqrt{9}\quad\text{True}$$

21. $\sqrt{3x+9}=2\sqrt{x}$

$$\left(\sqrt{3x+9}\right)^2=\left(2\sqrt{x}\right)^2$$
$$3x+9=4x$$
$$9=x$$

Check: $\sqrt{3x+9}=2\sqrt{x}$

$$\sqrt{3(9)+9}=2\sqrt{9}$$
$$\sqrt{36}=2\cdot 3$$
$$6=6\quad\text{True}$$

23. $\sqrt{4x-5}=\sqrt{x+9}$

$$\left(\sqrt{4x-5}\right)^2=\left(\sqrt{x+9}\right)^2$$
$$4x-5=x+9$$
$$3x-5=9$$
$$3x=14$$
$$x=\frac{14}{3}$$

Check: $\sqrt{4x-5}=\sqrt{x+9}$

$$\sqrt{4\left(\frac{14}{3}\right)-5}=\sqrt{\left(\frac{14}{3}\right)+9}$$
$$\sqrt{\frac{41}{3}}=\sqrt{\frac{41}{3}}\quad\text{True}$$

25. $3\sqrt{x}=\sqrt{x+8}$

$$(3\sqrt{x})^2=(\sqrt{x+8})^2$$
$$9x=x+8$$
$$8x=8$$
$$x=1$$

Check: $3\sqrt{x}=\sqrt{x+8}$

$$3\sqrt{1}=\sqrt{1+8}$$
$$3(1)=\sqrt{9}$$
$$3=3\quad\text{True}$$

27. $4\sqrt{x}=x+3$

$$(4\sqrt{x})^2=(x+3)^2$$
$$16x=x^2+6x+9$$
$$0=x^2-10x+9$$
$$0=(x-9)(x-1)$$
$$x-9=0\quad\text{or}\quad x-1=0$$
$$x=9\quad\text{or}\quad x=1$$

Check: $x = 9$ $\quad$ $x = 1$

$4\sqrt{x} = x + 3$ $\quad$ $4\sqrt{x} = x + 3$

$4\sqrt{9} = 9 + 3$ $\quad$ $4\sqrt{1} = 1 + 3$

$4 \cdot 3 = 12$ $\quad$ $4 \cdot 1 = 4$

$12 = 12$ True $\quad$ $4 = 4$ True

The solutions are 9 and 1

29. $\sqrt{2x-3} = 2\sqrt{3x-2}$

$$\left(\sqrt{2x-3}\right)^2 = \left(2\sqrt{3x-2}\right)^2$$
$$2x - 3 = 4(3x-2)$$
$$2x - 3 = 12x - 8$$
$$-3 = 10x - 8$$
$$5 = 10x$$
$$\frac{1}{2} = x$$

Check: $\sqrt{2x-3} = 2\sqrt{3x-2}$

$$\sqrt{2\left(\frac{1}{2}\right)-3} = 2\sqrt{3\left(\frac{1}{2}\right)-2}$$
$$\sqrt{-2} = 2\sqrt{-\frac{1}{2}} \quad \text{False}$$

$\frac{1}{2}$ is an extraneous root. There is no solution.

31. $\sqrt{x^2+5} = x + 5$

$$\left(\sqrt{x^2+5}\right)^2 = (x+5)^2$$
$$x^2 + 5 = x^2 + 10x + 25$$
$$5 = 10x + 25$$
$$-20 = 10x$$
$$-2 = x$$

Check: $\sqrt{x^2+5} = x + 5$

$$\sqrt{(-2)^2+5} = -2 + 5$$
$$\sqrt{9} = 3$$
$$3 = 3 \quad \text{True}$$

33. $5 + \sqrt{x-5} = x$

$$5 - 5 + \sqrt{x-5} = x - 5$$
$$\sqrt{x-5} = x - 5$$
$$\left(\sqrt{x-5}\right)^2 = (x-5)^2$$
$$x - 5 = x^2 - 10x + 25$$
$$0 = x^2 - 11x + 30$$
$$0 = (x-5)(x-6)$$

$x - 5 = 0$ or $x - 6 = 0$

$x = 5$ or $x = 6$

Check: $x = 5$ $\quad$ $x = 6$

$5 + \sqrt{x-5} = x$ $\quad$ $5 + \sqrt{x-5} = x$

$5 + \sqrt{5-5} = 5$ $\quad$ $5 + \sqrt{6-5} = 6$

$5 + \sqrt{0} = 5$ $\quad$ $5 + \sqrt{1} = 6$

$5 = 5$ True $\quad$ $5 + 1 = 6$

$6 = 6$ True

The solutions are 5 and 6.

35. $\sqrt{8-7x} = x - 2$

$$\left(\sqrt{8-7x}\right)^2 = (x-2)^2$$
$$8 - 7x = x^2 - 4x + 4$$
$$8 = x^2 + 3x + 4$$
$$0 = x^2 + 3x - 4$$
$$0 = (x+4)(x-1)$$

$x + 4 = 0$ or $x - 1 = 0$

$x = -4$ or $x = 1$

Check: $x = -4$

$\sqrt{8-7x} = x-2$

$\sqrt{8-7(-4)} = -4-2$

$\sqrt{8+28} = -6$

$\sqrt{36} = -6$

$6 = -6$ False

Check $x = 1$

$\sqrt{8-7x} = x-2$

$\sqrt{8-7(1)} = 1-2$

$\sqrt{8-7} = -1$

$\sqrt{1} = -1$

$1 = -1$ False

Both $x = -4$ and $x = -1$ are extraneous roots. There is no real number solution.

37. An equation that contains a variable in a radicand.

39. There may be extraneous roots.

Cumulative Review Exercises: 9.5

41. $3x - 2y = 6$ Ordered Pair

Let $x = 0$, then $y = -3$ (0, –3)

Let $y = 0$, then $x = 2$ (2, 0)

$y = 2x - 4$ Ordered Pair

Let $x = 0$, then $y = -4$ (0, –4)

Let $y = 0$, then $x = 2$ (2, 0)

Graph both equations.

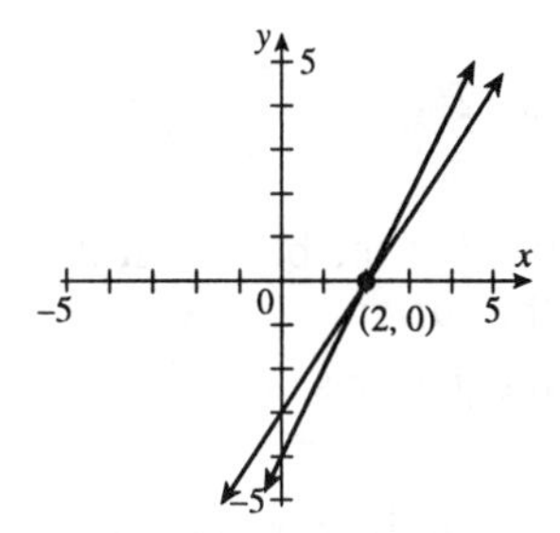

The graphs intersect at the point (2, 0).

Check:	$3x - 2y = 6$	$y = 2x - 4$
	$3(2) - 2(0) = 6$	$0 = 2(2) - 4$
	$6 - 0 = 6$	$0 = 4 - 4$
	$6 = 6$ True	$0 = 0$ True

The solution is (2, 0).

42. Substitute $2x - 4$ for y in the first equation.

$3x - 2y = 6$

$3x - 2(2x - 4) = 6$

$3x - 4x + 8 = 6$

$-x + 8 = 6$

$-x = -2$

$x = 2$

Solve for y.

$y = 2x - 4$

$y = 2(2) - 4$

$y = 4 - 4 = 0$

The solution is (2, 0).

43. Align the x and y terms on the left side of the equation $y = 2x - 4$

$$-2x + y = -4$$

$3x - 2y = 6$ gives $3x - 2y = 6$
$2[-2x + y = -4]$ $\quad \underline{-4x + 2y = -8}$

$$-x + 0 = -2$$
$$x = 2$$

Solve for y.

$y = 2x - 4$

$y = 2(2) - 4$

$y = 4 - 4 = 0$

The solution is (2, 0).

44. Let b = the speed of the boat in still water, c = speed of the current.
The system of equations is

$b + c = 12$

Speed of boat with the current = 12 mph.

$b - c = 4$

Speed of boat against current = 4 mph.

$$b + c = 12$$
$$\underline{b - c = 4}$$
$$2b \quad = 16$$
$$b \quad = 8$$

Solve for c.

$b + c = 12$

$8 + c = 12$

$c = 4$

Speed of boat in still water = 8 mph, speed of the current = 4 mph.

Group Activity/Challenge Problems

1. $\sqrt{x} + 2 = \sqrt{x + 16}$

$(\sqrt{x} + 2)^2 = (\sqrt{x + 16})^2$

$x + 4\sqrt{x} + 4 = x + 16$

$4\sqrt{x} = 12$

$(4\sqrt{x})^2 = (12)^2$

$16x = 144$

$x = 9$

3. $\sqrt{x + 7} = 5 - \sqrt{x - 8}$

$\left(\sqrt{x + 7}\right)^2 = \left(5 - \sqrt{x - 8}\right)^2$

$x + 7 = 25 - 10\sqrt{x - 8} + x - 8$

$10\sqrt{x - 8} = 25 + x - 8 - x - 7$

$10\sqrt{x - 8} = 10$

$\sqrt{x - 8} = 1$

$\left(\sqrt{x - 8}\right)^2 = 1^2$

$x - 8 = 1$

$x = 9$

5. **(a)**

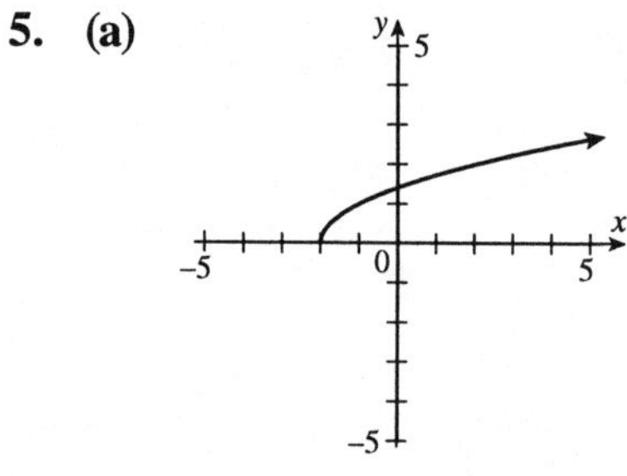

(b) The graph is not linear.

(c) The graph is a function.

Exercise Set 9.6

1. $a^2 + b^2 = c^2$

$x^2 + 5^2 = (12)^2$

$x^2 + 25 = 144$

$x^2 = 119$

$x = \sqrt{119} \approx 10.91$

3. $a^2 + b^2 = c^2$

$(10)^2 + 8^2 = x^2$

$100 + 64 = x^2$

$164 = x^2$

$x = \sqrt{164} \approx 12.81$

5. $a^2 + b^2 = c^2$

$(15)^2 + x^2 = (20)^2$

$225 + x^2 = 400$

$x^2 = 175$

$x = \sqrt{175} \approx 13.23$

7. $a^2 + b^2 = c^2$

$6^2 + \left(\sqrt{5}\right)^2 = x^2$

$36 + 5 = x^2$

$41 = x^2$

$x = \sqrt{41} \approx 6.40$

9. $a^2 + b^2 = c^2$

$(12)^2 + \left(\sqrt{5}\right)^2 = x^2$

$144 + 5 = x^2$

$149 = x^2$

$x = \sqrt{149} \approx 12.21$

11. $a^2 + b^2 = c^2$

$4^2 + x^2 = (12)^2$

$16 + x^2 = 144$

$x^2 = 128$

$x = \sqrt{128} \approx 11.31$

13.

53.3
d
120

$a^2 + b^2 = c^2$

$(120)^2 + (53.3)^2 = x^2$

$14400 + 2840.89 = x^2$

$17{,}240.89 = x^2$

$x = \sqrt{17{,}240.89}$

The length of the diagonal is $\sqrt{17{,}240.89} \approx 131.30$ yards

15.

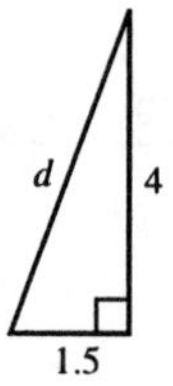

$$a^2 + b^2 = c^2$$
$$4^2 + (1.5)^2 = d^2$$
$$16 + 2.25 = d^2$$
$$18.25 = d^2$$
$$d = \sqrt{18.25}$$

The wire must be $\sqrt{18.25} \approx 4.27$ meters long.

17. $d = \sqrt{(x_2 - x_1)^2 + (y_2 - y_1)^2}$
$$= \sqrt{[-1-(-4)]^2 + (4-3)^2}$$
$$= \sqrt{(-1+4)^2 + (4-3)^2}$$
$$= \sqrt{3^2 + 1^2}$$
$$= \sqrt{10} \approx 3.16$$

19. $d = \sqrt{(x_2 - x_1)^2 + (y_2 - y_1)^2}$
$$= \sqrt{[4-(-8)]^2 + (-8-4)^2}$$
$$= \sqrt{(4+8)^2 + (-12)^2}$$
$$= \sqrt{(12)^2 + (-12)^2}$$
$$= \sqrt{144 + 144}$$
$$= \sqrt{288} \approx 16.97$$

21. $A = s^2$
$$144 = s^2$$
$$s = \sqrt{144} = 12 \text{ in.}$$
The side of the square is 12 inches.

23. $A = \pi r^2$
$$80 = (3.14)r^2$$
$$\frac{80}{3.14} = r^2$$
$$r^2 \approx 25.48$$
$$r = \sqrt{25.48} \text{ ft}$$
The radius of the circle is $\sqrt{25.48} \approx 5.05$ feet

25. $a^2 + b^2 = c^2$
$$9^2 + (12)^2 = d^2$$
$$81 + 144 = d^2$$
$$225 = d^2$$
$$d = \sqrt{225} = 15 \text{ in.}$$
The length of the diagonal is 15 inches.

27. $v = \sqrt{2gh}$
$$= \sqrt{2(32)(80)}$$
$$= \sqrt{5120} \approx 71.55 \text{ ft/sec}$$
The velocity of the lamp after falling 80 feet is approximately 71.55 feet per second.

29. $v = \sqrt{2gh}$
$$= \sqrt{2(32)(1250)}$$
$$= \sqrt{80{,}000} \approx 282.84 \text{ ft/sec}$$
The velocity of the olive when it strikes the ground is approximately 282.84 feet per second.

31. $T = 2\pi\sqrt{L/32}$

$= 2(3.14)\sqrt{\dfrac{40}{32}}$

$= 6.28\sqrt{1.25} \approx 7.02$ sec

The period of a 40 foot pendulum is approximately 7.02 seconds.

33. $T = 2\pi\sqrt{L/32}$

$= 2(3.14)\sqrt{\dfrac{10}{32}}$

$= 6.28\sqrt{0.3125} \approx 3.51$ sec

The period of a 10 foot pendulum is approximately 3.51 seconds.

35. $N = 0.2\left(\sqrt{R}\right)^3$

$N = 0.2\left(\sqrt{1418}\right)^3$

$\approx 0.2(37.7)^3$

$\approx 0.2(53396.7)$

$\approx 10{,}679.34$

There are approximately 10,679.34 Earth days in the year of Saturn.

37. $v_e = \sqrt{2gR}$

$= \sqrt{2(9.75)(6{,}370{,}000)}$

$= \sqrt{124{,}215{,}000} \approx 11{,}145.18$ m/sec

The escape velocity for Earth is approximately 11,145.18 meters per second.

Cumulative Review Exercises: 9.6

38. $2(x+3) < 4x - 6$

$2x + 6 < 4x - 6$

$6 < 2x - 6$

$12 < 2x$

$6 < x$ or $x > 6$

39. $(4x^{-4}y^3)^{-1} = (4)^{-1}(x^{-4})^{-1}(y^3)^{-1}$

$= (4^{-1})(x^4)(y^{-3}) = \dfrac{x^4}{4y^3}$

40. $x^3 - 2x^2 - 6x + 4 - (3x^3 - 6x^2 + 8)$

$= x^3 - 2x^2 - 6x + 4 - 3x^3 + 6x^2 - 8$

$= x^3 - 3x^3 + 6x^2 - 2x^2 - 6x + 4 - 8$

$= -2x^3 + 4x^2 - 6x - 4$

41. $\dfrac{5x^4 - 9x^3 + 6x^2 - 4x - 3}{3x^2}$

$= \dfrac{5x^4}{3x^2} - \dfrac{9x^3}{3x^2} + \dfrac{6x^2}{3x^2} - \dfrac{4x}{3x^2} - \dfrac{3}{3x^2}$

$= \dfrac{5}{3}x^2 - 3x + 2 - \dfrac{4}{3x} - \dfrac{1}{x^2}$

Group Activity/Challenge Problems

1. Let the width be w, then the length is $w + 3$.

$w^2 + (w+3)^2 = (15)^2$

$w^2 + w^2 + 6w + 9 = 225$

$2w^2 + 6w - 216 = 0$

$w^2 + 3w - 108 = 0$

$(w-9)(w+12) = 0$

$w = 9$ or $w = -12$

Since the measurement must be positive, $w = 9$. Therefore the width is 9 inches and the length is $9 + 3 = 12$ inches.

3. $T = 2\pi\sqrt{\frac{L}{32}}$

$2 = 2\pi\sqrt{\frac{L}{32}}$

$\frac{2}{2\pi} = \frac{\sqrt{L}}{\sqrt{32}}$

$\frac{\sqrt{32}}{\pi} = \sqrt{L}$

$\left(\frac{\sqrt{32}}{\pi}\right)^2 = \left(\sqrt{L}\right)^2$

$\frac{32}{\pi^2} = L$

$\frac{32}{(3.14)^2} = L$

$3.25 = L$

The length is approximately 3.25 ft.

Exercise Set 9.7

1. $\sqrt[3]{8} = 2$ since $2^3 = 8$

3. $\sqrt[3]{8} = -2$ since $(-2)^3 = 8$

5. $\sqrt[4]{16} = 2$ since $2^4 = 16$

7. $\sqrt[4]{81} = 3$ since $3^4 = 81$

9. $\sqrt[3]{-1} = -1$ since $(-1)^3 = -1$

11. $\sqrt[3]{64} = 4$ since $4^3 = 64$

13. $\sqrt[3]{54} = \sqrt[3]{27 \cdot 2} = \sqrt[3]{27}\ \sqrt[3]{2} = 3\sqrt[3]{2}$

15. $\sqrt[3]{16} = \sqrt[3]{8 \cdot 2} = \sqrt[3]{8}\ \sqrt[3]{2} = 2\sqrt[3]{2}$

17. $\sqrt[3]{81} = \sqrt[3]{27 \cdot 3} = \sqrt[3]{27}\ \sqrt[3]{3} = 3\sqrt[3]{3}$

19. $\sqrt[4]{32} = \sqrt[4]{16 \cdot 2} = \sqrt[4]{16}\ \sqrt[4]{2} = 2\sqrt[4]{2}$

21. $\sqrt[3]{40} = \sqrt[3]{8 \cdot 5} = \sqrt[3]{8}\ \sqrt[3]{5} = 2\sqrt[3]{5}$

23. $\sqrt[3]{x^3} = x^{3/3} = x^1 = x$

25. $\sqrt[4]{y^{12}} = y^{12/4} = y^3$

27. $\sqrt[3]{x^{12}} = x^{12/3} = x^4$

29. $\sqrt[4]{y^4} = y^{4/4} = y^1 = y$

31. $\sqrt[3]{x^{15}} = x^{15/3} = x^5$

33. $8^{4/3} = \left(\sqrt[3]{8}\right)^4 = 2^4 = 16$

35. $16^{3/4} = \left(\sqrt[4]{16}\right)^3 = 2^3 = 8$

37. $1^{5/3} = \left(\sqrt[3]{1}\right)^5 = 1^5 = 1$

39. $9^{3/2} = \left(\sqrt[2]{9}\right)^3 = 3^3 = 27$

41. $16^{3/4} = \left(\sqrt[4]{16}\right)^3 = 2^3 = 8$

43. $125^{4/3} = \left(\sqrt[3]{125}\right)^4 = 5^4 = 625$

45. $27^{-2/3} = \dfrac{1}{27^{2/3}} = \dfrac{1}{\left(\sqrt[3]{27}\right)^2} = \dfrac{1}{3^2} = \dfrac{1}{9}$

47. $8^{-5/3} = \dfrac{1}{8^{5/3}} = \dfrac{1}{\left(\sqrt[3]{8}\right)^5} = \dfrac{1}{2^5} = \dfrac{1}{32}$

49. $\sqrt[3]{x^7} = x^{7/3}$

51. $\sqrt[3]{x^4} = x^{4/3}$

53. $\sqrt[4]{y^{15}} = y^{15/4}$

55. $\sqrt[4]{y^{21}} = y^{21/4}$

57. $\sqrt[3]{x} \cdot \sqrt[3]{x} = x^{1/3} \cdot x^{1/3}$

$= x^{(1/3)+(1/3)}$

$= x^{2/3}$

59. $\sqrt[4]{x^2} \cdot \sqrt[4]{x^2} = x^{2/4} \cdot x^{2/4}$

$= x^{1/2} \cdot x^{1/2}$

$= x$

61. $\left(\sqrt[3]{x^2}\right)^6 = \left(x^{2/3}\right)^6 = x^4$

63. $\left(\sqrt[4]{x^2}\right)^4 = \left(x^{2/4}\right)^4$

$= \left(x^{1/2}\right)^4 = x^2$

65. For $x = 8$, $\left(\sqrt[3]{x}\right)^2 = \left(\sqrt[3]{8}\right)^2 = 2^2 = 4$

For $x = 8$, $\left(\sqrt[3]{x^2}\right) = \left(\sqrt[3]{8^2}\right) = \sqrt[3]{64} = 4$

67. **(b)** $x^{5/8} = \sqrt[8]{x^5}$ or $\left(\sqrt[8]{x}\right)^5$

Cumulative Review Exercises: 9.7

69. Substitute 2 for x and -4 for y.

$-x^2 + 4xy - 6 = -(2)^2 + 4(2)(-4) - 6$

$= -4 - 32 - 6 = -42$

70. $3x^2 - 28x + 32 = 3x^2 - 24x - 4x + 32$

$= 3x(x-8) - 4(x-8) = (3x-4)(x-8)$

71. $y = \frac{2}{3}x - 4$ Ordered Pair

Let $x = 0$, then $y = \frac{2}{3}(0) - 4 = -4$ $(0, -4)$

Let $y = 0$, then $0 = \frac{2}{3}x - 4$

$4 = \frac{2}{3}x$

$6 = x$ $(6, 0)$

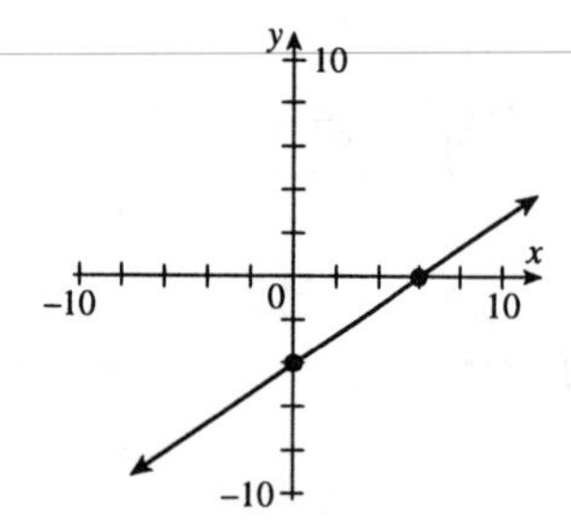

72. $\sqrt{\frac{64x^3y^7}{2x^4}} = \sqrt{\frac{32y^7}{x}} = \frac{\sqrt{32y^7}}{\sqrt{x}}$

$= \frac{\sqrt{16y^6}\sqrt{2y}}{\sqrt{x}} = \frac{4y^3\sqrt{2y}}{\sqrt{x}}$

$= \frac{4y^3\sqrt{2y}}{\sqrt{x}} \cdot \frac{\sqrt{x}}{\sqrt{x}} = \frac{4y^3\sqrt{2xy}}{x}$

Group Activity/Challenge Problems

1. $\sqrt[3]{xy} \cdot \sqrt[3]{x^2y^2} = \sqrt[3]{x^3y^3} = xy$

3. $\sqrt[4]{32} - \sqrt[4]{2} = \sqrt[4]{16 \cdot 2} - \sqrt[4]{2}$

$= \sqrt[4]{16} \cdot \sqrt[4]{2} - \sqrt[4]{2} = \sqrt[4]{2}\left(\sqrt[4]{16} - 1\right)$

$= \sqrt[4]{2}(2-1) = \sqrt[4]{2}$

5. **(a)** Individual answer.

(b) $\frac{1}{\sqrt[3]{2}} = \frac{1 \cdot \sqrt[3]{2^2}}{\sqrt[3]{2}\sqrt[3]{2^2}} = \frac{\sqrt[3]{4}}{2}$

(c) Individual answer

7. $\frac{1}{\sqrt[3]{x^2}} = \frac{1 \cdot \sqrt[3]{x}}{\sqrt[3]{x^2} \cdot \sqrt[3]{x}} = \frac{\sqrt[3]{x}}{x}$

9. $\frac{4}{\sqrt[3]{5y^4}} = \frac{4 \cdot \sqrt[3]{5^2y^2}}{\sqrt[3]{5y^4} \cdot \sqrt[3]{5^2y^2}} = \frac{4\sqrt[3]{25y^2}}{5y^2}$

Chapter 9 Review Exercises

1. $\sqrt{25} = 5$ since $5^2 = 25$

2. $\sqrt{6} = 6$ since $6^2 = 36$

3. $\sqrt{81} = 9$

Take the opposite of both sides to get $-\sqrt{81} = -9$

4. $\sqrt{8} = 8^{1/2}$

5. $\sqrt{26x} = (26x)^{1/2}$

6. $\sqrt{20xy^2} = (20xy^2)^{1/2}$

7. $\sqrt{32} = \sqrt{16 \cdot 2} = \sqrt{16}\sqrt{2} = 4\sqrt{2}$

8. $\sqrt{44} = \sqrt{4 \cdot 11} = \sqrt{4}\sqrt{11} = 2\sqrt{11}$

9. $\sqrt{45x^5y^4} = \sqrt{9x^4y^4}\sqrt{5x} = 3x^2y^2\sqrt{5x}$

10. $\sqrt{125x^4y^6} = \sqrt{25x^4y^6}\sqrt{5} = 5x^2y^3\sqrt{5}$

11. $\sqrt{15x^5yz^3} = \sqrt{x^4z^2}\sqrt{15xyz} = x^2z\sqrt{15xyz}$

12. $\sqrt{48ab^4c^5} = \sqrt{16b^4c^4}\sqrt{3ac} = 4b^2c^2\sqrt{3ac}$

13. $\sqrt{8}\cdot\sqrt{12} = \sqrt{96} = \sqrt{16\cdot 6} = \sqrt{16}\sqrt{6} = 4\sqrt{6}$

14. $\sqrt{5x}\sqrt{5x} = \sqrt{25x^2} = 5x$

15. $\sqrt{18x}\cdot\sqrt{2xy} = \sqrt{36x^2y} = \sqrt{36x^2}\sqrt{y}$
$= 6x\sqrt{y}$

16. $\sqrt{25x^2y}\cdot\sqrt{3y} = \sqrt{75x^2y^2} = \sqrt{25x^2y^2}\sqrt{3}$
$= 5xy\sqrt{3}$

17. $\sqrt{20xy^4}\cdot\sqrt{5xy^3} = \sqrt{100x^2y^7}$
$= \sqrt{100x^2y^6}\sqrt{y} = 10xy^3\sqrt{y}$

18. $\sqrt{8x^3y}\cdot\sqrt{3y^4} = \sqrt{24x^3y^5} = \sqrt{4x^2y^4}\sqrt{6xy}$
$= 2xy^2\sqrt{6xy}$

19. $\dfrac{\sqrt{32}}{\sqrt{2}} = \sqrt{\dfrac{32}{2}} = \sqrt{16} = 4$

20. $\sqrt{\dfrac{10}{250}} = \sqrt{\dfrac{1}{25}} = \dfrac{\sqrt{1}}{\sqrt{25}} = \dfrac{1}{5}$

21. $\sqrt{\dfrac{7}{28}} = \sqrt{\dfrac{1}{4}} = \dfrac{\sqrt{1}}{\sqrt{4}} = \dfrac{1}{2}$

22. $\dfrac{3}{\sqrt{5}} = \dfrac{3}{\sqrt{5}}\cdot\dfrac{\sqrt{5}}{\sqrt{5}} = \dfrac{3\sqrt{5}}{5}$

23. $\sqrt{\dfrac{5x}{12}} = \dfrac{\sqrt{5x}}{\sqrt{12}} = \dfrac{\sqrt{5x}}{\sqrt{4}\sqrt{3}} = \dfrac{\sqrt{5x}}{2\sqrt{3}}$
$= \dfrac{\sqrt{5x}}{2\sqrt{3}}\cdot\dfrac{\sqrt{3}}{\sqrt{3}} = \dfrac{\sqrt{15x}}{2\cdot 3} = \dfrac{\sqrt{15x}}{6}$

24. $\sqrt{\dfrac{x}{6}} = \dfrac{\sqrt{x}}{\sqrt{6}} = \dfrac{\sqrt{x}}{\sqrt{6}}\cdot\dfrac{\sqrt{6}}{\sqrt{6}} = \dfrac{\sqrt{6x}}{6}$

25. $\sqrt{\dfrac{x^2}{2}} = \dfrac{\sqrt{x^2}}{\sqrt{2}} = \dfrac{x}{\sqrt{2}} = \dfrac{x}{\sqrt{2}}\cdot\dfrac{\sqrt{2}}{\sqrt{2}} = \dfrac{x\sqrt{2}}{2}$

26. $\sqrt{\dfrac{x^5}{8}} = \dfrac{\sqrt{x^5}}{\sqrt{8}} = \dfrac{\sqrt{x^4}\sqrt{x}}{\sqrt{4}\sqrt{2}} = \dfrac{x^2\sqrt{x}}{2\sqrt{2}}$
$= \dfrac{x^2\sqrt{x}}{2\sqrt{2}}\cdot\dfrac{\sqrt{2}}{\sqrt{2}} = \dfrac{x^2\sqrt{2x}}{4}$

27. $\sqrt{\dfrac{60xy^5}{4x^5y^3}} = \sqrt{\dfrac{15y^2}{x^4}} = \dfrac{\sqrt{15y^2}}{\sqrt{x^4}}$
$= \dfrac{\sqrt{y^2}\sqrt{15}}{\sqrt{x^4}} = \dfrac{y\sqrt{15}}{x^2}$

28. $\sqrt{\dfrac{30x^4y}{15x^2y^4}} = \sqrt{\dfrac{2x^2}{y^3}} = \dfrac{\sqrt{2x^2}}{\sqrt{y^3}} = \dfrac{\sqrt{2}\sqrt{x^2}}{\sqrt{y^2}\sqrt{y}}$
$= \dfrac{x\sqrt{2}}{y\sqrt{y}} = \dfrac{x\sqrt{2}}{y\sqrt{y}}\cdot\dfrac{\sqrt{y}}{\sqrt{y}} = \dfrac{x\sqrt{2y}}{y^2}$

29. $\dfrac{\sqrt{90}}{\sqrt{8x^3y^2}} = \dfrac{\sqrt{9}\sqrt{10}}{\sqrt{4x^2y^2}\sqrt{2x}} = \dfrac{3\sqrt{10}}{2xy\sqrt{2x}}$

$= \dfrac{3\sqrt{10}}{2xy\sqrt{2x}} \cdot \dfrac{\sqrt{2x}}{\sqrt{2x}} = \dfrac{3\sqrt{20x}}{2xy \cdot 2x} = \dfrac{3\sqrt{4}\sqrt{5x}}{4x^2y}$

$= \dfrac{3 \cdot 2\sqrt{5x}}{4x^2y} = \dfrac{3\sqrt{5x}}{2x^2y}$

30. $\dfrac{\sqrt{2x^4yz^4}}{\sqrt{7x^5yz^2}} = \sqrt{\dfrac{2x^4yz^4}{7x^5yz^2}} = \sqrt{\dfrac{2z^2}{7x}} = \dfrac{\sqrt{2z^2}}{\sqrt{7x}}$

$= \dfrac{\sqrt{2}\sqrt{z^2}}{\sqrt{7x}} = \dfrac{z\sqrt{2}}{\sqrt{7x}} = \dfrac{z\sqrt{2}}{\sqrt{7x}} \cdot \dfrac{\sqrt{7x}}{\sqrt{7x}} = \dfrac{z\sqrt{14x}}{7x}$

31. $\dfrac{3}{1+\sqrt{2}} = \dfrac{3}{1+\sqrt{2}} \cdot \dfrac{1-\sqrt{2}}{1-\sqrt{2}} = \dfrac{3(1-\sqrt{2})}{1-2}$

$= \dfrac{3(1-\sqrt{2})}{-1} = -3(1-\sqrt{2})$

32. $\dfrac{5}{3-\sqrt{6}} = \dfrac{5}{3-\sqrt{6}} \cdot \dfrac{3+\sqrt{6}}{3+\sqrt{6}} = \dfrac{5(3+\sqrt{6})}{9-6}$

$= \dfrac{15+5\sqrt{6}}{3}$

33. $\dfrac{\sqrt{3}}{2+\sqrt{x}} = \dfrac{\sqrt{3}}{2+\sqrt{x}} \cdot \dfrac{2-\sqrt{x}}{2-\sqrt{x}} = \dfrac{\sqrt{3}(2-\sqrt{x})}{4-x}$

$= \dfrac{2\sqrt{3}-\sqrt{3x}}{4-x}$

34. $\dfrac{2}{\sqrt{x}-5} = \dfrac{2}{\sqrt{x}-5} \cdot \dfrac{\sqrt{x}+5}{\sqrt{x}+5} = \dfrac{2(\sqrt{x}+5)}{x-25}$

$= \dfrac{2\sqrt{x}+10}{x-25}$

35. $\dfrac{\sqrt{5}}{\sqrt{x}+\sqrt{3}} = \dfrac{\sqrt{5}}{\sqrt{x}+\sqrt{3}} \cdot \dfrac{\sqrt{x}-\sqrt{3}}{\sqrt{x}-\sqrt{3}}$

$= \dfrac{\sqrt{5}(\sqrt{x}-\sqrt{3})}{x-3} = \dfrac{\sqrt{5x}-\sqrt{15}}{x-3}$

36. $6\sqrt{3} - 2\sqrt{3} = (6-2)\sqrt{3} = 4\sqrt{3}$

37. $6\sqrt{2} - 8\sqrt{2} + \sqrt{2} = (6-8+1)\sqrt{2} = -1\sqrt{2}$

$= -\sqrt{2}$

38. $3\sqrt{x} - 5\sqrt{x} = (3-5)\sqrt{x} = -2\sqrt{x}$

39. $\sqrt{x} + 3\sqrt{x} - 4\sqrt{x} = (1+3-4)\sqrt{x}$

$= 0\sqrt{x} = 0$

40. $\sqrt{8} - \sqrt{2} = \sqrt{4}\sqrt{2} - \sqrt{2} = 2\sqrt{2} - \sqrt{2}$

$= (2-1)\sqrt{2} = 1\sqrt{2} = \sqrt{2}$

41. $7\sqrt{40} - 2\sqrt{10} = 7\sqrt{4}\sqrt{10} - 2\sqrt{10}$

$= 7 \cdot 2\sqrt{10} - 2\sqrt{10} = 14\sqrt{10} - 2\sqrt{10}$

$= (14-2)\sqrt{10} = 12\sqrt{10}$

42. $2\sqrt{98} - 4\sqrt{72} = 2\sqrt{49}\sqrt{2} - 4\sqrt{36}\sqrt{2}$

$= 2 \cdot 7\sqrt{2} - 4 \cdot 6\sqrt{2} = 14\sqrt{2} - 24\sqrt{2}$

$= (14-24)\sqrt{2} = -10\sqrt{2}$

43. $3\sqrt{18} + 5\sqrt{50} - 2\sqrt{32}$

$= 3\sqrt{9}\sqrt{2} + 5\sqrt{25}\sqrt{2} - 2\sqrt{16}\sqrt{2}$

$= 3 \cdot 3\sqrt{2} + 5 \cdot 5\sqrt{2} - 2 \cdot 4\sqrt{2}$

$= 9\sqrt{2} + 25\sqrt{2} - 8\sqrt{2}$

$= (9+25-8)\sqrt{2} = 26\sqrt{2}$

44. $4\sqrt{27}+5\sqrt{80}+2\sqrt{12}$

$=4\sqrt{9}\sqrt{3}+5\sqrt{16}\sqrt{5}+2\sqrt{4}\sqrt{3}$

$=4\cdot 3\sqrt{3}+5\cdot 4\sqrt{5}+2\cdot 2\sqrt{3}$

$=12\sqrt{3}+20\sqrt{5}+4\sqrt{3}$

$=(12+4)\sqrt{3}+20\sqrt{5}$

$=16\sqrt{3}+20\sqrt{5}$

45. $\sqrt{x}=9$

$(\sqrt{x})^2=9^2$

$x=81$

Check: $\sqrt{x}=9$

$\sqrt{81}=9$

$9=9$ True

46. $\sqrt{x}=-2$

$(\sqrt{x})^2=(-2)^2$

$x=4$

Check: $\sqrt{x}=-2$

$\sqrt{4}=-2$

$2=-2$ False

4 is an extraneous root. There is no solution.

47. $\sqrt{x-3}=6$

$(\sqrt{x-3})^2=6^2$

$x-3=36$

$x=39$

Check: $\sqrt{x-3}=6$

$\sqrt{39-3}=6$

$\sqrt{36}=6$

$6=6$ True

48. $\sqrt{3x+1}=5$

$(\sqrt{3x+1})^2=5^2$

$3x+1=25$

$3x=24$

$x=8$

Check: $\sqrt{3x+1}=5$

$\sqrt{3(8)+1}=5$

$\sqrt{25}=5$

$5=5$ True

49. $\sqrt{2x+4}=\sqrt{3x-5}$

$\left(\sqrt{2x+4}\right)^2=\left(\sqrt{3x-5}\right)^2$

$2x+4=3x-5$

$4=x-5$

$9=x$

Check: $\sqrt{2x+4}=\sqrt{3x-5}$

$\sqrt{2(9)+4}=\sqrt{3(9)-5}$

$\sqrt{22}=\sqrt{22}$ True

50. $4\sqrt{x}-x=4$

$4\sqrt{x}-x+x=x+4$

$4\sqrt{x}=x+4$

$\left(4\sqrt{x}\right)^2=(x+4)^2$

$16x=x^2+8x+16$

$0=x^2-8x+16$

$0=(x-4)^2$

$x-4=0$

$x=4$

Check: $4\sqrt{x} - x = 4$

$4\sqrt{4} - 4 = 4$

$4 \cdot 2 - 4 = 4$

$4 = 4$ True

51. $\sqrt{x^2 + 4} = x + 2$

$\left(\sqrt{x^2 + 4}\right)^2 = (x + 2)^2$

$x^2 + 4 = x^2 + 4x + 4$

$4 = 4x + 4$

$0 = 4x$

$0 = x$

Check: $\sqrt{x^2 + 4} = x + 2$

$\sqrt{0^2 + 4} = 0 + 2$

$\sqrt{4} = 2$

$2 = 2$ True

52. $\sqrt{3x + 5} - \sqrt{5x - 9} = 0$

$\sqrt{3x + 5} = \sqrt{5x - 9}$

$\left(\sqrt{3x + 5}\right)^2 = \left(\sqrt{5x - 9}\right)^2$

$3x + 5 = 5x - 9$

$3x + 14 = 5x$

$14 = 2x$

$7 = x$

Check $\sqrt{3x + 5} - \sqrt{5x - 9} = 0$

$\sqrt{3(7) + 5} - \sqrt{5(7) - 9} = 0$

$\sqrt{26} - \sqrt{26} = 0$

$\sqrt{26} = \sqrt{26}$ True

53. $3\sqrt{2x + 3} = 9$

$\left(3\sqrt{2x + 3}\right)^2 = 9^2$

$9(2x + 3) = 81$

$2x + 3 = 9$

$2x = 6$

$x = 3$

Check: $3\sqrt{2x + 3} = 9$

$3\sqrt{2(3) + 3} = 9$

$3\sqrt{9} = 9$

$3 \cdot 3 = 9$

$9 = 9$ True

54. $a^2 + b^2 = c^2$

$8^2 + 6^2 = x^2$

$64 + 36 = x^2$

$100 = x^2$

$x = \sqrt{100} = 10$

55. $a^2 + b^2 = c^2$

$x^2 + 9^2 = 13^2$

$x^2 + 81 = 169$

$x^2 = 88$

$x = \sqrt{88} \approx 9.38$

56. $a^2 + b^2 = c^2$

$x^2 + (\sqrt{3})^2 = (\sqrt{15})^2$

$x^2 + 3 = 15$

$x^2 = 12$

$x = \sqrt{12} \approx 3.46$

57. $a^2 + b^2 = c^2$

$6^2 + 5^2 = x^2$

$36 + 25 = x^2$

$61 = x^2$

$x = \sqrt{61} \approx 7.81$

58.

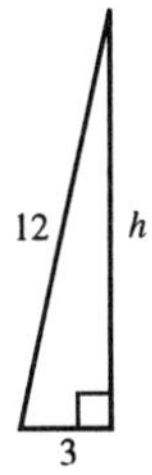

$a^2 + b^2 = c^2$

$h^2 + 3^2 = 12^2$

$h^2 + 9 = 144$

$h^2 = 135$

$h = \sqrt{135}$

The height of the ladder on the house is $\sqrt{135} \approx 11.62$ feet.

59.

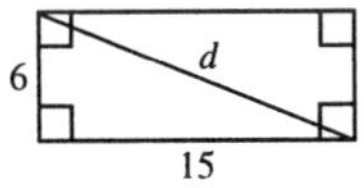

$a^2 + b^2 = c^2$

$15^2 + 6^2 = d^2$

$225 + 36 = d^2$

$261 = d^2$

$d = \sqrt{261}$

The length of the diagonal is $\sqrt{261} \approx$ 16.16 inches.

60. $d = \sqrt{(x_2 - x_1)^2 + (y_2 - y_1)^2}$

$= \sqrt{[3-(-5)]^2 + (-2-4)^2}$

$= \sqrt{(3+5)^2 + (-6)^2}$

$= \sqrt{8^2 + (-6)^2}$

$= \sqrt{64 + 36}$

$= \sqrt{100} = 10$

61. $d = \sqrt{(x_2 - x_1)^2 + (y_2 - y_1)^2}$

$= \sqrt{(-6-6)^2 + (8-5)^2}$

$= \sqrt{(-12)^2 + (3)^2}$

$= \sqrt{144 + 9}$

$= \sqrt{153} \approx 12.37$

62. $A = s^2$

$121 = s^2$

$s = \sqrt{121} = 11$

The side of the square is 11 feet.

63. $\sqrt[3]{8} = 2$ since $2^3 = 8$

64. $\sqrt[3]{-27} = -3$ since $(-3)^3 = -27$

65. $\sqrt[4]{16} = 2$ since $2^4 = 16$

66. $\sqrt[3]{16} = \sqrt[3]{8 \cdot 2} = \sqrt[3]{8}\,\sqrt[3]{2} = 2\sqrt[3]{2}$

67. $\sqrt[3]{24} = \sqrt[3]{8 \cdot 3} = \sqrt[3]{8}\,\sqrt[3]{3} = 2\sqrt[3]{3}$

68. $\sqrt[4]{32} = \sqrt[4]{16 \cdot 2} = \sqrt[4]{16}\,\sqrt[4]{2} = 2\sqrt[4]{2}$

69. $\sqrt[3]{48} = \sqrt[3]{8 \cdot 6} = \sqrt[3]{8}\,\sqrt[3]{6} = 2\sqrt[3]{6}$

70. $\sqrt[3]{54} = \sqrt[3]{27 \cdot 2} = \sqrt[3]{27}\,\sqrt[3]{2} = 3\sqrt[3]{2}$

71. $\sqrt[4]{96} = \sqrt[4]{16 \cdot 6} = \sqrt[4]{16}\,\sqrt[4]{6} = 2\sqrt[4]{6}$

72. $\sqrt[3]{x^{15}} = x^{15/3} = x^5$

73. $\sqrt[3]{x^{12}} = x^{12/3} = x^4$

74. $\sqrt[4]{y^{16}} = y^{16/4} = y^4$

75. $8^{2/3} = (\sqrt[3]{8})^2 = 2^2 = 4$

76. $16^{1/2} = \sqrt{16} = 4$

77. $27^{-2/3} = \dfrac{1}{27^{2/3}} = \dfrac{1}{(\sqrt[3]{27})^2} = \dfrac{1}{3^2} = \dfrac{1}{9}$

78. $64^{2/3} = (\sqrt[3]{64})^2 = 4^2 = 16$

79. $16^{-3/4} = \dfrac{1}{16^{3/4}} = \dfrac{1}{(\sqrt[4]{16})^3} = \dfrac{1}{8}$

80. $25^{3/2} = (\sqrt{25})^3 = 5^3 = 125$

81. $\sqrt[3]{x^5} = x^{5/3}$

82. $\sqrt[3]{x^{10}} = x^{10/3}$

83. $\sqrt[4]{y^9} = y^{9/4}$

84. $\sqrt{x^5} = x^{5/2}$

85. $\sqrt{y^{11}} = y^{11/2}$

86. $\sqrt[4]{x^7} = x^{7/4}$

87. $\sqrt{x} \cdot \sqrt{x} = x^{1/2} \cdot x^{1/2} = x^{1/2+1/2} = x^1 = x$

88. $\sqrt[3]{x} \cdot \sqrt[3]{x} = x^{1/3} \cdot x^{1/3} = x^{1/3+1/3} = x^{2/3}$
$= \sqrt[3]{x^2}$

89. $\sqrt[3]{x^2} \cdot \sqrt[3]{x^7} = x^{2/3} \cdot x^{7/3} = x^{2/3+7/3}$
$= x^{9/3} = x^3$

90. $\sqrt[4]{x^2} \cdot \sqrt[4]{x^6} = x^{2/4} \cdot x^{6/4}$
$= x^{2/4+6/4} = x^{1/2+3/2} = x^{4/2} = x^2$

91. $\left(\sqrt[3]{x^3}\right)^2 = \left(x^{3/3}\right)^2 = (x^1)^2 = x^2$

92. $(\sqrt[3]{x^2})^3 = (x^{2/3})^3 = x^2$

93. $(\sqrt[4]{x^2})^6 = (x^{2/4})^6 = (x^{1/2})^6 = x^3$

94. $(\sqrt[4]{x^3})^8 = (x^{3/4})^8 = x^6$

Chapter 9 Practice Test

1. $\sqrt{3xy} = (3xy)^{1/2}$

2. $\sqrt{(x+3)^2} = (x+3)^{2/2} = (x+3)^1 = x+3$

3. $\sqrt{96} = \sqrt{16 \cdot 6} = \sqrt{16}\sqrt{6} = 4\sqrt{6}$

4. $\sqrt{12x^2} = \sqrt{4x^2}\sqrt{3} = 2x\sqrt{3}$

5. $\sqrt{32x^4y^5} = \sqrt{16x^4y^4}\sqrt{2y} = 4x^2y^2\sqrt{2y}$

6. $\sqrt{8x^2y} \cdot \sqrt{6xy} = \sqrt{48x^3y^2}$
$= \sqrt{16x^2y^2}\sqrt{3x}$
$= 4xy\sqrt{3x}$

7. $\sqrt{15xy^2} \cdot \sqrt{5x^3y^3} = \sqrt{75x^4y^5}$
$= \sqrt{25x^4y^4}\sqrt{3y}$
$= 5x^2y^2\sqrt{3y}$

8. $\sqrt{\dfrac{5}{125}} = \sqrt{\dfrac{1}{25}} = \dfrac{\sqrt{1}}{\sqrt{25}} = \dfrac{1}{5}$

9. $\dfrac{\sqrt{3xy^2}}{\sqrt{48x^3}} = \sqrt{\dfrac{3xy^2}{48x^3}} = \sqrt{\dfrac{y^2}{16x^2}}$
$= \dfrac{\sqrt{y^2}}{\sqrt{16x^2}} = \dfrac{y}{4x}$

10. $\dfrac{1}{\sqrt{2}} = \dfrac{1}{\sqrt{2}} \cdot \dfrac{\sqrt{2}}{\sqrt{2}} = \dfrac{\sqrt{2}}{2}$

11. $\sqrt{\dfrac{4x}{5}} = \dfrac{\sqrt{4x}}{\sqrt{5}} = \dfrac{\sqrt{4}\sqrt{x}}{\sqrt{5}} = \dfrac{2\sqrt{x}}{\sqrt{5}}$
$= \dfrac{2\sqrt{x}}{\sqrt{5}} \cdot \dfrac{\sqrt{5}}{\sqrt{5}} = \dfrac{2\sqrt{5x}}{5}$

12. $\sqrt{\dfrac{40x^2y^5}{6x^3y^7}} = \sqrt{\dfrac{20}{3xy^2}} = \dfrac{\sqrt{4}\sqrt{5}}{\sqrt{y^2}\sqrt{3x}}$
$= \dfrac{2\sqrt{5}}{y\sqrt{3x}} \cdot \dfrac{\sqrt{3x}}{\sqrt{3x}} = \dfrac{2\sqrt{15x}}{3xy}$

13. $\dfrac{3}{2+\sqrt{5}} = \dfrac{3}{2+\sqrt{5}} \cdot \dfrac{2-\sqrt{5}}{2-\sqrt{5}} = \dfrac{3(2-\sqrt{5})}{4-5}$
$= \dfrac{3(2-\sqrt{5})}{-1} = -3(2-\sqrt{5})$

14. $\dfrac{6}{\sqrt{x}-3} = \dfrac{6}{\sqrt{x}-3} \cdot \dfrac{\sqrt{x}+3}{\sqrt{x}+3} = \dfrac{6(\sqrt{x}+3)}{x-9}$
$= \dfrac{6\sqrt{x}+18}{x-9}$

15. $\sqrt{48} + \sqrt{75} + 2\sqrt{3}$
$= \sqrt{16 \cdot 3} + \sqrt{25 \cdot 3} + 2\sqrt{3}$
$= 4\sqrt{3} + 5\sqrt{3} + 2\sqrt{3} = 11\sqrt{3}$

16. $4\sqrt{x} - 6\sqrt{x} - \sqrt{x} = (4-6-1)\sqrt{x} = -3\sqrt{x}$

17. $\sqrt{x+5}=9$

$\left(\sqrt{x+5}\right)^2=9^2$

$x+5=81$

$x=76$

Check: $\sqrt{x+5}=9$

$\sqrt{76+5}=9$

$\sqrt{81}=9$

$9=9$ True

18. $2\sqrt{x-4}+4=x$

$2\sqrt{x-4}+4-4=x-4$

$2\sqrt{x-4}=x-4$

$\left(2\sqrt{x-4}\right)^2=(x-4)^2$

$4(x-4)=x^2-8x+16$

$4x-16=x^2-8x+16$

$0=x^2-12x+32$

$0=(x-4)(x-8)$

$x-4=0$ or $x-8=0$

$x=4$ or $x=8$

Check: $x=4$

$2\sqrt{x-4}+4=x$

$2\sqrt{4-4}+4=4$

$2\sqrt{0}+4=4$

$4=4$ True

Check: $x=8$

$2\sqrt{x-4}+4=x$

$2\sqrt{8-4}+4=8$

$2\sqrt{4}+4=8$

$2\cdot 2+4=8$

$8=8$ True

The solutions are 4 and 8.

19. $a^2+b^2=c^2$

$9^2+5^2=x^2$

$81+25=x^2$

$106=x^2$

$x=\sqrt{106}\approx 10.30$

20. $d=\sqrt{(x_2-x_1)^2+(y_2-y_1)^2}$

$=\sqrt{[5-(-2)]^2+[1-(-4)]^2}$

$=\sqrt{(5+2)^2+(1+4)^2}$

$=\sqrt{7^2+5^2}$

$=\sqrt{49+25}$

$=\sqrt{74}\approx 8.60$

21. $27^{-4/3}=\dfrac{1}{27^{4/3}}=\dfrac{1}{\left(\sqrt[3]{27}\right)^4}=\dfrac{1}{3^4}=\dfrac{1}{81}$

22. $\sqrt[3]{x^4}\cdot\sqrt[3]{x^{11}}=x^{4/3}\cdot x^{11/3}=x^{(4/3)+(11/3)}$

$=x^{15/3}=x^5$

CHAPTER 10

Exercise Set 10.1

1. $x^2 = 16$

$x = \pm\sqrt{16}$

$x = 4, -4$

3. $x^2 = 100$

$x = \pm\sqrt{100}$

$x = 10, -10$

5. $y^2 = 36$

$y = \pm\sqrt{36}$

$y = 6, -6$

7. $x^2 = 10$

$x = \sqrt{10}, -\sqrt{10}$

9. $x^2 = 8$

$x = \pm\sqrt{8}$

$x = \pm\sqrt{4}\sqrt{2}$

$x = \pm 2\sqrt{2}$

$x = 2\sqrt{2}, -2\sqrt{2}$

11. $3x^2 = 12$

$x^2 = 4$

$x = \pm\sqrt{4}$

$x = 2, -2$

13. $2w^2 = 34$

$w^2 = 17$

$w = \sqrt{17}, -\sqrt{17}$

15. $2x^2 + 1 = 19$

$2x^2 = 18$

$x^2 = 9$

$x = \pm\sqrt{9}$

$x = 3, -3$

17. $4w^2 - 3 = 12$

$4w^2 = 15$

$w^2 = \frac{15}{4}$

$w = \pm\sqrt{\frac{15}{4}}$

$w = \pm\frac{\sqrt{15}}{\sqrt{4}}$

$w = \frac{\sqrt{15}}{2}, \frac{-\sqrt{15}}{2}$

19. $5x^2 - 9 = 30$

$5x^2 = 39$

$x^2 = \frac{39}{5}$

$x = \pm\sqrt{\frac{39}{5}}$

$x = \pm\frac{\sqrt{39}}{\sqrt{5}}$

$x = \pm\frac{\sqrt{39}}{\sqrt{5}} \cdot \frac{\sqrt{5}}{\sqrt{5}}$

$x = \frac{\sqrt{195}}{5}, \frac{-\sqrt{195}}{5}$

21. $(x+1)^2 = 4$

$$x+1 = \pm\sqrt{4}$$
$$x+1 = \pm 2$$
$$x = -1 \pm 2$$
$$x = -1+2 \quad \text{or} \quad x = -1-2$$
$$x = 1 \quad \text{or} \quad x = -3$$

The solutions are 1 and –3.

23. $(x-3)^2 = 16$

$$x-3 = \pm\sqrt{16}$$
$$x-3 = \pm 4$$
$$x = 3 \pm 4$$
$$x = 3+4 \quad \text{or} \quad x = 3-4$$
$$x = 7 \quad \text{or} \quad x = -1$$

The solutions are 7 and –1.

25. $(x+4)^2 = 36$

$$x+4 = \pm\sqrt{36}$$
$$x+4 = \pm 6$$
$$x = -4 \pm 6$$
$$x = -4+6 \quad \text{or} \quad x = -4-6$$
$$x = 2 \quad \text{or} \quad x = -10$$

The solutions are 2 and –10.

27. $(x-1)^2 = 12$

$$x-1 = \pm\sqrt{12}$$
$$x-1 = \pm\sqrt{4}\sqrt{3}$$
$$x-1 = \pm 2\sqrt{3}$$
$$x = 1 \pm 2\sqrt{3}$$

The solutions are $1+2\sqrt{3},\ 1-2\sqrt{3}$

29. $(x+6)^2 = 20$

$$x+6 = \pm\sqrt{20}$$
$$x+6 = \pm\sqrt{4}\sqrt{5}$$
$$x+6 = \pm 2\sqrt{5}$$
$$x = -6 \pm 2\sqrt{5}$$

The solutions are $-6+2\sqrt{5},\ -6-2\sqrt{5}$.

31. $(x+2)^2 = 25$

$$x+2 = \pm\sqrt{25}$$
$$x+2 = \pm 5$$
$$x = -2 \pm 5$$
$$x = -2+5 \quad \text{or} \quad x = -2-5$$
$$x = 3 \quad \text{or} \quad x = -7$$

The solutions are 3 and –7.

33. $(x-9)^2 = 100$

$$x-9 = \pm\sqrt{100}$$
$$x-9 = \pm 10$$
$$x = 9 \pm 10$$
$$x = 9+10 \quad \text{or} \quad x = 9-10$$
$$x = 19 \quad \text{or} \quad x = -1$$

The solutions are 19 and –1.

35. $(2x+3)^2 = 18$

$$2x+3 = \pm\sqrt{18}$$
$$2x+3 = \pm\sqrt{9}\sqrt{2}$$
$$2x+3 = \pm 3\sqrt{2}$$
$$2x = -3 \pm 3\sqrt{2}$$
$$x = \frac{-3 \pm 3\sqrt{2}}{2}$$

The solutions are $\frac{-3+3\sqrt{2}}{2}$ and $\frac{-3-3\sqrt{2}}{2}$.

37. $(4x+1)^2 = 20$

$$4x+1 = \pm\sqrt{20}$$
$$4x+1 = \pm\sqrt{4}\sqrt{5}$$
$$4x+1 = \pm 2\sqrt{5}$$
$$4x = -1 \pm 2\sqrt{5}$$
$$x = \frac{-1 \pm 2\sqrt{5}}{4}$$

The solutions are $\frac{-1+2\sqrt{5}}{4}$ and $\frac{-1-2\sqrt{5}}{4}$.

39. $(2x-6)^2 = 18$

$$2x-6 = \pm\sqrt{18}$$
$$2x-6 = \pm\sqrt{9}\sqrt{2}$$
$$2x-6 = \pm 3\sqrt{2}$$
$$2x = 6 \pm 3\sqrt{2}$$
$$x = \frac{6 \pm 3\sqrt{2}}{2}$$

The solutions are $\frac{6+3\sqrt{2}}{2}$ and $\frac{6-3\sqrt{2}}{2}$.

41. Let x = width of rectangle; then $2x$ = length of rectangle.

Area = length · width

$$80 = (2x)x$$
$$80 = 2x^2$$
$$x^2 = 40$$
$$x = \pm\sqrt{40}$$
$$x = \pm\sqrt{4}\sqrt{10}$$
$$x = \pm 2\sqrt{10}$$

Since the width cannot be negative, the width is $2\sqrt{10} \approx 6.32$ feet.

The length is $2(2\sqrt{10}) = 4\sqrt{10} \approx 12.65$ feet.

43. $x^2 = 36$; other answers are possible

45. $x^2 - 9 = 27$; need an equation equivalent to $x^2 = 36$.

47. $ax^2 + bx + c = 0,\ a \neq 0$

Cumulative Review Exercises: 10.1

49. $4x^2 - 10x - 24$

$$= 2(2x^2 - 5x - 12)$$
$$= 2(2x^2 - 8x + 3x - 12)$$
$$= 2[2x(x-4) + 3(x-4)]$$
$$= 2[(2x+3)(x-4)]$$
$$= 2(2x+3)(x-4)$$

50. $\frac{3x-7}{x-4}=\frac{5}{x-4}$

$(3x-7)(x-4)=5(x-4)$

$3x^2-19x+28=5x-20$

$3x^2-24x+48=0$

$3(x^2-8x+16)=0$

$3(x-4)(x-4)=0$

$x-4=0$

$x=4$

Check: $x=4$

$\frac{3x-7}{x-4}=\frac{5}{x-4}$

$\frac{3(4)-7}{4-4}=\frac{5}{4-4}$

$\frac{5}{0}=\frac{5}{0}$

Since $\frac{5}{0}$ is not a real number, 4 is an extraneous solution. The equation has no solution.

52. $m=\frac{y_2-y_1}{x_2-x_1}=\frac{3-(-1)}{1-0}=\frac{3+1}{1}=4$

The y-intercept is –1. Thus, the equation of the line is $y=4x-1$.

51. Let x = principal invested at 6%,

y = principal invested at 8%.

Then:

Interest from 6% account in one year

$=x(1)(0.06)=0.06x$

Interest from 8% account in one year

$=y(1)(0.08)=0.08y$

Thus:

$x+y=10{,}000$

Total prinicipal invested = $10,000

$0.06x+0.08y=760$

Total interest = $760

$-6[x+y=10{,}000]$

$100[0.06x+0.08y=760]$

gives

$-6x-6y=-60{,}000$

$\underline{6x+8y=76{,}000}$

$2y=16{,}000$

$y=8{,}000$

Solve for x:

$x+y=10{,}000$

$x+8000=10{,}000$

$x=2000$

$2000 was in the 6% account and $8000 was in the 8% account.

Group Activity/Challenge Problems

1. $A=s^2$

$\sqrt{A}=\sqrt{s^2}$

$\sqrt{A}=s$

3. $A=\pi r^2$

$\frac{A}{\pi}=\frac{\pi r^2}{\pi}$

$\frac{A}{\pi}=r^2$

$\sqrt{\frac{A}{\pi}}=\sqrt{r^2}$

$\sqrt{\frac{A}{\pi}}=r$

5. $I = \frac{k}{d^2}$

$d^2 I = \frac{k}{d^2}(d^2)$

$\frac{d^2 I}{I} = \frac{k}{I}$

$d^2 = \frac{k}{I}$

$\sqrt{d^2} = \sqrt{\frac{k}{I}}$

$d = \sqrt{\frac{k}{I}}$

6. $A = p(1+r)^2$

$\frac{A}{p} = \frac{p(1+r)^2}{p}$

$\frac{A}{p} = (1+r)^2$

$\sqrt{\frac{A}{p}} = \sqrt{(1+r)^2}$

$\sqrt{\frac{A}{p}} = 1+r$

$\sqrt{\frac{A}{p}} - 1 = 1+r-1$

$\sqrt{\frac{A}{p}} - 1 = r$

Exercise Set 10.2

1. $x^2 + 2x - 3 = 0$

$x^2 + 2x = 3$

$x^2 + 2x + 1 = 3 + 1$

$(x+1)^2 = 4$

$x + 1 = \pm\sqrt{4}$

$x + 1 = \pm 2$

$x = -1 \pm 2$

$x = -1 + 2$ or $x = -1 - 2$

$x = 1$ or $x = -3$

The solutions are 1 and –3.

3. $x^2 - 4x - 5 = 0$

$x^2 - 4x = 5$

$x^2 - 4x + 4 = 5 + 4$

$(x-2)^2 = 9$

$x - 2 = \pm\sqrt{9}$

$x - 2 = \pm 3$

$x = 2 \pm 3$

$x = 2 + 3$ or $x = 2 - 3$

$x = 5$ or $x = -1$

The solutions are 5 and –1.

5. $x^2+3x+2=0$

$x^2+3x=-2$

$x^2+3x+\frac{9}{4}=-2+\frac{9}{4}$

$\left(x+\frac{3}{2}\right)^2=-\frac{8}{4}+\frac{9}{4}$

$\left(x+\frac{3}{2}\right)^2=\frac{1}{4}$

$x+\frac{3}{2}=\pm\sqrt{\frac{1}{4}}$

$x+\frac{3}{2}=\pm\frac{1}{2}$

$x=-\frac{3}{2}\pm\frac{1}{2}$

$x=-\frac{3}{2}+\frac{1}{2}$ or $x=-\frac{3}{2}-\frac{1}{2}$

$x=-1$ or $x=-2$

The solutions are –2 and –1.

7. $x^2-2x-8=0$

$(x-4)(x+2)=0$

$x-4=0$ and $x+2=0$

$x=4$ and $x=-2$

The solutions are 4 and –2.

9. $x^2=-6x-9$

$x^2+6x=-9$

$x^2+6x+9=-9+9$

$(x+3)^2=0$

$x+3=\pm\sqrt{0}$

$x+3=\pm 0$

$x=-3\pm 0$

$x=-3$

The solution is –3.

11. $x^2=-5x-6$

$x^2+5x=-6$

$x^2+5x+\frac{25}{4}=-6+\frac{25}{4}$

$\left(x+\frac{5}{2}\right)^2=-\frac{24}{4}+\frac{25}{4}$

$\left(x+\frac{5}{2}\right)^2=\frac{1}{4}$

$x+\frac{5}{2}=\pm\sqrt{\frac{1}{4}}$

$x+\frac{5}{2}=\pm\frac{1}{2}$

$x=-\frac{5}{2}\pm\frac{1}{2}$

$x=-\frac{5}{2}+\frac{1}{2}$ or $x=-\frac{5}{2}-\frac{1}{2}$

$x=-\frac{4}{2}=-2$ or $x=-\frac{6}{2}=-3$

The solutions are –2, –3.

13. $x^2+9x+18=0$

$x^2+9x=-18$

$x^2+9x+\frac{81}{4}=-18+\frac{81}{4}$

$\left(x+\frac{9}{2}\right)^2=-\frac{72}{4}+\frac{81}{4}$

$\left(x+\frac{9}{2}\right)^2=\frac{9}{4}$

$x+\frac{9}{2}=\pm\sqrt{\frac{9}{4}}$

$x+\frac{9}{2}=\pm\frac{3}{2}$

$x=-\frac{9}{2}\pm\frac{3}{2}$

$$x=-\frac{9}{2}+\frac{3}{2} \quad \text{or} \quad x=-\frac{9}{2}-\frac{3}{2}$$

$$x=-3 \quad \text{or} \quad x=-6$$

The solutions are –3 and –6.

15.

$$x^2 = 15x - 56$$

$$x^2 - 15x = -56$$

$$x^2 - 15 + \frac{225}{4} = -56 + \frac{225}{4}$$

$$\left(x-\frac{15}{2}\right)^2 = -\frac{224}{4}+\frac{225}{4}$$

$$\left(x-\frac{15}{2}\right)^2 = \frac{1}{4}$$

$$x-\frac{15}{2} = \pm\sqrt{\frac{1}{4}}$$

$$x-\frac{15}{2} = \pm\frac{1}{2}$$

$$x = \frac{15}{2}\pm\frac{1}{2}$$

$$x=\frac{15}{2}+\frac{1}{2} \quad \text{or} \quad x=\frac{15}{2}-\frac{1}{2}$$

$$x=8 \quad \text{or} \quad x=7$$

The solutions are 7 and 8.

17.

$$-4x = -x^2 + 12$$

$$x^2 - 4x = 12$$

$$x^2 - 4x + 4 = 12 + 4$$

$$(x-2)^2 = 16$$

$$x-2 = \pm\sqrt{16}$$

$$x-2 = \pm 4$$

$$x = 2 \pm 4$$

$$x=2+4 \quad \text{or} \quad x=2-4$$

$$x=6 \quad \text{or} \quad x=-2$$

The solutions are 6 and –2.

19. $x^2 + 2x - 6 = 0$

$$x^2 + 2x = 6$$

$$x^2 + 2x + 1 = 6 + 1$$

$$(x+1)^2 = 7$$

$$x+1 = \pm\sqrt{7}$$

$$x = -1 \pm \sqrt{7}$$

$$x=-1+\sqrt{7} \quad \text{or} \quad x=-1-\sqrt{7}$$

The solutions are $-1+\sqrt{7}$ and $-1-\sqrt{7}$.

21.

$$6x + 6 = -x^2$$

$$x^2 + 6x + 6 = 0$$

$$x^2 + 6x = -6$$

$$x^2 + 6x + 9 = -6 + 9$$

$$(x+3)^2 = 3$$

$$x+3 = \pm\sqrt{3}$$

$$x = -3 \pm \sqrt{3}$$

The solutions are $-3+\sqrt{3}$ and $-3-\sqrt{3}$.

23.

$$-x^2 + 5x = -8$$

$$-1(-x^2 + 5x) = (-1)(-8)$$

$$x^2 - 5x = 8$$

$$x^2 - 5x + \frac{25}{4} = 8 + \frac{25}{4}$$

$$\left(x-\frac{5}{2}\right)^2 = \frac{32}{4}+\frac{25}{4}$$

$$\left(x-\frac{5}{2}\right)^2 = \frac{57}{4}$$

$$x-\frac{5}{2} = \pm\sqrt{\frac{57}{4}}$$

$$x-\frac{5}{2} = \frac{\pm\sqrt{57}}{2}$$

$$x = \frac{5}{2}\pm\frac{\sqrt{57}}{2}$$

The solutions are $\frac{5}{2}+\frac{\sqrt{57}}{2}$ and $\frac{5}{2}-\frac{\sqrt{57}}{2}$

25. $2x^2+4x-6=0$

$$\frac{1}{2}(2x^2+4x-6)=\frac{1}{2}(0)$$
$$x^2+2x-3=0$$
$$x^2+2x=3$$
$$x^2+2x+1=3+1$$
$$(x+1)^2=4$$
$$x+1=\pm\sqrt{4}$$
$$x+1=\pm 2$$
$$x=-1\pm 2$$
$$x=-1+2 \quad \text{or} \quad x=-1-2$$
$$x=1 \quad \text{or} \quad x=-3$$

The solutions are 1 and –3.

27. $2x^2+18x+4=0$

$$\frac{1}{2}(2x^2+18x+4)=\frac{1}{2}(0)$$
$$x^2+9x+2=0$$
$$x^2+9x=-2$$
$$x^2+9x+\frac{81}{4}=-2+\frac{81}{4}$$
$$\left(x+\frac{9}{2}\right)^2=-\frac{8}{4}+\frac{81}{4}$$
$$\left(x+\frac{9}{2}\right)^2=\frac{73}{4}$$
$$x+\frac{9}{2}=\pm\sqrt{\frac{73}{4}}$$
$$x+\frac{9}{2}=\frac{\pm\sqrt{73}}{2}$$
$$x=-\frac{9}{2}\pm\frac{\sqrt{73}}{2}$$

The solutions are $\frac{-9+\sqrt{73}}{2}$ and $\frac{-9-\sqrt{73}}{2}$.

29. $3x^2+33x+72=0$

$$\frac{1}{3}(3x^2+33x+72)=\frac{1}{3}(0)$$
$$x^2+11x+24=0$$
$$x^2+11x=-24$$
$$x^2+11x+\frac{121}{4}=-24+\frac{121}{4}$$
$$\left(x+\frac{11}{2}\right)^2=-\frac{96}{4}+\frac{121}{4}$$
$$\left(x+\frac{11}{2}\right)^2=\frac{25}{4}$$
$$x+\frac{11}{2}=\pm\sqrt{\frac{25}{4}}$$
$$x+\frac{11}{2}=\pm\frac{5}{2}$$
$$x=-\frac{11}{2}\pm\frac{5}{2}$$
$$x=-\frac{11}{2}+\frac{5}{2} \quad \text{or} \quad x=-\frac{11}{2}-\frac{5}{2}$$
$$x=-3 \quad \text{or} \quad x=-8$$

The solutions are –3 and –8.

31. $2x^2+10x-3=0$

$$\frac{1}{2}(2x^2+10x-3)=\frac{1}{2}(0)$$
$$x^2+5x-\frac{3}{2}=0$$
$$x^2+5x=\frac{3}{2}$$
$$x^2+5x+\frac{25}{4}=\frac{3}{2}+\frac{25}{4}$$
$$\left(x+\frac{5}{2}\right)^2=\frac{31}{4}$$
$$x+\frac{5}{2}=\pm\sqrt{\frac{31}{4}}$$
$$x+\frac{5}{2}=\frac{\pm\sqrt{31}}{2}$$
$$x=-\frac{5}{2}\pm\frac{\sqrt{31}}{2}$$

The solutions are $-\frac{5}{2}+\frac{\sqrt{31}}{2}$ and $-\frac{5}{2}-\frac{\sqrt{31}}{2}$.

33. $3x^2 + 6x = 6$

$\frac{1}{3}(3x^2 + 6x) = \frac{1}{3}(6)$

$x^2 + 2x = 2$

$x^2 + 2x + 1 = 2 + 1$

$(x+1)^2 = 3$

$x + 1 = \pm\sqrt{3}$

$x = -1 \pm \sqrt{3}$

The solutions are $-1 + \sqrt{3}$ and $-1 - \sqrt{3}$.

35. $x^2 + 4x = 0$

$x^2 + 4x + 4 = 0 + 4$

$(x+2)^2 = 4$

$x + 2 = \pm\sqrt{4}$

$x + 2 = \pm 2$

$x = -2 \pm 2$

$x = -2 + 2$ or $x = -2 - 2$

$x = 0$ or $x = -4$

The solutions are 0 and –4.

37. $2x^2 - 4x = 0$

$\frac{1}{2}(2x^2 - 4x) = \frac{1}{2}(0)$

$x^2 - 2x = 0$

$x^2 - 2x + 1 = 0 + 1$

$(x-1)^2 = 1$

$x - 1 = \pm\sqrt{1}$

$x - 1 = \pm 1$

$x = 1 \pm 1$

$x = 1 + 1$ or $x = 1 - 1$

$x = 2$ or $x = 0$

The solutions are 0 and 2.

39. $x^2 + 3x = 4$

$x^2 + 3x + \frac{9}{4} = 4 + \frac{9}{4}$

$\left(x + \frac{3}{2}\right)^2 = \frac{16}{4} + \frac{9}{4}$

$\left(x + \frac{3}{2}\right)^2 = \frac{25}{4}$

$x + \frac{3}{2} = \pm\sqrt{\frac{25}{4}}$

$x + \frac{3}{2} = \pm\frac{5}{2}$

$x = -\frac{3}{2} \pm \frac{5}{2}$

$x = -\frac{3}{2} + \frac{5}{2}$ or $x = -\frac{3}{2} - \frac{5}{2}$

$x = 1$ or $x = -4$

The possible numbers are 1 and –4.

41. $(x+3)^2 = 9$

$x + 3 = \pm\sqrt{9}$

$x + 3 = \pm 3$

$x = -3 \pm 3$

$x = -3 + 3$ or $x = -3 - 3$

$x = 0$ or $x = -6$

The possible numbers are 0 and –6.

43. $xy = 21$ The product of the numbers is 21

$y = x + 4$ The larger number is 4 greater than the smaller.

Substitute $x + 4$ for y in the first equation.

$xy = 21$

$x(x+4) = 21$

$x^2 + 4x = 21$

$x^2 + 4x + 4 = 21 + 4$

$(x+2)^2 = 25$

$x+2 = \pm\sqrt{25}$

$x+2 = \pm 5$

$x = -2 \pm 5$

$x = -2+5$ or $x = -2-5$

$x = 3$ or $x = -7$

Since the numbers are positive, $x = 3$
Solve for y.

$y = x+4 = 3+4 = 7$

The numbers are 3 and 7.

45. The constant is the square of half the coefficient of the x-term.

Cumulative Review Exercises: 10.2

47. $\dfrac{x^2}{x^2-x-6} - \dfrac{x-2}{x-3}$

$= \dfrac{x^2}{(x-3)(x+2)} - \dfrac{x-2}{x-3}$

$= \dfrac{x^2}{(x-3)(x+2)} - \dfrac{(x-2)}{(x-3)} \cdot \dfrac{(x+2)}{(x+2)}$

$= \dfrac{x^2}{(x-3)(x+2)} - \dfrac{(x^2-4)}{(x-3)(x+2)}$

$= \dfrac{4}{(x-3)(x+2)}$

48. Write the equations in slope-intercept form and compare the slopes. If the slopes are the same and the y-intercepts are different, the equations represent parallel lines.

49. $3x-4y=6$ gives $3x-4y = 6$

$4[2x+y=8]$ $\quad \underline{8x+4y=32}$

$11x = 38$

$x = \dfrac{38}{11}$

Solve for y.

$3x-4y=6$

$3\left(\dfrac{38}{11}\right) - 4y = 6$

$\dfrac{114}{11} - 4y = 6$

$-4y = -\dfrac{48}{11}$

$y = \dfrac{12}{11}$

The solution is $\left(\dfrac{38}{11}, \dfrac{12}{11}\right)$

50. $\sqrt{2x+3} = 2x-3$

$\left(\sqrt{2x+3}\right)^2 = (2x-3)^2$

$2x+3 = 4x^2 - 12x + 9$

$0 = 4x^2 - 14x + 6$

$0 = 2(2x-1)(x-3)$

$2x-1=0$ or $x-3=0$

$x = \dfrac{1}{2}$ or $x = 3$

Check: $x = \dfrac{1}{2}$

$\sqrt{2x+3} = 2x-3$

$\sqrt{2\left(\dfrac{1}{2}\right)+3} = 2\left(\dfrac{1}{2}\right) - 3$

$\sqrt{4} = -2$

$2 = -2$ False

Check: $x = 3$

$\sqrt{2x+3} = 2x-3$

$\sqrt{2(3)+3} = 2(3)-3$

$\sqrt{9} = 3$

$3 = 3$ True

$\dfrac{1}{2}$ is an extraneous root. The solution is 3.

Group Activity/Challenge Problems

1. $x^2+\frac{3}{5}x-\frac{1}{2}=0$

$x^2+\frac{3}{5}x=\frac{1}{2}$

$x^2+\frac{3}{5}x+\frac{9}{100}=\frac{1}{2}+\frac{9}{100}$

$\left(x+\frac{3}{10}\right)^2=\frac{59}{100}$

$x+\frac{3}{10}=\pm\sqrt{\frac{59}{100}}$

$x=\frac{-3\pm\sqrt{59}}{10}$

Answer: $\frac{-3-\sqrt{59}}{10}, \frac{-3+\sqrt{59}}{10}$

3. $3x^2+\frac{1}{2}x=4$

$x^2+\frac{1}{6}x=\frac{4}{3}$

$x+\frac{1}{6}x+\frac{1}{144}=\frac{4}{3}+\frac{1}{144}$

$\left(x+\frac{1}{12}\right)^2=\frac{193}{144}$

$x+\frac{1}{12}=\pm\sqrt{\frac{193}{144}}$

$x=-\frac{1}{12}\pm\frac{\sqrt{193}}{12}$

$x=\frac{-1\pm\sqrt{193}}{12}$

Answer: $\frac{-1-\sqrt{193}}{12}, \frac{-1+\sqrt{193}}{12}$

5. $-5.26x^2+7.89x+15.78=0$

$x^2-1.5x-3=0$

$x^2-1.5x=3$

$x^2-1.5x+\frac{9}{16}=3+\frac{9}{16}$

$(x-0.75)^2=\frac{57}{16}$

$x-0.75=\pm\sqrt{\frac{57}{16}}$

$x=0.75\pm\sqrt{3.5625}$

Answer: $0.75-\sqrt{3.5625}, 0.75+\sqrt{3.5625}$

Exercise Set 10.3

1. $b^2-4ac=(3)^2-4(1)(-5)$

$=9+20=29$

Since the discriminant is positive, this equation has the two distinct real number solutions.

3. $b^2-4ac=(-4)^2-4(3)(7)$

$=16-84=-68$

Since the discriminant is negative, this equation has no real number solutions.

5. $b^2-4ac=(3)^2-4(5)(-7)$

$=9+140=149$

Since the discriminant is positive, this equation has two distinct real number solutions.

7. $4x^2 - 24x = -36$

$4x^2 - 24x + 36 = 0$

$b^2 - 4ac = (-24)^2 - 4(4)(36)$

$= 576 - 576 = 0$

Since the discriminant is zero, this equation has one real number solution.

9. $b^2 - 4ac = (-8)^2 - 4(1)(5)$

$= 64 - 20 = 44$

Since the discriminant is positive, this equation has two distinct real number solutions.

11. $b^2 - 4ac = (5)^2 - 4(-3)(-8)$

$= 25 - 96 = -71$

Since the discriminant is negative, this equation has no real number solutions.

13. $b^2 - 4ac = (7)^2 - 4(1)(-3)$

$= 49 + 12 = 61$

Since the discriminant is positive, this equation has two distinct real number solutions.

15. $b^2 - 4ac = 0^2 - 4(4)(-9) = 144$

Since the discriminant is positive, this equation has two distinct real number solutions.

17. $a = 1,\ b = -3,\ c = 2$

$x =$ Quadratic Formula

$$= \frac{-(-3) \pm \sqrt{(-3)^2 - 4(1)(2)}}{2(1)}$$

$$= \frac{3 \pm \sqrt{9 - 8}}{2}$$

$$= \frac{3 \pm \sqrt{1}}{2}$$

$$= \frac{3 \pm 1}{2}$$

$x = \frac{3+1}{2}$ or $x = \frac{3-1}{2}$

$x = 2$ or $x = 1$

The solutions are 2 and 1.

19. $a = 1,\ b = -9,\ c = 20$

$$x = \frac{-b \pm \sqrt{b^2 - 4ac}}{2a}$$

$$= \frac{-(-9) \pm \sqrt{(-9)^2 - 4(1)(20)}}{2(1)}$$

$$= \frac{9 \pm \sqrt{81 - 80}}{2}$$

$$= \frac{9 \pm \sqrt{1}}{2}$$

$$= \frac{9 \pm 1}{2}$$

$x = \frac{9+1}{2}$ or $x = \frac{9-1}{2}$

$x = 5$ $\qquad x = 4$

The solutions are 5 and 4.

21. $a = 1,\ b = 5,\ c = -24$

$$x = \frac{-b \pm \sqrt{b^2 - 4ac}}{2a}$$
$$= \frac{-5 \pm \sqrt{(5)^2 - 4(1)(-24)}}{2(1)}$$
$$= \frac{-5 \pm \sqrt{25 + 96}}{2}$$
$$= \frac{-5 \pm \sqrt{121}}{2}$$
$$= \frac{-5 \pm 11}{2}$$
$$x = \frac{-5 + 11}{2} \quad \text{or} \quad x = \frac{-5 - 11}{2}$$
$$x = 3 \qquad x = -8$$

The solutions are 3 and –8.

23. Write in standard form.

$$x^2 - 13x + 36 = 0$$
$$a = 1,\ b = -13,\ c = 36$$
$$x = \frac{-b \pm \sqrt{b^2 - 4ac}}{2a}$$
$$= \frac{-(-13) \pm \sqrt{(-13)^2 - 4(1)(36)}}{2(1)}$$
$$= \frac{13 \pm \sqrt{169 - 144}}{2}$$
$$= \frac{13 \pm \sqrt{25}}{2}$$
$$= \frac{13 \pm 5}{2}$$
$$x = \frac{13 + 5}{2} \quad \text{or} \quad x = \frac{13 - 5}{2}$$
$$x = 9 \quad \text{or} \quad x = 4$$

The solutions are 9 and 4.

25. $a = 1,\ b = 0,\ c = -25$

$$x = \frac{-b \pm \sqrt{b^2 - 4ac}}{2a}$$
$$= \frac{-(0) \pm \sqrt{(0)^2 - 4(1)(-25)}}{2(1)}$$
$$= \frac{\pm\sqrt{100}}{2}$$
$$= \pm\frac{10}{2} = \pm 5$$

The solutions are 5 and –5.

27. $a = 1,\ b = -3,\ c = 0$

$$x = \frac{-b \pm \sqrt{b^2 - 4ac}}{2a}$$
$$= \frac{-(-3) \pm \sqrt{(-3)^2 - 4(1)(0)}}{2(1)}$$
$$= \frac{3 \pm \sqrt{9}}{2}$$
$$= \frac{3 \pm 3}{2}$$
$$x = \frac{3 + 3}{2} \quad \text{or} \quad x = \frac{3 - 3}{2}$$
$$x = 3 \quad \text{or} \quad x = 0$$

The solutions are 3 and 0.

29. $a = 1,\ b = -7,\ c = 12$

$$p = \frac{-b \pm \sqrt{b^2 - 4ac}}{2a}$$
$$= \frac{-(-7) \pm \sqrt{(-7)^2 - 4(1)(12)}}{2(1)}$$
$$= \frac{7 \pm \sqrt{49 - 48}}{2}$$
$$= \frac{7 \pm \sqrt{1}}{2}$$
$$= \frac{7 \pm 1}{2}$$

$p = \frac{7+1}{2}$ or $p = \frac{7-1}{2}$

$p = 4$ or $p = 3$

The solutions are 4 and 3.

31. $a = 2,\ b = -7,\ c = 4$

$$y = \frac{-b \pm \sqrt{b^2 - 4ac}}{2a}$$
$$= \frac{-(-7) \pm \sqrt{(-7)^2 - 4(2)(4)}}{2(2)}$$
$$= \frac{7 \pm \sqrt{49 - 32}}{4}$$
$$= \frac{7 \pm \sqrt{17}}{4}$$

The solutions are $\frac{7+\sqrt{17}}{4}$ and $\frac{7-\sqrt{17}}{4}$.

33. Write in standard form:

$6x^2 + x - 1 = 0$

$a = 6,\ b = 1,\ c = -1$

$$x = \frac{-b \pm \sqrt{b^2 - 4ac}}{2a}$$
$$= \frac{-1 \pm \sqrt{(1)^2 - 4(6)(-1)}}{2(6)}$$
$$= \frac{-1 \pm \sqrt{1 + 24}}{12}$$
$$= \frac{-1 \pm \sqrt{25}}{12}$$
$$= \frac{-1 \pm 5}{12}$$

$x = \frac{-1+5}{12}$ or $x = \frac{-1-5}{12}$

$x = \frac{1}{3}$ or $x = -\frac{1}{2}$

The solutions are $\frac{1}{3}$ and $-\frac{1}{2}$.

35. $a = 2,\ b = -4,\ c = -1$

$$x = \frac{-b \pm \sqrt{b^2 - 4ac}}{2a}$$
$$= \frac{-(-4) \pm \sqrt{(-4)^2 - 4(2)(-1)}}{2(2)}$$
$$= \frac{4 \pm \sqrt{16 + 8}}{4}$$
$$= \frac{4 \pm \sqrt{24}}{4}$$
$$= \frac{4 \pm 2\sqrt{6}}{4}$$
$$= \frac{2(2 \pm \sqrt{6})}{4}$$
$$= \frac{2 \pm \sqrt{6}}{2}$$

The solutions are $\frac{2+\sqrt{6}}{2}$ and $\frac{2-\sqrt{6}}{2}$.

37. $a = 2,\ b = -4,\ c = 3$

$$s = \frac{-b \pm \sqrt{b^2 - 4ac}}{2a}$$
$$= \frac{-(-4) \pm \sqrt{(-4)^2 - 4(2)(3)}}{2(2)}$$
$$= \frac{4 \pm \sqrt{16 - 24}}{4}$$
$$= \frac{4 \pm \sqrt{-8}}{4}$$

Since $\sqrt{-8}$ is not a real number, this equation has no real number solution.

39. Write in standard form:

$4x^2 - x - 5 = 0$

$a = 4,\ b = -1,\ c = -5$

$$x = \frac{-b \pm \sqrt{b^2 - 4ac}}{2a}$$
$$= \frac{-(-1) \pm \sqrt{(-1)^2 - 4(4)(-5)}}{2(4)}$$
$$= \frac{1 \pm \sqrt{1 + 80}}{8}$$
$$= \frac{1 \pm \sqrt{81}}{8}$$
$$= \frac{1 \pm 9}{8}$$

$x = \frac{1 + 9}{8}$ or $x = \frac{1 - 9}{8}$

$x = \frac{5}{4}$ $\qquad x = -1$

The solutions are $\frac{5}{4}$ and -1.

41. Write in standard form:

$2x^2 - 7x - 9 = 0$

$a = 2,\ b = -7,\ c = -9$

$$x = \frac{-b \pm \sqrt{b^2 - 4ac}}{2a}$$
$$= \frac{-(-7) \pm \sqrt{(-7)^2 - 4(2)(-9)}}{2(2)}$$
$$= \frac{7 \pm \sqrt{49 + 72}}{4}$$
$$= \frac{7 \pm \sqrt{121}}{4}$$
$$= \frac{7 \pm 11}{4}$$

$x = \frac{7 + 11}{4}$ or $x = \frac{7 - 11}{4}$

$x = \frac{9}{2}$ or $x = -1$

The solutions are $\frac{9}{2}$ and -1.

43. $a = -2,\ b = 11,\ c = -15$

$$x = \frac{-b \pm \sqrt{b^2 - 4ac}}{2a}$$
$$= \frac{-11 \pm \sqrt{(11)^2 - 4(-2)(-15)}}{2(-2)}$$
$$= \frac{-11 \pm \sqrt{121 - 120}}{-4}$$
$$= \frac{-11 \pm \sqrt{1}}{-4}$$
$$= \frac{-11 \pm 1}{-4}$$

$x = \frac{-11 + 1}{-4}$ or $x = \frac{-11 - 1}{-4}$

$x = \frac{5}{2}$ or $x = 3$

The solutions are $\frac{5}{2}$ and 3.

45. Let x = the smaller integer

Then $x + 1$ = the larger integer

$x(x+1) = 20$

$x^2 + x = 20$

$x^2 + x - 20 = 0$

$a = 1,\ b = 1,\ c = -20$

$$x = \frac{-b \pm \sqrt{b^2 - 4ac}}{2a}$$

$$= \frac{-1 \pm \sqrt{1^2 - 4(1)(-20)}}{2(1)}$$

$$= \frac{-1 \pm \sqrt{1 + 80}}{2}$$

$$= \frac{-1 \pm \sqrt{81}}{2}$$

$$= \frac{-1 \pm 9}{2}$$

$x = \frac{-1+9}{2}$ or $x = \frac{-1-9}{2}$

$x = 4$ $\quad$ $x = -5$

Since the numbers are positive, $x = 4$. The numbers are 4 and $4 + 1 = 5$.

47. Let w = width of rectangle

Then $2w - 3$ = length of rectangle

Area = length · width

$20 = (2w - 3)w$

$20 = 2w^2 - 3w$

$0 = 2w^2 - 3w - 20$

$a = 2,\ b = -3,\ c = -20$

$$x = \frac{-b \pm \sqrt{b^2 - 4ac}}{2a}$$

$$= \frac{-(-3) \pm \sqrt{(-3)^2 - 4(2)(-20)}}{2(2)}$$

$$= \frac{3 \pm \sqrt{9 + 160}}{4}$$

$$= \frac{3 \pm \sqrt{169}}{4}$$

$$= \frac{3 \pm 13}{4}$$

$x = \frac{3+13}{4}$ or $x = \frac{3-13}{4}$

$x = 4$ or $x = -\frac{5}{2}$

Since the width is positive, $w = 4$. The width is 4 feet and the length is $2(4) - 3 = 5$ feet.

49. Let x = width of strip

Area of garden = $l \cdot w = (30)(40)$

$= 1200$ square feet

Area of the garden plus border = $l \cdot w$

$= (2x + 30)(2x + 40) = 4x^2 + 140x + 1200$

Area of border = Area of garden plus border – area of garden

$= 4x^2 + 140x + 1200 - 1200$

$= 4x^2 + 140x$

Area of border = 296 square feet

$296 = 4x^2 + 140x$

$4x^2 + 140x - 296 = 0$

$4(x^2 + 35x - 74) = 0$

$a = 1,\ b = 35,\ c = -74$

$x =$ Quadratic Formula

$$= \frac{-35 \pm \sqrt{(35)^2 - 4(1)(-74)}}{2(1)}$$

$$= \frac{-35 \pm \sqrt{1225 + 296}}{2}$$

$$= \frac{-35 \pm \sqrt{1521}}{2}$$

$$= \frac{-35 \pm 39}{2}$$

$x = \frac{-35 + 39}{2}$ or $x = \frac{-35 - 39}{2}$

$x = 2$ or $x = -37$

Since lengths are positive, $x = 2$

The width of the uniform strip will be 2 feet.

51. **(a)** none (discriminant negative)

(b) one (discriminant equal to zero)

(c) two (discriminant positive)

Cumulative Review Exercises: 10.3

54. **(a)** $x^2 - 13x + 42 = 0$

$(x - 6)(x - 7) = 0$

$x - 6 = 0$ or $x - 7 = 0$

$x = 6$ or $x = 7$

(b) $x^2 - 13x + 42 = 0$

$x^2 - 13x = -42$

$$x^2 - 13x + \frac{169}{4} = -42 + \frac{169}{4}$$

$$\left(x - \frac{13}{2}\right)^2 = -\frac{168}{4} + \frac{169}{4}$$

$$\left(x - \frac{13}{2}\right)^2 = \frac{1}{4}$$

$$x - \frac{13}{2} = \pm\sqrt{\frac{1}{4}}$$

$$x - \frac{13}{2} = \pm\frac{1}{2}$$

$$x = \frac{13}{2} \pm \frac{1}{2}$$

$x = \frac{13}{2} + \frac{1}{2}$ or $x = \frac{13}{2} - \frac{1}{2}$

$x = 7$ or $x = 6$

(c) $a = 1,\ b = -13,\ c = 42$

$$x = \frac{-b \pm \sqrt{b^2 - 4ac}}{2a}$$

$$= \frac{-(-13) \pm \sqrt{(-13)^2 - 4(1)(42)}}{2(1)}$$

$$= \frac{13 \pm \sqrt{169 - 168}}{2}$$

$$= \frac{13 \pm \sqrt{1}}{2}$$

$$= \frac{13 \pm 1}{2}$$

$x = \frac{13 + 1}{2}$ or $= \frac{13 - 1}{2}$

$x = 7$ or $x = 6$

55. **(a)** $6x^2 + 11x - 35 = 0$

$(2x+7)(3x-5) = 0$

$2x + 7 = 0$ or $3x - 5 = 0$

$x = -\frac{7}{2}$ or $x = \frac{5}{3}$

(b) $6x^2 + 11x - 35 = 0$

$\frac{1}{6}(6x^2 + 11x - 35) = \frac{1}{6}(0)$

$x^2 + \frac{11}{6}x - \frac{35}{6} = 0$

$x^2 + \frac{11}{6}x = \frac{35}{6}$

$x^2 + \frac{11}{6}x + \frac{121}{144} = \frac{35}{6} + \frac{121}{144}$

$\left(x + \frac{11}{12}\right)^2 = \frac{840}{144} + \frac{121}{144}$

$\left(x + \frac{11}{12}\right)^2 = \frac{961}{144}$

$x + \frac{11}{12} = \pm\sqrt{\frac{961}{144}}$

$x + \frac{11}{12} = \pm\frac{31}{12}$

$x = -\frac{11}{12} \pm \frac{31}{12}$

$x = -\frac{11}{12} + \frac{31}{12}$ or $x = -\frac{11}{12} - \frac{31}{12}$

$x = \frac{5}{3}$ or $x = -\frac{7}{2}$

(c) $a = 6,\ b = 11,\ c = -35$

$$x = \frac{-b \pm \sqrt{b^2 - 4ac}}{2a}$$

$$= \frac{-11 \pm \sqrt{(11)^2 - 4(6)(-35)}}{2(6)}$$

$$= \frac{-11 \pm \sqrt{121 + 840}}{12}$$

$$= \frac{-11 \pm \sqrt{961}}{12}$$

$$= \frac{-11 \pm 31}{12}$$

$x = \frac{-11 + 31}{12}$ or $= \frac{-11 - 31}{12}$

$x = \frac{5}{3}$ or $x = -\frac{7}{2}$

56. **(a)** Cannot be solved by factoring.

(b) $2x^2 + 3x - 4 = 0$

$\frac{1}{2}(2x^2 + 3x - 4) = \frac{1}{2}(0)$

$x^2 + \frac{3}{2}x - 2 = 0$

$x^2 + \frac{3}{2}x + \frac{9}{16} = 2 + \frac{9}{16}$

$\left(x + \frac{3}{4}\right)^2 = \frac{41}{16}$

$x + \frac{3}{4} = \pm\sqrt{\frac{41}{16}}$

$x + \frac{3}{4} = \pm\frac{\sqrt{41}}{4}$

$x = -\frac{3}{4} \pm \frac{\sqrt{41}}{4}$

The solutions are $x = \frac{-3 + \sqrt{41}}{4}$

and $x = \frac{-3 - \sqrt{41}}{4}$.

(c) $a = 2, b = 3, c = -4$

$$x = \frac{-b \pm \sqrt{b^2 - 4ac}}{2a}$$
$$= \frac{-3 \pm \sqrt{3^2 - 4(2)(-4)}}{2(2)}$$
$$= \frac{-3 \pm \sqrt{9 + 32}}{4}$$
$$= \frac{-3 \pm \sqrt{41}}{4}$$

The solutions are $x = \frac{-3 + \sqrt{41}}{4}$ and $x = \frac{-3 - \sqrt{41}}{4}$.

57. (a) $6x^2 = 54$

$6x^2 - 54 = 0$

$6(x + 3)(x - 3) = 0$

$x + 3 = 0$ or $x - 3 = 0$

$x = -3$ or $x = 3$

(b) $6x^2 = 54$

$\frac{1}{6}(6x^2) = \frac{1}{6}(54)$

$x^2 = 9$

$x^2 = \pm\sqrt{9}$

$x = \pm 3$

The solutions are 3 and –3.

(c) $6x^2 = 54$

$6x^2 - 54 = 0$

$\frac{1}{6}(6x^2 - 54) = \frac{1}{6}(0)$

$x^2 - 9 = 0$

$a = 1, b = 0, c = -9$

$$x = \frac{-b \pm \sqrt{b^2 - 4ac}}{2a}$$
$$= \frac{-0 \pm \sqrt{0^2 - 4(1)(-9)}}{2(1)}$$
$$= \frac{\pm\sqrt{36}}{2}$$
$$= \frac{\pm 6}{2} = \pm 3$$

The solutions are 3 and – 3.

Group Activity/Challenge Problems

1. (a)

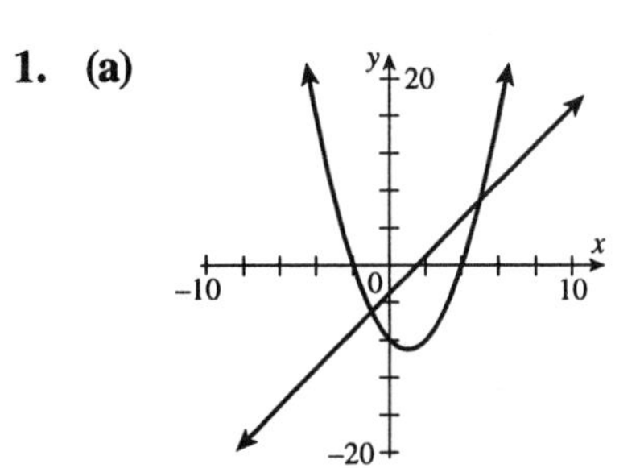

Answer: (–1, –5), (5, 7)

(b) $y = x^2 - 2x - 8, \ y = 2x - 3$

$x^2 - 2x - 8 = 2x - 3$

$x^2 - 4x - 5 = 0$

$(x - 5)(x + 1) = 0$

$x - 5 = 0$ or $x + 1 = 0$

$x = 5$ and $x = -1$

(c) Yes

(d) $x = -1; \ y = x^2 - 2x - 8$

$y = (-1)^2 - 2(-1) - 8 = 1 + 2 - 8 = -5$

$x = 5; \ y = x^2 - 2x - 8$

$y = 5^2 - 2(5) - 8 = 25 - 10 - 8 = 7$

Answer: (–1, 5), (5, 7)

2. $x^2 + 6x + c = 0$

$$\frac{-6 \pm \sqrt{36-4c}}{2} = 0$$

$36 - 4c = 0$

$c = 9$

(a) $c < 9$

(b) $c = 9$

(c) $c > 9$

3. $2x^2 + 3x + c = 0$

$$\frac{-3 \pm \sqrt{9-8c}}{4} = 0$$

$9 - 8c = 0$

$c = \frac{9}{8}$

(a) $c < \frac{9}{8}$

(b) $c = \frac{9}{8}$

(c) $c > \frac{9}{8}$

5. $400 = l + 2w;\ l = 400 - 2w$

$15{,}000 = lw;\ 15{,}000 = (400 - 2w)w$

$15000 = 400w - 2w^2$

$2w^2 - 400w + 15{,}000 = 0$

$w^2 - 200w + 7500 = 0$

$(w - 150)(w - 50) = 0$

$w = 150$ or $w = 50$

Therefore, the width is 150 and length is $(400 - 2(150)) = 100$ feet, or the width is 50 and the length is $(400 - 2(50)) = 300$ feet.

Exercise Set 10.4

1. $a = 1, b = 2, c = -7$

$$x = \frac{-b}{2a} = \frac{-2}{2(1)} = -1$$

The axis of symmetry is $x = -1$

Find the y coordinate of the vertex:

$y = x^2 + 2x - 7$

$y = (-1)^2 + 2(-1) - 7$

$y = 1 - 2 - 7 = -8$

The vertex is $(-1, -8)$

Since $a > 0$, the parabola opens up.

3. $a = -1, b = 5, c = -6$

$$x = \frac{-b}{2a} = \frac{-5}{2(-1)} = \frac{5}{2}$$

The axis of symmetry is $x = \frac{5}{2}$

Find the y coordinate of the vertex:

$y = -x^2 + 5x - 6$

$$y = -\left(\frac{5}{2}\right)^2 + 5\left(\frac{5}{2}\right) - 6$$

$$y = -\frac{25}{4} + \frac{25}{2} - 6 = \frac{1}{4}$$

The vertex is $\left(\frac{5}{2}, \frac{1}{4}\right)$.

Since $a < 0$, the parabola opens down.

5. $a=-3,\ b=5,\ c=8$

$x=\frac{-b}{2a}=\frac{-5}{2(-3)}=\frac{5}{6}$

The axis of symmetry is $x=\frac{5}{6}$

Find the y coordinate of the vertex:

$y=-3x^2+5x+8$

$y=-3\left(\frac{5}{6}\right)^2+5\left(\frac{5}{6}\right)+8$

$y=-\frac{25}{12}+\frac{25}{6}+8$

$y=\frac{121}{12}$

The vertex is $\left(\frac{5}{6},\frac{121}{12}\right)$.

Since $a<0$, the parabola opens down.

7. $a=-4,\ b=-8,\ c=-12$

$x=\frac{-b}{2a}=\frac{-(-8)}{2(-4)}=-1$

The axis of symmetry is $x=-1$.

Find the y coordinate of the vertex:

$y=-4x^2-8x-12$

$y=-4(-1)^2-8(-1)-12$

$y=-4+8-12=-8$

The vertex is $(-1,-8)$.

Since $a<0$, the parabola opens down.

9. $a=3,\ b=-2,\ c=2$

$x=\frac{-b}{2a}=\frac{-(-2)}{2(3)}=\frac{1}{3}$

The axis of symmetry is $x=\frac{1}{3}$

Find the y coordinate of the vertex:

$y=3x^2-2x+2$

$y=3\left(\frac{1}{3}\right)^2-2\left(\frac{1}{3}\right)+2$

$y=\left(\frac{1}{3}\right)-\frac{2}{3}+2=\frac{5}{3}$

The vertex is $\left(\frac{1}{3},\frac{5}{3}\right)$.

Since $a>0$, the parabola opens up.

11. $a=4,\ b=12,\ c=-5$

$x=\frac{-b}{2a}=\frac{-12}{2(4)}=-\frac{3}{2}$

The axis of symmetry is $x=-\frac{3}{2}$.

Find the y coordinate of the vertex:

$y=4x^2+12x-5$

$y=4\left(-\frac{3}{2}\right)^2+12\left(-\frac{3}{2}\right)-5$

$y=9-18-5=-14$

The vertex is $\left(-\frac{3}{2},-14\right)$.

Since $a>0$, the parabola opens up.

13. $a = 1,\ b = 0,\ c = -1$

Since $a > 0$, the parabola opens up.

Axis of symmetry: $x = \dfrac{-b}{2a}$

$$x = \frac{-0}{2(1)} = 0$$

y coordinate of vertex:

$y = x^2 - 1$

$y = (0)^2 - 1 = -1$

The vertex is at $(0, -1)$

$$y = x^2 - 1$$

Let $x = 1$, $y = 1^2 - 1 = 0$

Let $x = 2$, $y = 2^2 - 1 = 3$

Let $x = 3$, $y = 3^2 - 1 = 8$

x	y
1	0
2	3
3	8

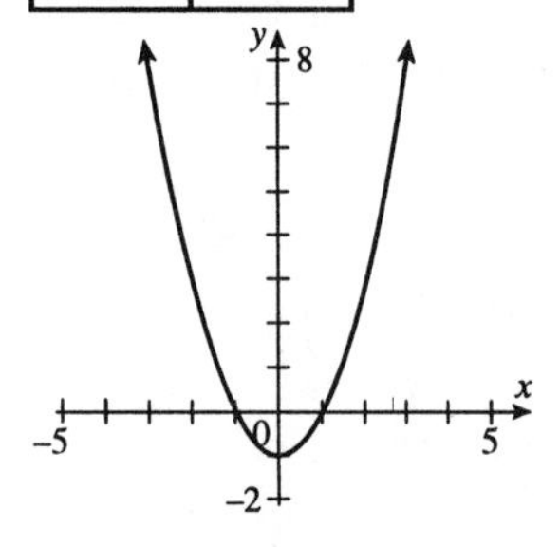

The x-intercepts are 1 and -1

15. $a = -1,\ b = 0,\ c = 3$

Since $a < 0$, the parabola opens down.

Axis of symmetry: $x = \dfrac{-b}{2a}$

$$x = \frac{-0}{2(-1)} = 0$$

y coordinate of vertex:

$y = -x^2 + 3$

$y = -(0)^2 + 3 = 3$

The vertex is at $(0, 3)$

$$y = -x^2 + 3$$

Let $x = 1$, $y = -1^2 + 3 = 2$

Let $x = 2$, $y = -2^2 + 3 = -1$

x	y
1	2
2	-1

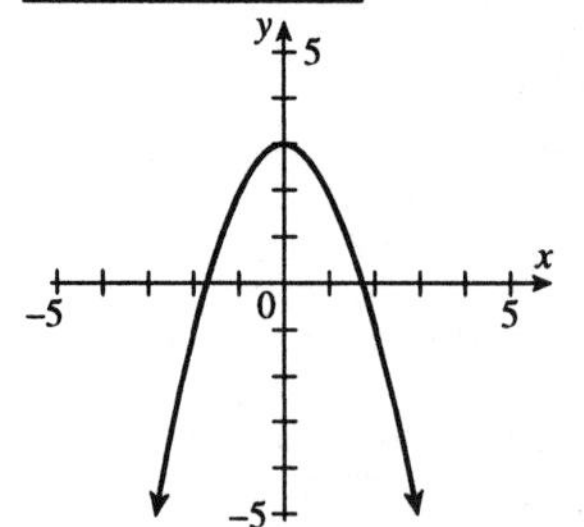

To find x-intercepts, set $y = 0$

$0 = -x^2 + 3$

$x^2 = 3$

$x = \pm\sqrt{3}$

The x-intercepts are $\sqrt{3}$ and $-\sqrt{3}$

17. $a = 1, b = 2, c = 3$

Since $a > 0$, the parabola opens up.

Axis of symmetry: $x = \frac{-b}{2a}$

$$x = \frac{-2}{2(1)} = -1$$

y coordinate of vertex:

$y = x^2 + 2x + 3$

$y = (-1)^2 + 2(-1) + 3$

$y = 1 - 2 + 3 = 2$

The vertex is at $(-1, 2)$

$$y = x^2 + 2x + 3$$

Let $x = 0$, $y = 0^2 + 2(0) + 3 = 3$

Let $x = 1$, $y = 1^2 + 2(1) + 3 = 6$

x	y
0	3
1	6

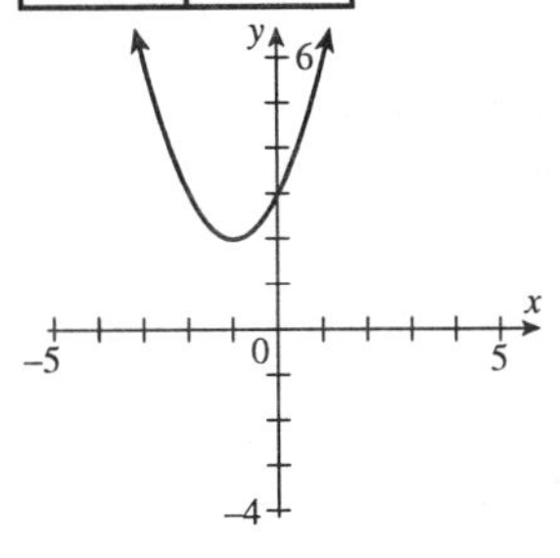

There are no x-intercepts.

19. $a = 1, b = 2, c = -15$

Since $a > 0$, the parabola opens up.

Axis of symmetry: $x = \frac{-b}{2a}$

$$x = \frac{-2}{2(1)} = -1$$

y coordinate of vertex:

$y = x^2 + 2x - 15$

$y = (-1)^2 + 2(-1) - 15$

$y = 1 - 2 - 15 = -16$

The vertex is at $(-1, -16)$

$$y = x^2 + 2x - 15$$

Let $x = 0$, $y = 0^2 + 2(0) - 15 = -15$

Let $x = 1$, $y = 1^2 + 2(1) - 15 = -12$

Let $x = 2$, $y = 2^2 + 2(2) - 15 = -7$

Let $x = 3$, $y = 3^2 + 2(3) - 15 = 0$

Let $x = 4$, $y = 4^2 + 2(4) - 15 = 9$

x	y
0	–15
1	–12
2	–7
3	0
4	9

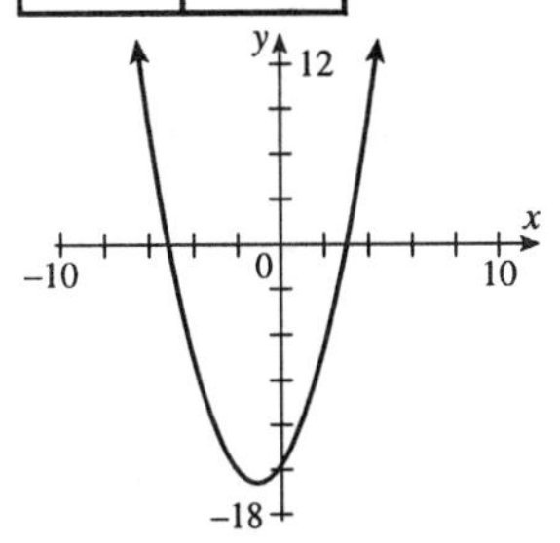

The x-intercepts are 3 and –5.

21. $a=-1$, $b=4$, $c=-5$

Since $a<0$, the parabola opens down.

Axis of symmetry: $x=\dfrac{-b}{2a}$

$$x=\frac{-4}{2(-1)}=2$$

y coordinate of vertex:

$y=-x^2+4x-5$

$y=-2^2+4(2)-5$

$y=-4+8-5=-1$

The vertex is at $(2,-1)$.

$$y=-x^2+4x-5$$

Let $x=3$ $y=-3^2+4(3)-5=-2$

Let $x=4$ $y=-4^2+4(4)-5=-5$

x	y
3	–2
4	–5

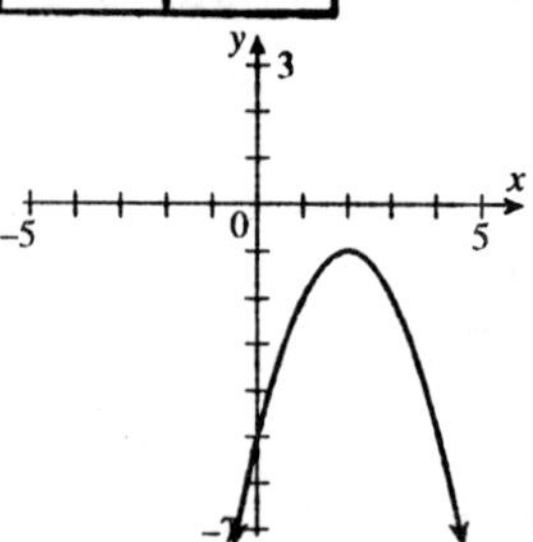

There are no x-intercepts.

23. $a=1$, $b=-1$, $c=-12$

Since $a>0$, the parabola opens up.

Axis of symmetry: $x=\dfrac{-b}{2a}$

$$x=\frac{-(-1)}{2(1)}=\frac{1}{2}$$

y coordinate of the vertex:

$y=x^2-x-12$

$y=\left(\dfrac{1}{2}\right)^2-\dfrac{1}{2}-12$

$y=\dfrac{1}{4}-\dfrac{1}{2}-12=-\dfrac{49}{4}$

The vertex is at $\left(\dfrac{1}{2},-\dfrac{49}{4}\right)$.

$$y=x^2-x-12$$

Let $x=1$ $y=1^2-1-12=-12$

Let $x=2$ $y=2^2-2-12=-10$

Let $x=3$ $y=3^2-3-12=-6$

Let $x=4$ $y=4^2-4-12=0$

x	y
1	–12
2	–10
3	–6
4	0

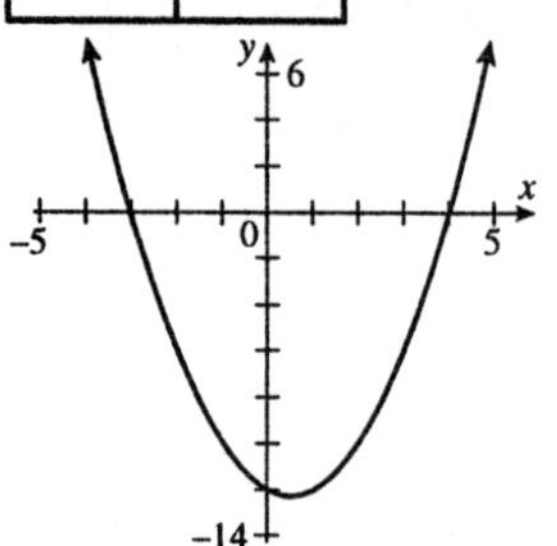

The x-intercepts are 4 and –3.

25. $a = 1,\ b = -6,\ c = 9$

Since $a > 0$, the parabola opens up.

Axis of symmetry: $x = \dfrac{-b}{2a}$

$$x = \frac{-(-6)}{2(1)} = 3$$

y coordinate of the vertex:

$y = x^2 - 6x + 9$

$y = 3^2 - 6(3) + 9$

$y = 9 - 18 + 9 = 0$

The vertex is at (3, 0)

$y = x^2 - 6x + 9$

Let $x = 4$ $\quad y = 4^2 - 6(4) + 9 = 1$

Let $x = 5$ $\quad y = 5^2 - 6(5) + 9 = 4$

Let $x = 6$ $\quad y = 6^2 - 6(6) + 9 = 9$

x	y
4	1
5	4
6	9

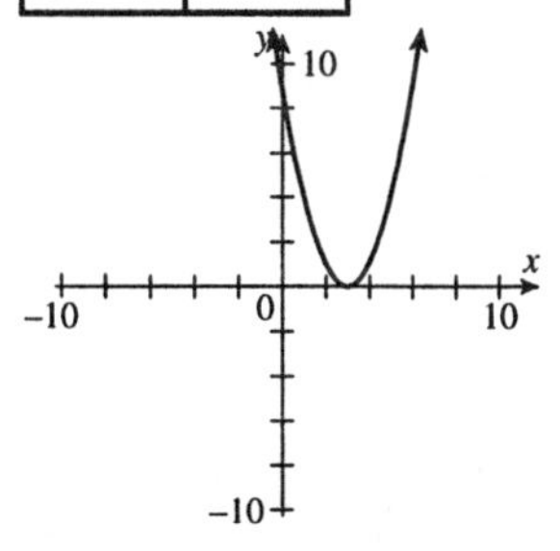

The x-intercept is 3.

27. $a = -1,\ b = 5,\ c = 0$

Since $a < 0$, the parabola opens down.

Axis of symmetry: $x = \dfrac{-b}{2a}$

$$x = \frac{-5}{2(-1)} = \frac{5}{2}$$

y coordinate of the vertex:

$y = -x^2 + 5x$

$y = -\left(\frac{5}{2}\right)^2 + 5\left(\frac{5}{2}\right)$

$y = -\frac{25}{4} + \frac{25}{2} = \frac{25}{4}$

The vertex is at $\left(\frac{5}{2}, \frac{25}{4}\right)$

$y = -x^2 + 5x$

Let $x = 3$ $\quad y = -3^2 + 5(3) = 6$

Let $x = 4$ $\quad y = -4^2 + 5(4) = 4$

Let $x = 5$ $\quad y = -5^2 + 5(5) = 0$

x	y
3	6
4	4
5	0

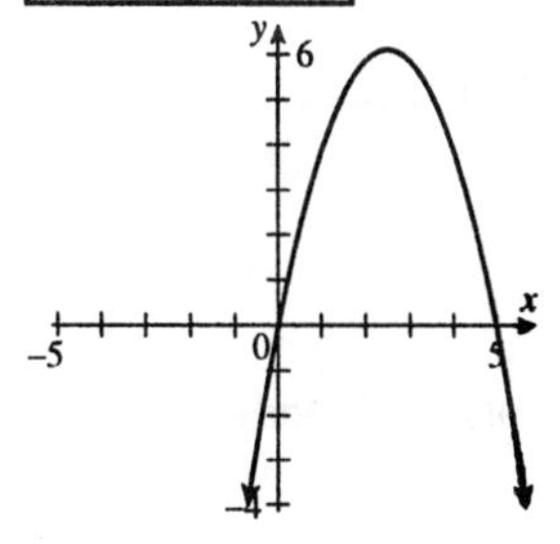

The x-intercepts are 0 and 5.

29. $a = 2,\ b = -6,\ c = 4$

Since $a > 0$, the parabola opens up.

Axis of symmetry: $x = \dfrac{-b}{2a}$

$$x = \frac{-(-6)}{2(2)} = \frac{3}{2}$$

y coordinate of the vertex:

$y = 2x^2 - 6x + 4$

$y = 2\left(\dfrac{3}{2}\right)^2 - 6\left(\dfrac{3}{2}\right) + 4$

$y = \dfrac{9}{2} - 9 + 4 = -\dfrac{1}{2}$

The vertex is at $\left(\dfrac{3}{2}, -\dfrac{1}{2}\right)$.

$$y = 2x^2 - 6x + 4$$

Let $x = 2$ $\quad y = 2(2)^2 - 6(2) + 4 = 0$

Let $x = 3$ $\quad y = 2(3)^2 - 6(3) + 4 = 4$

x	y
2	0
3	4

The x-intercepts are 1 and 2.

31. $a = -1,\ b = 11,\ c = -28$

Since $a < 0$, the parabola opens down.

Axis of symmetry: $x = \dfrac{-b}{2a}$

$$x = \frac{-11}{2(-1)} = \frac{11}{2}$$

y coordinate of the vertex:

$y = -x^2 + 11x - 28$

$y = -\left(\dfrac{11}{2}\right)^2 + 11\left(\dfrac{11}{2}\right) - 28$

$y = -\dfrac{121}{4} + \dfrac{121}{2} - 28 = \dfrac{9}{4}$

The vertex is at $\left(\dfrac{11}{2}, \dfrac{9}{4}\right)$.

$$y = -x^2 + 11x - 28$$

Let $x = 6$ $\quad y = -6^2 + 11(6) - 28 = 2$

Let $x = 7$ $\quad y = -7^2 + 11(7) - 28 = 0$

Let $x = 8$ $\quad y = -8^2 + 11(8) - 28 = -4$

x	y
6	2
7	0
8	−4

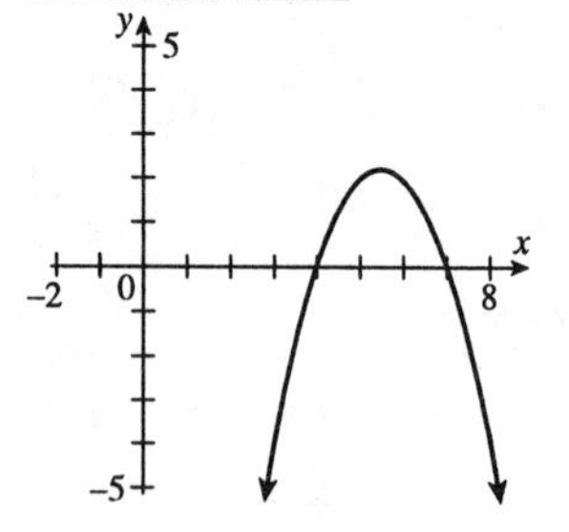

The x-intercepts are 4 and 7.

33. $a = 1, b = -2, c = -15$

Since $a > 0$, the parabola opens up.

Axis of symmetry: $x = \frac{-b}{2a}$

$$x = \frac{-(-2)}{2(1)} = 1$$

y coordinate of the vertex:

$y = x^2 - 2x - 15$

$y = 1^2 - 2(1) - 15$

$y = 1 - 2 - 15 = -16$

The vertex is at $(1, -16)$

$$y = x^2 - 2x - 15$$

Let $x = 2$ $\quad y = 2^2 - 2(2) - 15 = -15$

Let $x = 3$ $\quad y = 3^2 - 2(3) - 15 = -12$

Let $x = 4$ $\quad y = 4^2 - 2(4) - 15 = -7$

Let $x = 5$ $\quad y = 5^2 - 2(5) - 15 = 0$

x	y
2	–15
3	–12
4	–7
5	0

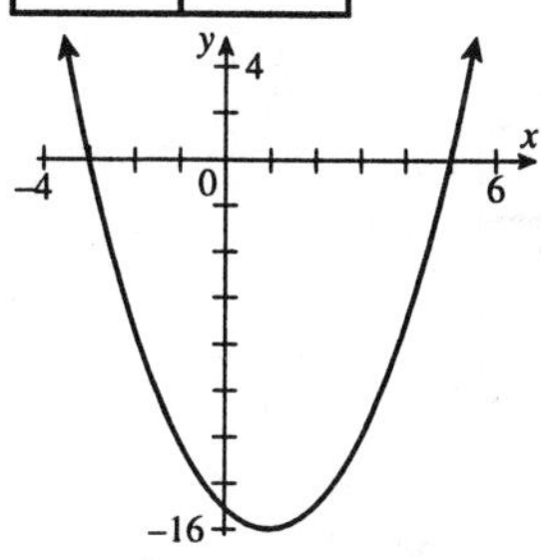

The x-intercepts are –3 and 5.

35. $a = -2, b = 7, c = -3$

Since $a < 0$, the parabola opens down.

Axis of symmetry: $x = \frac{-b}{2a}$

$$x = \frac{-7}{2(-2)} = \frac{7}{4}$$

y coordinate of the vertex:

$y = -2x^2 + 7x - 3$

$y = -2\left(\frac{7}{4}\right)^2 + 7\left(\frac{7}{4}\right) - 3$

$y = -\frac{49}{8} + \frac{49}{4} - 3 = \frac{25}{8}$

The vertex is at $\left(\frac{7}{4}, \frac{25}{8}\right)$

$$y = -2x^2 + 7x - 3$$

Let $x = 1$ $\quad y = -2(1)^2 + 7(1) - 3 = 2$

Let $x = 0$ $\quad y = -2(0)^2 + 7(0) - 3 = -3$

Let $x = 2$ $\quad y = -2(2)^2 + 7(2) - 3 = 3$

Let $x = 3$ $\quad y = -2(3)^2 + 7(3) - 3 = 0$

Let $x = 4$ $\quad y = -2(4)^2 + 7(4) - 3 = -7$

x	y
2	3
3	0
4	–7
1	2
0	–3

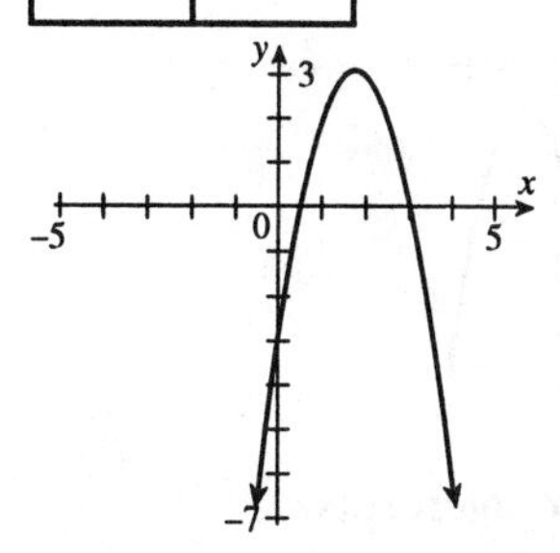

The x-intercepts are 3 and 0.5

37. $a=-2, b=3, c=-2$

Since $a<0$, the parabola opens down.

Axis of symmetry: $x=\frac{-b}{2a}$

$$x=\frac{-3}{2(-2)}=\frac{3}{4}$$

y coordinate of the vertex:

$y=-2x^2+3x-2$

$y=-2\left(\frac{3}{4}\right)^2+3\left(\frac{3}{4}\right)-2$

$y=-\frac{9}{8}+\frac{9}{4}-2=-\frac{7}{8}$

The vertex is at $\left(\frac{3}{4},-\frac{7}{8}\right)$

$$y=-2x^2+3x-2$$

Let $x=-1$ $\quad y=-2(-1)^2+3(-1)-2$
$\quad =-7$

Let $x=0$ $\quad y=-2(0)^2+3(1)-2=-2$

Let $x=1$ $\quad y=-2(1)^2+3(1)-2=-1$

Let $x=2$ $\quad y=-2(2)^2+3(2)-2=-4$

x	y
−1	−7
0	−2
1	−1
2	−4

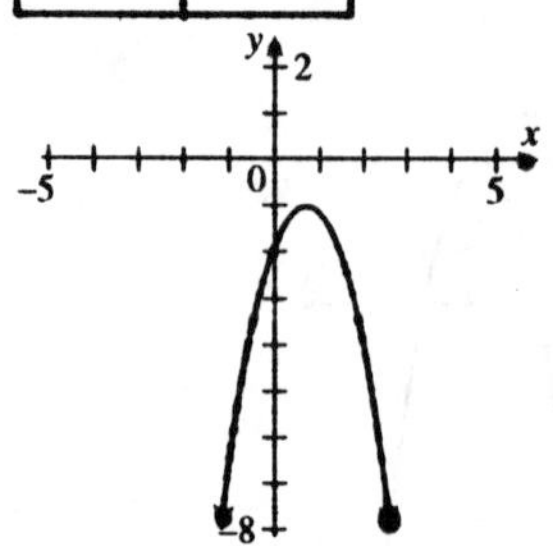

The are no x-intercepts.

39. $a=2, b=-1, c=-15$

Since $a>0$, the parabola opens up.

Axis of symmetry: $x=\frac{-b}{2a}$

$$x=\frac{-(-1)}{2(2)}=\frac{1}{4}$$

y coordinate of the vertex:

$y=2x^2-x-15$

$y=2\left(\frac{1}{4}\right)^2-\frac{1}{4}-15$

$y=\frac{1}{8}-\frac{1}{4}-15=-\frac{121}{8}$

The vertex is at $\left(\frac{1}{4},-\frac{121}{8}\right)$

$$y=2x^2-x-15$$

Let $x=-2$ $\quad y=2(-2)^2-(-2)-15=-5$

Let $x=-1$ $\quad y=2(-1)^2-(-1)-15=-12$

Let $x=1$ $\quad y=2(1)^2-1-15=-14$

Let $x=2$ $\quad y=2(2)^2-2-15=-9$

Let $x=3$ $\quad y=2(3)^2-3-15=0$

x	y
−2	−5
−1	−12
1	−14
2	−9
3	0

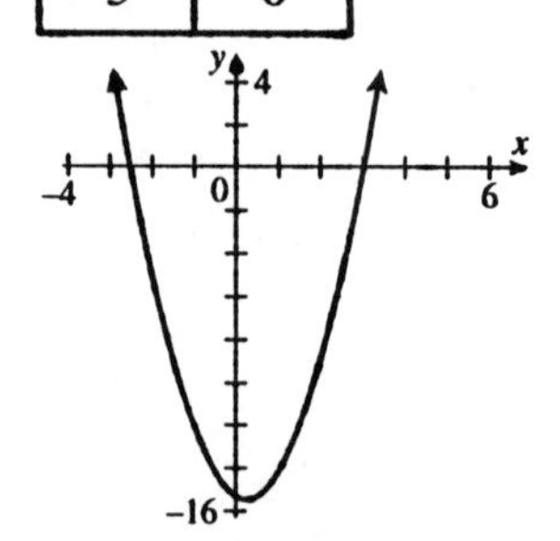

The are x-intercepts are 3 and $-\frac{5}{2}$

41. $b^2 - 4ac = (-6)^2 - 4(3)(4)$
$= 36 - 48 = -12$

Since the discriminant is negative, there will be no x -intercepts.

43. $b^2 - 4ac = (-6)^2 - 4(4)(-7)$
$= 36 + 112 = 148$

Since the discriminant is positive, there will be two x -intercepts.

45. $b^2 - 4ac = 2^2 - 4(0.1)(-3)$
$= 4 + 1.2 = 5.2$

Since the discriminant is positive, there will be two x -intercepts.

47. $b^2 - 4ac = 0^2 - 4\left(-\frac{1}{2}\right)\left(-\frac{3}{4}\right)$
$= 0 - \frac{3}{2} = -\frac{3}{2}$

Since the discriminant is negative, there will be no x -intercepts.

49. $\left(\frac{-b}{2a}, \frac{4ac - b^2}{4a}\right)$

51. (a) The values of x where the graph crosses the x axis

(b) Let $y = 0$ and solve the resulting equation for x.

53. (a) $x = -\frac{b}{2a}$

(b) The axis of symmetry.

55. Two; the vertex is below the x axis and the parabola opens upward.

57. None; the vertex is below the x axis and the parabola opens downward.

Cumulative Review Exercises: 10.4

59. $\frac{3}{x+3} - \frac{x-2}{x-4}$

$= \frac{3}{x+3} \cdot \frac{x-4}{x-4} - \frac{x-2}{x-4} \cdot \frac{x+3}{x+3}$

$= \frac{3x - 12}{(x+3)(x-4)} - \frac{(x^2 + x - 6)}{(x+3)(x-4)}$

$= \frac{-x^2 + 2x - 6}{(x+3)(x-4)}$

60. $\frac{1}{3}(x+6) = 3 - \frac{1}{4}(x-5)$

$\frac{1}{3}x + 2 = 3 - \frac{1}{4}x + \frac{5}{4}$

$\frac{1}{3}x + 2 = \frac{17}{4} - \frac{1}{4}x$

$\frac{1}{3}x = \frac{9}{4} - \frac{1}{4}x$

$\frac{7}{12}x = \frac{9}{4}$

$x = \frac{12}{7} \cdot \frac{9}{4} = \frac{27}{7}$

61. $4x - 6y = 20$ **Ordered Pair**

Let $x = 0$, then $y = -\frac{10}{3}$ $\left(0, -\frac{10}{3}\right)$

Let $y = 0$, then $x = 5$ $(5, 0)$

Plot the points and draw the line.

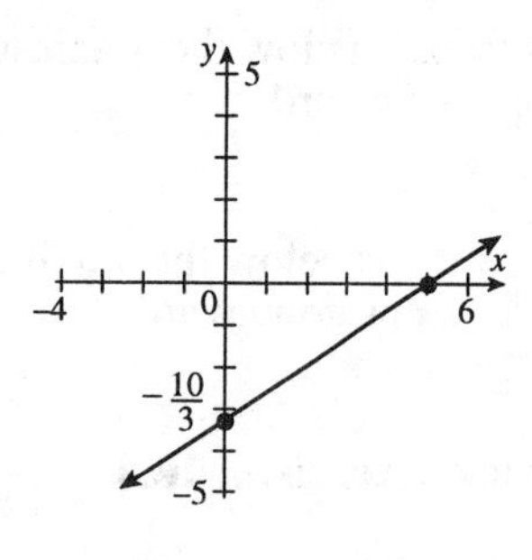

62. Draw the line $y = 2$ (horizontal line). Since the inequality symbol is <, draw a dashed line.

Check: (0, 0)

$y < 2$

$0 < 2$ True

Since the origin does satisfy the inequality, all the points on the same side of the line as the origin satisfy the inequality.

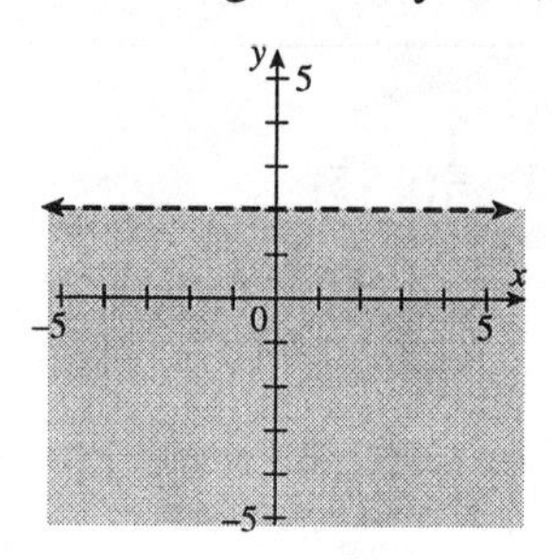

Group Activity/Challenge Problems

2.

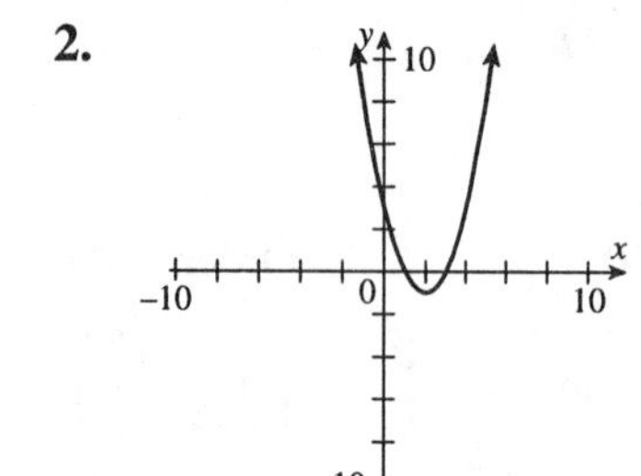

Chapter 10 Review Exercises

1. $x^2 = 25$

$x = \pm\sqrt{25}$

$x = \pm 5$

The solutions are 5 and –5.

2. $x^2 = 8$

$x = \pm\sqrt{8}$

$x = \pm 2\sqrt{2}$

The solutions are $2\sqrt{2}$ and $-2\sqrt{2}$.

3. $2x^2 = 12$

$\frac{1}{2}(2x^2) = \frac{1}{2}(12)$

$x^2 = 6$

$x = \pm\sqrt{6}$

The solutions are $\sqrt{6}$ and $-\sqrt{6}$.

4. $x^2 + 3 = 9$

$x^2 = 6$

$x = \pm\sqrt{6}$

The solutions are $\sqrt{6}$ and $-\sqrt{6}$.

5. $x^2 - 4 = 16$

$x^2 = 20$

$x = \pm\sqrt{20}$

$x = \pm 2\sqrt{5}$

The solutions are $2\sqrt{5}$ and $-2\sqrt{5}$.

6. $2x^2 - 4 = 10$

$2x^2 = 14$

$x^2 = 7$

$x = \pm\sqrt{7}$

The solutions are $\sqrt{7}$ and $-\sqrt{7}$.

7. $3x^2 + 8 = 32$

$3x^2 = 24$

$x^2 = 8$

$x = \pm\sqrt{8}$

$x = \pm 2\sqrt{2}$

The solutions are $2\sqrt{2}$ and $-2\sqrt{2}$.

8. $(x-3)^2 = 12$

$x - 3 = \pm\sqrt{12}$

$x - 3 = \pm 2\sqrt{3}$

$x = 3 \pm 2\sqrt{3}$

The solutions are $3 + 2\sqrt{3}$ and $3 - 2\sqrt{3}$.

9. $(2x+4)^2 = 30$

$2x + 4 = \pm\sqrt{30}$

$2x = -4 \pm \sqrt{30}$

$x = \dfrac{-4 \pm \sqrt{30}}{2}$

The solutions are $\dfrac{-4+\sqrt{30}}{2}$ and $\dfrac{-4-\sqrt{30}}{2}$.

10. $(3x-5)^2 = 50$

$3x - 5 = \pm\sqrt{50}$

$3x - 5 = \pm 5\sqrt{2}$

$3x = 5 \pm 5\sqrt{2}$

$x = \dfrac{5 \pm 5\sqrt{2}}{3}$

The solutions are $\dfrac{5+5\sqrt{2}}{3}$ and $\dfrac{5-5\sqrt{2}}{3}$.

11. $x^2 - 10x + 16 = 0$

$x^2 - 10x = -16$

$x^2 - 10x + 25 = -16 + 25$

$(x-5)^2 = 9$

$x - 5 = \pm\sqrt{9}$

$x - 5 = \pm 3$

$x = 5 \pm 3$

$x = 5 + 3$ or $x = 5 - 3$

$x = 8$ or $x = 2$

12. $x^2 - 8x + 15 = 0$

$x^2 - 8x = -15$

$x^2 - 8x + 16 = -15 + 16$

$(x-4)^2 = 1$

$x - 4 = \pm\sqrt{1}$

$x - 4 = \pm 1$

$x = 4 \pm 1$

$x = 4 + 1$ or $x = 4 - 1$

$x = 5$ or $x = 3$

13. $x^2 - 14x + 13 = 0$

$x^2 - 14x = -13$

$x^2 - 14x + 49 = -13 + 49$

$(x-7)^2 = 36$

$x - 7 = \pm\sqrt{36}$

$x - 7 = \pm 6$

$x = 7 \pm 6$

$x = 7 + 6$ or $x = 7 - 6$

$x = 13$ or $x = 1$

14. $x^2 + x - 6 = 0$

$x^2 + x = 6$

$x^2 + x + \frac{1}{4} = 6 + \frac{1}{4}$

$\left(x + \frac{1}{2}\right)^2 = \frac{25}{4}$

$x + \frac{1}{2} = \pm\sqrt{\frac{25}{4}}$

$x + \frac{1}{2} = \pm\frac{5}{2}$

$x = -\frac{1}{2} \pm \frac{5}{2}$

$x = -\frac{1}{2} + \frac{5}{2}$ or $x = -\frac{1}{2} - \frac{5}{2}$

$x = 2$ or $x = -3$

15. $x^2 - 3x - 54 = 0$

$x^2 - 3x = 54$

$x^2 - 3x + \frac{9}{4} = 54 + \frac{9}{4}$

$\left(x - \frac{3}{2}\right)^2 = \frac{225}{4}$

$x - \frac{3}{2} = \pm\sqrt{\frac{225}{4}}$

$x - \frac{3}{2} = \pm\frac{15}{2}$

$x = \frac{3}{2} \pm \frac{15}{2}$

$x = \frac{3}{2} + \frac{15}{2}$ or $x = \frac{3}{2} - \frac{15}{2}$

$x = 9$ or $x = -6$

16. $x^2 = -5x + 6$

$x^2 + 5x = 6$

$x^2 + 5x + \frac{25}{4} = 6 + \frac{25}{4}$

$\left(x + \frac{5}{2}\right)^2 = \frac{49}{4}$

$x + \frac{5}{2} = \pm\sqrt{\frac{49}{4}}$

$x + \frac{5}{2} = \pm\frac{7}{2}$

$x = -\frac{5}{2} \pm \frac{7}{2}$

$x = -\frac{5}{2} + \frac{7}{2}$ or $x = -\frac{5}{2} - \frac{7}{2}$

$x = 1$ or $x = -6$

17. $x^2 + 2x - 5 = 0$

$x^2 + 2x = 5$

$x^2 + 2x + 1 = 5 + 1$

$(x+1)^2 = 6$

$x + 1 = \pm\sqrt{6}$

$x = -1 \pm \sqrt{6}$

The solutions are $-1 + \sqrt{6}$ and $-1 - \sqrt{6}$.

18. $x^2 - 3x - 8 = 0$

$x^2 - 3x = 8$

$x^2 - 3x + \frac{9}{4} = 8 + \frac{9}{4}$

$\left(x - \frac{3}{2}\right)^2 = \frac{41}{4}$

$x - \frac{3}{2} = \pm\sqrt{\frac{41}{4}}$

$x - \frac{3}{2} = \pm\frac{\sqrt{41}}{2}$

$x = \frac{3}{2} \pm \frac{\sqrt{41}}{2}$

$x = \frac{3}{2} + \frac{\sqrt{41}}{2}$ or $x = \frac{3}{2} - \frac{\sqrt{41}}{2}$

The solutions are $\frac{3}{2} + \frac{\sqrt{41}}{2}$ and $\frac{3}{2} - \frac{\sqrt{41}}{2}$.

19. $2x^2 - 8x = 64$

$\frac{1}{2}(2x^2 - 8x) = \frac{1}{2}(64)$

$x^2 - 4x = 32$

$x^2 - 4x + 4 = 32 + 4$

$(x - 2)^2 = 36$

$x - 2 = \pm\sqrt{36}$

$x - 2 = \pm 6$

$x = 2 + 6$ or $x = 2 - 6$

$x = 8$ or $x = -4$

20. $2x^2 - 4x = 30$

$\frac{1}{2}(2x^2 - 4x) = \frac{1}{2}(30)$

$x^2 - 2x = 15$

$x^2 - 2x + 1 = 15 + 1$

$(x - 1)^2 = 16$

$x - 1 = \pm\sqrt{16}$

$x - 1 = \pm 4$

$x = 1 \pm 4$

$x = 1 + 4$ or $x = 1 - 4$

$x = 5$ or $x = -3$

21. $4x^2 + 2x - 12 = 0$

$\frac{1}{4}(4x^2 + 2x - 12) = \frac{1}{4}(0)$

$x^2 + \frac{1}{2}x - 3 = 0$

$x^2 + \frac{1}{2}x = 3$

$x^2 + \frac{1}{2}x + \frac{1}{16} = 3 + \frac{1}{16}$

$\left(x + \frac{1}{4}\right)^2 = \frac{49}{16}$

$x + \frac{1}{4} = \pm\sqrt{\frac{49}{16}}$

$x + \frac{1}{4} = \pm\frac{7}{4}$

$x = -\frac{1}{4} \pm \frac{7}{4}$

$x = -\frac{1}{4} + \frac{7}{4}$ or $x = -\frac{1}{4} - \frac{7}{4}$

$x = \frac{3}{2}$ or $x = -2$

22. $6x^2 - 19x + 15 = 0$

$\frac{1}{6}(6x^2 - 19x + 15) = \frac{1}{6}(0)$

$x^2 - \frac{19}{6}x + \frac{5}{2} = 0$

$x^2 - \frac{19}{6}x = -\frac{5}{2}$

$x^2 - \frac{19}{6}x + \frac{361}{144} = -\frac{5}{2} + \frac{361}{144}$

$\left(x - \frac{19}{12}\right)^2 = \frac{1}{144}$

$x - \frac{19}{12} = \pm\sqrt{\frac{1}{144}}$

$x - \frac{19}{12} = \pm\frac{1}{12}$

$x = \frac{19}{12} \pm \frac{1}{12}$

$x = \frac{19}{12} + \frac{1}{12}$ or $x = \frac{19}{12} - \frac{1}{12}$

$x = \frac{5}{3}$ or $x = \frac{3}{2}$

23. $b^2 - 4ac = (-4)^2 - 4(3)(-20) = 16 + 240$

$= 256$

Since the discriminant is positive, there are two real number solutions.

24. Write in standard form.

$-3x^2 + 4x - 9 = 0$

$b^2 - 4ac = (4)^2 - 4(-3)(-9) = 16 - 108$

$= -92$

Since the discriminant is negative, there is no real number solution.

25. $b^2 - 4ac = 6^2 - 4(2)(7) = 36 - 56 = -20$

Since the discriminant is negative, there is no real number solution.

26. $b^2 - 4ac = (-1)^2 - 4(1)(8) = 1 - 32$

$= -31$

Since the discriminant is negative, there is no real number solution.

27. Write in standard form.

$x^2 - 12x + 36 = 0$

$b^2 - 4ac = (-12)^2 - 4(1)(36) = 144 - 144$

$= 0$

Since the discriminant is zero, there is one real number solution.

28. $b^2 - 4ac = (-4)^2 - 4(3)(5) = 16 - 60$

$= -44$

Since the discriminant is negative, there is no real number solution.

29. $b^2 - 4ac = (-4)^2 - 4(-3)(8) = 16 + 96$

$= 112$

Since the discriminant is positive, there are two real number solutions.

30. $b^2 - 4ac = (-9)^2 - 4(1)(6) = 81 - 24 = 57$

Since the discriminant is positive, there are two real number solutions.

31. $a = 1,\ b = -9,\ c = 14$

$x = \frac{-b \pm \sqrt{b^2 - 4ac}}{2a}$

$= \frac{-(-9) \pm \sqrt{(-9)^2 - 4(1)(14)}}{2(1)}$

$= \frac{9 \pm \sqrt{25}}{2}$

$= \frac{9 \pm 5}{2}$

$x = \frac{9 + 5}{2}$ or $x = \frac{9 - 5}{2}$

$x = 7$ or $x = 2$

32. $a = 1, b = 7, c = -30$

$$x = \frac{-b \pm \sqrt{b^2 - 4ac}}{2a}$$
$$= \frac{-7 \pm \sqrt{7^2 - 4(1)(-30)}}{2(1)}$$
$$= \frac{-7 \pm \sqrt{49 + 120}}{2}$$
$$= \frac{-7 \pm \sqrt{169}}{2}$$
$$= \frac{-7 \pm 13}{2}$$
$$x = \frac{-7 + 13}{2} \text{ or } x = \frac{-7 - 13}{2}$$
$$x = 3 \text{ or } x = -10$$

33. Write in standard form.

$x^2 - 7x + 10 = 0$

$a = 1, b = -7, c = 10$

$$x = \frac{-b \pm \sqrt{b^2 - 4ac}}{2a}$$
$$= \frac{-(-7) \pm \sqrt{(-7)^2 - 4(1)(10)}}{2(1)}$$
$$= \frac{7 \pm \sqrt{49 - 40}}{2}$$
$$= \frac{7 \pm \sqrt{9}}{2}$$
$$= \frac{7 \pm 3}{2}$$
$$x = \frac{7 + 3}{2} \text{ or } x = \frac{7 - 3}{2}$$
$$x = 5 \text{ or } x = 2$$

34. Write in standard form.

$5x^2 - 7x - 6 = 0$

$a = 5, b = -7, c = -6$

$$x = \frac{-b \pm \sqrt{b^2 - 4ac}}{2a}$$
$$= \frac{-(-7) \pm \sqrt{(-7)^2 - 4(5)(-6)}}{2(5)}$$
$$= \frac{7 \pm \sqrt{49 + 120}}{10}$$
$$= \frac{7 \pm \sqrt{169}}{10}$$
$$= \frac{7 \pm 13}{10}$$
$$x = \frac{7 + 13}{10} \text{ or } x = \frac{7 - 13}{10}$$
$$x = 2 \text{ or } x = -\frac{3}{5}$$

35. Write in standard form.

$x^2 - 18 = 7x$

$x^2 - 7x - 18 = 0$

$a = 1, b = -7, c = -18$

$$x = \frac{-b \pm \sqrt{b^2 - 4ac}}{2a}$$
$$= \frac{-(-7) \pm \sqrt{(-7)^2 - 4(1)(-18)}}{2(1)}$$
$$= \frac{7 \pm \sqrt{49 + 72}}{2}$$
$$= \frac{7 \pm \sqrt{121}}{2}$$
$$= \frac{7 \pm 11}{2}$$
$$x = \frac{7 + 11}{2} \text{ or } x = \frac{7 - 11}{2}$$
$$x = 9 \text{ or } x = -2$$

36. $a = 1, b = -1, c = -30$

$$x = \frac{-b \pm \sqrt{b^2 - 4ac}}{2a}$$
$$= \frac{-(-1) \pm \sqrt{(-1)^2 - 4(1)(-30)}}{2(1)}$$
$$= \frac{1 \pm \sqrt{1+120}}{2}$$
$$= \frac{1 \pm \sqrt{121}}{2}$$
$$= \frac{1 \pm 11}{2}$$
$$x = \frac{1+11}{2} \text{ or } x = \frac{1-11}{2}$$
$$x = 6 \text{ or } x = -5$$

37. $a = 6, b = 1, c = -15$

$$x = \frac{-b \pm \sqrt{b^2 - 4ac}}{2a}$$
$$= \frac{-1 \pm \sqrt{1^2 - 4(6)(-15)}}{2(6)}$$
$$= \frac{-1 \pm \sqrt{1+360}}{12}$$
$$= \frac{-1 \pm \sqrt{361}}{12}$$
$$= \frac{-1 \pm 19}{12}$$
$$x = \frac{-1+19}{12} \text{ or } x = \frac{-1-19}{12}$$
$$x = \frac{3}{2} \text{ or } x = -\frac{5}{3}$$

38. $a = 2, b = 4, c = -3$

$$x = \frac{-b \pm \sqrt{b^2 - 4ac}}{2a}$$
$$= \frac{-4 \pm \sqrt{4^2 - 4(2)(-3)}}{2(2)}$$
$$= \frac{-4 \pm \sqrt{16+24}}{4}$$
$$= \frac{-4 \pm \sqrt{40}}{4}$$
$$= \frac{-4 \pm 2\sqrt{10}}{4}$$
$$= \frac{2\left(-2 \pm \sqrt{10}\right)}{4}$$
$$= \frac{-2 \pm \sqrt{10}}{2}$$

The solutions are $\frac{-2+\sqrt{10}}{2}$ and $\frac{-2-\sqrt{10}}{2}$.

39. $a = -2, b = 3, c = 6$

$$x = \frac{-b \pm \sqrt{b^2 - 4ac}}{2a}$$
$$= \frac{-3 \pm \sqrt{3^2 - 4(-2)(6)}}{2(-2)}$$
$$= \frac{-3 \pm \sqrt{9+48}}{-4}$$
$$= \frac{-3 \pm \sqrt{57}}{-4}$$
$$= \frac{3 \pm \sqrt{57}}{4}$$

The solutions are $\frac{3+\sqrt{57}}{4}$ and $\frac{3-\sqrt{57}}{4}$.

40. $a = 1, b = -6, c = 7$

$$x = \frac{-b \pm \sqrt{b^2 - 4ac}}{2a}$$
$$= \frac{-(-6) \pm \sqrt{(-6)^2 - 4(1)(7)}}{2(1)}$$
$$= \frac{6 \pm \sqrt{36 - 28}}{2}$$
$$= \frac{6 \pm \sqrt{8}}{2}$$
$$= \frac{6 \pm 2\sqrt{2}}{2}$$
$$= \frac{2\left(3 \pm \sqrt{2}\right)}{2}$$
$$= 3 \pm \sqrt{2}$$

The solutions are $3 + \sqrt{2}$ and $3 - \sqrt{2}$.

41. $a = 3, b = -4, c = 6$

$$x = \frac{-b \pm \sqrt{b^2 - 4ac}}{2a}$$
$$= \frac{-(-4) \pm \sqrt{(-4)^2 - 4(3)(6)}}{2(3)}$$
$$= \frac{4 \pm \sqrt{16 - 72}}{6}$$
$$= \frac{4 \pm \sqrt{-56}}{6}$$

Since $\sqrt{-56}$ is not a real number, there is no real number solution.

42. $a = 3, b = -6, c = -8$

$$x = \frac{-b \pm \sqrt{b^2 - 4ac}}{2a}$$
$$= \frac{-(-6) \pm \sqrt{(-6)^2 - 4(3)(-8)}}{2(3)}$$
$$= \frac{6 \pm \sqrt{36 + 96}}{6}$$
$$= \frac{6 \pm \sqrt{132}}{6}$$
$$= \frac{6 \pm 2\sqrt{33}}{6}$$
$$= \frac{2\left(3 \pm \sqrt{33}\right)}{6}$$
$$= \frac{3 \pm \sqrt{33}}{3}$$

The solutions are $\frac{3 + \sqrt{33}}{3}$ and $\frac{3 - \sqrt{33}}{3}$.

43. $a = 2, b = 3, c = 0$

$$x = \frac{-b \pm \sqrt{b^2 - 4ac}}{2a}$$
$$= \frac{-3 \pm \sqrt{3^2 - 4(2)(0)}}{2(2)}$$
$$= \frac{-3 \pm \sqrt{9}}{4}$$
$$= \frac{-3 \pm 3}{4}$$
$$x = \frac{-3 + 3}{4} \text{ or } \frac{-3 - 3}{4}$$
$$x = 0 \text{ or } x = -\frac{3}{2}$$

44. $a = 2, b = -5, c = 0$

$$x = \frac{-b \pm \sqrt{b^2 - 4ac}}{2a}$$

$$= \frac{-(-5) \pm \sqrt{(-5)^2 - 4(2)(0)}}{2(2)}$$

$$= \frac{5 \pm \sqrt{25}}{4}$$

$$= \frac{5 \pm 5}{4}$$

$$x = \frac{5+5}{4} \text{ or } \frac{5-5}{4}$$

$$x = \frac{5}{2} \text{ or } x = 0$$

45. $x^2 - 11x + 24 = 0$

$(x-8)(x-3) = 0$

$x - 8 = 0$ or $x - 3 = 0$

$x = 8$ or $x = 3$

46. $x^2 - 16x + 63 = 0$

$(x-7)(x-9) = 0$

$x - 7 = 0$ or $x - 9 = 0$

$x = 7$ or $x = 9$

47. $x^2 = -3x + 40$

$x^2 + 3x - 40 = 0$

$(x-5)(x+8) = 0$

$x - 5 = 0$ or $x + 8 = 0$

$x = 5$ or $x = -8$

48. $x^2 + 6x = 27$

$x^2 + 6x - 27 = 0$

$(x-3)(x+9) = 0$

$x - 3 = 0$ or $x + 9 = 0$

$x = 3$ or $x = -9$

49. $x^2 - 4x - 60 = 0$

$(x+6)(x-10) = 0$

$x + 6 = 0$ or $x - 10 = 0$

$x = -6$ or $x = 10$

50. $x^2 - x - 42 = 0$

$(x-7)(x+6) = 0$

$x - 7 = 0$ or $x + 6 = 0$

$x = 7$ or $x = -6$

51. $x^2 + 11x - 12 = 0$

$(x+12)(x-1) = 0$

$x + 12 = 0$ or $x - 1 = 0$

$x = -12$ or $x = 1$

52. $x^2 = 25$

$x = \pm\sqrt{25}$

$x = \pm 5$

$x = 5$ or $x = -5$

53. $x^2 + 6x = 0$

$x(x+6) = 0$

$x = 0$ or $x + 6 = 0$

$x = 0$ or $x = -6$

54. $2x^2 + 5x = 3$

$2x^2 + 5x - 3 = 0$

$(2x-1)(x+3) = 0$

$2x - 1 = 0$ or $x + 3 = 0$

$x = \frac{1}{2}$ or $x = -3$

55. $2x^2 = 9x - 10$

$2x^2 - 9x + 10 = 0$

$(2x-5)(x-2) = 0$

$2x - 5 = 0$ or $x - 2 = 0$

$x = \frac{5}{2}$ or $x = 2$

56. $6x^2 + 5x = 6$

$6x^2 + 5x - 6 = 0$

$(2x+3)(3x-2) = 0$

$2x + 3 = 0$ or $3x - 2 = 0$

$x = -\frac{3}{2}$ or $x = \frac{2}{3}$

57. $x^2 + 3x - 6 = 0$

$a = 1,\ b = 3,\ c = -6$

$$x = \frac{-b \pm \sqrt{b^2 - 4ac}}{2a}$$

$$= \frac{-3 \pm \sqrt{3^2 - 4(1)(-6)}}{2(1)}$$

$$= \frac{-3 \pm \sqrt{9 + 24}}{2}$$

$$= \frac{-3 \pm \sqrt{33}}{2}$$

The solutions are $\frac{-3+\sqrt{33}}{2}$ and $\frac{-3-\sqrt{33}}{2}$.

58. $3x^2 - 11x + 10 = 0$

$(3x-5)(x-2) = 0$

$3x - 5 = 0$ or $x - 2 = 0$

$x = \frac{5}{3}$ or $x = 2$

59. $-3x^2 - 5x + 8 = 0$

$-1(-3x^2 - 5x + 8) = (-1)(0)$

$3x^2 + 5x - 8 = 0$

$(3x+8)(x-1) = 0$

$3x + 8 = 0$ or $x - 1 = 0$

$x = -\frac{8}{3}$ or $x = 1$

60. $-2x^2 + 6x = -9$

$2x^2 - 6x - 9 = 0$

$a = 2,\ b = -6,\ c = -9$

$$x = \frac{-b \pm \sqrt{b^2 - 4ac}}{2a}$$

$$= \frac{-(-6) \pm \sqrt{(-6)^2 - 4(2)(-9)}}{2(2)}$$

$$= \frac{6 \pm \sqrt{36 + 72}}{4}$$

$$= \frac{6 \pm \sqrt{108}}{4}$$

$$= \frac{6 \pm 6\sqrt{3}}{4}$$

$$= \frac{2\left(3 \pm 3\sqrt{3}\right)}{4}$$

$$= \frac{3 \pm 3\sqrt{3}}{2}$$

The solutions are $\frac{3+3\sqrt{3}}{2}$ and $\frac{3-3\sqrt{3}}{2}$.

61. $2x^2 - 5x = 0$

$x(2x-5) = 0$

$x = 0$ or $2x - 5 = 0$

$x = 0$ or $x = \frac{5}{2}$

62. $3x^2 + 5x = 0$

$x(3x + 5) = 0$

$x = 0$ or $3x + 5 = 0$

$x = 0$ or $x = -\frac{5}{3}$

63. $a = 1, b = -2, c = -3$

$x = \frac{-b}{2a} = \frac{-(-2)}{2(1)} = 1$

The axis of symmetry is $x = 1$.

y-coordinate of the vertex:

$y = x^2 - 2x - 3$

$y = 1^2 - 2(1) - 3$

$y = 1 - 2 - 3 = -4$

The vertex is $(1, -4)$.

Since $a > 0$, the parabola opens up.

64. $a = 1, b = -10, c = 24$

$x = \frac{-b}{2a} = \frac{-(-10)}{2(1)} = 5$

The axis of symmetry is $x = 5$.

y-coordinate of the vertex:

$y = x^2 - 10x + 24$

$y = 5^2 - 10(5) + 24$

$y = 25 - 50 + 24 = -1$

The vertex is $(5, -1)$.

Since $a > 0$, the parabola opens up.

65. $a = 1, b = 7, c = 12$

$x = \frac{-b}{2a} = \frac{-7}{2(1)} = -\frac{7}{2}$

The axis of symmetry is $x = -\frac{7}{2}$.

y-coordinate of the vertex:

$y = x^2 + 7x + 12$

$y = \left(-\frac{7}{2}\right)^2 + 7\left(-\frac{7}{2}\right) + 12$

$y = \frac{49}{4} - \frac{49}{2} + 12 = -\frac{1}{4}$

The vertex is $\left(-\frac{7}{2}, -\frac{1}{4}\right)$.

Since $a > 0$, the parabola opens up.

66. $a = -1, b = -2, c = 15$

$x = \frac{-b}{2a} = \frac{-(-2)}{2(-1)} = -1$

The axis of symmetry is $x = -1$.

y-coordinate of the vertex:

$y = -x^2 - 2x + 15$

$y = -(-1)^2 - 2(-1) + 15$

$y = -1 + 2 + 15 = 16$

The vertex is $(-1, 16)$.

Since $a < 0$, the parabola opens down.

67. $a = 1, b = -3, c = 0$

$x = \frac{-b}{2a} = \frac{-(-3)}{2(1)} = \frac{3}{2}$

The axis of symmetry is $x = \frac{3}{2}$.

y-coordinate of the vertex:

$y = x^2 - 3x$

$y = \left(\frac{3}{2}\right)^2 - 3\left(\frac{3}{2}\right)$

$y = \frac{9}{4} - \frac{9}{2} = -\frac{9}{4}$

The vertex is $\left(\frac{3}{2}, -\frac{9}{4}\right)$.

Since $a > 0$, the parabola opens up.

68. $a = 2,\ b = 7,\ c = 3$

$$x = \frac{-b}{2a} = \frac{-7}{2(2)} = -\frac{7}{4}$$

The axis of symmetry is $x = -\frac{7}{4}$.

y-coordinate of the vertex:

$y = 2x^2 + 7x + 3$

$$y = 2\left(-\frac{7}{4}\right)^2 + 7\left(-\frac{7}{4}\right) + 3$$

$$y = \frac{49}{8} - \frac{49}{4} + 3 = -\frac{25}{8}$$

The vertex is $\left(-\frac{7}{4}, -\frac{25}{8}\right)$.

Since $a > 0$, the parabola opens up.

69. $a = -1,\ b = 0,\ c = -8$

$$x = \frac{-b}{2a} = \frac{-0}{2(-1)} = 0$$

The axis of symmetry is $x = 0$.

y-coordinate of the vertex:

$y = -x^2 - 8$

$y = -0^2 - 8 = -8$

The vertex is $(0, -8)$.

Since $a < 0$, the parabola opens down.

70. $a = -4,\ b = 8,\ c = 5$

$$x = \frac{-b}{2a} = \frac{-8}{2(-4)} = 1$$

The axis of symmetry is $x = 1$.

y-coordinate of the vertex:

$y = -4x^2 + 8x + 5$

$y = -4(1)^2 + 8(1) + 5$

$y = -4 + 8 + 5 = 9$

The vertex is $(1, 9)$.

Since $a < 0$, the parabola opens down.

71. $a = -1,\ b = -1,\ c = 20$

$$x = \frac{-b}{2a} = \frac{-(-1)}{2(-1)} = -\frac{1}{2}$$

The axis of symmetry is $x = -\frac{1}{2}$.

y-coordinate of the vertex:

$y = -x^2 - x + 20$

$$y = -\left(-\frac{1}{2}\right)^2 - \left(-\frac{1}{2}\right) + 20$$

$$y = -\frac{1}{4} + \frac{1}{2} + 20 = \frac{81}{4}$$

The vertex is $\left(-\frac{1}{2}, \frac{81}{4}\right)$.

Since $a < 0$, the parabola opens down.

72. $a = 3,\ b = 5,\ c = -8$

$$x = \frac{-b}{2a} = \frac{-5}{2(3)} = -\frac{5}{6}$$

The axis of symmetry is $x = -\frac{5}{6}$.

y-coordinate of the vertex:

$y = 3x^2 + 5x - 8$

$$y = 3\left(-\frac{5}{6}\right)^2 + 5\left(-\frac{5}{6}\right) - 8$$

$$y = \frac{25}{12} - \frac{25}{6} - 8 = -\frac{121}{12}$$

The vertex is $\left(-\frac{5}{6}, -\frac{121}{12}\right)$.

Since $a > 0$, the parabola opens up.

73. $a = 1,\ b = 6,\ c = 0$

Since $a > 0$, the parabola opens up.

The axis of symmetry is

$$x = \frac{-b}{2a} = \frac{-6}{2(1)} = -3$$

y-coordinate of the vertex:

$y = x^2 + 6x$

$y = (-3)^2 + 6(-3)$

$y = 9 - 18 = -9$

The vertex is $(-3, -9)$.

$y = x^2 + 6x$

Let $x = -2$ $\quad y = (-2)^2 + 6(-2) = -8$

Let $x = -1$ $\quad y = (-1)^2 + 6(-1) = -5$

Let $x = 0$ $\quad y = (0)^2 + 6(0) = 0$

Let $x = 1$ $\quad y = (1)^2 + 6(1) = 7$

x	y
–2	–8
–1	–5
0	0
1	7

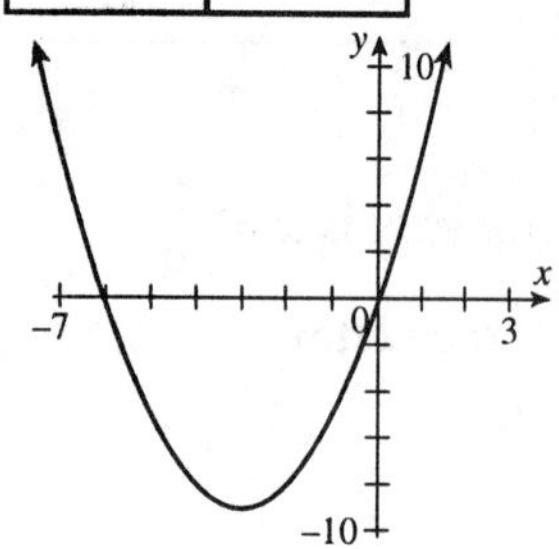

The x-intercepts are 0 and –6.

74. $a = -2,\ b = 0,\ c = 8$

Since $a < 0$, the parabola opens down.

The axis of symmetry is

$$x = \frac{-b}{2a} = \frac{-0}{2(-2)} = 0$$

y-coordinate of the vertex:

$y = -2x^2 + 8$

$y = -2(0)^2 + 8 = 8$

The vertex is $(0, 8)$.

$y = -2x^2 + 8$

Let $x = 1$ $\quad y = -2(1)^2 + 8 = 6$

Let $x = 2$ $\quad y = -2(2)^2 + 8 = 0$

x	y
1	6
2	0

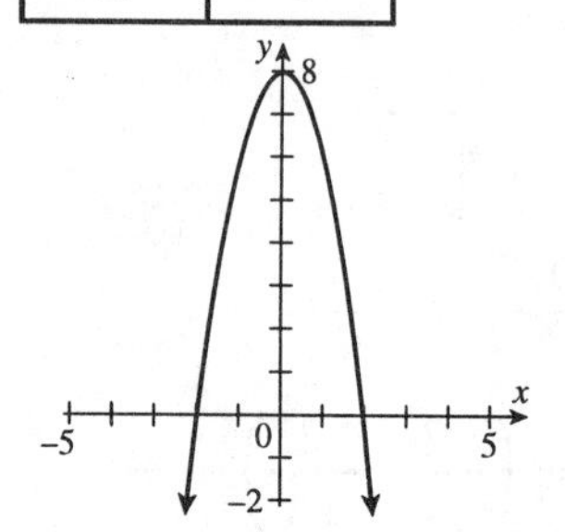

The x-intercepts are 2 and –2.

75. $a = 1,\ b = 2,\ c = -8$

Since $a > 0$, the parabola opens up.

The axis of symmetry is

$$x = \frac{-b}{2a} = \frac{-2}{2(1)} = -1$$

y-coordinate of the vertex:

$y = x^2 + 2x - 8$

$y = (-1)^2 + 2(-1) - 8$

$y = 1 - 2 - 8 = -9$

The vertex is $(-1, -9)$.

$y = x^2 + 2x - 8$

Let $x = 0$ $\quad y = 0^2 + 2(0) - 8 = -8$

Let $x = 1$ $\quad y = 1^2 + 2(1) - 8 = -5$

Let $x = 2$ $\quad y = 2^2 + 2(2) - 8 = 0$

x	y
0	–8
1	–5
2	0

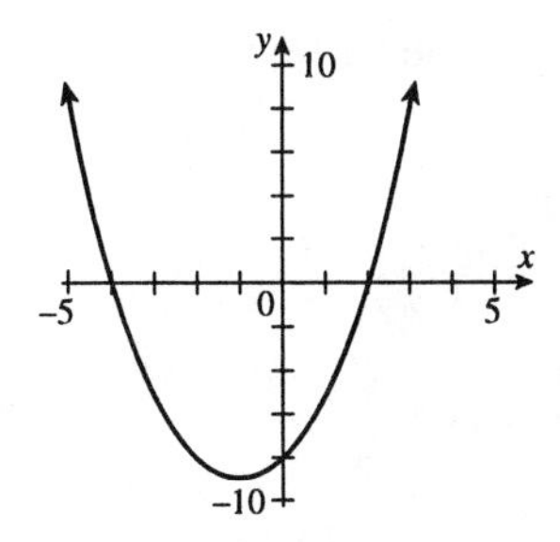

The x-intercepts are -4 and 2.

76. $a = 1, b = -1, c = -2$

Since $a > 0$, the parabola opens up.

The axis of symmetry is

$$x = \frac{-b}{2a} = \frac{-(-1)}{2(1)} = \frac{1}{2}$$

y-coordinate of the vertex:

$$y = x^2 - x - 2$$

$$y = \left(\frac{1}{2}\right)^2 - \frac{1}{2} - 2$$

$$y = \frac{1}{4} - \frac{1}{2} - 2 = -\frac{9}{4}$$

The vertex is $\left(\frac{1}{2}, -\frac{9}{4}\right)$.

$y = x^2 - x - 2$

Let $x = 1$ $\quad y = 1^2 - 1 - 2 = -2$

Let $x = 2$ $\quad y = 2^2 - 2 - 2 = 0$

x	y
1	−2
2	0

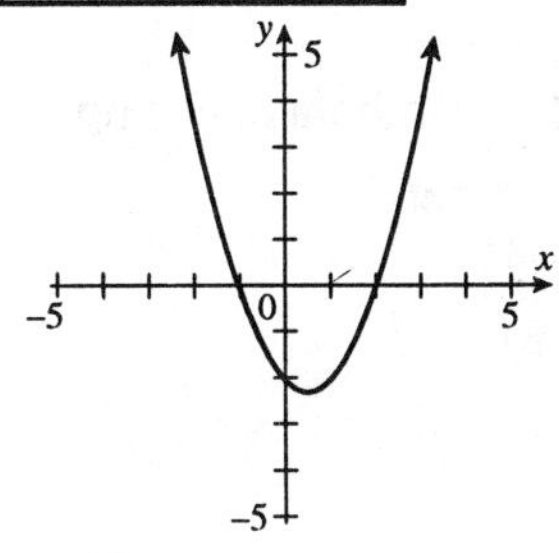

The x-intercepts are -1 and 2.

77. $a = 1, b = 5, c = 4$

Since $a > 0$, the parabola opens up.

The axis of symmetry is

$$x = \frac{-b}{2a} = \frac{-5}{2(1)} = -\frac{5}{2}$$

y-coordinate of the vertex:

$$y = x^2 + 5x + 4$$

$$y = \left(-\frac{5}{2}\right)^2 + 5\left(-\frac{5}{2}\right) + 4$$

$$y = \frac{25}{4} - \frac{25}{2} + 4 = -\frac{9}{4}$$

The vertex is $\left(-\frac{5}{2}, -\frac{9}{4}\right)$.

$y = x^2 + 5x + 4$

Let $x = -2$ $\quad y = (-2)^2 + 5(-2) + 4 = -2$

Let $x = -1$ $\quad y = (-1)^2 + 5(-1) + 4 = 0$

Let $x = 0$ $\quad y = (0)^2 + 5(0) + 4 = 4$

x	y
−2	−2
−1	0
0	4

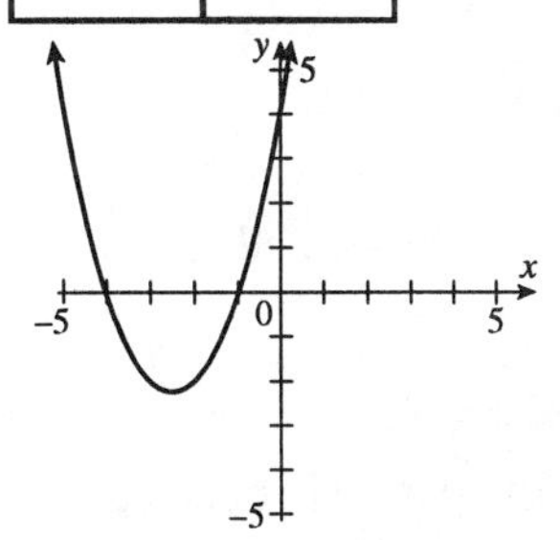

The x-intercepts are -1 and -4.

78. $a = 1,\ b = 4,\ c = 3$

Since $a > 0$, the parabola opens up.

The axis of symmetry is

$$x = \frac{-b}{2a} = \frac{-4}{2(1)} = -2$$

y-coordinate of the vertex:

$y = x^2 + 4x + 3$

$y = (-2)^2 + 4(-2) + 3$

$y = 4 - 8 + 3 = -1$

The vertex is $(-2, -1)$.

$y = x^2 + 4x + 3$

Let $x = -1$ $\quad y = (-1)^2 + 4(-1) + 3 = 0$

Let $x = 0$ $\quad y = (0)^2 + 4(0) + 3 = 3$

x	y
–1	0
0	3

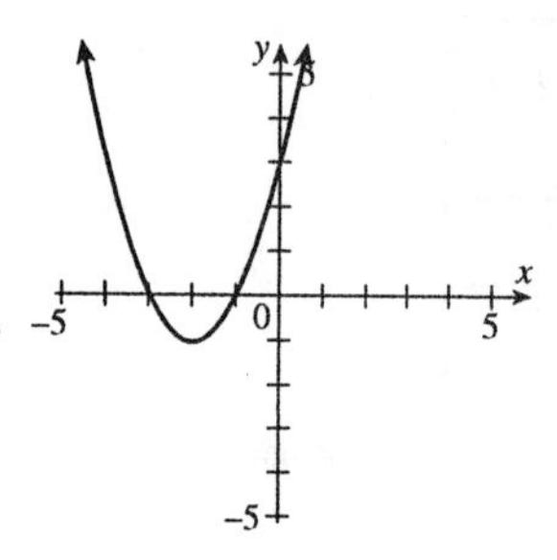

The x-intercepts are –1 and –3.

79. $a = -2,\ b = 3,\ c = -2$

Since $a < 0$, the parabola opens down.

The axis of symmetry is

$$x = \frac{-b}{2a} = \frac{-3}{2(-2)} = \frac{3}{4}$$

y-coordinate of the vertex:

$y = -2x^2 + 3x - 2$

$$y = -2\left(\frac{3}{4}\right)^2 + 3\left(\frac{3}{4}\right) - 2$$

$$y = -\frac{9}{8} + \frac{9}{4} - 2 = -\frac{7}{8}$$

The vertex is $\left(\frac{3}{4}, -\frac{7}{8}\right)$.

$y = -2x^2 + 3x - 2$

Let $x = -1$ $\quad y = -2(-1)^2 + 3(-1) - 2 = -7$

Let $x = 0$ $\quad y = -2(0)^2 + 3(0) - 2 = -2$

Let $x = 1$ $\quad y = -2(1)^2 + 3(1) - 2 = -1$

Let $x = 2$ $\quad y = -2(2)^2 + 3(2) - 2 = -4$

x	y
–1	–7
0	–2
1	–1
2	–4

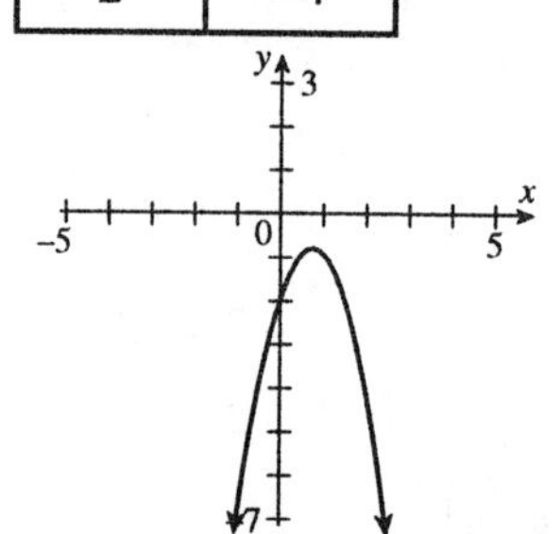

There are no x-intercepts.

80. $a = 3,\ b = -4,\ c = 1$

Since $a > 0$, the parabola opens up.

The axis of symmetry is

$$x = \frac{-b}{2a} = \frac{-(-4)}{2(3)} = \frac{2}{3}$$

y-coordinate of the vertex:

$y = 3x^2 - 4x + 1$

$$y = 3\left(\frac{2}{3}\right)^2 - 4\left(\frac{2}{3}\right) + 1$$

$$y = \frac{4}{3} - \frac{8}{3} + 1 = -\frac{1}{3}$$

The vertex is $\left(\frac{2}{3}, -\frac{1}{3}\right)$.

$y = 3x^2 - 4x + 1$

Let $x = 0$ $\quad y = 3(0)^2 - 4(0) + 1 = 1$

Let $x = 1$ $\quad y = 3(1)^2 - 4(1) + 1 = 0$

Let $x = 2$ $\quad y = 3(2)^2 - 4(2) + 1 = 5$

x	y
0	1
1	0
2	5

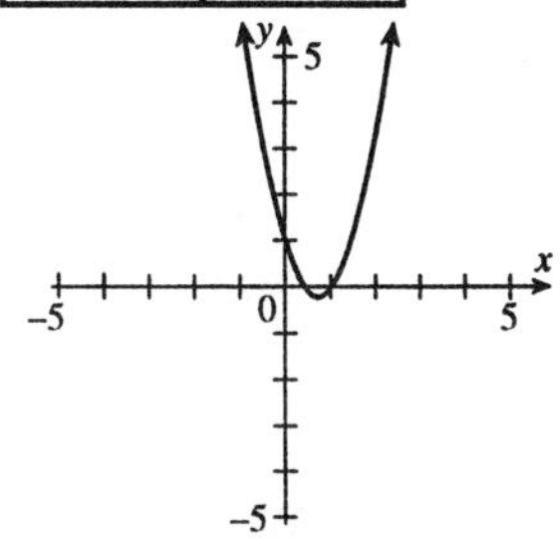

To find the x-intercepts, set y to 0:

$0 = 3x^2 - 4x + 1$

$0 = (3x - 1)(x - 1)$

$x = 1$ or $x = \frac{1}{3}$

81. $a = 4$, $b = -8$, $c = 6$

Since $a > 0$, the parabola opens up.

The axis of symmetry is

$$x = \frac{-b}{2a} = \frac{-(-8)}{2(4)} = 1$$

y-coordinate of the vertex:

$y = 4x^2 - 8x + 6$

$y = 4(1)^2 - 8(1) + 6$

$y = 4 - 8 + 6 = 2$

The vertex is (1, 2).

$y = 4x^2 - 8x + 6$

Let $x = 2$ $\quad y = 4(2)^2 - 8(2) + 6 = 6$

x	y
2	6

There are no x-intercepts.

82. $a = -3$, $b = -14$, $c = 5$

Since $a < 0$, the parabola opens down.

The axis of symmetry is

$$x = \frac{-b}{2a} = \frac{-(-14)}{2(-3)} = -\frac{7}{3}$$

y-coordinate of the vertex:

$y = -3x^2 - 14x + 5$

$$y = -3\left(-\frac{7}{3}\right)^2 - 14\left(-\frac{7}{3}\right) + 5$$

$$y = -\frac{49}{3} + \frac{98}{3} + 5 = \frac{64}{3}$$

The vertex is $\left(-\frac{7}{3}, \frac{64}{3}\right)$.

$y = -3x^2 - 14x + 5$

Let $x = -5$ $\quad y = -3(-5)^2 - 14(-5) + 5 = 0$

Let $x = -4$ $\quad y = -3(-4)^2 - 14(-4) + 5 = 13$

Let $x = -3$ $\quad y = -3(-3)^2 - 14(-3) + 5 = 20$

Let $x = -2$ $\quad y = -3(-2)^2 - 14(-2) + 5 = 21$

Let $x = -1$ $\quad y = -3(-1)^2 - 14(-1) + 5 = 16$

Let $x = 0$ $\quad y = -3(0)^2 - 14(0) + 5 = 5$

Let $x = 1$ $\quad y = -3(1)^2 - 14(1) + 5 = -12$

x	y
–5	0
–4	13
–3	20
–2	21
–1	16
0	5
1	–12

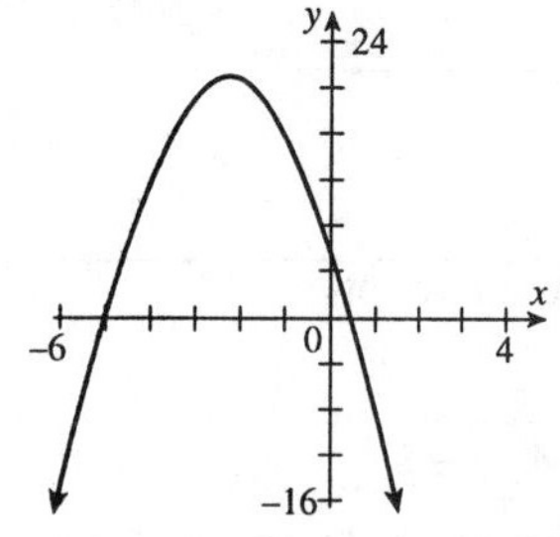

To find the x-intercepts, set y to 0:

$0 = -3x^2 - 14x + 5$

$0 = -(3x - 1)(x + 5)$

$x = -5$ or $x = \dfrac{1}{3}$

83. $a = -1,\ b = -6,\ c = -4$

Since $a < 0$, the parabola opens down.

The axis of symmetry is

$x = \dfrac{-b}{2a} = \dfrac{-(-6)}{2(-1)} = -3$

y-coordinate of the vertex:

$y = -x^2 - 6x - 4$

$y = -(-3)^2 - 6(-3) - 4$

$y = -9 + 18 - 4 = 5$

The vertex is $(-3, 5)$.

$y = -x^2 - 6x - 4$

Let $x = -2$ $\quad y = -(-2)^2 - 6(-2) - 4 = 4$

Let $x = -1$ $\quad y = -(-1)^2 - 6(-1) - 4 = 1$

Let $x = 0$ $\quad y = -(0)^2 - 6(0) - 4 = -4$

x	y
–2	4
–1	1
0	–4

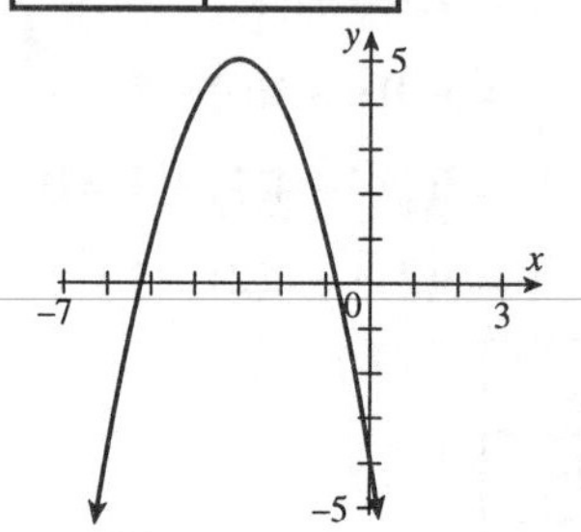

To find the x-intercepts, set y to 0:

$0 = -x^2 - 6x - 4$

$$x = \frac{-b \pm \sqrt{b^2 - 4ac}}{2a}$$

$$= \frac{-(-6) \pm \sqrt{(-6)^2 - 4(-1)(-4)}}{2(-1)}$$

$$= \frac{6 \pm \sqrt{36 - 16}}{-2} = \frac{6 \pm \sqrt{20}}{-2} = -3 \pm \sqrt{5}$$

The x-intercepts are $-3 - \sqrt{5}$ and $-3 + \sqrt{5}$.

84. $a = 2,\ b = 5,\ c = -12$

Since $a > 0$, the parabola opens up.

The axis of symmetry is

$x = \dfrac{-b}{2a} = \dfrac{-5}{2(2)} = -\dfrac{5}{4}$

y-coordinate of the vertex:

$y = 2x^2 + 5x - 12$

$y = 2\left(-\frac{5}{4}\right)^2 + 5\left(-\frac{5}{4}\right) - 12$

$y = \frac{25}{8} - \frac{25}{4} - 12 = -\frac{121}{8}$

The vertex is $\left(-\frac{5}{4}, -\frac{121}{8}\right)$.

$y = 2x^2 + 5x - 12$

Let $x = -4$ $\quad y = 2(-4)^2 + 5(-4) - 12 = 0$

Let $x = -3$ $\quad y = 2(-3)^2 + 5(-3) - 12 = -9$

Let $x = -2$ $\quad y = 2(-2)^2 + 5(-2) - 12 = -14$

Let $x = -1$ $\quad y = 2(-1)^2 + 5(-1) - 12 = -15$

Let $x = 0$ $\quad y = 2(0)^2 + 5(0) - 12 = -12$

Let $x = 1$ $\quad y = 2(1)^2 + 5(1) - 12 = -5$

x	y
–4	0
–3	–9
–2	–14
–1	–15
0	–12
1	–5

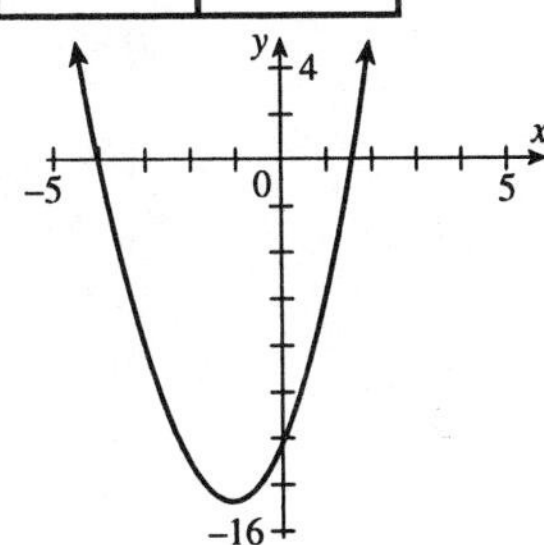

To find the x-intercepts, set y to 0:

$0 = 2x^2 + 5x - 12$

$0 = (2x - 3)(x + 4)$

$x = \frac{3}{2}$ or $x = -4$

85. Let the two integers be x and y. Then,

$xy = 88$ The product is 88.

$y = x + 3$ The larger is 3 greater than the smaller.

Substitute $x + 3$ for y in the first equation.

$x(x + 3) = 88$

$x^2 + 3x = 88$

$x^2 + 3x + \frac{9}{4} = 88 + \frac{9}{4}$

$\left(x + \frac{3}{2}\right)^2 = \frac{361}{4}$

$x + \frac{3}{2} = \pm\sqrt{\frac{361}{4}}$

$x + \frac{3}{2} = \pm\frac{19}{2}$

$x = -\frac{3}{2} \pm \frac{19}{2}$

Since the numbers are positive,

$x = -\frac{3}{2} + \frac{19}{2} = 8$ and $y = 8 + 3 = 11$.

86. Let w = width of rectangle.

Then $2w - 5$ = length of rectangle

Area = length · width

$63 = w(2w - 5)$

$2w^2 - 5w = 63$

$2w^2 - 5w - 63 = 0$

$(2w + 9)(w - 7) = 0$

$2w + 9 = 0$ or $w - 7 = 0$

$w = -\frac{9}{2}$ or $w = 7$

Since w is positive, the width is 7 feet and the length is $2(7) - 5 = 9$ feet.

Chapter 10 Practice Test

1. $x^2 + 1 = 21$

$x^2 = 20$

$x = \pm\sqrt{20}$

$x = \pm 2\sqrt{5}$

The solutions are $2\sqrt{5}$ and $-2\sqrt{5}$.

2. $(2x-3)^2 = 35$

$2x - 3 = \pm\sqrt{35}$

$2x = 3 \pm \sqrt{35}$

$x = \dfrac{3 \pm \sqrt{35}}{2}$

The solutions are $\dfrac{3+\sqrt{35}}{2}$ and $\dfrac{3-\sqrt{35}}{2}$.

3. $x^2 - 4x = 60$

$x^2 - 4x + 4 = 60 + 4$

$(x-2)^2 = 64$

$x - 2 = \pm\sqrt{64}$

$x - 2 = \pm 8$

$x = 2 \pm 8$

$x = 2 + 8$ or $x = 2 - 8$

$x = 10$ or $x = -6$

4. $x^2 = -x + 12$

$x^2 + x = 12$

$x^2 + x + \dfrac{1}{4} = 12 + \dfrac{1}{4}$

$\left(x + \dfrac{1}{2}\right)^2 = \dfrac{49}{4}$

$x + \dfrac{1}{2} = \pm\sqrt{\dfrac{49}{4}}$

$x + \dfrac{1}{2} = \pm\dfrac{7}{2}$

$x = -\dfrac{1}{2} \pm \dfrac{7}{2}$

$x = -\dfrac{1}{2} + \dfrac{7}{2}$ or $x = -\dfrac{1}{2} - \dfrac{7}{2}$

$x = 3$ or $x = -4$

5. $a = 1,\ b = -5,\ c = -6$

$$x = \frac{-b \pm \sqrt{b^2 - 4ac}}{2a}$$

$$= \frac{-(-5) \pm \sqrt{(-5)^2 - 4(1)(-6)}}{2(1)}$$

$$= \frac{5 \pm \sqrt{25 + 24}}{2}$$

$$= \frac{5 \pm \sqrt{49}}{2}$$

$$= \frac{5 \pm 7}{2}$$

$x = \dfrac{5+7}{2}$ or $x = \dfrac{5-7}{2}$

$x = 6$ or $x = -1$

6. $2x^2 + 5 = -8x$

$2x^2 + 8x + 5 = 0$

$a = 2,\ b = 8,\ c = 5$

$$x = \frac{-b \pm \sqrt{b^2 - 4ac}}{2a}$$

$$= \frac{-8 \pm \sqrt{8^2 - 4(2)(5)}}{2(2)}$$

$$= \frac{-8 \pm \sqrt{64 - 40}}{4}$$

$$= \frac{-8 \pm \sqrt{24}}{4}$$

$$= \frac{-8 \pm 2\sqrt{6}}{4}$$

$$= \frac{-4 \pm \sqrt{6}}{2}$$

The solutions are $\frac{-4 + \sqrt{6}}{2}$ and $\frac{-4 - \sqrt{6}}{2}$.

7. $3x^2 - 5x = 0$

$x(3x - 5) = 0$

$x = 0$ or $3x - 5 = 0$

$x = 0$ or $x = \frac{5}{3}$

8. $2x^2 + 9x = 5$

$2x^2 + 9x - 5 = 0$

$(2x - 1)(x + 5) = 0$

$2x - 1 = 0$ or $x + 5 = 0$

$x = \frac{1}{2}$ or $x = -5$

9. $b^2 - 4ac = (-4)^2 - 4(3)(2) = 16 - 24 = -8$

Since the discriminant is negative, the equation has no real solution.

10. $a = -1,\ b = 3,\ c = 8$

Axis of symmetry: $x = \frac{-b}{2a} = \frac{-3}{2(-1)} = \frac{3}{2}$

The axis of symmetry is $x = \frac{3}{2}$.

y-coordinate of vertex:

$y = -x^2 + 3x + 8$

$y = -\left(\frac{3}{2}\right)^2 + 3\left(\frac{3}{2}\right) + 8$

$y = -\frac{9}{4} + \frac{9}{2} + 8 = \frac{41}{4}$

The vertex is $\left(\frac{3}{2}, \frac{41}{4}\right)$.

Since $a < 0$, the parabola opens down.

11. $a = 1,\ b = 2,\ c = -8$

Since $a > 0$, the parabola opens up.

Axis of symmetry: $x = \frac{-b}{2a} = \frac{-2}{2(1)} = -1$

y-coordinate of vertex:

$y = x^2 + 2x - 8$

$y = (-1)^2 + 2(-1) - 8$

$y = 1 - 2 - 8 = -9$

The vertex is $(-1, -9)$.

	$y = x^2 + 2x - 8$
Let $x = 0$	$y = 0^2 + 2(0) - 8 = -8$
Let $x = 1$	$y = 1^2 + 2(1) - 8 = -5$
Let $x = 2$	$y = 2^2 + 2(2) - 8 = 0$
Let $x = 3$	$y = 3^2 + 2(3) - 8 = 7$

x	y
0	–8
1	–5
2	0
3	7

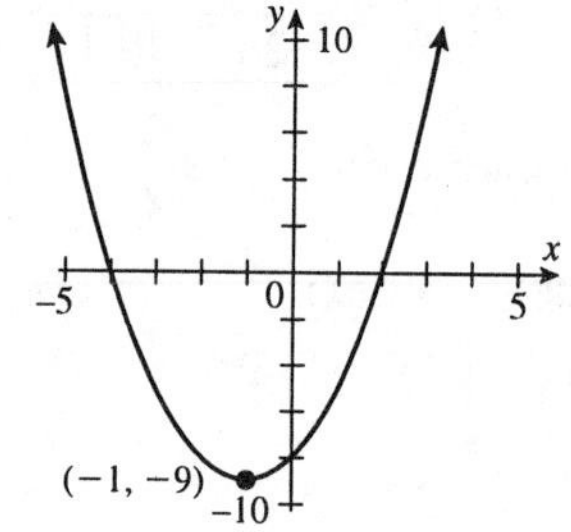

The x-intercepts are –4 and 2.

12. $a = -1,\ b = 6,\ c = -9$

Since $a < 0$, the parabola opens down.

Axis of symmetry: $x = \frac{-b}{2a} = \frac{-6}{2(-1)} = 3$

y-coordinate of vertex:

$y = -x^2 + 6x - 9$

$y = -3^2 + 6(3) - 9$

$y = -9 + 18 - 9 = 0$

The vertex is (3, 0).

$y = -x^2 + 6x - 9$

Let $x = 0$ $\quad y = -0^2 + 6(0) - 9 = -9$

Let $x = 1$ $\quad y = -1^2 + 6(1) - 9 = -4$

Let $x = 2$ $\quad y = -2^2 + 6(2) - 9 = -1$

x	y
0	–9
1	–4
2	–1

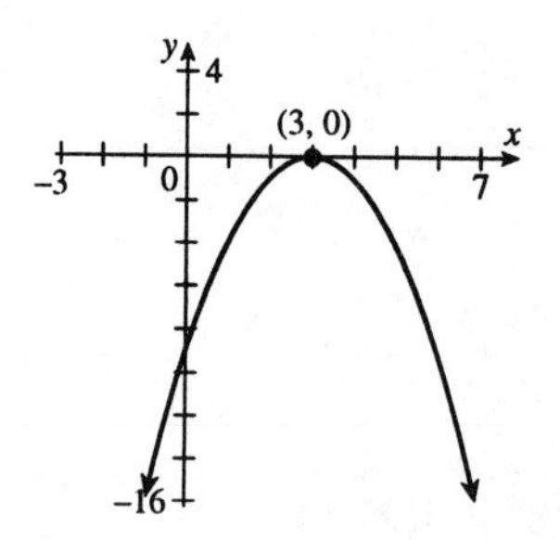

The x-intercept is 3.

13. Let w = width of the rectangle

Then $3w + 1$ = length of the rectangle

Area = length · width

$30 = (3w + 1)w$

$3w^2 + w = 30$

$3w^2 + w - 30 = 0$

$(3w + 10)(w - 3) = 0$

$3w + 10 = 0$ or $w - 3 = 0$

$w = -\frac{10}{3}$ or $w = 3$

Since w is positive, the width is 3 feet and the length is $3(3) + 1 = 10$ feet.

Chapter 10 Cumulative Review Test

1. Substitute –3 for x and 4 for y.

$-x^2y + y^2 - 3xy$

$= -(-3)^2(4) + (4)^2 - 3(-3)(4)$

$= -36 + 16 + 36 = 16$

2. $\frac{1}{4}x + \frac{3}{5}x = \frac{1}{3}(x + 2)$

$\frac{1}{4}x + \frac{3}{5}x = \frac{1}{3}x + \frac{2}{3}$

$\frac{17}{20}x = \frac{1}{3}x + \frac{2}{3}$

$\frac{17}{20}x - \frac{1}{3}x = \frac{2}{3}$

$$\frac{51}{60}x - \frac{20}{60}x = \frac{2}{3}$$

$$\frac{31}{60}x = \frac{2}{3}$$

$$x = \frac{40}{31}$$

3. $\frac{x}{8} = \frac{2}{3}$

$3x = (8)(2)$

$3x = 16$

$x = \frac{16}{3}$ or $5\frac{1}{3}$

The length of side x is $5\frac{1}{3}$ inches.

4. $2(x-3) \le 6x - 5$

$2x - 6 \le 6x - 5$

$2x - 1 \le 6x$

$-1 \le 4x$

$-\frac{1}{4} \le x$

$x \ge -\frac{1}{4}$

-5 0 5

5. $A = \frac{m+n+P}{3}$

$3A = m + n + P$

$3A - m = n + P$

$3A - m - n = P$

6. $(6x^2y^4)^3(2x^4y^5)^2 = (216x^6y^{12})(4x^8y^{10})$

$= 864x^{14}y^{22}$

7.

$$\begin{array}{r} x+4 \\ x+2\overline{)x^2+6x+5} \\ \underline{-x^2-2x} \\ 4x+5 \\ \underline{-4x-8} \\ -3 \end{array}$$

$$\frac{x^2+6x+5}{x+2} = x + 4 - \frac{3}{x+2}$$

8. $2x^2 - 3xy - 4xy + 6y^2$

$= x(2x - 3y) - 2y(2x - 3y)$

$= (x - 2y)(2x - 3y)$

9. $4x^2 - 14x - 8$

$= 2(2x^2 - 7x - 4)$

$= 2[2x^2 + x - 8x - 4]$

$= 2[x(2x + 1) - 4(2x + 1)]$

$= 2(x - 4)(2x + 1)$

10. $\frac{4}{a^2 - 16} + \frac{2}{(a-4)^2}$

$$= \frac{4}{(a-4)(a+4)} + \frac{2}{(a-4)(a-4)}$$

$$= \frac{4}{(a-4)(a+4)} \cdot \frac{a-4}{a-4} + \frac{2}{(a-4)(a-4)} \cdot \frac{a+4}{a+4}$$

$$= \frac{4a-16}{(a+4)(a-4)^2} + \frac{2a+8}{(a+4)(a-4)^2}$$

$$= \frac{6a-8}{(a+4)(a-4)^2}$$

11. $x+\frac{48}{x}=14$

$x\left[x+\frac{48}{x}\right]=14\cdot x$

$x^2+48=14x$

$x^2-14x+48=0$

$(x-6)(x-8)=0$

$x-6=0$ or $x-8=0$

$x=6$ or $x=8$

12. $3x+5y=10$ — Ordered Pair

Let $x=0$, then $y=2$ — $(0, 2)$

Let $x=5$, then $y=-1$ — $(5, -1)$

Plot the points and draw the line.

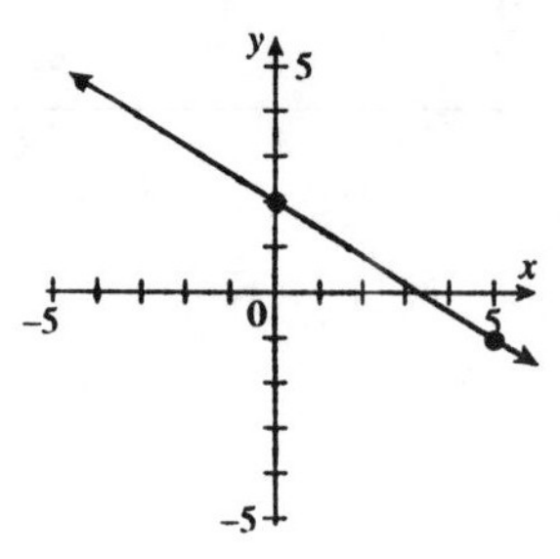

13. $4[3x-4y=12]$ gives $12x-16y=48$

$-3[4x-3y=6]$ $\quad \underline{-12x+9y=-18}$

$-7y=30$

$y=-\frac{30}{7}$

Solve for x.

$3x-4y=12$

$3x-4\left(-\frac{30}{7}\right)=12$

$3x+\frac{120}{7}=12$

$3x=-\frac{36}{7}$

$x=-\frac{12}{7}$

The solution is $\left(-\frac{12}{7}, -\frac{30}{7}\right)$

14. $\sqrt{\frac{3x^2y^3}{54x}}=\sqrt{\frac{xy^3}{18}}=\frac{\sqrt{xy^3}}{\sqrt{18}}$

$=\frac{\sqrt{y^2}\sqrt{xy}}{\sqrt{9}\sqrt{2}}=\frac{y\sqrt{xy}}{3\sqrt{2}}$

$=\frac{y\sqrt{xy}}{3\sqrt{2}}\cdot\frac{\sqrt{2}}{\sqrt{2}}=\frac{y\sqrt{2xy}}{6}$

15. $2\sqrt{28}-3\sqrt{7}+\sqrt{63}$

$=2\sqrt{4}\sqrt{7}-3\sqrt{7}+\sqrt{9}\sqrt{7}$

$=2\cdot2\sqrt{7}-3\sqrt{7}+3\sqrt{7}$

$=4\sqrt{7}-3\sqrt{7}+3\sqrt{7}$

$=4\sqrt{7}$

16. $\sqrt{x^2+5}=x+1$

$\left(\sqrt{x^2+5}\right)^2=(x+1)^2$

$x^2+5=x^2+2x+1$

$5=2x+1$

$4=2x$

$2=x$

17. $a = 2$, $b = -4$, $c = -5$

$$x = \frac{-b \pm \sqrt{b^2 - 4ac}}{2a}$$
$$= \frac{-(-4) \pm \sqrt{(-4)^2 - 4(2)(-5)}}{2(2)}$$
$$= \frac{4 \pm \sqrt{16 + 40}}{4}$$
$$= \frac{4 \pm \sqrt{56}}{4}$$
$$= \frac{4 \pm 2\sqrt{14}}{4}$$
$$= \frac{2 \pm \sqrt{14}}{2}$$

The solutions are $\frac{2 + \sqrt{14}}{2}$ and $\frac{2 - \sqrt{14}}{2}$.

18. $\frac{500 \text{ square feet}}{4 \text{ pounds fertilizer}} = \frac{3200 \text{ square feet}}{x \text{ pounds fertilizer}}$

$$\frac{500}{4} = \frac{3200}{x}$$
$$500x = 4 \cdot 3200$$
$$500x = 12{,}800$$
$$x = 25.6$$

25.6 pounds of fertilizer are needed for 3,200 square feet of lawn.

19. Let w = width of rectangle

Then $3w - 3$ = length of rectangle

$$P = 2l + 2w$$
$$74 = 2(3w - 3) + 2w$$
$$74 = 6w - 6 + 2w$$
$$80 = 8w$$
$$10 = w$$

The width is 10 feet and the length is $3(10) - 3 = 27$ feet.

20. Let w = her walking speed

Then $w + 3$ = her jogging speed

Time to walk 2 miles = $\frac{2}{w}$

Time to jog 2 miles = $\frac{2}{w+3}$

Time to walk 2 miles + time to jog 2 miles = 1 hour.

$$\frac{2}{w} + \frac{2}{w+3} = 1$$
$$(w+3)\left[\frac{2}{w} + \frac{2}{w+3}\right] = 1(w+3)w$$
$$2(w+3) + 2w = w^2 + 3w$$
$$2w + 6 + 2w = w^2 + 3w$$
$$w^2 - w - 6 = 0$$
$$(w-3)(w+2) = 0$$
$$w - 3 = 0 \text{ or } w + 2 = 0$$
$$w = 3 \text{ or } w = -2$$

Since w is positive, her walking speed is 3 mph and her jogging speed is $3 + 3 = 6$ mph.